The table has been permuted to bring groups VIIA and IA together, eliminating VIIIA (the noble gases). This maximizes the number of blocks of elements with similar biotic functions.

Found in a few special proteins.

Structural elements in walls, skeletons, shells and so on.

Rare and special, functions largely unknown.

All isotopes are radioactive.

( ) Indicates mass number of longest-lived or best-known isotopes.

The small numbers at far left show electron distribution in preceding noble gas.

C,H,N,O,P,S: the fundamental six.

Activators and cofactors of enzymes and other functional proteins.

Electrolytes: principal cations and anions which neutralize large molecules. Also function as activators of enzymes and other proteins.

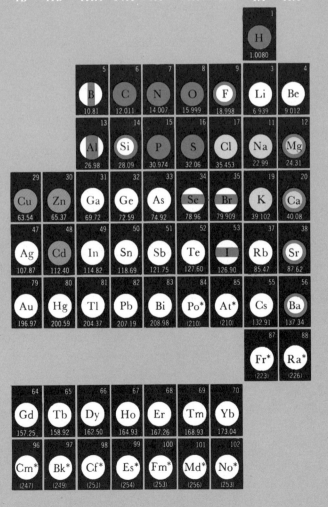

biological principles

biological principles

burton s. guttman

university of kentucky

w. a. benjamin, inc.
new york
1971

**biological
principles**

Copyright © 1971
by W. A. Benjamin, Inc.
All rights reserved
Standard Book Number 8053–3573–0
(Clothbound Edition)
Library of Congress Catalog
Card Number 79–157567
Manufactured in the
United States of America
1 2 3 4 5 K 4 3 2 1
The author and publisher are
pleased to acknowledge
the assistance of
Wladislaw Finne, who
designed the text and
the cover,
and of F. W. Taylor and
his associates, who
drew the illustrations.
W. A. Benjamin, Inc.
New York, New York 10016

for
paddy, erica,
carrie, lindy,
and meg

Somehow, we must find a way to communicate the true
essence of today's science to our young people.
Somehow, thinking must again be made popular,
and great thinking synonymous with high adventure.
In biology, for example, the contemporary scene is
exciting and dramatic, swift-paced and paradoxical. . . .
The new biology is being spoken everywhere and
by everyone—by intellectual giants and bunglers,
by physicists and cosmologists on sabbaticals,
in the laboratories of Cambridge and Iowa State,
in the journals of microbiology, at the summer seminars
in physical chemistry at Woods Hole, in the
third-class cabins of Fulbright scholars, at
the International Congresses of Biochemistry, in
the wards and corridors of great medical schools,
at the Cold Spring Harbor symposia on ''the gene'' or on
''nerve function,'' next to the lead-shielded monsters
of the United States Atomic Energy Commission,
on the docks of the Oceanographic Institute.
The living organism today has as many profiles as
viewers, as many dimensions as measuring rods.
Its image should be an image of adventure.

—William S. Beck,
Modern Science and the Nature of Life
(Harcourt, Brace, Jovanovich, New York, 1957)

preface

This book is an introduction to the principles of biology. It is designed for the kind of course generally given for sophomores who intend to major in biology or to enter one of the applied biology professions such as medicine or dentistry. It is not designed for those students who want to study biology as part of their general education. A background of general chemistry and at least college algebra is assumed; most students will have had a little calculus and will be taking organic chemistry along with this course.

I have tried to accomplish several things at once in this book. First, I have tried to present a picture of biology as it really is in the 1970s; this is not a unique goal, and all courses of this type certainly try to present as modern a view of the science as possible. The second goal is to present biology within a general conceptual framework that can serve as a foundation for the student's whole lifetime of learning. I have tried to weave together the major facts, hypotheses, analytical methods, and principles of biology to create this particular viewpoint, and the whole book is organized around it. Finally, I have tried to create a book that is pedagogically correct, so all of these concepts are presented in a way that is really useful for students who are approaching the subject for the first time at this level.

The book is not based on a parochial point of view; it is not the view that biology is "nothing but chemistry" or "nothing but anatomy," but rather a broadly integrative outlook that happens to center primarily on genetic concepts: reproduction, inheritance, mutation, selection, and evolution. This viewpoint has been discussed several times by Norman Horowitz, but he himself points out that the first clear statement of it came in a 1929 paper by H. J. Muller, "The Gene as the Basis of Life," (*Proc. Int. Cong. Plant Sci.* 1:897–921). Regardless of priorities, it is a very valuable concept and it serves nicely as the foundation for other concepts. Donald L. D. Caspar pointed out that the principle of natural selection coupled with the physical principle of least energy leads directly to the principles of biomolecular organization—subassembly and self-assembly of symmetrical structures. Other ideas, such as Peter Mitchell's view of vectorial metabolism, fit neatly into this framework and have been woven in throughout.

Superimposed upon all of this is the theme of energy flow through individual organisms and through the biosphere. Naturally, this creates an ecological viewpoint (what would biology be without it?) but, in spite of the gratifying renewal of interest in ecology today, I have not stressed this matter much. I hope that the students who read this book will be out campaigning and working to help save us from the impending disasters of pollution and overpopulation; but I assume that they will be the biologists, physicians, wildlife managers, etc. of the future—in other words, the experts upon whom much of our future is going to depend—and I assume that their primary task right now is to acquire the solid foundation in biology that they will need in a few years. Many of these students will enter the health professions, and there are some major health problems that we will have to solve soon along with the problems of pollution; for these students, information that gives them a firm understanding of how organisms operate at the levels emphasized here is as "relevant" as information about population ecology is to students who will be businessmen and history teachers.

This book is an introduction; it concentrates on one set of themes that provide an essential background for further study in all areas of biology. To keep the book relatively concise and limit it to a size that students can easily carry around, I have stopped short of many fascinating topics, particularly phenomena peculiar to multicellular organisms and to populations. Many of these topics simply require more careful development in a later course, but the backgrounds for them are here. For example, the background to a discussion of differentiation is set by thorough treatments of regulatory mechanisms, cell surface phenomena, and so on. Similarly, population dynamics and population genetics require more mathematics than students in this course are expected to have, but the background for these topics is presented here. The student who finishes this course should be able to take up these other topics later with some real understanding.

However, I specifically reject any artificial separation of the topics in this book as merely "molecular biology," in contrast to an "organismic biology" that emphasizes multicellular organisms. We happen to consider the simplest systems first, and, of course, these are primarily unicellular and subcellular, but they are no less "organismic" than cats and worms. I have not avoided more complex organisms when they illustrate some important point; because students are alive, it is often simplest to call their attention to something they already know about themselves. But by concentrating on simple systems, we get much clearer insights into biotic systems as a whole and we achieve a very useful and satisfying picture of the science. Topics usually assigned to traditional categories such as genetics, cytology, cell physiology, and ecology can all be unified into a single course, with considerable advantages for both students and instructors. Instructors in more advanced courses can then assume this kind of background and go on into more interesting topics.

Many topics traditionally taught in biochemistry courses are also included here. Biochemistry courses for seniors or first-year graduate students usually have two main emphases. Their "bio-" part is partially presented here because it is essential to understand the metabolism of organisms, their general chemical structure, and the general significance of metabolism and structure. Once this is understood, future courses can concentrate on the "-chemistry," including enzyme kinetics and reaction mechanisms. These topics do not belong here, both because they obscure the main biological principles and because they require more physics and chemistry than one can assume at this level.

For pedagogical reasons, the major themes of the book are repeated three times, and many of them are repeated a fourth time. With each repetition, the student's understanding should become greater; if everything were piled on at one time, the biological framework would be buried in details. Part I is an introduction and a broad overview of

biology. In Chapter 1 we look at the general structure of the biosphere and the molecular and cellular organization of organisms. Chapter 2 outlines the basic phenomena
of growth, mutation, and evolution, while introducing bacteria and viruses as experimental tools. In Chapter 3 we first examine the concept of an enzyme and take a very
broad look at metabolism; we also take up a few basic concepts of thermodynamics
and consider how organisms obtain free energy and use it for growth. Chapter 4 lays
the foundation for understanding organisms as genetic systems; it introduces the
informational relationships between nucleic acids and proteins and outlines some
general features of organisms and some mechanisms of evolution. Then in Chapter 5
we introduce the important tools of genetic analysis through mapping and complementation studies.

In Part II we review most of this material in more detail, with emphasis on procaryotic
cells. Chapter 6 is a closer look at biomolecular structure and Chapter 7 is a closer look
at some general properties of cells, especially the properties of cell boundaries. In
Chapter 8 we see how bacteria are built and we focus a little more sharply on metabolism, with emphasis on the Krebs cycle. In Chapter 9 we see how the principles of
genetic analysis can be used to explore bacteria, and in Chapter 10 we again look at
information transfer in the cell and see how proteins are made and how their synthesis
is regulated.

In Part III we take up these themes for a third time, with emphasis on eucaryotic
organisms. Chapter 11 presents a detailed examination of eucaryotic cells, including
a discussion of the complex movements seen in these cells. Chapter 12 discusses
cell cycles in general, with some emphasis on the mitotic cycle and the eucaryotic
nuclear apparatus. Then in Chapter 13 we look at some of the major types of eucaryotes,
their lines of evolution, and their reproductive cycles. With this background, eucaryotic
(Mendelian) genetics can be discussed in Chapter 14.

Finally, in Part IV we take a series of topics whose general significance has already
been discussed and we consider them in more detail. In Chapter 15 we consider proteins
as large, multifunctional molecules and examine their interactions with small ligands;
here we see how proteins can function as enzymes, as translocators, and as antibodies,
and we introduce the concept of allosteric interactions. In Chapter 16 we try to develop
a realistic model of cell membrane structure, and we see how membranes can behave
in electron transport and energy conservation and how nerve cells may function in
response to small ligands. Chapter 17 is primarily a collection of topics from which
instructors can pick and choose to illustrate interesting points; this compendium of
pathways is used to illustrate various regulatory mechanisms, interesting features
of molecular and cellular organization, and a few examples of metabolic diseases.
In Chapter 18 we examine some general features of regulation and consider briefly
how hormones may be superimposed upon other regulatory mechanisms. Chapter 19
briefly considers the structure of cell surfaces and provides a basis for understanding
cell-to-cell interactions and some aspects of development. Chapter 20 discusses the
viruses, because they are intrinsically interesting, because they are important in com-
municable diseases and probably in cancer, and because they illustrate some features
of regulation and self-assembly so well. Chapter 21 then ties off some loose ends.

The emphasis in this book is on major principles and concepts, not on minutiae; I hope
the book will be used in this spirit. Even though we present many facts in courses

where this approach is used, students are not expected to regurgitate these facts, not even the details of experiments—or even worse, the names of experimenters. We expect them to develop the ability to think about biology, and to test them, we ask them to solve new problems of the kind that are scattered through this book. Many of these problems relate to imaginary extraterrestrial systems, for if we are really learning the principles of biology, those principles should be applicable to any organisms, anywhere. If it is given in the right spirit, this kind of exam can be fun for both students and instructors. We also expect students to develop a feeling for sizes in the biological universe, but rather than making them memorize numbers we ask them to solve quantitative problems; a student who has calculated sizes, numbers, and rates will remember these things more easily than one who has memorized, and he will have information that he can apply to other problems in the future.

In writing this book, I have tried to convey some of the excitement of contemporary biology, because it may well be the most exciting of all modern sciences. Many topics are developed semihistorically to show how science operates at its best. A good working scientist must have a historical sense; he knows what others have done before him and generally what they are doing simultaneously, and he asks questions within the framework of current concepts and developments. He interprets his results in the light of current theories and models, in order to exclude them or force their revision. Science proceeds very rapidly and forcefully through what John Platt has called "strong inference," whereby alternative models are set up and well-designed, critical experiments are performed to distinguish between them. Modern biology is full of such elegant and revealing experiments, and many of them are presented in enough detail so the student can see science at its best and understand its logic. I think it is important to see our current views of the biological world as stages in an intellectual history, rather than truths engraved on golden tablets, and to see that science is a social phenomenon that revolves around real people. I apologize to those who feel that there are too many names in the book, but I would rather lean too heavily in this direction than in the opposite direction of letting it appear that discoveries have come from anonymous sources who know the truth in some mysterious way. I also apologize to anyone who has been part of an intellectual history and does not find his name here; I have tried to reconstruct histories as well as possible, but I have undoubtedly missed some people in the process.

Although many topics are covered in this book, there are certainly many other themes that could have been woven into the framework I have used. For example, I am sure that the whole theme of reactions of organisms to light could have been treated more generally. I would be grateful for any suggestions for similar topics that could be included in a second edition.

I have been developing this book for several years. A preliminary edition entitled General Biology was published in 1967. In spite of its weaknesses, it was classroom tested by several instructors and read thoroughly by others; comments by them and their students have been extremely valuable in shaping this revision. Many other people have contributed criticisms, ideas, teaching strategies, and encouragement; I particularly want to thank Sid Bernhard, Ted Borun, Ted Cox, Grace Donnelly, Bob Edgar, John and Jette Foss, Norman Guttman, Norman Horowitz, Sharon Hotchkiss, Cy Levinthal, Harmon McAllister, David Mehlman, Aaron Novick, William Purves, Tom Roszman,

Frank Stahl, and Gerald Selzer. My wife Shelley proofread the manuscript and helped eliminate some of my worst gibberish and nonsense. My daughter Erica is responsible for the name *groodies* in Chapter 5. I also want to thank many of the present and former staff of W. A. Benjamin for all their help. However, I am the only one who can be held responsible for weaknesses and errors that remain, and I hope anyone with suggestions for future improvements will bring them to my attention.

Burton S. Guttman

Lexington, Kentucky
April, 1970

to the student

There was a time when the only path to higher education was to sit in the cold, drafty hall of a Renaissance university and listen to the Master lecture. The Master was truly a master; he knew essentially everything that could be said about his subject, and the notes a student made during the lecture became a valuable treatise to be cherished and studied. Things have changed a lot since them. In the first place, the Master doesn't know everything. It is impossible for one man to know everything in biology in the 1970s, and none of your instructors would pretend to be more than a guide. Moreover, the attitude of contemporary science is that the methods of experimentation and analysis are much more important than any particular results, and your instructors will probably emphasize these much more than a mere collection of facts.

However, the most important change is in our ability to convey information efficiently to large numbers of students. The great advantage you have over your Renaissance counterpart lies in books like this, made quickly and inexpensively by modern machines, with drawings by professional artists that are much more informative than anything you can copy from the blackboard in two minutes. It therefore behooves you to get most of your information from this book and from other references—preferably before you go into the classroom, so you will be able to follow the lecture better or participate in a discussion.

Biology textbooks frequently suffer from a disease of informational inflammation; there are more trivial facts in biology than anyone knows what to do with, and they have a tendency to fill up textbooks. So the good student takes his thick book and wades through it paragraph by paragraph until he comes to meaty, significant sentences, and then he underlines these. In this book, I have tried to concentrate on concepts, on models, on enlightening experiments, and generally on ways of *thinking* about biology, and I have tried to do this in such a way that you will want to underline most of the book—and so you will have to underline none of it. It would be a crime if you were expected to memorize all the details in this book; I have gone into the details of most things in order to bring out some major take-home lessons, and these are the things you ought to learn. When someone's experiments are described in detail, it is not because you ought to memorize the names of biologists or learn all of those details; experiments are the ways we learn about the world, and I want you to see how our concepts have developed and what the evidence for some concept is. Moreover, some experiments are such models of clarity that they ought to be studied as examples of scientific work at its best.

To help ensure that you understand the major concepts, there are many exercises throughout the text. Work on these as you come to them; they will help you to find out what you do and do not understand. Some of these exercises require that you calculate some simple quantities, such as the number of molecules of a substance in a cell. These are not trivial exercises; you are already familiar with numbers in the world you have to live in, but it is time for you to become familiar with a larger world. Just as you know that people may weigh 100–200 lb and cats may weigh 10–20 lb, you must learn that some cells may weigh 10^{-12} g and other may weigh 10^{-8} g. Other exercises refer to organisms on other planets; if we are really studying the principles of biology, you should be able to apply those principles to radically different biological systems, and you can also learn a lot by making up your own problems about imaginary systems. (To allay the fears of one reviewer, let me assure you that these are all science *fiction* problems.) But work on these problems systematically; there is no point in going ahead until you understand what has come before. And you will often need the information in the exercises to follow the text and solve later problems. (The answers are in the back of the book.)

Please be sure that you go through the Appendixes—at least 1 and 2—early enough to do you some good. The words you ought to add to your vocabulary are in boldface type; Appendix 1 will help you to learn them. The math required in most of biology is not very difficult, but it is not always covered in standard college courses, so Appendix 2 has been added to cover some essentials.

symbols and abbreviations

A	area; absorbancy
B	bacteria (mass or number); cells, in general
C	percent concentration, g/100 ml
c	concentration, g/liter
D	diffusion constant; number of doublings
d	deviation
E	energy; enzyme/cell
e	Napierian base $= 2.718\ldots$
$\mathscr{E}$	electromotive force (emf)
$\mathscr{F}$	Faraday constant
f	coefficient of friction
G	Gibbs free energy
g	growth-yield constant
H	enthalpy
h	Planck's constant $= 6.624 \times 10^{-34}$ joule sec
I	information; light intensity; inhibitor
j	axial ratio of molecule
K	equilibrium constant
k	rate constant; Boltzmann's constant $= 1.38 \times 10^{-23}$ joules/deg
l	length
M	molarity
M	molecular weight
m	mass; number of mutants
N	negentropy; Avogadro's number $= 6.023 \times 10^{23}$ molecules/mole (or daltons/g)
n	refractive index; number
O	osmolarity
P	probability
p	pressure
Q	heat
R	frequency of recombination; gas constant
S	entropy; coefficient of coincidence
S	Svedberg unit
s	sedimentation coefficient
T	absolute temperature; transmittancy
t	time
V	volume
v	velocity
W	work
w	flow rate of chemostat
X	mole fraction
Z	enzyme/ml
z	charge in valence units
α	growth-rate constant
Δ	difference or change

δ specific refraction increment

ϵ molar extinction coefficient

η viscosity

λ wavelength

μ chemical potential; mutation rate

μ electrochemical potential

ν frequency; Simha viscosity coefficient

Π osmotic pressure

π 3.14159 . . .

ρ density

Σ sum

τ generation time

Φ optical retardation

ϕ volume fraction; phase angle; osmotic correction

χ^2 goodness of fit

ψ electric potential

ω angular velocity

Contents

PART  ONE

organisms, molecules, and cells

Somewhere in the universe, at a time and place we do not know, a cloud of hydrogen and other gases begins to take shape. As the eons pass, it condenses under its own gravitation and collects more material, until it becomes unstable and breaks up into smaller and smaller masses. Eventually there are billions of small, dense clouds that become hot enough to initiate nuclear reactions and start to burn with tremendous heat and light. The cloud has become a galaxy of billions of stars.

Many of these stars have thrown off small planets during their growth. The atmospheres of these primitive planetary systems (principally ammonia, hydrogen, methane, and water) capture some of the intense energy of their stars, condense into complex molecules with carbon backbones, and aggregate on the bits of dust that are still drifting about. Some of these aggregates finally fall on planets far out in space, where the sunlight has little warmth; here, even the light gases are frozen and nothing more happens that is of interest to us. Other aggregates fall too close to their stars and burn up. But some fall on planets in between, where the temperatures are moderate and water remains liquid; and here, using the energy of their stars and the raw materials around them, they evolve into large, complex structures with a unique property: They can arrange smaller carbon compounds to make replicas of themselves—they are self-reproducing. The changes come much faster now, as the self-reproducing systems increase in size and complexity and multiply rapidly, so that when these planets have circled their suns millions of times more, their surfaces are covered with a multitude of complex systems.

The Greeks had a word for it. They called these structures *bios*—life. There are living things all around you; you cannot escape them, if only because you are alive too. Were you to travel around the universe you would certainly find them on other planets in other planetary systems, and in other galaxies. They are as much a part of the universe as gases and minerals, gravitation and light, electricity and magnetism.

You belong to a class of living things called man. Man is a curious animal; he spends a large part of his time looking at the universe around him and wondering about it. He probes and examines, experiments and thinks, exchanges his thoughts with others of his kind, and then goes back to probe and examine some more. He does all this in response to curiosity. It is a human trait, and needs no more justification than eating or sleeping. All of this work, and the knowledge that comes of it, is called science. By now, you have presumably studied some of the sciences, such as physics and chemistry, that deal with the most fundamental properties of matter. With this background, you are ready to examine some of these curious bits of matter that we say are "living." The science that deals with them is biology, which takes its name from the Greek word for life.

As you know from your own experience, there are many different kinds of living things on this planet. There are at least a million species of animals and about a quarter of a million species of plants, along with a host of microscopic forms. No one can learn everything there is to know about them all and it would be foolish to try, although you can derive a great deal of pleasure from learning about those that you are likely to come across on camping or hiking trips or while traveling around the country. If you have learned to identify a few hundred birds or plants, you are doing very well. In any case, you would not have gotten this far in your education if you did not already know the names of many different organisms and have some knowledge of how they are built. We shall assume this kind of common-sense knowledge as a basis for this course. Although we shall describe many organisms in more detail, we shall not try to catalog them all, nor shall we dwell upon the differences between them. Our time will be devoted primarily to a discussion of their most general properties, and you will see that there is more than enough general information to fill this book. More specialized information about specific groups of organisms will have to come from more advanced courses.

1-1 A BROAD VIEW OF THE WORLD

There would be no life on our planet or anywhere else without a star like the sun to supply energy. To twentieth century sophisticates like ourselves, it is obvious that all of the biological activity we see around us requires energy, but we sometimes forget that the sun is the source. Let's think about this for a minute. We get our own energy by eating; a large part of our food consists of cows, chickens, fish, oysters, and other animals. They in turn get their energy by eating plants and smaller animals. The fish eats little fish and water plants. The oyster lies in its bed sucking in a stream of sea water, from which it extracts a variety of tiny plants and animals. These little animals, in turn, eat smaller organisms. You can see how complicated the picture can become; if we take a very small, limited environment and draw a line from every organism to all of the other organisms it uses for food, we find a complicated maze of **food chains.**

Every one of these food chains eventually comes back to some plant. The chain may be very short: Man eats cow, cow eats grass. Or it may be quite devious, but the "grass" always appear in some form, whether it be the leaves of a maple tree, the plants in your local fishing hole, or the tiny algae that fill the oceans. If we stand back a few hundred miles and look at our planet with a satellite's eye, much of the land we see is green. We have hacked and burned and abused that green surface, but we are beginning to learn (hopefully) that we dare not hack it much more, for it is the basis of our existence. The green is due to a molecule called **chlorophyll;** this molecule is our link with the universe, for it allows green plants to capture some of the sun's energy and to begin all of the marvelous chemical transformations that lead to the oyster, the fish, the cow, and you and me.

Our planet weighs about 6.6×10^{21} tons. It is convenient to divide it into a number of regions that form more or less concentric layers; there is an *atmosphere* of air, a *hydrosphere* of water, a *lithosphere* of rock, and a **biosphere** of all of the organisms living on its surface. The biosphere weighs relatively little in comparison with the rest, probably about 1.7×10^{13} tons, but it is responsible for an enormous share of the transformation of matter on earth. There is a constant flux of mass and energy through the biosphere, and the source of this energy is the sun. How much energy is involved? Moyer D. Thomas has collected some useful numbers; many of them are estimates, but they agree well with one another and give an order-of-magnitude picture. About 1.3×10^{24} cal reach our atmosphere each year from the sun, but the atmosphere filters out half of this, so 6.5×10^{23} cal/year reaches the surface. Sixty percent of this energy is infrared or ultraviolet light, which plants cannot usefully absorb, so only 2.6×10^{23} cal/year can be used. Of this amount, 1.8×10^{23} cal fall on the oceans and 0.8×10^{23} cal on land; but only about half of the land is covered with vegetation, leaving 0.4×10^{23} cal/year that fall on land plants. This is divided between 44 million km^2 of forest land and 27 million km^2 of farm land, and if the latter lies dormant about a third of the year, only 0.25×10^{23} cal/year are absorbed by land plants.

What do the green plants do with all this energy? They use it to remove carbon dioxide (CO_2) from the atmosphere and **photosynthesize** organic compounds with it— that is, the carbon compounds that are typical of organisms. The efficiency of photosynthesis is about 2%—only 2% of the light energy is converted by the plant into chemical energy. About 10^{10} cal are required to convert 1 ton of carbon from CO_2 to organic material, so each year about 5×10^{10} tons of carbon are converted on land and about 18×10^{10} tons in the ocean. Other estimates agree reasonably well with these figures. In essence, then, about 200 billion tons of carbon are removed from the atmosphere yearly, and much of this is then sent on its way through a series of food chains and is converted into animal tissue or returned to the atmosphere.

While we can easily recite such numbers as 200 billion tons, we do not really understand them. In fact, one of the major problems of understanding biology is that we have to deal with numbers of this size at one time and then a moment later with tiny fractions of a microgram. We can come to grips with this problem by focusing on a part of the world that is more familiar to us.

1-2 THE ECOSYSTEM

A tree stands in the forest, its green leaves reaching up to the sky to capture the sunlight. The sugar the leaves make becomes a sweet sap carried throughout the tree to support its growth; but it also supports many other creatures. Tiny sucking insects suck sap out of the leaves. Some of them, the aphids, are milked of their sap by ants; others are eaten by larger insects. A small nest high in the tree holds a family of warblers, who feed on the insects. A Cooper's hawk in the forest may occasionally add one of the warblers to its diet of mice and rabbits. A few caterpillars munch on the leaves and these add to the birds' diet. Farther down the trunk, a pair of red-headed woodpeckers have made their nest in a hole where a limb was ripped off by an old storm; they occasionally forage the trunk and keep it clear of carpenter ants and bark beetles. And among the roots of the tree, a family of mice have built a warm burrow. The tree is not just an organism by itself; it is a small world of organisms.

Our tree is healthy and growing, but nearby stands the bare trunk of another that has not done so well. Beetles and rabbits have gnawed at its bark and cut the vessels that carry its sap, so the tree has died. Already its base is weakened by a growth of mushrooms that are beginning to push into the heart of the trunk and soften the once strong wood. One day a strong gust of wind suddenly snaps the trunk in two and adds another log to the forest floor.

In the moist soil, surrounded by a thick growth of woodland flowers, the log gradually becomes wet and soft. Now a variety of soil bacteria and molds move in and start to reduce the firm wood to a soft pulp. The large mushrooms spread out over the log; they extend an unseen network of rootlets deep into the wood and add to its decay. The carpenter ants and termites move in and fill the wood with burrows. Water rises into the burrows, carrying more molds and bacteria with it, and soon the bottom of the log is hardly distinguishable from the soil on which it lies. When winter comes, the water in the log freezes and thaws again and again, opening larger cracks in the wood. In a few years, all that remains are a few chips of brown wood; the log is now soil, and the seeds dropped by healthy trees above it are already starting to take hold in it, to grow into new trees and begin the story again.

We do not fully understand this story if we only know about the tree or the hawk or the mushroom, for each of them is only a small part of the whole. The entire scene we have described here is an **ecosystem;** it is a complex of organisms that live with one another and on one another, that supply food for each other, and that generally form a network of food chains and mutual dependencies. In the final analysis, it is the ecosystem as a whole that we must try to understand. If for no other reason, we must understand it because we are part of it, and our survival depends upon understanding our place in it. However, we do not understand the ecosystem as a whole any more than we understand the meaning of 200 billion tons. The whole system—the forest, the pond, the ocean—may be an object of great beauty and a source of pleasure, but we can only comprehend it by trying to dissect it into pieces small enough to be analyzed and by understanding them and then fitting them back together. The rest of this book is primarily an attempt to understand the pieces, although we will try to show their relationships to one another as often as possible.

It is convenient to divide the ecosystem into four components. First, there is the *physical environment;* this is the set of all nonbiological components, including the water, the air, and the mineral environment. Second, there are *producer* organisms, mostly plants but including many bacteria, which remove CO_2 from the physical environment and convert it into the organic compounds of their own structure. Since these organisms feed themselves and are to this extent independent of the rest of the ecosystem, they are said to be **autotrophic.** Third, there are the *consumer* organisms, primarily animals, that live by eating the producer organisms and one another, and thus constitute the major part of the food chains. Fourth, there are *decomposer* organisms, primarily molds and bacteria, that break the bodies of the other organisms back into soil, CO_2, and so on. Since the latter two groups live off the food created originally by the producers, they are said to be **heterotrophic.**

At this point, we are in danger of skipping one of the first roadblocks to understanding. We cannot really understand the food relationships between all these organisms unless we first understand how they are built, what sort of compounds they are made of, and what their food must consist of. We begin by asking what organisms are made of.

We will have to leave the forest and go into a well-equipped laboratory, but we will take with us a few specimens to analyze, such as a fresh bit of animal tissue or a plant. To begin, we weigh the specimen and then put it into a warm oven to drive off all of the free water in it. Its weight decreases sharply at first and then levels off, indicating that as much water as possible has been removed. This experiment reveals that most of an organism's mass is water. The **dry weight** that remains behind is typically about 10–30% of the **wet weight,** although organisms that have heavy supporting structures (such as snail shells, vertebrate skeletons, and the wood of plants) have more solid mass.

Now we place the dried specimen in a dish over a Bunsen burner and carefully incinerate it. If this is done properly, all of the organic compounds in the body will be oxidized to CO_2 and water, leaving behind only an inorganic ash. Over 90% of the dry weight can typically be burned off, so organic compounds are clearly the major non-aqueous components of an organism. Again, of course, the inorganic ash is a larger percentage of the total if there is a solid shell or skeleton, since these are built largely of calcium, phosphorus, and silicon.

For the next experiment, we put a variety of biological samples through some modern analytical procedures to determine how much they contain of each element. The exact composition varies from organism to organism, but some interesting patterns emerge, shown in Fig. 1-1. Ninety-five percent or more of each sample can be accounted for by four elements: hydrogen, carbon, nitrogen, and oxygen. This is not surprising, since hydrogen and oxygen form water, which we already know to be a major component, and the four elements together are the chief components of organic compounds. For every atom of carbon, there are usually one to three atoms of both oxygen and hydrogen and 0.1 atom of nitrogen or slightly less. There is a group of elements sometimes known as **macronutrients** in the range from about 10^{-3} to 10^{-1} on the same scale; it includes sodium, magnesium, phosphorus, sulfur, chlorine, potassium, calcium, and sometimes silicon and iodine. In the range from about 10^{-6} to 10^{-3} there is a group of **micronutrients** or **trace elements** that includes fluorine, cobalt, nickel, manganese, iron, copper, zinc, molybdenum, and sometimes boron and selenium. We can also find small traces of practically every other known element in these samples, but since there are small amounts of these substances everywhere, we can hardly expect them to be absent; they rarely have any specific biological function. However, no matter how carefully and critically we examine our data, we can find no evidence for any element that is unique to organisms. (Would we have called it bionium?) Figure 1-1 also shows the composition of the universe as a whole and that of soils, for comparison; it should be clear that in their elementary composition, organisms are not particularly different from the rest of the universe. Of course, it is not surprising that the composition of soil should be so much like that of organisms in general, since soil is largely made of decaying organisms.

The flow of mass through the biosphere must be a reshuffling of atoms such as carbon, oxygen, hydrogen, nitrogen, and all of the minor components. No organism has a very different elemental composition from any other. When the warbler eats a caterpillar, it is largely because it needs the elements found in the caterpillar's structure in order to make more of its own structure. At this point, it makes sense to pursue this

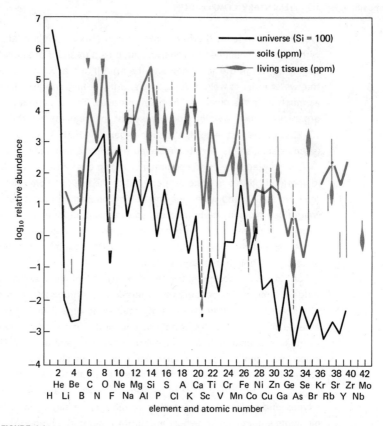

FIGURE 1-1

*Relative abundance of the elements in the universe and in soils, compared with
their abundance in organisms. The width of each vertical bar reflects the number
of samples in the range.*

matter of structure a little further and try to find out generally how the warbler and the
caterpillar are built.

1-4 DESCENDING THE SCALE

By this time, you should be quite familiar with a biological world that includes yourself
and many other objects not very different in size. From your previous education, you
are probably quite familiar with many components inside your own body, such as your
heart, liver, intestines, lungs, and eyes. You know generally where these organs are and
what they do. These are all objects that are large enough to see easily, even if you
personally don't relish the idea of following the medical student's course of study and
examining them in detail. You are also familiar with some fairly small organisms that
are only a few millimeters long, and it is not hard for you to understand the largest
organisms, such as the redwood trees, which reach heights of 365 ft, and the giant

blue whales, which may be 100 ft long. We can easily explore this world, and you have probably already had the experience of dissecting some common animals to see how their hearts beat or how food moves through their intestines.

However, there is a much smaller world, with which you must become familiar and in which you must eventually come to feel as much at home as in the world you now know. You know about objects that are measured by millimeters or centimeters; you must learn about those that are measured by micrometers and nanometers, which are a thousand to a million times smaller. We can begin by gradually descending the scale from a point with which we are familiar, primarily by using a microscope.

The microscope increases our **resolving power.** Two objects are said to be resolved if one can see that they are not a single object. Obviously, the closer we get to something, the more its structure becomes resolved, but the limit of resolution of our eyes is about 0.2 mm. To see objects that are smaller and closer together, we have to use microscopes such as the one shown in Fig. 1-2, which has a resolving power of

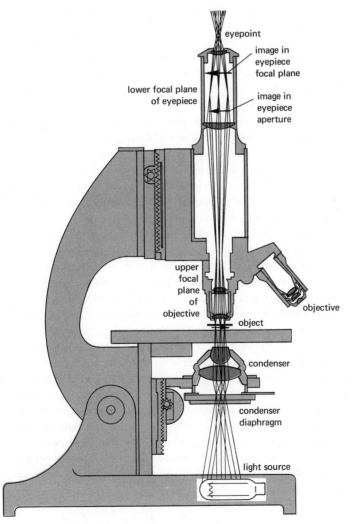

FIGURE 1-2
A compound microscope, showing the light path through the instrument.

about 0.2 μm. We won't use its full power immediately; since it has several lenses with different magnifications, we can gradually descend the scale.

Figure 1-3 shows the trachea or windpipe taken from a small mammal. The whole object is a few millimeters in diameter and can be handled easily. To see how it is built, we have to put it through a regime of **fixation** and **staining.** This piece of tissue would soon disintegrate during all the manipulations it will be put through if it were not treated with a preservative such as formalin (formaldehyde) or osmic acid; these reagents fix the tissue by binding its components into more stable structures (although they may also create artifacts that were not present in the pristine state). The tissue is then embedded in paraffin to make it rigid and is cut into very thin slices with the sharp blade of an instrument called a *microtome.* The slices are then stained with various dyes that have more affinity for some materials than others, so that some structures stand out very sharply in contrast to others.

Figure 1-4 shows a slice through the trachea at such low magnification that the whole cross section can be seen. Notice that the trachea has several more or less concentric layers; to be more technical, we really ought to consider each of the layers a distinct *tissue,* even though we use the word tissue more generally for the whole structure. At this level, we are looking at differences over a few tenths of a millimeter, or, to change units, a few hundred micrometers. Now we can increase the magnification of the microscope and look at finer details, as shown in Fig. 1-5, and we can see that the tissue is really divided into many small compartments, called **cells.** Each cell is somewhat rectangular and has a large **nucleus** near its center. Inside some of the nuclei you can see a prominent, dark **nucleolus.** We have now come down by another factor of 10, and we are looking at cells whose dimensions are around 10–30 μm and even at some of their components, which may be only 1–2 μm in length. This is an important and interesting size range; let's explore it for a while.

Figure 1-6 shows the cells in a thin slice of plant tissue. These cells are even more prominent than those of the animal tissue, primarily because they have heavy **walls** around them. They contain nuclei and many small green bodies called **chloroplasts** which contain chlorophyll. Cells like these were first seen in 1665 by Robert Hooke, who found the little compartments in a slice of cork and gave them the name cell that we still use. However, cork is simply the light, dry bark of the cork oak tree, and Hooke was really looking at the dry walls, not at their contents. He did not really understand the significance of his observations; others who came after him toyed with microscopes and perhaps saw some internal cell structures, but they could not extend their knowledge very far because of the poor quality of their microscopes. There are two kinds of distortion in optical instruments, both of which plagued these early microscopists. **Spherical aberration** is the distortion of shape that is obvious with any crude lens, such as a clear glass ball or a cheap magnifier. **Chromatic aberration** is the separation of light waves with different wavelengths; since the path that any ray of light takes through a lens system depends upon its wavelength, a crude system of lenses disperses white light into a spectrum and objects viewed in the system appear to have colored rings around them. The more lenses introduced into a poor system, the worse these distortions become. In fact, the best early observations were made with simple microscopes, rather than with compound microscopes with several lenses such as we use today.

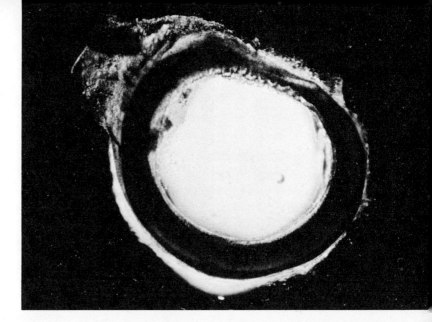

FIGURE 1-3

A view down a rat's windpipe. The tube is about 2 mm in diameter, but even though this picture is greatly enlarged, you cannot see much except the tubular shape and a bit of ragged tissue on the side. (Courtesy of Grace M. Donnelly.)

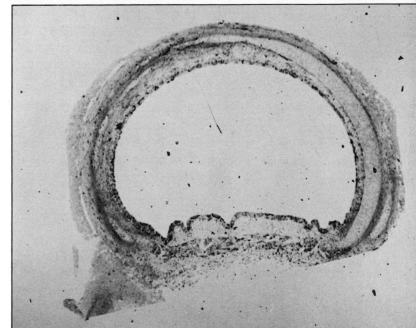

FIGURE 1-4

A thin slice has been taken through the windpipe, and although the magnification is no higher than before, you can see layers of tissue and a hint of single cells. (Courtesy of Grace M. Donnelly.)

FIGURE 1-5

With increasing magnification, you can now see several types of individual cells with nuclei, nucleoli, and a brushy border of cilia. (Courtesy of Grace M. Donnelly.)

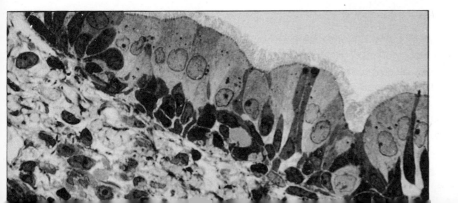

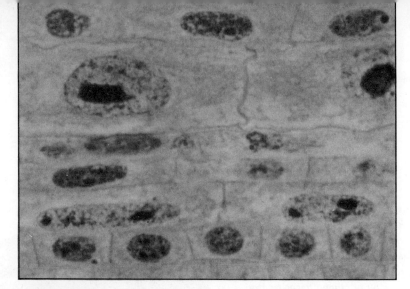

├──10 μm──┤

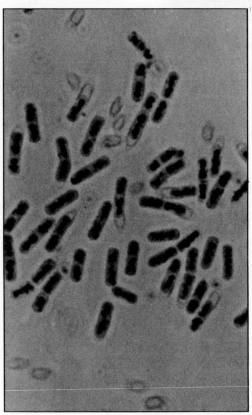

FIGURE 1-6
Cells in an onion root tip. Cell walls are clearly visible and each cell contains a large nucleus, many of which contain black nucleoli. (Courtesy of Stephen D. Smith.)

FIGURE 1-7
Bacterial cells (Bacillus subtilis) *treated with HCl-Giemsa stain to show their nuclear bodies. (Courtesy of Terry Korotzer.)*

However, in the early nineteenth century the remedies for these distortions were discovered and good compound microscopes with better resolution and higher magnification were developed. With these, some internal features of cells could be discerned. In 1831, Robert Brown discovered the nucleus, four years after his discovery of the thermal agitation of small particles that we now call Brownian motion. Gabriel Valentine discovered the nucleolus in 1836, but little else could be seen in cells for a long time.

Other observers saw cells in many tissues; in every case, these cells could be defined by some kind of limiting envelope that separated them from one another, whether they had heavy walls or not. Gradually the significance of these observations began to dawn on a number of people. For example, in 1824 René Dutrochet could write that "all organic tissues are actually globular cells of exceeding smallness, which appear to be united only by simple adhesive forces." Finally, in 1838, Matthias J. Schleiden wrote a new textbook of botany in which he strongly championed the view that cells are universal plant structures, and in the next year Theodor Schwann published some of his observations on animal tissues and extended Schleiden's generalization to them. Schleiden and Schwann are generally given credit for the so-called cell theory, which states in its simplest form that cells are universal units of *structure*. You will find historians of science today trying to achieve a delicate balance in their evaluation of these men; in their roles as publicity agents, if nothing else, Schleiden and Schwann certainly performed an enormous service by saying clearly and forcefully what many of their contemporaries were thinking. But they expressed other ideas, particularly about the origin of cells, that were dead wrong. They believed that new cells either grow out of the nuclei of old ones or are formed from material between existing cells. It was not until some years later (1859), when Rudolf Virchow established clearly that all new cells arise through the division of old ones, that cells were seen as universal units of *reproduction* as well as structure.

One of Hooke's contemporaries, Antony van Leeuwenhoek, used a simple microscope to examine some natural waters and discovered a world of tiny "animalcules." Among them were some bacteria like those shown in Fig. 1-7. We can now see that these are also cells, even though they are usually only a few micrometers long and the cells of plants and animals are generally about 10 times as long. Leeuwenhoek and others who followed him found many other little organisms swimming about in pond water, and it gradually became clear that Schleiden and Schwann's generalization must be extended to them, too. Thus by the middle of the nineteenth century the generalization could reasonably be made that all organisms have a cellular structure, even though some consist of only a single cell. We will not try to define cells any more carefully at this time because of the pitfalls involved in framing strict dictionary definitions. It is sufficient to say that cells are quite small, that they have definite boundaries, and that they generally contain nuclei. As we go along, we will see that they all have a great deal more in common.

1-5 SOME CLASSES OF ORGANISMS

Our world has been divided from ancient times into animal, vegetable, and mineral kingdoms, and none of us has any trouble distinguishing these from one another on the basis of some common-sense criteria. In the first place, we can distinguish the third from the other two because we know intuitively what it is to be alive. We feel life within ourselves, we enjoy it, and we hope the condition will persist for a while. We ascribe life to other things because they exhibit some kind of activity, such as barking, or growing on leftover food, or turning green in the spring. And you certainly know what I mean if I say that I live in America, or my uncle is alive, or the old gray goose is dead.

However, there is no technical meaning for terms like "life" and "alive" that is understood and agreed upon by all biologists. There have been many arguments about this point and many attempts have been made to find simple desiderata that will clearly distinguish all living things from all nonliving things. The standard solution is to list characteristics such as the ability to grow, sensitivity to stimuli, or the ability to move, and say that something is alive if it has most of these features, but this has never been very satisfactory.

It is most reasonable to admit that we simply don't know yet what it means to be alive in the sense in which we feel life; in a way, all of this book is an attempt to clearly define the characteristics of living things, but we will not try to give technical meanings to words like "life" and "alive." We will only use these terms in their colloquial sense from now on. For technical purposes, we will talk about *organisms,* which we will try to define in the next few chapters. More generally, we will talk about **biotic systems,** which may be whole organisms, or their parts, or populations of similar organisms, or a whole ecosystem made up of very different organisms. (System, after all, is just a general term for any part of the universe that we choose to isolate and study.) We are more likely to get a clear view of biology by taking this more modest approach to it.

We certainly use some common-sense criteria to distinguish plants from animals. We think of plants as **sessile** organisms—that is, they stay in one place—that get their nourishment from the air and the soil, while animals are **motile** organisms—they move around independently—that get their nourishment by eating something else. As we have seen, plants are usually green, but we include the fungi, such as mushrooms and molds, among the plants in spite of their color. People who live near the ocean are familiar with a variety of seaweeds that are generally red or brown, and which they also think of as plants.

We have seen that there are some differences between plant cells, which have heavy walls and generally contain chloroplasts, and animal cells, which generally have only thin envelopes and are not colored. However, both plant and animal cells are very active. Fresh, growing plant cells generally show a **cyclotic motion (cyclosis)** in which their chloroplasts and other cell contents move around and around the cell perimeter. Animal cells can be seen squirming around and sending out little extensions here and there, although sometimes time-lapse photography is required to show this activity.

The plants and animals we have been looking at are **multicellular**—they are built of many cells. They are also **differentiated,** so that each organism consists of several different cell types. For example, an adult human consists of about 10^{14} cells of 90–100 different types: muscle cells, nerve cells, blood cells, bone cells, liver cells, and so on. Each one has a distinctive size and shape, and their names imply that they are specialized for different tasks, such as contracting, carrying messages, or supporting. The organism as a whole is an integrated complex of specialized cells.

The same thing is true of a plant, although there are fewer distinct cell types. However, we find a different situation in the fungi and the large red and brown seaweeds. These organisms are also multicellular, but all of the cells look essentially alike and are undifferentiated (except for some reproductive cells, which we will ignore for the moment). These organisms are more like colonies of similar cells than highly integrated multicellular organisms.

The microorganisms that were discovered as microscopes improved were always classified as either plants or animals. Some, which we commonly call *algae,* are

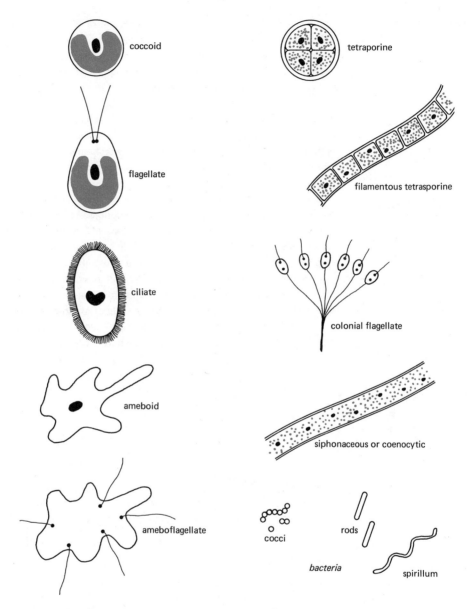

FIGURE 1-8
A variety of microorganisms.

colored green, yellow, red, or brown and generally resemble plants, so they were called *protophyta*—"first plants." Others look more like animals cells, are typically active, and often eat solid food, so they were called *protozoa*—"first animals." The tiny fungi, such as yeasts and molds, were generally included in the protophyta, as were the bacteria, which resemble fungi but are much smaller. Some of these organisms are shown in Fig. 1-8.

As more organisms like algae and protozoa were discovered, it became clear that the old division of the organic world into plants and animals was artificial, for too many of the new species had some of the characteristics of both. In 1866, Ernst Haeckel proposed a new classification with a third group, *protista,* to include the protozoa, many of the algae, some small fungi, the sponges, and others. More modern versions of this system have been published, but we are just entering an era in which the really definitive criteria for classifying organisms are being discovered. It would be misleading to write down a modern classification now, so we will continue to use terms like algae, protozoa, and fungi in a nontechnical way, just as we use words like fish, worm, and tree. We will not try to set up a complete, technical classification of organisms.

We do have to make one other distinction between different kinds of organisms, but for this we must improve our powers of resolution and come down the size scale a little more. The light microscope reveals details of a few tenths of a micrometer or, to switch units again, a few hundred nanometers. To see structures that are measured in nanometers or tens of nanometers we have to use the electron microscope, shown schematically in Fig. 1-9, which uses a beam of electrons instead of light and magnets to focus the beam instead of lenses. The microscope looks at electron density, which is normally quite uniform in a cell made mostly of organic compounds, so the cell must be stained with heavy metals, using such reagents as osmium tetroxide, OsO_4, and lead hydroxide, $Pb(OH)_2$. The regions where these metals have been deposited

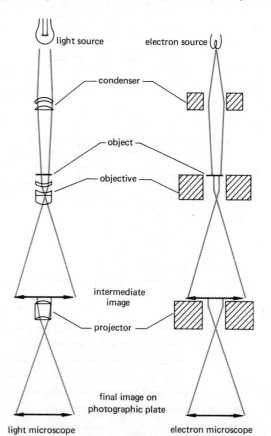

FIGURE 1-9

A schematic view of an electron microscope, comparing its optical path with that of a light microscope.

light source   electron source

condenser

object

objective

intermediate
image

projector

final image on
photographic plate

light microscope   electron microscope

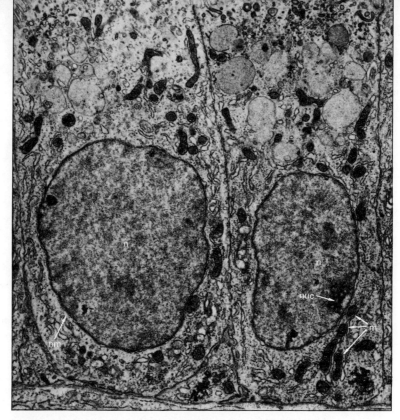

FIGURE 1-10
*A low-power electron micrograph (enlarged 4700 times) of cells in a rat yolk sac,
showing nuclei (n), nuclear membrane (nm), nucleolus (nuc), and mitochondria (m).
This material was fixed in 2% OsO₄ and stained with Pb(OH)₂. (Courtesy of
Roger O. Lambson.)*

absorb electrons and make dark areas on a photograph. It is always important
to keep in mind that you are looking at such a pattern of electron density and not at
the cell structure directly; as with light microscopy, we are often uncertain about the
relationship between what we see and what was originally there.

A typical animal cell reveals an enormous amount of internal structure when
properly fixed and stained (Fig. 1-10). The nucleus can be seen as a region that is
delimited from the rest of the cell by a **nuclear envelope.** Dark material generally
called **chromatin** fills much of the nucleus and prominent nucleoli are often visible. A
large part of the cell is filled with membranes. There are large bodies called
mitochondria (singular **mitochondrion**) scattered through the cell, and these have a
system of internal membranes. Other structures can be identified now and then,
but these are the most common ones.

However, if we examine some bacterial cells that have been prepared in the same
way we can see that they are much simpler (Fig. 1-11). Very few internal membranes
are ever seen and there is never a nuclear envelope to separate a nucleus from
the rest of the cell. There is a diffuse central mass (or more generally two to four such
masses) called the **nuclear body.** There are no mitochondria; in fact, the entire
bacterium is no larger than a single mitochondrion from an animal cell.

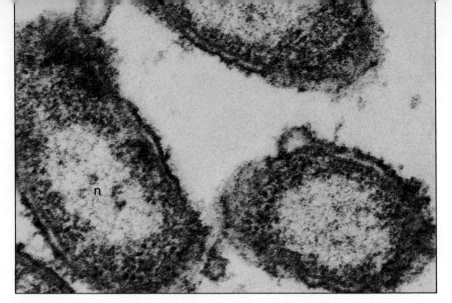

├────── 0.5 μm ──────┤

FIGURE 1-11
Electron micrograph of bacterial cells (Hemophilus influenzae) *showing their simple internal structure. n = nuclear body. This picture is enlarged about 70,000 times. (Courtesy of David C. White.)*

Roger Y. Stanier and C. B. van Niel noted the great difference in size and organization between bacterial cells and most other cells. They called the organization of bacteria and some of their relatives **procaryotic** (prokaryotic) and the organization of other cells **eucaryotic** (eukaryotic). Individual cells may be called **procells** and **eucells.** The chief differences between them are (1) size and (2) the presence or absence of a nucleus and associated internal membranes. We will find many other differences as we go along.

1-6 ORGANIC COMPOUNDS AND POLYMERS

We turn now to a different view of biological structure. Since the organic compounds are the most definitive and characteristic molecules in organisms, we will separate the major classes of these compounds and see what they look like.

Organic compounds are fundamentally chains of carbon atoms; they can reach enormous lengths because the carbon-carbon bond is so strong. Carbon atoms generally form four equal bonds to other atoms, which may include hydrogen, nitrogen, oxygen, phosphorus, and sulfur (the molecules made from these elements are called CHNOPS compounds) so the number of possible organic compounds is truly enormous.

In the basic hydrocarbon chain (Fig. 1-12), the C—C bonds generally alternate in direction; we will often represent hydrocarbon chains by zigzag lines

where each peak represents a —CH_2— group. Double and triple bonds within hydrocarbons cause slight distortions from this pattern. The chains may branch and they may form closed rings, with or without other atoms, which may be saturated or unsaturated; the rings that are most important for us are shown in Table 1-1.

FIGURE 1-12

Model of the basic hydrocarbon chain. The dark units in the core are carbon atoms; the white balls are hydrogens.

TABLE 1-1

some organic ring compounds having biological importance

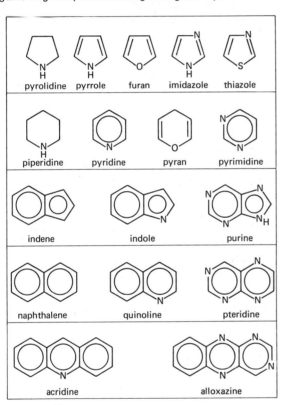

pyrolidine   pyrrole   furan   imidazole   thiazole

piperidine   pyridine   pyran   pyrimidine

indene   indole   purine

naphthalene   quinoline   pteridine

acridine   alloxazine

saturated ring   cyclohexane

unsaturated ring   =   +

electron smear
abbreviation

for two or more
resonating forms

benzene

Since it is easy to grow small organisms like bacteria, let us begin with a suspension of these cells. We first add 5% trichloroacetic acid (TCA) and let them sit on ice for an hour. The TCA separates soluble material from an insoluble residue that can be collected by centrifugation at moderate speeds. The soluble fraction is extremely complex and contains many organic compounds and inorganic ions; 8–10% of all the carbon and 20% of all the phosphorus in the cell can typically be found here. These compounds are very small, and therefore they do not sediment in the centrifuge. But we will not try to analyze the soluble fraction further at this time.

The insoluble residue can be fractionated again. First, we treat it with ethanol or an ethanol-ether mixture; everything that will dissolve in these organic solvents is called **lipid.** In bacteria, this constitutes 10–15% of the dry weight and consists mostly of hydrocarbons, including a variety of **fatty acids,** which are hydrocarbons with a **carboxyl group** at one end. There may also be rather brightly colored hydrocarbons in this fraction.

$$-C\overset{O}{\underset{OH}{\diagup\!\!\!\diagup}} \rightleftharpoons -C\overset{O}{\underset{O^-}{\diagup\!\!\!\diagup}} + H^+ \qquad CH_3CH_2CH_2CH_2\cdots CH_2C\overset{O}{\underset{OH}{\diagup\!\!\!\diagup}}$$

carboxyl group, uncharged and charged   fatty acid

The insoluble residue that remains after lipid extraction can now be divided in two by adding 5% TCA again and heating it to 90°C for 30 minutes. A fraction of this material amounting to 15–20% of the dry weight of the cells dissolves in the hot TCA; it is acidic, contains a large amount of the cell phosphorus, and is called **nucleic acid,** since it constitutes a large share of the material inside the cell nucleus. The fraction that is insoluble in hot TCA amounts to 50% of the dry weight, contains almost all of the cell sulfur, and is called **protein.** Proteins and nucleic acids can both be dissolved in various salt solutions and their properties analyzed further.

One of the most striking characteristics of proteins and nucleic acids is their size. A solution of either of them can be spun very rapidly in an analytical ultracentrifuge, which can reach a speed of nearly 60,000 rpm and exert enormous gravitational forces (nearly 10^6 times g). Since the speed with which an object falls in a frictional gravitational field is a function of its mass and shape, the sizes of large molecules can be measured by subjecting them to these great forces and measuring their sedimentation rates with an optical system (see Appendix 3). The molecular weights of proteins and nucleic acids range from about 15,000 up to several million daltons (a dalton is 1 atomic mass unit, the mass of a hydrogen atom); in contrast, most of the organic compounds in the soluble cold TCA fraction weigh no more than a few hundred daltons. However, whenever the large molecules are subjected to a strong acid or strong base, they can always be decomposed into the smaller molecules.

The cell proteins may be incubated with hot 6-N HCl for several hours to break them into α-**amino acids,** which have both a carboxyl and an **amino group** on the

$$-NH_2 + H^+ \rightleftharpoons -NH_3^+ \qquad R-\overset{\overset{\displaystyle NH_3^+}{|}}{\underset{\underset{\displaystyle H}{|}}{C}}-C\overset{O}{\underset{O^-}{\diagup\!\!\!\diagup}}$$

amino group, uncharged and charged   amino acid

α carbon (C_α). Their general formula is given in the margin, where R represents some unspecified radical. Two amino acids can be bonded together as follows:

$$\overset{R_1}{\underset{H}{^+H_3N\cdot C\cdot COO^-}} + \overset{R_2}{\underset{H}{^+H_3N\cdot C\cdot COO^-}} \rightarrow \overset{R_1\quad H\;\; R_2}{\underset{H\;\; \overset{\|}{O}\;\; H}{^+H_3N\cdot C\cdot C\cdot N\cdot C\cdot COO^-}} + H_2O$$

The (C=O)—NH structure is a **peptide linkage** and the whole compound is a **dipeptide.** Since the dipeptide still has a free amino and a free carboxyl group, a third amino acid could be added at either end to make a *tri*peptide; this process could be repeated indefinitely to make an *oligo*peptide, with a *few* amino acids, and finally a *poly*peptide, with *many* amino acids. Notice that each peptide bond is synthesized by a process of **dehydrolysis,** in which one molecule of water is removed; the portion of the amino acid that remains in the polypeptide is a **residue.** The treatment with HCl causes **hydrolysis** of the protein by adding water molecules and breaking the peptide bonds.

The protein is called a **polymer** (literally, *many parts,* from the Greek) and its constituent amino acids are **monomers;** the number of monomers in a polymer is its **degree of polymerization (DP).** An oligomer is a short polymer. A homopolymer consists of identical monomers, but most biological polymers are heteropolymers, consisting of several very similar monomers. By somewhat arbitrary conventions, a polypeptide is generally called a protein only when its DP is greater than about 40–50 (about 5000 daltons), although no one would argue strongly about the distinction between a large polypeptide and a small protein.

aldehyde group

keto group

hydroxyl group

sugars (ring forms)

A variable amount of the organic matter consists of **sugars;** these compounds are **aldehydes** and **ketones** with several **hydroxyl** groups that have a formula such as $CH_2OH(CHOH)_nCHO$, where *n* is usually 3 or 4; these sugars generally exist as rings, as shown in the margin. A single sugar molecule or **monosaccharide** can be bonded to another sugar by a **glycosidic bond:**

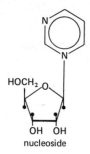

nucleoside

This makes a **disaccharide,** and again, oligosaccharides and polysaccharides can be built. These compounds appear in several of the fractions that can be separated from cells.

When the nucleic acid fraction of the cell is hydrolyzed, monomers called **nucleotides** and **nucleosides** appear. A nucleoside is a sugar bonded to some large ring compound, and a nucleotide has a phosphate on the sugar as well. Two nucleotides can be joined through phosphodiester bonds to make a dinucleotide:

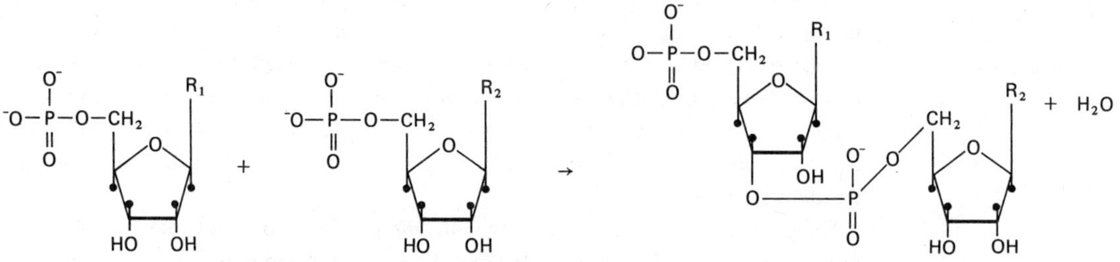

This process can be continued to make polynucleotides, which are the nucleic acids. Several nucleotides and dinucleotides are important components of the soluble fraction of the cell.

As you have probably learned by now, many organic compounds contain asymmetrical centers. A carbon atom usually forms four equally spaced bonds directed toward the corners of a tetrahedron, with the carbon nucleus in the center. If the carbon is attached to four different groups, it is asymmetrical because there are two different ways to arrange these groups. The two possible molecules are mirror images of one another and are called **enantiomers;** they are also called **optical isomers,** because they interact differently with polarized light, one rotating the light to the right and the other to the left.

One of the simplest optically active molecules is glyceraldehyde, the enantiomers of which are shown in Fig. 1-13 in both perspective and two-dimensional representations. The designations D and L are arbitrary; they can be extended to several series of compounds related to glyceraldehyde, including the sugars and amino acids. All of this is important here simply because the monomers of most biological polymers belong to one series or the other, but not to both. All amino acids found in proteins

FIGURE 1-13

Perspective and projection view of optical isomers.

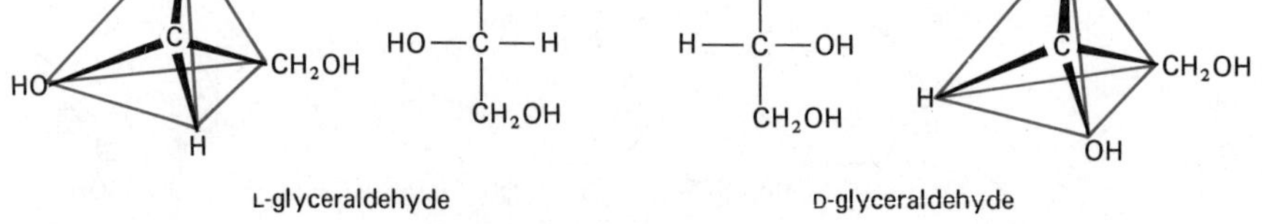

L-glyceraldehyde   D-glyceraldehyde

belong to the L series; the D acids are only found in special places. The D series of sugars are most often found in polysaccharides, although the L series is not excluded.

Why should biotic systems be made of polymers like these? The answer must be premised on the idea that nature is a good architect and that it is advantageous for organisms to be well designed. A human architect contemplating some new buildings does not generally plan on making each one out of a huge block of concrete molded on the spot or an enormous piece of steel cast as a whole. It is much more efficient to use many small units—bricks, girders, wooden panels, etc.—that are designed so they can be combined in many ways. Then a large number of different buildings can be constructed by varying the arrangement of the subunits.

The proteins, nucleic acids, and polysaccharides easily account for 80 % or more of the organic material in a cell. Polymeric construction is clearly very economical because so much of an organism can be made out of similar or identical subunits, using only a small number of distinct chemical reactions. Through polymerization, an organism can easily make the three basic structures it needs: *strings, branches,* and *sheets.*

The strings are **linear** polymers like polypeptides and polynucleotides. Amino acids and nucleotides both have two groups that participate in polymerization, so they form linear molecules in which all of the bonds between monomers are identical and can be made by repetition of the same chemical reaction:

A linear polymer has a **polarity;** for example, a polypeptide can be assigned an arbitrary polarity from the end with a free amino group to the end with a free carboxyl group.

Polysaccharides are typical of **branched** polymers. Some polysaccharides are linear, but since each sugar has several hydroxyl groups that can participate in bonding, a single sugar may be bonded to three others to make a branch point:

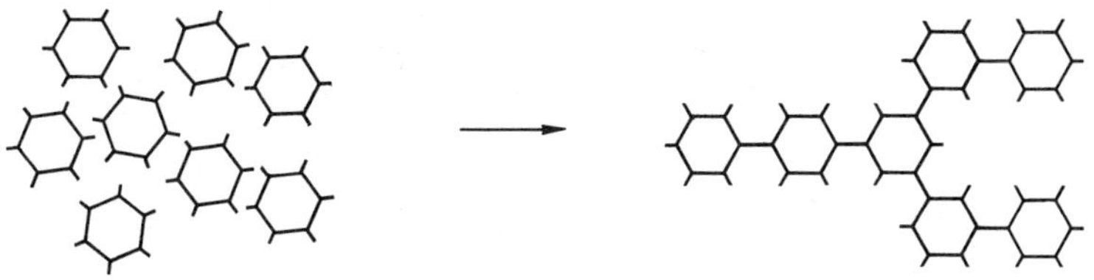

All of the bonds in such a polymer are not necessarily identical, but nevertheless, only a few distinct chemical reactions would be necessary to make this structure.

We will see later that there are more complex monomeric units that can participate in one type of bond in one direction and another type of bond in a perpendicular direction, to make a **network** polymer:

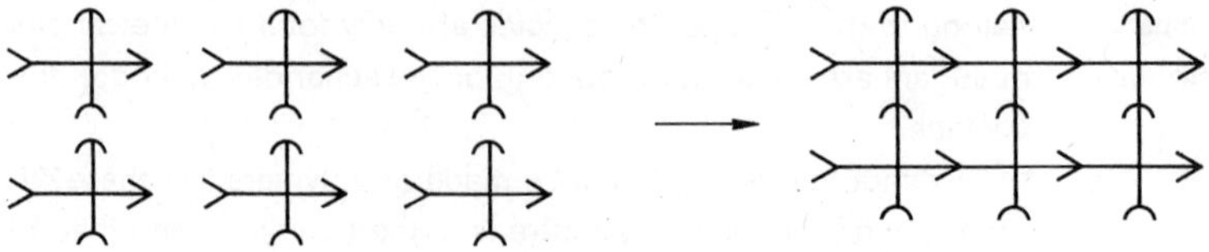

The best examples of sheets like this are found in bacterial cell walls.

Polymeric construction is also a way of achieving enormous variety. Proteins consist of 20 different amino acids, most nucleic acids of five different nucleotides, and polysaccharides of about 25 sugars. Since these monomers can be combined in any sequence, the number of possible polymers is enormous. For example, in a small protein with DP $= 150$ each position can be occupied by any one of 20 amino acids, so there are 20^{150} different proteins of this size. This number is more than astronomical; it is beyond comprehension. It should be obvious that the biotic diversity we observe can be easily accounted for by differences between the polymers of which organisms are made. Even if every organism were made of millions of different proteins, the number of organisms that have existed on our planet would be an insignificant fraction of the number of possible organisms.

exercise 1-1 Your local toy store carries "pop" beads that are amusing and instructive for children and biologists. Each bead has a knob on one side and a hole on the other into which a knob will fit snugly, so the beads can be strung together. What kind of pop-bead polymers could you make from the following:
(a) Beads as described, all of them red?
(b) Beads as described, colored red, blue, yellow, and green?
(c) Beads with two knobs and one hole each?
(d) Beads with two knobs and two holes each?
(e) Beads with two knobs and no holes each?

exercise 1-2 Which atoms in the following compounds are asymmetrical?

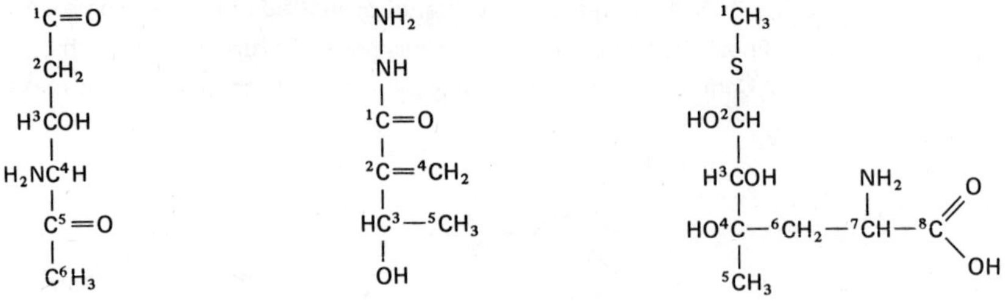

1-7 WATER

If 70–90% of most tissue consists of water, then the properties of water must be very important for biotic systems and we ought to understand them. Figures 1-14a and b show some physical properties of compounds with the formula RH_n; each series (points connected by a line) is based on elements from one column of the periodic table. Notice that the properties of most compounds change regularly as a function of the molecular weight of R, except when R is oxygen, nitrogen, or fluorine; water,

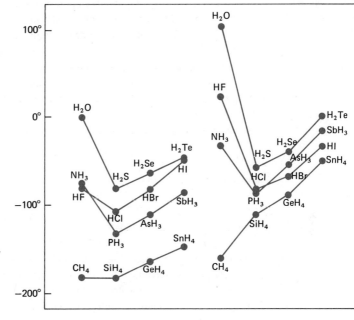

FIGURE 1-14

Properties of isoelectronic sequences of hydride molecules. (a) Melting points (left) and boiling points (right). (b) Heats of vaporization.

(a)

(b)

ammonia, and hydrofluoric acid melt and boil at higher temperatures than one would predict and with considerably greater changes in energy. This suggests that these compounds have unique bonds that require additional energy for disruption, and in fact, N, O, and F atoms, among others, form **hydrogen bonds** with one another. In these bonds, a proton (hydrogen ion) is held between two more negative atoms. The hydrogen is carried into the bond by a donor atom, D, and becomes linked to an acceptor atom, A, so the bond may be represented by —D—H · · · A. The bond strength increases as the positive charge on the donor and the negative charge on the acceptor increase. In order of increasing strength, some typical donor groups are

$$SH < \text{—OH} < \text{—NH} < HF$$

and some typical acceptors are

$$Cl^- < F^- < N: < C{=\mathrel{\mkern-3mu}=}O < \text{—OH} < O^-$$

Hydrogen bond energies typically range from about 3 to 6 kcal/mole.

The crystal structure of ice is shown in Fig. 1-15. The oxygens are negative and each one is simultaneously the donor for two hydrogen bonds and the acceptor for two more, so it is surrounded by four hydrogens and the basic unit of the crystal is a tetrahedron. In ice, such a structure continues indefinitely, while water probably consists of small crystalline regions ("icebergs") surrounded by more random regions. As water is heated, it becomes more random, but the high heat of vaporization indicates that even at 100°C there are still many hydrogen bonds that must be destroyed during boiling. The energy of a hydrogen bond in water is only 4.5 kcal/mole, compared with 110 kcal/mole for the covalent O—H bond, but this small energy is enough to account for some of water's unusual properties.

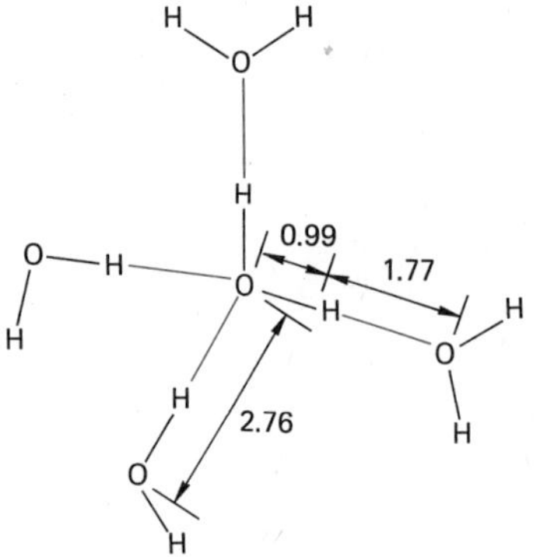

FIGURE 1-15
The crystal structure of water (distances in Å).

When water is heated, part of the heat energy goes into breaking hydrogen bonds; when it is cooled, energy can only be removed by making hydrogen bonds. Because of the large amount of heat required to change the properties of water significantly, three of its thermal parameters are unusually high:

1. **Heat capacity,** defined as the amount of heat required to raise the temperature of a substance by 1°C. By definition, the heat capacity of water is 1 cal/g, while the heat capacities of most organic compounds are close to 0.5 cal/g.

2. **Heat of vaporization,** defined as the amount of heat required to vaporize 1 g of a substance. The heat of vaporization of water is 540 cal/g at 100°C, in contrast to values of less than 100 cal/g for most organic compounds.

3. **Heat of fusion,** defined as the amount of heat that must be removed from a substance to freeze 1 g. The heat of fusion of water is 80 cal/g, which is about twice the value for most organic compounds.

Because of its open, hydrogen-bonded structure, water is one of the few substances that is less dense in its solid than in its liquid form. It also comes as close to the ideal of a universal solvent as any alchemist could wish; more substances will dissolve in it than in any other liquid. Most salts become highly dissociated in water and are therefore very soluble in it; this is due to water's very high dielectric constant, which is a measure of the force between two electostatic charges in a given medium compared to the force between them in a vacuum. The higher the dielectric constant, the weaker is the force; thus, anions and cations in water attract each other less strongly than in most other media, so they have a greater tendency to remain in solution rather than aggregating and precipitating out.

Water is truly a unique substance. Strong hydrogen-bonded structures are formed in ammonia and hydrofluoric acid as well, but these compounds only form chains or rings of hydrogen-bonded atoms, not the extensive three-dimensional crystals found in water. Some years ago, Lawrence J. Henderson pointed out just how uniquely suitable water is for biotic processes. The main points of his thesis are worth summarizing here.

First, organisms are susceptible to heat and must cool themselves. An organism living in the water is protected by the high heat capacity of the medium, which can absorb a lot of heat with little change in temperature. An organism living on land must cool itself, but conserve its internal fluid. Since water has such a high heat of vaporization, the evaporation of a small amount dissipates a lot of heat. An organism built largely of some other liquid would have to use much more to effect the same cooling.

Second, organisms are susceptible to cold and must maintain their temperatures. Again, the high heat capacity of the liquid protects them against severe temperature changes. Furthermore, it is harder to solidify water than most other liquids, so that an organism made of water is more protected against freezing.

It is also significant that water is one of the few compounds whose solid phase is less dense than its liquid phase. Ice floats on top of lakes and rivers, and when spring comes it melts, leaving the surface free and clear. Henderson pointed out that if ice were more dense than water it would sink to the bottoms of lakes and rivers, so that eventually the whole body of water might freeze and in the spring it would only melt a little at the top. No organisms could pass the winter below the ice as they now do; it is unlikely that aquatic organisms could survive, and there might well be no organisms at all on earth or any other planet.

Finally, it is important that water is such a good solvent. Organisms must carry out many chemical reactions in solution, particularly on organic compounds that are soluble in water. The hydrocarbons that are not water soluble separate out and form parts of the membranes that separate one aqueous compartment of a cell from another.

As the biologists of about 1840 came to realize that the cell is a universal unit of structure, they began to conceive of its contents as having a certain unit also. Schleiden called the contents of the plant cell "plant slime" and Felix Dujardin said that the insides of animal cells and protozoa consists of "sarcode." In 1846, Hugo von Mohl concluded that the two substances are identical and united them under the term "protoplasm" that Johannes Purkinje had used to denote the fundamental material of the animal embryo. "Protoplasm" became "the physical basis of life," in the words of Thomas Huxley, the great popularizer of biology, and in the elucidation of its properties lay the secrets of life itself.

In the late nineteenth and early twentieth centuries, many people tried to understand the structure of "protoplasm." In 1861, Lionel Beale distinguished the "formative" or "living" matter of the cell from the "formed material," such as starch granules, crystals, and oil droplets, that obviously could not be considered living in themselves. But many theories developed about the structure of the "living" part; some people thought it was basically made of little fibers, others that it was a complex network, and still others that it was basically a foam, depending upon the pictures that their microscopes revealed. These theories were laid to rest when investigations of microscopic techniques showed that many of the structures seen in cells were merely artifacts of fixation and staining.

Early in this century, colloid chemistry was developed and then the most popular view was that "protoplasm" is a complex colloid consisting mostly of proteins where the secret of life was somehow buried. This view could be maintained until the rise of modern biochemistry, when it became possible to identify all of the individual proteins, nucleic acids, and other components in the cell as distinct chemical entities, none of which had any magical properties. At the same time, the first really good electron micrographs of cells started to appear, and from this point on very fine details of cell structure could actually be seen.

We have identified many distinct chemical compounds in cells and we have dissected them down to a resolution of a few angstrom units (1 Å = 10^{-8} cm). We have found nothing that we can call "life" and nothing that fits the concept of "protoplasm." There is no "unitary substance of life;" there are many substances, each with its own role in the total structure and economy of the organism. The region of the cell that the old mapmakers would label *terra incognita* has all but disappeared. Like the Cheshire Cat, the more we look at the "protoplasm," the more it disappears and now only its mocking smile is left behind, laughing at us for being silly enough to think that a smile can exist without a cat.

The terms "nucleoplasm" and "cytoplasm" used to be used for the "protoplasm" of the nucleus and the extranuclear region, respectively. "Cytoplasm" is still used regularly to designate the region outside of the nucleus, with no intention of conjuring up any demon concepts from the past. However, we will use the term **cytosome** for the extranuclear region whenever possible.

There is clearly a fluid portion in all cells, and we will call this the **cytosol.** This term is generally used for the cell fluid that is outside of the nucleus, chloroplasts, mitochondria, and other such compartments, but there is really a constant exchange of materials between these regions. Cytosol is a good word for the fluid, aqueous portion of the cell, regardless of where it is, but the word must not be allowed to imply that there is only a single kind of aqueous phase with an invariant composition. The composition of the cytosol depends upon the reactions that are taking place in the cell and it undoubtedly varies enormously from cell to cell and from one part of a cell to another.

Electron microscopists often use the terms *hyaloplasm* or *ground substance* for any clear, apparently structureless regions of a cell. You should recognize these words; there is nothing wrong with them, as long as you don't trick yourself into thinking of the cytosol, hyaloplasm, or anything similar as a kind of modern substitute for "protoplasm." Conceptually, there is all the difference in the world.

SUPPLEMENTARY EXERCISES

exercise 1-3   Calculate the number of different linear polymers with various degrees of polymerization that can be made in each case.

(a) How many dimers (DP = 2) can be made out of four different kinds of monomers?

(b) How many different trimers (DP = 3) can be made out of five different kinds of monomers?

(c) How many polymers of DP = 100 can be made out of 10 different kinds of monomers?

exercise 1-4   Idealize a bacterium as a cylinder 1 μm in diameter and 2 μm long.

(a) Calculate its surface area in square micrometers.

(b) Suppose the surface of the cell is covered with a network polymer whose monomers are 10 Å $\times$ 10 Å. How many such monomers would it take to cover the cell?

exercise 1-5   The green leaf I hold in my hand looks uniformly green, but when I look at a section under the microscope I see that the green is confined to isolated chloroplasts in a clear background. What has the microscope done to enhance my view of the leaf's structure?

exercise 1-6   Assuming that a useful protein could be built either with L-amino acids or D-amino acids, but not a mixture, would you ever expect to find an ecosystem in which some organisms are built out of one type and some out of the other? Why, or why not?

exercise 1-7   Let's think for a moment about the soil to which the organisms in our ecosystem eventually return when they decay. A good, rich soil might have about 10^{10} bacteria, 10^4 algae, 10^4 protozoa, and 10^7 fungal masses per gram. If we say that each bacterium weighs 2.5×10^{-12} g, the algae and protozoa weigh 10^4 times this, and the fungi weigh 10^3 times this, roughly how much of the soil consists of microorganisms? Does this explain why the distribution of elements in soil is so much like that in the biosphere?

exercise 1-8   A wet bacterium weighs 2.5×10^{-12} g.

(a) If it is 90% water, what is its dry weight?

(b) What is the "molecular weight" of a dry bacterium in daltons?

(c) If half the dry weight of the cell is protein and the average protein weighs 3×10^4 daltons, how many protein molecules are there in the cell?

Biologists spend a lot of time looking at both light and electron micrographs. After some years of experience, it is easy to interpret a two-dimensional picture in realistic, three-dimensional terms, but it is sometimes hard for a student with little experience to do the same. Let's see if we can condense this experience into a short lesson on the elements of the art.

When we pick a piece of soft tissue to be examined, we usually excise very tiny blocks, no more than a few millimeters in any dimension, and drop them into a fixative such as formalin or alcohol. If we used larger pieces, the fixative would diffuse into the center too slowly, so parts of the tissue would be poorly fixed and the whole thing would be uneven. The piece is then embedded in something solid enough to hold it while it is being cut. The standard for many years has been paraffin; the tissue is saturated with xylene and then transferred to warm, melted paraffin, which penetrates completely, so that when the paraffin is cooled suddenly every part of the specimen is well supported. Electron microscopy requires a stronger supporting material, so a plastic such as epoxy resin is used. The tissue is saturated with a solvent in which the monomers of the plastic are soluble; the monomers are then introduced and finally a catalyst is added to polymerize them into a solid.

The block of tissue is then sectioned, and this is where three dimensions are reduced to two. The slices taken for light microscopy can be as thin as 1 μm, although 5 μm or more is a more common thickness, and for electron microscopy, sections of only 20 nm can be cut. When we focus on such a thin slice, we are looking at an essentially two-dimensional field.

The easiest way to see what happens in sectioning is to look at an oversimplified animal that has a gut and one blood vessel. We'll make some slices through this animal and look at them face on:

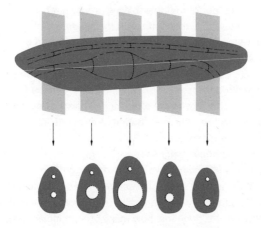

In each section, the gut is a rather large, rounded space and the blood vessel is a small, round space. We can see that the blood vessel is quite uniform all along the length of the animal, but the gut is small at both ends and enlarges in the middle, as a kind of stomach.

When pieced together, these serial sections really produce an excellent picture of the animal. The process is similar to creating a moving picture from a series of still pictures flashed on the screen in rapid succession. Incidentally, one reason we can do this so easily in microscopic work is that the slices come off the microtome as a long ribbon, so each section is in its proper position and we merely have to avoid breaking the ribbon indiscriminately.

Now, what do we see when we look at a cell? We can idealize everything in the cell as either a solid rod, a ball, or a sheet. Suppose there are solid rods running through the cell; thin rods will generally be called *fibers* and very thin ones *fibrils.* A section through a bunch of fibers will show dots:

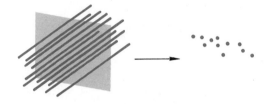

You can tell that you are looking at fibers either because you examine serial sections and can follow the pattern of dots from one section to another, or because you make a cut at right angles and see the fibers in *longitudinal section:*

On the other hand, if you were looking at little particles (balls) that are smaller than a section's thickness, you would only see the dots isolated in the sections and there would be no continuity from one to another.

Most of the objects we see in cells are sheets (membranes). These take very different shapes, and it is important to recognize them. First, take a sheet by itself; in cross section, this will appear as a layer or a series of layers in the best preparations:

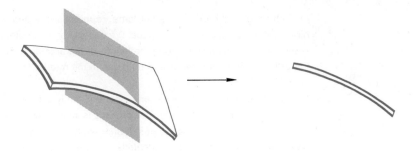

More rarely, a sheet will be cut somewhat tangentially, so you may see its edge and then a part of its face fading off out of the section:

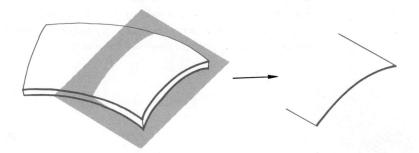

Suppose a membrane rounds up to make a little pocket or *vesicle;* a section through a vesicle will look round:

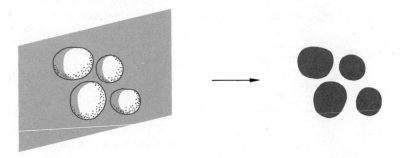

In serial sections, the diameters of vesicles will increase and decrease. However, you will also encounter *tubules* — particularly some called microtubules — whose dimensions do not change from one section to another:

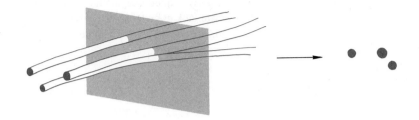

An interesting structure sometimes encountered is a *bleb,* where a sheet is forming a pocket that has not separated completely but looks as if it is going to at any moment:

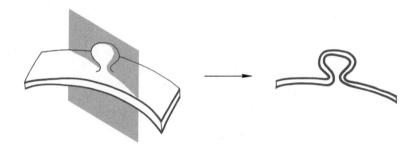

Of course, we may be running the clock in the wrong direction, and it is just as likely that the little pocket used to be a vesicle and has been caught in the act of making a connection to the larger sheet.

A very common pattern that you will see is a *lamellar* arrangement of sheets, all more or less parallel to one another; in cross section, lamellae will look like parallel layers:

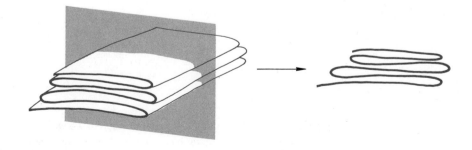

Clearly, a good imagination is important to a microscopist, and as you look at pictures in this book, you should try to see in your mind's eye the three-dimensional objects they represent.

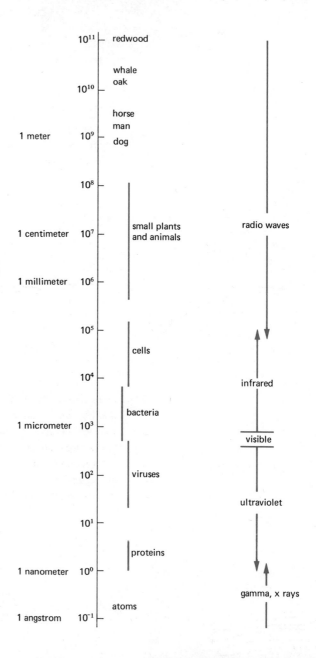

You will also have to acquire a concept of size, for we will have to discuss an enormous range of objects. Everything from a millimeter upwards should be relatively familiar. With your unaided eye, you can just barely see some of the largest protozoa, which are a few tenths of a millimeter long, but below this point you will need some guidelines. We will give the dimensions of cells in micrometers, and you should use 10 μm as a reference point for an average eucaryotic cell and 1 μm for an average bacterium. We will generally give the dimensions of subcellular structures in nanometers (nm), and here you will have to acquire your own set of guidelines as you learn more about cells. To help you, we will put some dimensions on micrographs; it would be wise to keep a small centimeter ruler handy for measuring small structures.

We may also give the factor by which each picture is enlarged through all the microscopic and photographic processes between the object and the printed page. Suppose a photograph is designated X20,000; since 1 mm = 10^6 nm, each millimeter on the page must correspond to $10^6 \div 2 \times 10^4 = 50$ nm. In general, 10^6 divided by the enlargement factor will give you the number of nanometers per millimeter.

We will generally give molecular dimensions in angstrom units. Atomic dimensions are in the range of 1–2 Å, and many proteins are little particles approximately 30–60 Å in diameter. The typical biological membrane is about 75 Å wide.

growth,
mutation,
and evolution

In this chapter we consider a few elementary experiments and observations that illustrate the most distinctively biological phenomena, those associated primarily with growth. This will provide the foundation for a more general and theoretical discussion.

2-1 CELL GROWTH AND PROLIFERATION

Because they are relatively simple, bacteria make excellent experimental subjects for biology. We will concentrate on the common, widely used bacterium *Escherichia coli,* which lives in the human intestine and is easily isolated from sewage. *E. coli* will grow in a **nutrient medium** such as a simple phosphate buffer containing $MgSO_4$, NH_4Cl, and a sugar such as glucose or lactose. We can put a single cell in this medium and observe its activities with a microscope. The cell appears to do nothing at first, but after a while we can clearly see that it has grown longer. Soon a division appears about midway and the cell very quickly divides transversely into two cells, which separate and continue to grow until each of them again divides in two. The **division time** t_D in this case might be about an hour, so if we record the number of cells at time zero B_0 and the number B after t hours have elapsed, we find:

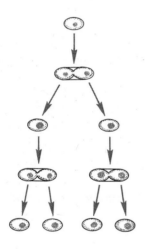

$$t = 0, \quad B = B_0$$
$$t = 1, \quad B = 2B_0$$
$$t = 2, \quad B = 2^2 B_0$$
$$t = 3, \quad B = 2^3 B_0$$
$$t = n, \quad B = 2^n B_0$$

More generally, for any division time t_D,

$$B = 2^{t/t_D} B_0 = 2^D B_0 \qquad (2\text{-}1)$$

where $D = t/t_D$ is the number of divisions. It should be obvious that a **clone** of cells, all of the cells derived from a single cell by repeated divisions, can become enormous in a short time, since growth is exponential.

Suppose we take a more dense culture of bacteria, with about 10^7 cells/ml, and shake it in a flask to keep the cells well aerated. We can measure growth in this case by counting cells with a microscope, by measuring the turbidity of the medium with an optical instrument, or by weighing the cells in a small sample. Whatever measure we use, a graph of the number of cells or the bacterial mass plotted against time will look much like Fig. 2-1, where we use a logarithmic scale so that most of the points fall on a straight line. At first there is a brief **lag period** during which each cell increases somewhat in mass,

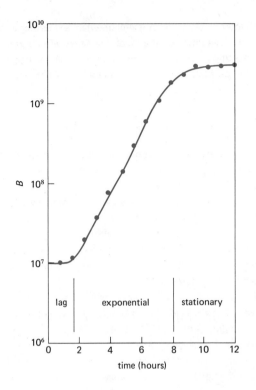

FIGURE 2-1
The normal bacterial growth curve.

but there is no increase in the number of cells. The culture soon enters the **exponential phase** (sometimes carelessly called the "logarithmic" phase) during which all of the cells divide at regular intervals; the divisions are not all synchronized, however, so at every instant there are some cells dividing and both the number of cells and the mass of cells increase smoothly, rather than in steps. The growth of the culture is well described by

$$\frac{dB}{dt} = \alpha B \tag{2-2}$$

where α is the growth-rate constant. Upon integrating equation 2-2 we obtain

$$\ln B/B_0 = \alpha(t - t_0)$$

or, rewriting this in exponential form, with $t_0 = 0$,

$$B = B_0 e^{\alpha t} \tag{2-3}$$

Equations 2-1 and 2-3 are both good descriptions of exponential growth; the latter is more generally useful because it refers to an increase in mass, independent of any change in the number of cells. Notice that the division time defined by equation 2-1 is a *doubling time* for E. coli (and for all other known bacteria) since the cells divide after their mass has doubled. For contrast, consider the (hypothetical) Martian bacterium *Martibacillus novellus,* which divides into *three* cells at each division; a sensible equation for its growth would be $B = B_0 3^{t/t_D}$; in this case t_D is a *tripling time.* However, equation 2-3 is still a useful description of growth in this case; in this equation, α

clearly has the dimensions of inverse time, so we can define a time $\tau = 1/\alpha$, commonly called the **generation time.**

The implications of equation 2-2 are important; the equation says that the rate at which the bacteria grow is proportional to the number of bacteria already present. This exponential growth is the same as the growth of money invested with a bank that compounds it continuously (except that the interest rate is 100% for the bacteria and this is hard to get commercially). In investment, every penny earned in interest is immediately reinvested to earn more; in growth, every bacterium that is made immediately starts to grow to make more bacteria.

exercise 2-1 Calculate the growth rates, α, for the following cultures:
(a) $B_0 = 5 \times 10^6$ cells/ml; B at 8 hours $= 10^9$ cells/ml.
(b) $B_0 = 3.6 \times 10^4$ cells/ml; B at 10 hours $= 1.8 \times 10^8$ cells/ml.

exercise 2-2 From equations 2-1 and 2-3, calculate the relationship between doubling time and generation time.

exercise 2-3 Solve equation 2-1 for t_D so that doubling times can be calculated easily using $\log_{10}$ values. Remember that $\log_{10} 2 = 0.3$. Use the resulting equation to obtain t_D for cultures that grow
(a) from 2×10^5 to 2×10^6 in 8 hours;
(b) from 3×10^5 to 10^7 in 8 hours;
(c) from 7×10^5 to 5×10^8 in 5 hours;
(d) from 1.5×10^6 to 9×10^7 in 4 hours;

exercise 2-4 Bacteria of strain A have a generation time of 1 hour while bacteria of strain B have a generation time of $\frac{1}{2}$ hour. If you start a culture with N cells of type B and $10N$ of type A, how long will it take the faster growing cells to catch up with the slower ones?

After a considerable exponential increase, the population levels off and enters the **stationary phase.** This can happen for a variety of reasons; the culture may not be getting enough air, the medium may have become too acid, or the cells may have exhausted the supply of some essential nutrient. The population approaches a limit B_l, which is often called the **carrying capacity** for this particular organism in this environment; growth of the culture may be described by

$$\frac{dB}{dt} = \alpha B(1 - \frac{B}{B_l})$$

so the growth curve levels off gradually. The limitation on growth is always imposed by the environment; that is, the growth potential of the population is much greater than the carrying capacity. We could start the culture growing again simply by diluting the bacteria into fresh medium or otherwise adjusting the conditions so they are again optimal.

For the moment, let us assume that growth has stopped because the cells have exhausted the supply of sugar we put in originally. The sugar is the only carbon compound in the medium and is called the **carbon source.** Since organisms are made largely of carbon, it would be reasonable to assume that the bacteria have used the sugar for growth and have transformed it into their own mass. This idea can be tested easily by using sugar made with radioactive ^{14}C instead of ^{12}C; the radioisotope serves as a **tracer** to indicate the fate of the sugar molecules. In fact, bacteria grown on ^{14}C sugar are highly radioactive, and about a third of the carbon in the sugar appears in the cellular material. Similar experiments can be done with other tracers, such as ^{32}P in phosphate and ^{35}S in sulfate; these experiments reveal that bacteria take all of these elements from the growth medium and transform them into cell material.

Using conditions where the carbon source is the limiting factor, we can grow bacteria on various concentrations of sugar and measure the mass of bacteria, B, after all growth has stopped. B is always proportional to the amount of sugar. In general, if N is the amount of any limiting nutrient in the medium, whatever it may be, we find that $B = gN$, where g is the **growth-yield constant.** This result has at least one practical application; if we know the value of g for a given medium and type of bacteria, we can set the final yield of cells at any desired value.

exercise 2-5 C. B. van Niel grew some purple bacteria on varying amounts of acetate, with these results:

ACETATE(mg/ml)	TOTAL GROWTH (mg/ml)
0.5	0.18
1.0	0.36
2.0	0.70
3.0	1.12

Calculate the growth-yield constant for each case; is this the expected result? How do these data look when plotted on a graph?

Since the sugar in the medium is so different in structure from most of the organic material in the bacteria (amino acids, nucleotides, etc.), the transformation from one to the other must involve many chemical reactions. It is important to realize from the beginning that growth is not a simple process and that the increase in mass or number of cells is the result of a multitude of processes, many of which we still do not understand. We can get a better idea of the magnitude of this transformation by growing a pretty, red-purple bacterium called *Chromatium,* which is found in many muddy soils. *Chromatium* requires a 1%-NaCl medium in phosphate buffer with a few other salts, including $Na_2S_2O_3$, Na_2S, and $NaHCO_3$. Rather than aerating these cells, as we did with *E. coli,* we keep them in a tightly stoppered vessel to exclude air, and shine a strong light on them. These bacteria grow a little more slowly than *E. coli,* but eventually they produce a dense culture. Their composition is not much different from *E. coli,* yet all of their organic compounds are made from the bicarbonate in the medium. This is certainly a spectacular transformation.

Growth requires energy as well as matter. *E. coli* gets its energy by diverting a part of its carbon source into energy-yielding reactions, where a slow oxidation occurs. (This is one reason why only a third of the carbon actually appears as cell mass.) Because its energy source is chemical, *E. coli* is called a **chemosynthetic** or **chemergonic** organism. *Chromatium,* on the other hand, is a **photosynthetic** or

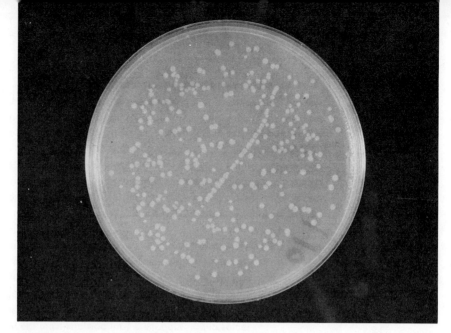

FIGURE 2-2
A petri plate of nutrient agar with colonies of E. coli.

photoergonic organism because it will grow only in the light; its red pigments absorb light and begin a series of reactions in which the light energy is converted into chemical energy.

Sometimes we must carefully distinguish the increase in mass of cells from the increase in numbers of cells; we will then call the increase in mass *growth,* in a stricter sense, and the increase in numbers *proliferation.* Notice that a multicellular organism is actually a clone resulting from the repeated division of one original cell, such as a fertilized egg, and that its growth entails both proliferation and growth of cells. We should also recognize that the proliferation of bacterial cells is a kind of *reproduction,* since each cell is an individual organism. However, the reproduction of many other organisms is a much more complex process.

2-2 INHERITANCE AND MUTATION

The aspect of proliferation that we have glossed over is its regularity and uniformity. Every cell in a clone looks very much like every other cell and they are all made of essentially the same materials. Similarly, the most striking fact about reproduction in other organisms is that the new generation looks very much like the old one. Horses have small horses, not small cows or small turtles. The progeny of a pair of butterflies go through a number of stages during which they are not recognizable as butterflies, but eventually they grow up to look like their parents. This phenomenon is called **inheritance** and we say that offspring inherit their characteristics from their parents. We sometimes take heredity for granted, but it is an observation that any rational biology must explain. Somewhat paradoxically, the best way to investigate the mechanism of heredity is to search for cases where it breaks down, where organisms are produced that differ from their parents.

To do this, we need some tools called **Petri plates** (Fig. 2-2). These shallow plates are half-filled with a layer of nutrient medium containing about 2% *agar,* a polysaccharide derived from seaweed that makes a semisolid gel when it is dissolved in water, heated, and cooled. Separate clones of bacteria can be selected by spreading an appropriately diluted culture over the surface of the plate. When incubated for several hours at 37°C, each cell will grow into a large clone, but since the cells are relatively immobile on the agar surface, they will stay in place and each clone will form a distinct round **colony** on the plate. A liquid culture might be quite heterogeneous, but pure cultures can be obtained from it by isolating single cells and their progeny on plates. Every distinct type of bacterium isolated in this way is called a **strain.**

exercise 2-6   The number of **viable** ("live") **bacteria** in a culture can be determined by counting colonies on plates. A **serial dilution** is made by carrying aliquots *successively* from one dilution tube to the next and then plating from the last tube. For example, a culture is diluted by a factor of 100; this dilution is diluted by another factor of 100 and this, in turn, by a factor of 50. Suppose 0.1-ml portions from the last tube are spread on four plates; after incubation there are 73, 62, 83, and 76 colonies on the plates. How many bacteria were in each milliliter of the original culture?

Most *E. coli* can use the sugar lactose as a carbon source; since this is the usual natural condition, we call it the standard or **wild-type** condition and designate lactose users *lac⁺*. We can search for a strain of bacteria that cannot use lactose, designated *lac⁻*, by making plates that contain lactose, another carbon source such as glycerol, and a mixture of the dyes eosin and methylene blue (EMB), which makes the agar dark red. In fermenting lactose, *lac⁺* bacteria make acids which darken the dyes and even precipitate some of them, so *lac⁺* colonies are a very dark red on EMB-lactose plates. However, *lac⁻* colonies, which are growing only on the glycerol, are light pink. We can pick up bacteria from a pink colony with a sterile wire and be quite sure that we have a pure strain of *lac⁻* cells.

Both *lac⁺* and *lac⁻* bacteria can be grown indefinitely on glycerol, but when a portion of a culture is spread on EMB-lactose plates, the resulting colonies are uniformly *lac⁺* or *lac⁻*, respectively, almost without exception. In other words, the ability or inability to use lactose is a very stable characteristic that is inherited from one generation of cells to the next. The few exceptional cells that do occur are termed **mutants;** to start the *lac⁻* culture originally, we selected one of the rare clones of *lac⁻* cells out of the *lac⁺* population. And occasionally **back-mutants** or **revertants** occur as *lac⁺* cells among the *lac⁻* population.

Bacteria are sensitive to ultraviolet (uv) light and are rapidly killed by exposure to a germicidal lamp. Suppose we irradiate a *lac⁺* population with uv light for various times and plate a sample of the survivors on EMB-lactose plates. The number of survivors decreases exponentially with time of irradiation, but the fraction of *lac⁻* bacteria *increases.* This shows that the same general type of damage inflicted by uv light that kills cells also causes a conversion of *lac⁺* to *lac⁻* cells. These damages, which are still rather hypothetical, are called **mutations.** The nature of mutations, and of the hereditary process in general, will become clearer as we look for additional examples.

If many bacteria are spread on a plate, the clones they form grow together into a confluent layer called a **lawn,** instead of remaining separate colonies. We can now take a small sample of the same sewage from which *E. coli* is isolated, mix it with a few drops of chloroform to kill any bacteria in it, and then spread it on the surface of a plate with enough bacteria to form a lawn. After several hours of incubation, the lawn that grows is interrupted here and there by clear, round areas called **plaques** where there are very few bacteria (Fig. 2-3). The plaques are not identical; they differ in size, sharpness, and other details.

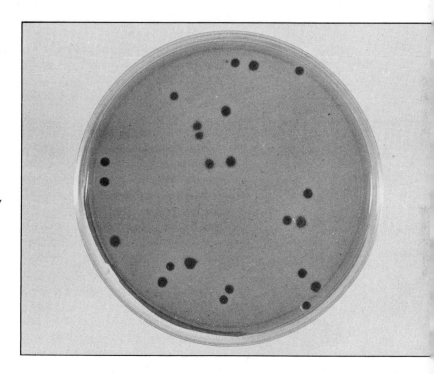

FIGURE 2-3
A Petri plate of nutrient agar whose surface is covered with a lawn of E. coli *interrupted by plaques of phage T4.*

If we pick up a bit of agar containing a plaque and mix it into a turbid culture of exponentially growing *E. coli,* the culture soon becomes clear and almost all of the bacteria disappear, leaving bits of debris that look like broken cells. Furthermore, the medium contains large numbers of the agents that produce plaques, as we can demonstrate by spreading a portion of it on a Petri plate with more cells. The agent that is killing the cells can clearly multiply in numbers and it appears to multiply at the expense of the cells. The men who first observed this phenomenon, principally Frederick Twort and Felix d'Herelle around 1920, called this agent **bacteriophage**—literally, bacteria eater; we generally shorten the name to **phage.**

We now know that phage belong to a class of objects called **viruses.** Viruses are of interest to biologists because they can multiply rapidly, but always at the expense of some organism; they are responsible for many diseases of plants and animals. Phage are simply bacterial viruses, so a phage infection is a disease of bacteria.

Although viruses have many of the characteristics of bacteria and other organisms, there are some important differences. They are much smaller than cells and will pass through porcelain filters whose pores will hold back even the smallest bacteria. Their structure is not cellular; each virus is a distinct little particle called a **virion** that is really more like a tiny crystal than a cell. However, using methods entirely analogous to those used for bacteria, we can isolate pure strains of virus from single plaques and determine the **titer** of a virus suspension—the number of virions per milliliter—by plating an appropriate dilution with susceptible cells and counting plaques. During the early 1940s, Max Delbrück and his school isolated seven types of phage designated T1 through T7 and initiated the work we will discuss here, using T4 as an example.

A stock of phage T4, like many other viruses, can be kept indefinitely in the cold without any significant decrease in titer, but also without any increase. The viruses by themselves are inert; they can reproduce themselves only by infecting an appropriate strain of cells that serve as their **host** (though "victim" would be a more appropriate term). Phage T4 multiplies on *E. coli* strain B. To start an infection, we mix 10^8 cells with 10^9 phage, using a **multiplicity of infection** (**moi:** the ratio of phage to cells) of 10 to ensure that almost every cell is infected by at least one phage. Then we dilute into a warm, well-aerated nutrient medium so there are only about a thousand infected cells per milliliter and at various times we plate 0.1 ml from this culture with enough *E. coli* B to form a lawn. The number of plaques per plate is shown in Fig. 2-4 as a function of time. For about 24 minutes (the **latent period**) there are only about 100 plaques per plate, the number we would expect if each one were made by a single infected cell. At 24 minutes, the number of plaques suddenly starts to rise; it levels off at about 200 times its initial value. (Another dilution is necessary at later times, because you cannot count 20,000 plaques on a plate.) This is because the cells in the nutrient medium suddenly begin to disintegrate at the end of the latent period and, on the average, each one liberates about 200 new phage virions. Thus, the plaques obtained before 24 minutes are due to infected cells that liberate their phage on the plate, while those obtained after 24 minutes are predominantly due to free phage. The disintegration of cells produced by the phage is called **lysis.**

Chloroform does not affect phage T4, but it destroys bacteria, both infected and uninfected. If we repeat the above experiment by blowing samples of infected cells into buffer with chloroform at various times, we obtain the other curve of Fig. 2-4. This procedure measures the number of *intracellular phage* at each time. The rather remarkable feature revealed by this experiment is a period from about 0 to 11 minutes, the **eclipse period,** during which no phage at all can be found! This is a very mysterious result; we cannot really explain it at this time, except to say that the phage have entered a noninfectious form as an obligatory step in reproduction.

Phage infections can now be observed by electron microscopy. The phage virions look like little tadpoles (Fig. 2-5) and they attach to the cell surfaces by the tips of their tails. The cells show no obvious external changes until the end of the latent period; then they literally burst open and scatter their phage particles. Viruses that infact plant and animal cells follow much the same course, except that the infections take hours rather than minutes and the cells disintegrate more gradually.

If we try to plate too many phage on a plate, all the plaques run together and the

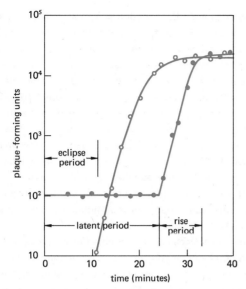

FIGURE 2-4
Single-step phage growth curves. The solid circles show lysis and release of phage, beginning at 23 minutes. The open circles show intracellular development of phage, beginning at 11 minutes. During the eclipse period, the phage are said to be in the **vegetative** *state, for they cannot be detected.*

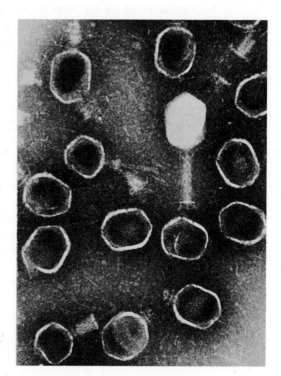

FIGURE 2-5
A T4 phage virion, surrounded by pieces of phage, mostly heads. (Electron micrograph by Alex Lielausis.)

plate is said to show *confluent lysis.* Even though almost all of the cells are lysed, a few colonies appear on the plate. Phage T4 will not grow on these cells; they are *mutant, phage-resistant* bacteria called B/4 (read "B bar four," the bar meaning "resistant to"). By picking colonies of B/4 and testing their progeny, we can again demonstrate that these cells carry a mutation that is a stable, inheritable characteristic.

Suppose we now plate a large number of T4 on B/4. We expect no plaques because the cells are resistant to the phage; however, a few plaques do occur. By picking these plaques and purifying the phage in them, we can demonstrate that these are *mutant phage,* designated T4*h* (referring to the fact that the phage can grow on a different *h*ost than the wild-type phage). This mutation is also a stable, inheritable characteristic; viruses, therefore, not only multiply, but also inherit specific characteristics.

We could go on indefinitely isolating other mutant bacteria and phage. There are mutant organisms everywhere; you already know of mutations in humans that account for such features as hair, skin, and eye color and for certain inherited diseases, such as diabetes and color blindness. Occasionally an albino robin appears and has its picture taken for *Life* magazine, and if you were to examine the fruit flies swarming around a dish of rotting apples on a hot day, you could find some that are a little different in eye color or some other feature. Mutations are recognized as inherited differences between individuals, and one must be careful to distinguish them from noninherited differences. For example, a color-blind man may have color-blind children, but if a man loses his sight in an accident, his children will not be affected directly. Many difficulties have occurred in biology because of the failure to distinguish carefully between mutational and nonmutational differences, particularly in microorganisms.

2-4 MUTATION RATE

To distinguish mutations from environmentally induced variations in bacteria, Salvador Luria and Max Delbrück set up an experiment that proved mutations actually occur in bacteria and permitted them to measure the rate at which the mutations occur. It is instructive to repeat part of this analysis.

If mutations are some kind of damage to the cell's hereditary apparatus and if they occur at random, then the probability of a mutation *per bacterium per unit time* should be constant; we call this probability the **mutation rate.** However, one may reasonably expect mutations to occur during the course of reproduction, so the faster bacteria are reproducing, the more likely they are to mutate. In this case, we should measure time in units of generations, or, to put it another way, mutation rate should be proportional to growth rate. If m is the number of mutations and μ is the mutation rate,

$$\frac{dm}{dt} = \mu \frac{dB}{dt}$$

$$= \mu \alpha B$$

Then

$$dm = \mu \alpha B_0 e^{\alpha t} dt \qquad (2\text{-}4)$$

and by integrating from an initial population of B_0 to a final population of B, the total number of mutations is

$$m = \mu(B - B_0) \qquad (2\text{-}5)$$

Note that this is the number of mutations, not the number of mutant cells; the original mutation creates a single mutant, which grows into a clone of mutants. We can count the number of mutants in a culture, but we must calculate the number of mutations that produced them.

Luria and Delbrück set up many identical tubes of nutrient medium, inoculated each one with the same small number of cells, and grew them identically until there were many cells per tube. They then assayed for the number of phage-resistant mutants. In some tubes, mutations occurred early, just by chance, and, because each original mutant could go through many doublings, there were large clones of mutants in these tubes. In other tubes, there were no mutations at all, while in most tubes, mutations occurred rather late in the growth period and there were only small clones of mutants. If there are any mutants at all in a tube, it is impossible to say how many mutations were actually responsible for them. But if mutations are randomly (Poisson) distributed among tubes (see Appendix II), then the fraction of tubes with no mutations (hence, no mutants) becomes a useful number. If m is the average number of mutations per tube, then the probability that no mutations occur in a tube is $P_0 = e^{-m}$. For example, if 37% of the tubes have no mutations, then $e^{-m} = 0.37$ and $m = 1$. The mutation rate can then be calculated from equation 2-5, since the inoculum size, B_0, is known and B can be determined.

In practice, there is a minimum population size B_n such that it is unlikely for mutations to occur in any smaller population, and this rather simple analysis is really valid only for a population growing from B_n to B. Luria and Delbrück made more sophisticated calculations to cover a wider range of cases. They also showed that, because of the randomness of mutation, the fluctuation in number of mutants from culture to culture was enormous. If the variations observed in bacterial populations were due directly to environmental influences (such as phage resistance being a consequence of exposure to phage) then the fluctuations from culture to culture should have been very small.

It may help to repeat this analysis for a *synchronized* culture where all the cells double simultaneously and we can measure time in units of D, the number of doublings. If B_D is the number of bacteria at the time of division, then we can write, by analogy with equation 2-4,

$$m_D = \mu B_D \qquad (2\text{-}6)$$

for the number of new mutations occurring at the time of division. Similarly, for the previous doublings $D - 1$, $D - 2$, and so on,

$$m_{D-1} = \mu B_{D-1} = \mu B_D/2$$

$$m_{D-2} = \mu B_{D-2} = \mu B_D/4$$

and so on, where the second equality follows from equation 2-1. Now for a sufficiently long series of terms,

$$1 + \frac{1}{2} + \frac{1}{4} + \frac{1}{8} + \cdots + \frac{1}{2^n} \approx 2$$

So if we consider enough doublings, the total number of mutations is approximately

$$m = \sum_n m_{D-n} = 2\mu B_D \tag{2-7}$$

This differs somewhat from equation 2-5; which one of these equations one uses depends upon taste and circumstances.

Now, how many mutants will arise from these mutations? Note that a mutation at doubling $D - n$ will yield 2^n mutants in the final population. Therefore the total contribution of each doubling will be 2^n times the number of mutations that occur at that time, m_{D-n}:

DOUBLING	m_{D-n}	2^n	TOTAL MUTANTS
D	μB_D	1	μB_D
$D - 1$	$\mu B_D/2$	2	μB_D
$D - 2$	$\mu B_D/4$	4	μB_D

It is clear that, on the average, every doubling contributes the same number of mutants, and after n doublings the total number of mutants is just $n\mu B_D$.

exercise 2-7 Consider 100 identical tubes of growth medium inoculated with 10^4 streptomycin-sensitive cells each and incubated until they contain 10^8 cells each. When they are assayed for streptomycin-resistant cells, 83 of the tubes are found to contain one or more of these mutants. Calculate the average number of mutations per tube and the mutation rate to drug resistance.

exercise 2-8 Consider 10^5 phage-sensitive cells inoculated into fresh medium and grown to 3×10^8 cells, where 10^{-5} of the cells in the final culture are resistant to phage. How many resistant cells will there be after one more doubling? What is the mutation rate?

2-5 THE CHEMOSTAT

We can measure mutation rates more directly and discover some other interesting properties of organisms by using a **chemostat** to grow a population of bacteria under constant conditions. The chemostat was invented by Aaron Novick and Leo Szilard, while simultaneously and independently Jacques Monod developed a similar device, the bactogen. As Fig. 2-6 shows, a bacterial culture is grown in a tube of volume V into which fresh medium is pumped at the rate of w ml/hour. A siphon removes excess medium and bacteria at the same rate, so the volume remains constant. The number of bacteria B is set by limiting the concentration of some essential nutrient whose growth-yield constant has been determined previously; this may be the carbon source, nitrogen source, or perhaps an amino acid that the bacteria cannot make for themselves. Let c be the concentration of this limiting compound.

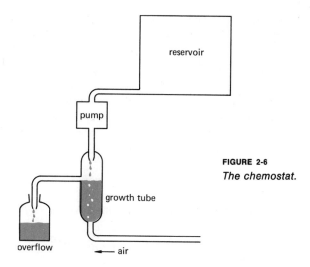

reservoir

pump

growth tube

overflow

← air

FIGURE 2-6

The chemostat.

The bacteria grow with a rate constant α and are washed out at the rate w/V, so their growth equation is

$$\frac{dB}{dt} = \alpha B - \frac{w}{V} B \qquad (2\text{-}8)$$

If the cell density is to remain constant, dB/dt must equal zero, and therefore

$$\alpha = \frac{w}{V} \qquad (2\text{-}9)$$

This condition is easily obtained if the washout rate w/V is set at any value less than the maximum growth rate, because α is a function of c, most conveniently described by

$$\alpha = \alpha_{max}\left(\frac{c}{c_h + c}\right) \qquad (2\text{-}10)$$

where c_h is a constant for a given nutrient and strain of cells; it is clearly the concentration of the nutrient for which α is half its maximum value. A graph of equation 2-10 shows that the growth rate is proportional to c when c is small and very quickly reaches a plateau as c increases—which merely means that the bacteria are limited by the known nutrient until they become limited by something else.

The bacteria initially grow up in the growth tube until they reach the density B set by the limiting nutrient; at this point they have reduced the nutrient to a very low value and their growth rate falls correspondingly. The flow rate is then set by adjusting w until w/V equals the desired value of α; fresh medium begins to flow into the tube and the concentration of the limiting nutrient reaches a low, steady-state value c such that the bacteria will grow at a rate $\alpha = \alpha(c) = w/V$. The system is self-stabilizing, for if B ever increases slightly, c will decrease, α will decrease, and B will decrease, while the opposite will happen if B ever decreases slightly. In fact, the chemostat operates without any such fluctuations.

exercise 2-9 What is the generation time of bacteria in a 21.2-ml chemostat tube if the flow rate is 9.6 ml/hour?

Mutations will occur naturally in any chemostat culture. Let us first consider mutations to phage resistance, since there is no reason why such mutants should grow any faster or slower than other cells when no phage are present. Mutations should occur at a rate $\mu \alpha B$, from equation 2-4; mutant cells grow at the rate α and are washed out at the rate w/V, just like other cells, so mutants should accumulate at the rate

$$\frac{dm}{dt} = \mu \alpha B + \alpha m - \frac{w}{V}m \qquad (2\text{-}11)$$

but the last two terms cancel each other, since $\alpha = w/V$, and

$$\frac{dm}{dt} = \mu \alpha B \qquad (2\text{-}12)$$

Since $m \ll B$, B is essentially constant and μ can be measured by sampling the growth tube at intervals, measuring the number of mutants, and plotting m/B against time (Fig. 2-7). The slope of the line is $\mu \alpha$.

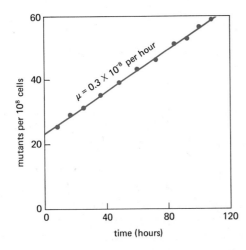

FIGURE 2-7
Accumulation of mutants in the chemostat.

When Novick actually measured the accumulation of mutants in various chemostat cultures growing with rather long generation times, he made the surprising discovery that the mutation rate per unit time was a constant and that it did not increase with increasing growth rate, as equation 2-12 predicts. Maurice Fox later found that mutation rate becomes a function of growth rate when generation times less than about 120 minutes are used. Novick also found that mutation rates depend very strongly on the limiting nutrient used, and therefore, presumably, on some internal conditions of the cells. While we may still think of mutations as a kind of damage to the cell's hereditary machinery, the processes leading to the damage are obviously not simple.

exercise 2-10 Once upon a time, Novick started to worry about his daily coffee consumption and performed the following experiment. He set up a chemostat with a generation time of 5.5 hours and inoculated it with *E. coli* B sensitive to phage T5. At intervals thereafter, he took samples from the growth tube and assayed for the number of B/5 mutants. After 70 hours, he added 150 mg of caffeine to each liter of the medium. He obtained the following results:

	TIME (HOURS)	MUTANTS PER 10^8 BACTERIA
	0	50
	25	75
	50	100
	65	115
caffeine →	80	200
	90	330
	100	460
	110	600

Plot these data on linear graph paper and calculate the mutation rates to phage resistance with and without caffeine. (Caffeine is a **mutagen** because it increases the spontaneous mutation rate.)

2-6 EVOLUTION

In the last experiment, we examined mutations to a characteristic that was neither a help nor a hinderance to the bacteria in the growth tube, but it is not hard to set up conditions under which certain mutations are distinctly beneficial. Suppose we use lactose, the sole carbon source, as the limiting nutrient. As before, the lactose concentration in the growth tube will remain at some low value c for which the growth rate is $\alpha(c)$. However, mutations occasionally occur which lead to the ability to grow at a greater growth rate $\alpha'c$. If a single such mutant occurs, it will produce a clone of cells that grow more rapidly than the original cells; the number of mutants increases under these conditions until the original cells are swamped and completely replaced by the mutants.

The process we have just described is one example of **biotic (organic) evolution.** We have put a population of organisms into an environment where they must *compete* for something essential and in this way we have *selected* those organisms that can grow most efficiently there. If the growth rate of the mutants is $\alpha' = s\alpha$, where $s > 1$, then s is called the **coefficient of selection** favoring the mutants. Whatever the basis of the selection, it is best to think of it as a race between slightly different individuals. The race is won by those individuals that can grow and reproduce the fastest and can therefore leave the most offspring in the next generation. The prize is the race itself—the ability to continue to grow and reproduce and to have a space to do it in.

The place where an organism lives is its **habitat** and the way it earns its living is its

niche. Its niche is the way it gets its food and energy — in other words, the way it fits into its ecosystem. The chemostat is an artificially simple system that has only a single niche, for all the organisms in it must live off the simple nutrients supplied in the medium. We began with bacteria that were able to occupy this niche, but they were soon replaced by cells that were better *adapted* and could use it more efficiently; if we kept the chemostat running, this population would probably be replaced by another that was still better adapted, and there is potentially no limit to this process. Evolution is a process in which well-adapted populations are created by continuing selection, and selection in turn depends upon mutation, which produces a supply of variants.

Notice the terminology we have been using; evolution is a property of populations, not individuals. Individuals change, through mutation and other processes, but the population as a whole evolves. In fact, it is the whole ecosystem that evolves, for no population lives in isolation. Each organism is eaten by some other organism, and the relationships between them are at least as complex as the various food chains we can draw. If some population fails to adapt to a stress in its environment and becomes extinct, the entire ecosystem may be affected; the extinct organisms were part of a series of food chains, and the chains will have to be restructured. Consider, for example, that when the wolves are systematically killed in an area, the deer herds increase dangerously and overeat their food supply, creating a series of new problems. Consider another simple food chain series:

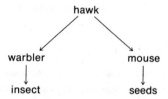

If there was a sudden stress applied to the warbler population and the warblers disappeared, the insect population would increase, the hawks would be forced to feed more heavily on the mice, and more seeds would germinate into plants. Each of these changes would produce other stresses, and the entire ecosystem might have to change dramatically. Natural populations are constantly being subjected to environmental stresses, but in general they adapt to these stresses and change, rather than becoming extinct, primarily because every population contains a reservoir of mutants that are better adapted to the new conditions.

It is important to understand that the relationships within the ecosystem are not simply a matter of one organism eating another. There are important relationships between pairs of organisms that may be only partially based on food requirements and may also be mutually beneficial; these relationships also have had long evolutionary histories of mutual adjustment. For example, **symbiosis** is a relationship that is beneficial to both parties; an excellent example is the relationship between termites and the little ciliated protozoa that live in their intestines. The termites can chew wood but cannot digest it; the ciliates can digest wood but cannot chew it. In cooperating, they feed each other. There are also **commensal** relationships in which one organism benefits and the other is not affected particularly; for example, the remora is a fish that attaches itself to sharks and other large animals and feeds on scraps of food that they miss. The remora does not hurt its benefactor; in a **parasitic** relationship, however, one organism definitely gains and the other loses, as when a tapeworm grows in the intestine of a large animal. The general classification of these relationships is shown in Table 2-1; there are many examples, and we cannot go into many of them here. For now, it is important to understand that such relationships form parts of the tangled web of interactions within which evolution occurs.

TABLE 2-1

EFFECT OF B ON A	EFFECT OF A ON B		
	NEGATIVE	NEUTRAL	POSITIVE
NEGATIVE	competition	amensalism	parasitism predation
NEUTRAL	amensalism	neutralism	commensalism
POSITIVE	parasitism predation	commensalism	mutualism

With the growth of biology during the last century, it became obvious to many observers that the organisms on this planet have evolved from one another and that they are all related in time, more or less as the twigs of a great tree are all connected to a central trunk. However, it required the genius of Charles Darwin to outline the general mechanism of evolution and to amass so much evidence in support of his theory that rational men could no longer doubt its general truth. Darwin recognized that as organisms tend to increase in numbers exponentially and the environment sets limitations on the number that can survive, there must be a competition among individuals. Since there are natural variations—we now say mutational differences— among individuals, some will be more fit than others and more likely to reproduce. Since Darwin's time, we have learned a great deal about the mechanism of selection and evolution, but the basic idea remains the same in spite of its modern dress. What we now call the **principle of natural selection** or **Darwin's principle** remains one of the most important and fundamental principles of biology.

The most convincing evidence for the evolution of one group of organisms into another comes from the fossil record. As long as there have been organisms on earth, they have died and sunk into environments such as soft mud where their bodies have decayed. Under favorable condition, their structures—particularly hard parts such as shells and skeletons—are preserved quite faithfully through replacement by mineral deposits. Somewhere close to your home there is almost certainly a bed of shale or limestone filled with the fossilized remains of organisms that lived millions of years ago. By carefully sequencing the rock strata and examining their fossil contents in detail, it has been possible to assemble excellent chronologies and to deduce the phylogeny of certain groups—that is, their evolutionary relationships to one another. Figure 2-8 shows the evolutionary sequences in a small group of animals where the relationships are quite obvious. In conjunction with studies of the anatomy of modern organisms and with certain biochemical studies that we will discuss later, the fossil

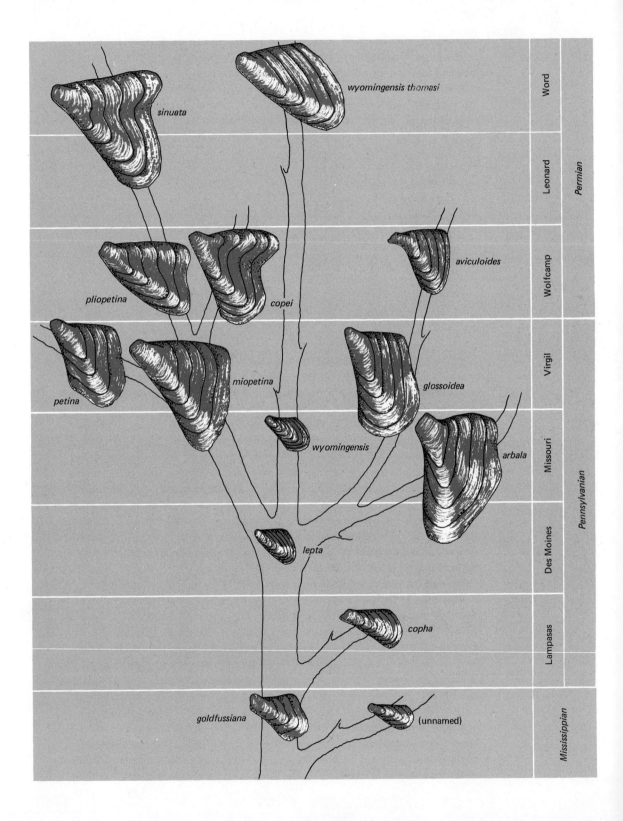

sinuata

wyomingensis thomasi

Word

Leonard

Permian

pliopetina

copei

aviculoides

Wolfcamp

petina

miopetina

glossoidea

Virgil

wyomingensis

arbala

Missouri

lepta

Des Moines

Pennsylvanian

copha

Lampasas

goldfussiana

(unnamed)

Mississippian

record gives us quite a good view of the relationships between various plants and animals. Unfortunately, fossils of most microorganisms are few and far between, since they do not preserve well, so some of the major lines of evolution are obscure.

Two excellent examples of contemporary evolution are worth mentioning at this point. The use of the insecticide DDT has increased enormously and dangerously since World War II. DDT was very effective at first against domestic pests such as houseflies, but its usefulness has been diminished by a relatively simple adaptation; many insects now have enzymes that remove a few Cl atoms from DDT and inactivate it. There has clearly been a selection of a few individuals who happened to have the necessary enzymes, though they may never have been used before DDT was invented; contemporary flies are the descendents of the survivors of early DDT sprayings. Unfortunately, many other organisms have not evolved mechanisms of resistance; DDT has been spread through food chains to birds and other animals and is doing serious damage.

A very obvious case of evolution has occurred among certain moths in heavily industrialized areas; H. B. D. Kettlewell has studied the striking case of *Biston betularia* in the vicinity of Manchester, England. The forest land around Manchester is typical of other areas where smokestacks are continually pouring out their industrial soot and blackening the trees. In 1848, a melanic (dark) form of *B. betularia* was first recorded in the area. While the light form of the moth is quite well camouflaged against the bark of unpolluted trees, the melanic form is practically invisible on a heavy layer of soot. Fifty years after the original report, the melanic form made up at least 99% of the population of this moth around Manchester. Its frequency in other areas is clearly a function of the amount of soot deposited; in the unpolluted areas, it is kept at a low level because it is so conspicuous. Kettlewell has observed birds feeding on moths in polluted and unpolluted areas, and there is little doubt that their ability to see the moths on the different backgrounds is responsible for color selection. In one experiment near Birmingham, Kettlewell captured 154 melanic and 73 light moths, marked them, and released them. He later captured 82 of the melanics and only 16 light moths. As an exercise, calculate the probability that a dark moth will be eaten compared with the probability that a light moth will be eaten in this region. The value of dark coloration is obvious.

It is wrong to think of evolution as a bloody struggle for survival, as some of the nineteenth century philosophers did. This concept has sometimes been used to justify ruthless and immoral behavior in human society. On the whole, the chemostat is a good model, for evolution in the growth tube is primarily a race to leave a maximum of offspring in the next generation. Predation, disease, and so on in natural populations are equivalent to the siphon that removes excess cells; the organisms that are selected are merely those that are eaten less often, eat better, stay healthier, and so forth. While some individuals must always die, the old picture of "nature red in tooth and claw" is a bit too dramatic.

FIGURE 2-8
Evolution in the genus Myalina, *an extinct group of bivalves that lived in the Mississippian through Permian periods, from about 260 to 200 million years ago. They are related to our modern mussels. Notice the diverging lines of evolution in the architecture of their shells. (Adapted from the work of Norman D. Newell.)*

2-7 SUMMARY

The most essential ideas of biology are contained in this chapter: growth and proliferation, inheritance, mutation, and evolution. Populations increase in numbers by using raw materials and energy from their environment. Random mutations occur in their systems of inheritance and lead to new types; the best of these are selected by their ability to reproduce more effectively, thus leading to evolution of the population. But all this goes on within a physical environment in which there are various sources of energy and a variety of ways to make a living; evolution is primarily a search for new ways to tap these energy sources—for niches that are not already occupied by other organisms. In Chapter 3 we will explore the concept of energy a little further and see just what an organism requires to make a living.

The very concept of evolution has had a profound effect on man in the last century. Man has been highly egocentric for most of his lifetime. Our ancestors believed that the universe revolved around the earth and that they had been placed on earth through a special act of creation. Astronomy shattered the geocentric view of the universe and showed us that the earth really occupies a rather obscure place in the cosmos. Darwin and his contemporaries showed that we have been shaped by the same forces that have produced all the other organisms around us. A people cannot face such cold realities without showing some scars; we are left with many questions with which the best of our philosophers still struggle. As if for security, we still cling to some child-like conceptions about what we are biologically, and we may become extinct unless we face realities here, too. We can watch a few bacteria grow exponentially into a few grams of packed cells and fail to recognize that our own population is growing according to the same kinetics. We can watch a herd of deer outgrow its food supply and forget the herds of people in India and China and South America who are in the same position. We do not link the springtime floods that destroy our cities to the forest fires, lumbering, and strip mining a few hundred miles upstream. The delusion that we are somehow above and beyond the rest of the biological world is temporarily comforting, but we may not survive to enjoy that comfort much longer. We dare not keep our illusions about the laws that govern our universe.

SUPPLEMENTARY EXERCISES

exercise 2-11 Census figures are available for the United States since 1800. Figure 2-9 shows the increase in the U.S. population since then, compared with the population of England and Wales. (a) Calculate the doubling time of the U.S. population before the Civil War. (b) Assuming that the graph has been drawn correctly for the past few decades (which included disturbances due to the Depression and World War II), what is our present doubling time? (c) What phase of the growth curve are we in? (d) What phase is England in? (e) Assuming that nothing drastic is done to limit populations, which country will see the greatest decrease in its living standards in the next few decades?

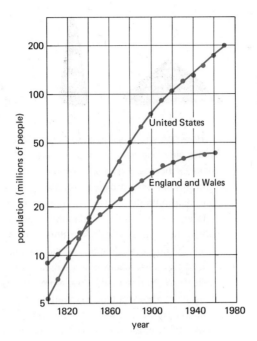

FIGURE 2-9
Growth curves for two human populations.

exercise 2-12 How would you select mutants resistant to the antibiotic penicillin?

exercise 2-13 Antibiotics, such as penicillin and streptomycin, are made by fungi and "higher" bacteria, apparently to kill "lower" bacteria in the soil around them. Briefly, what do you imagine is going on down in the soil? What kinds of mutants are being selected in each species?

exercise 2-14 Oil slicks from tankers and underwater drilling are becoming an increasingly common menace to the wildlife of our coasts. Remember that oil is made of organic compounds; how could you select bacteria that could be used to clean up oil messes?

exercise 2-15 Before the Europeans came to America, one of the food chains on the plains consisted of: coyotes → rodents → grasses. Some settlers brought sheep with them, which also fed on the grass. Some coyotes turned to eating sheep, perhaps because the rodent population was reduced by competition with the sheep; the settlers then started to shoot the coyotes. What effect did coyote hunting have on the sheep industry?

exercise 2-16 Suppose that in rocks that are 250 million years old you find snail fossils like 1 and 2, that you find fossils like 3 and 4 in 275-million-year-old rocks below them, and that farther down, in 300-million-year-old rocks you find only fossils like 5. What would you conclude about the course of evolution here?

1   2   3   4   5

How can you possibly determine that a rock stratum is 300 million years old? The best methods depend on a knowledge of radioactive decay schemes. Radioisotopes decay into stable products; in the atmosphere or in growing organisms, all isotopes can diffuse about and distribute themselves randomly, but once a rock has formed the "parent" isotope and its "daughter" isotopes are trapped in place. When the relative amounts of these two isotopes are measured with a mass spectrometer, the age of the rock can be determined.

Radioisotopes decay exponentially in the same way that clones of cells grow exponentially. If N_0 atoms of a radioactive isotope are present in a rock at time zero, after a time t a certain number N will remain and the rest will have decayed into D atoms of a stable "daughter" isotope: $N_0 \rightarrow N + D$. The decay occurs at a rate $dN/dt = -\lambda N$, so $N = N_0 e^{-\lambda t}$. Since

$$D = N_0 - N$$
$$= N e^{\lambda t} - N$$
$$= N(e^{\lambda t} - 1)$$

$$\frac{D}{N} + 1 = e^{\lambda t}$$

$$\ln\left(\frac{D}{N} + 1\right) = \lambda t$$

exercise 2-17 One of the best systems for studying very old rocks depends on the reaction $^{87}\text{Rb} \rightarrow {}^{87}\text{Sr} + \beta$, for which $\lambda = 1.39 \times 10^{-11}$ years^{-1}. Calculate the age of rocks for which $^{87}\text{Sr}/^{87}\text{Rb} = 0.0247$.

exercise 2-18 Another useful clock is provided by the decay of ^{40}K to ^{40}A. Actually, ^{40}K decays in two ways:

$$^{40}\text{K} \rightarrow {}^{40}\text{Ca} + \beta$$
$$^{40}\text{K} + e^o \rightarrow {}^{40}\text{A} + \gamma + \text{x ray}$$

Since the second reaction accounts for about 12% of the ^{40}K decay, we introduce a correction factor:

$$\lambda t = \ln\left(\frac{1.12D}{0.12N} + 1\right)$$

where $\lambda = 0.503 \times 10^{-9}$ year^{-1}. What is the age of rocks in which $^{40}\text{A}/^{40}\text{K} = 0.180$?

3

**energetics
and
metabolism**

The bacteria we discussed in Chapter 2 could make all their complex organic compounds from a carbon source, such as lactose, plus a few inorganic salts, such as phosphate and sulfate. It should be obvious that growth and reproduction entail many chemical reactions, which are collectively called metabolism. In this chapter we shall try to understand the principal features of **metabolism,** along with some thermodynamic and kinetic restrictions that govern biological processes just as they do nonbiological ones.

A. Enzymes and chemical reactions

3-1 ENZYME ACTIVITIES

Lactose is a disaccharide of glucose and galactose. A bacterium could easily use it for a carbon source by hydrolyzing it into its component monosaccharides:

$$\text{lactose} + H_2O \leftrightarrows \text{glucose} + \text{galactose}$$

From the elementary principle of mass action in chemical reactions, the rate of this process can be defined as

$$-\frac{d[\text{lactose}]}{dt} = k_1[\text{lactose}] \tag{3-1}$$

where brackets indicate the molar concentration of a substance and we ignore the concentration of water because the process occurs in aqueous solution. Even though this process does occur, it is so slow at pH 7 that it is hardly noticeable.

Lactose hydrolysis can be catalyzed by increasing the hydrogen ion concentration; a catalyst, you will recall, is any agent that changes the rate of a chemical reaction (generally increasing it) without being consumed in the reaction. In this case hydrogen ions do just that. The rate of the reaction increases as the pH decreases, but even at pH 2, k_1 is little more than 10^{-7} per second, which appears to be much too low to support the rapid bacterial growth we observed before, even if such a low pH were a realistic biological condition.

Perhaps the bacteria have an internal catalyst that is more efficient than mere hydrogen ions. To search for such a catalyst, we grow a large number of cells, break them by grinding with find sand or alumina (Al_2O_3) powder, and extract the broken cells with a buffer When a small amount of this cell extract is added to a lactose solution there is an immediate, rapid hydrolysis. The activity appears to be catalytic, for the active agent is not used up by the addition of lactose.

Many agents have been obtained from cells by similar methods that have the ability to catalyze chemical reactions—always, in each case, a rather specific reaction. They were once called "ferments," but later, because they could be extracted from yeast (zyme) the name **enzyme** became more common. In fact, every cell contains many different enzymes. They can be separated from one another and purified by using fractionation methods not too different from some we examined in Chapter 1, except that we generally try to be very gentle with the cell extract to keep its contents in their most pristine state. At each stage of the fractionation, we assay for enzyme activity and keep the fraction that contains the most catalyst. However, the first step in this work is to find the most convenient assay for the enzyme; in the case of the lactose hydrolytic activity, we need a substitute for the direct measurement of lactose hydrolysis.

Incidentally, we should also find a more convenient name for the enzyme. By convention, all enzyme names end in -ase (with very few exceptions from the dawn of biochemistry) and they are generally formed from the name of the enzyme's **substrate**—that is, the principal reactant that the enzyme acts upon; so a good name for this enzyme might be lactose hydrolase or simply lactase. In fact, the enzyme works on a number of compounds that are technically β-galactosides, so we will call it β-galactosidase.

3-2 SPECTROPHOTOMETRY

Some of the most important optical techniques depend upon the fact that compounds specifically absorb light of different frequencies. Absorption in different stretches of the electromagnetic spectrum reflect different types of changes. Light in the ultraviolet range, roughly below 350 nm in wavelength, is absorbed primarily in electron transitions. Figure 3-1 shows that if an electron can reside in a low-energy orbital with energy E_1 and in a higher orbital with energy E_2, it can be elevated from one to the other by absorbing light whose frequency is ν, where $\Delta E = E_2 - E_1 = h\nu$, where h is Planck's constant. If a compound is irradiated with light of this particular frequency, some of the light is absorbed in the electronic transition, and there will be one such absorption maximum for every characteristic transition. We generally plot the amount of light absorbed by a compound as a function of the frequency of the light or its wavelength $\lambda = c/\nu$, where c is the velocity of light; this produces a characteristic spectrum for each compound (Fig. 3-2) that is useful in identifying it and determining its concentration. Since we generally work with molar quantities of molecules, we define the energy of a mole of light—that is, an Avogadro's number of photons—as one **einstein;** when λ is measured in nanometers, this quantity is

$$E = \frac{28,600}{\lambda} \text{ kcal/mole} = \frac{1235}{\lambda} \text{ eV/mole}$$

[1 electron volt (eV) $= 1.6 \times 10^{-12}$ erg].

The infrared region of the spectrum, from about 2 to 15 μm, is also interesting and useful; light of this wavelength is absorbed primarily in transitions from one vibrational state to another. Every bond type has a characteristic absorption frequency associated with its stretching; these are generally not as sharp as the peaks of an electronic absorption spectrum because rotational transitions are superimposed on them, but

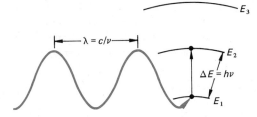

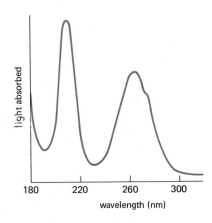

FIGURE 3-1

Light of the proper wavelength is absorbed in an electronic transition to a higher energy level.

FIGURE 3-2

The absorption spectrum of a nucleotide.

they are characteristic enough to identify many of the major functional groups in organic structures.

Spectrophotometers that measure light absorption are used for much of the quantitative work we will discuss. In these instruments, a solution containing the substance being studied at a concentration c moles/liter is placed in a cuvette, a small glass or quartz cell with a light path of l cm between optically clear faces. A light beam of intensity I_0 enters the solution and a beam of intensity I emerges; the instrument measures the **transmittancy** of the solution, $T = I/I_0$. If c and I_0 are not too great, the system obeys Beer's law, which states that

$$T = 10^{-\epsilon cl} \tag{3-2}$$

where ϵ is the **molar extinction coefficient** of the solute. The **absorbancy** of the solution, A, (sometimes also called **optical density**, OD) is defined as

$$A = -log_{10} T = \epsilon cl \tag{3-3}$$

The absorbancy, which can also be read on the spectrophotometer, is therefore directly proportional to the concentration of the absorbing solute. To determine the concentration of a solute, we use light that corresponds to one of its distinctive absorption maxima; values of ϵ for many compounds at characteristic wavelengths are now tabulated in reference books.

We can readily follow the course of a chemical reaction by using, for example, light that is strongly absorbed by one of the products but not by any of the reactants. To measure the activity of β-galactosidase we use the compound *o*-nitrophenyl-β-

galactoside (ONPG), which is colorless but yields on hydrolysis the yellow compound *o*-nitrophenol, which absorbs at 420 nm. The **enzyme activity** is defined as the rate at which *o*-nitrophenol is released, and the **specific enzyme activity** is the enzyme activity per unit of mass in the cell extract—for example, per microgram of protein.

3-3 FRACTIONATION AND CHROMATOGRAPHY

There are many good ways to separate enzyme activities, but we will select three which illustrate separations based on (1) size and shape, (2) solubility, and (3) electric charge.

A crude extract of broken cells can be separated easily on the basis of the size, shape, and density of its components by using **density gradient centrifugation.** Sucrose solutions are widely used for this purpose; Fig. 3-3 shows how a dense and a

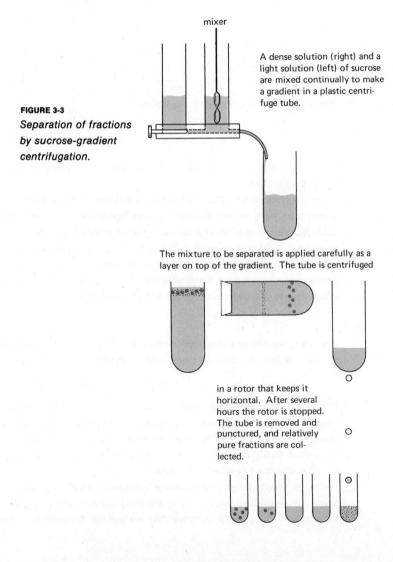

FIGURE 3-3

Separation of fractions by sucrose-gradient centrifugation.

mixer

A dense solution (right) and a light solution (left) of sucrose are mixed continually to make a gradient in a plastic centrifuge tube.

The mixture to be separated is applied carefully as a layer on top of the gradient. The tube is centrifuged

in a rotor that keeps it horizontal. After several hours the rotor is stopped. The tube is removed and punctured, and relatively pure fractions are collected.

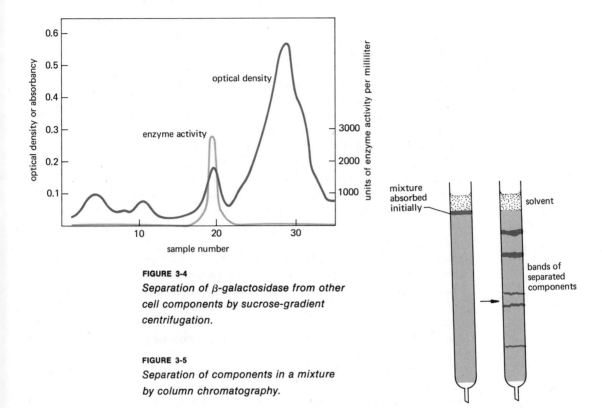

FIGURE 3-4

Separation of β-galactosidase from other cell components by sucrose-gradient centrifugation.

FIGURE 3-5

Separation of components in a mixture by column chromatography.

light solution can be mixed in a centrifuge tube to make a continuous gradient. A little cell extract is then layered on top and the tube is spun rapidly in a centrifuge on a special rotor that keeps it horizontal during the entire run. Various cell components separate from one another on the basis of their mass and shape, since the heaviest and most "streamlined" fall fastest, and also on the basis of their density, since the least dense tend to float. The tube is punctured at the bottom and samples are collected. Since we are interested in β-galactosidase, we take a small sample from each tube and assay for the enzyme. To get a measure of the mass of material in each tube, we determine the absorbancy of each sample at 280 nm, where proteins absorb strongly. The two measurements together yield the specific activity of each tube. In this case we are lucky, for the enzyme appears to be a fairly large compound that sediments more rapidly than most of the cell protein and therefore can be partially isolated (Fig. 3-4).

A second procedure, of which there are many variations, comes under the general heading of **chromatography.** All chromatographic techniques depend upon the fact that materials partition themselves in various ways between two distinct phases on the basis of chemical affinities, solubilities, and so on. One of the phases is a solid, such as a column of packed resin or a sheet of filter paper. The mixture of materials to be separated is applied to the solid phase; a solvent, or a combination of solvents, is then passed across the solid to elute different fractions from the mixture on the basis of their relative affinities for the solid and solubilities in the solvent (Fig. 3-5). For example, a column may be made from an ion exchange resin that consists of tiny beads that are very high polymers of a hydrocarbon such as styrene ($CH_2\!\!=\!\!CHC_5H_6$). The styrene residues make a *polystyrene* whose side chains are benzene rings that can be substituted with different acidic or basic groups. A column of these beads bathed in an aqueous solution presents an enormous number of these side chains. The various proteins and other components in the cell extract have different affinities for these groups. Those that only bind weakly will be readily eluted and will come off in an early sample, while those that are more tightly bound will come off later, or perhaps not at all unless a different solvent is used. Again, each fraction is assayed for enzyme activity and its optical density is determined. This is a very efficient way of getting relatively pure enzyme fractions.

A third major technique, **electrophoresis,** depends upon differences in charge on the surfaces of different compounds. The cell extract is generally applied to a piece of filter paper or a strip of gel made from a thick starch suspension. The solid is again bathed in an aqueous solution while a strong electric current passes through it. Neutral materials will be unaffected, but most substances in the cell extract have a distinct positive or negative charge, and they will move toward the cathode or the anode at rates that are functions of their charges. One can often achieve a remarkable degree of separation in this way. Furthermore, electrophoresis can be combined with chromatography. A sample can be applied to filter paper and run in one direction electrophoretically and then at right angles by chromatography; this often produces some very distinctive separations.

In all of these procedures, we find that the enzyme activity goes along with material that absorbs strongly at 280 nm — that is, with protein. Furthermore, there are some enzymes that specifically attack proteins; if one of these is added to a bit of extract with a high enzyme activity, the activity soon disappears as the protein is destroyed. The result strongly suggests that enzymes are proteins. If an enzyme is highly purified, it can be crystallized; with some effort, a pure crystalline protein that has a very high specific activity for ONPG hydrolysis can be obtained from our cell extracts.

In 1929, J. B. Sumner crystallized the first pure enzyme, urease, which degrades urea. This was the first proof that enzymes are proteins. Many other enzymes have been purified since and they have all proven to be proteins. Subject to later qualification, it is safe to say that all enzymes are proteins; but let us always remember that we rarely have pure protein preparations to work with, and enzyme activities are always defined operationally by the ability to catalyze specific reactions.

A common operation used in enzyme purification is **dialysis.** The cell extract or a fraction of it is placed in a tightly closed little bag made of sausage casing and left for several hours in a buffer solution. There are tiny pores in the casing which allow free diffusion of water and small molecules, but macromolecules such as proteins are trapped inside. This is a convenient way to get rid of unwanted small molecules, but early enzyme chemists frequently observed that dialysis also removed some components essential to enzyme activity but distinct from the enzyme itself. These **coenzymes** pass through the dialysis tubing and enzyme activity is restored only when they are added back to the protein, the **apoenzyme,** to make a complete enzymatic unit, the **holoenzyme** (Fig. 3-6). Coenzymes are small organic molecules; many of them, along with certain very similar compounds, are the vitamins so essential to the nutrition of many animals, including man. When certain ions, particularly transition metals such as iron and zinc, are removed by dialysis, the enzyme activity also disappears. Many enzymes require certain metals for their activity and these are spoken of as **activators** rather than coenzymes.

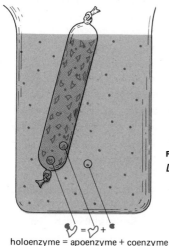

FIGURE 3-6
Dialysis separates small molecules from large ones.

holoenzyme = apoenzyme + coenzyme

3-4 CHEMICAL EQUILIBRIUM

If a lactose solution is left long enough, with or without a catalyst, it eventually reaches a point where no further hydrolysis of lactose can be detected. The reaction as written in equation 3-1 is reversible, and the reverse reaction is also taking place at the rate

$$\frac{d[\text{lactose}]}{dt} = k_{-1}[\text{glucose}][\text{galactose}] \tag{3-4}$$

When the rate of lactose hydrolysis, equation 3-1, is equal to the rate of lactose synthesis, equation 3-4, the overall composition of the solution will not change and the system is at *equilibrium*. We define an equilibrium constant

$$K = \frac{k_1}{k_{-1}} = \frac{[\text{glucose}][\text{galactose}]}{[\text{lactose}]} \tag{3-5}$$

that characterizes the reaction very well. In general, for the chemical reaction $aA + bB \rightleftarrows cC + dD$, where the lowercase letters are stoichiometric coefficients, the equilibrium constant is

$$K = \frac{[C]^c[D]^d}{[A]^a[B]^b} \tag{3-6}$$

Equilibrium is simply the point where the forward and reverse reaction rates are equal; the reactions are still going on, but there is no net change in the concentrations of the reactants. Furthermore, if we have a cyclical process, then at equilibrium the forward and reverse rates of all three reactions must be equal. The flow of matter one way around the cycle must equal the flow the other way, and the system is not at equilibrium if there is a flow in only one direction, even though the concentrations of all three components remain constant.

No matter how much enzyme or other catalyst we add to a chemical system at equilibrium, the concentrations of the reactants do not change. The catalyst speeds both the forward and reverse reactions, but their ratio, the equilibrium constant, remains the same. There is an important lesson here; from the definition of equilibrium it should be obvious that a growing, functioning cell or organism cannot be at equilibrium. And if it were at equilibrium, its enzymes would do it no good at all. The only reason that enzymes have any value in a cell is that the reactions that occur within it are quite far removed from equilibrium; there is a constant flow of material through the cell, and the enzymes increase the rate of this flow. However, much of the cell can be in a steady state, where the concentrations of all reactants remain constant while there is a unidirectional flow through the system. The chemostat is an example of a steady state; fresh nutrient solution runs in at the same rate that cells are siphoned out, but the concentrations of all cellular materials within the growth tube itself remain constant. An even simpler example is a river that retains its shape and its depth at every point, but exists in this condition only because there is a steady flow of water through it. We will have other occasions later to think of cells in their steady-state conditions.

The lactose-hydrolysis reaction illustrates one other point. The equilibrium constant of the reaction is enormous; the equilibrium condition is one where the lactose concentration is only about 10^{-6} that of the glucose and galactose. From what we have just said, there is no enzyme or other catalyst that we could add to the solution to drive the reaction toward lactose synthesis, and yet the synthesis of compounds such as lactose, polysaccharides, proteins, and nucleic acids must occur regularly and rapidly in all cells. So we can see that there is another problem that a biotic system must solve; it is not enough to have enzymes that drive reactions rapidly if there are some other physical limitations that prevent it from going far enough beyond its equilibrium point to be useful. So we must now think about some of these other limitations and find out how cells can synthesize complex materials that would exist in minute quantities in an equilibrium mixture.

exercise 3-1   Calculate the equilibrium constant for $2AB \rightleftarrows A^+ + B^- + CD$, if at equilibrium $[AB] = 0.1\ M$, $[A^+] = [B^-] = 5\ M$, and $[CD] = 0.04\ M$.

B. Energy and equilibrium

The purpose of the following sections is to provide a minimal understanding of some basic physical concepts, not in the depth and mathematical formality possible in a course devoted to these topics, but simply at an intuitive level. Anyone who is dissatisfied with equations without formal proofs will have to relieve his frustrations with a more advanced chemistry text.

3-5 HEAT AND ENERGY

Around the beginning of the nineteenth century, Benjamin Thompson (Count Rumford) was supervising the boring of cannons for the King of Bavaria. He was impressed by the enormous amount of heat produced during the drilling and suggested that the mechanical energy was being converted into heat; he even measured the heat equivalent to 1 horsepower hour and was not far from the modern value. But the scientific world did not become convinced of the equivalence of heat and mechanical energy until 1849, when James Joule proved that paddles beating vigorously in a water bath would raise the temperature of the water; he found that mechanical and electrical energy could be converted to heat, with a constant ratio between the various forms of energy. In modern units, 1 cal of heat is 4.184 joules.

The line of inquiry started by Thompson and Joule finds its modern expression in the first law of thermodynamics. We say that every system is characterized by an internal energy E which can change either because of the transfer of heat across the boundaries of the system or because of work performed:

change in energy	=	heat added to the system	−	work done by the system
ΔE		Q	−	W	(3-7)

According to the first law, energy is *conserved;* it depends only upon the state of the system and not upon the path taken to reach that state. If the system goes from state 1 to state 2, its energy changes by $\Delta E = E_2 - E_1$; if it then goes back to state 1, there is another energy change $\Delta E = E_1 - E_2$, so the total change in energy is zero. Even though the system may have undergone many changes, no energy has been created or lost; when it returns to its initial state its energy is the same as it was before.

While work is not solely a function of the state of the system, it can be expressed in terms of other functions that are. Work can be expressed as the product of a force f acting through a distance s, and there is some expression of the form

$$dW = f \, ds \tag{3-8}$$

for every kind of work (electrical, gravitational, etc.). Figure 3-7 shows that mechanical work can be expressed in terms of a pressure p acting through a volume change dV:

$$dW = p \, dV \tag{3-9}$$

The total work done in changing from V_1 to V_2 is just

$$W = \int_{V_1}^{V_2} p \, dV \tag{3-10}$$

To fix this idea a little more strongly, you might attempt the following exercise.

exercise 3-2 For an ideal gas, $pV = nRT$. By substituting in equation 3-10, derive an

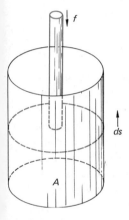

FIGURE 3-7
A cylinder of cross-sectional area A *performing work against an external force* f.

expression for the work done by an ideal gas in changing from V_1 to V_2 at constant temperature and pressure. Calculate the work at 300°K for the expansion of 1 mole of gas from 1.5 to 4.5 liters.

The conditions of constant temperature and pressure are particularly useful for understanding some biotic systems, because organisms really operate under these restrictions. In this case, the energy can be changed by adding heat, which produces a corresponding change in volume:

$$\Delta E = E_2 - E_1 = \Delta Q - p(V_2 - V_1)$$
$$\Delta Q = (E_2 + pV_2) - (E_1 + pV_1) \tag{3-11}$$

The combination $E + pV$ occurs so often that it is given the name **enthalpy** and denoted by H. Equation 3-11 then becomes

$$\Delta Q = H_2 - H_1 = \Delta H \tag{3-12}$$

The enthalpy is a useful function because at constant pressure it is just the heat added to or subtracted from a system. Every chemical reaction, for example, can be written as

$$A \rightleftarrows B + \Delta H \tag{3-13}$$

When this reaction proceeds toward the right, with the liberation of heat, ΔH is *negative* and the process is **exothermic.** When the reaction proceeds toward the left, heat must be added, ΔH is *positive,* and the process is **endothermic.**

On the basis of intuition and some experience, we are generally inclined to consider a chemical reaction such as equation 3-13 as spontaneous when it liberates heat and proceeds toward the right, while we feel that it is not spontaneous when heat must be applied to drive it toward the left. Nineteenth century chemists had the same feeling and performed many experiments that convinced them that they were right. By 1878, Marcellin Berthelot was confident enough to state that all chemical reactions will proceed in the direction in which they liberate the most heat. Since heat removed from a system decreases its internal energy, this is in accordance with the general principle that, other things being equal, every system will tend to achieve its condition of minimum energy.

However, our intuition is not quite good enough and Berthelot was wrong. Although the drive toward minimum energy is important and most reactions do proceed exothermically, there are other reactions that are endothermic but still spontaneous. This is because we have ignored a second important tendency that sometimes outweighs the tendency toward minimum enthalpy; we shall consider this next.

3-6 PROBABILITY AND ENTROPY

A second feeling we have about the universe is that if we wait for a while, things will get worse. The natural tendency of most things is to run down, get dirty, wear out, and get mixed up. It has been said that the feeling that the universe is running toward a state of greater disorder is the basis of our intuition about the direction of time; we would certainly all agree that a spot of ink placed on the surface of a glass of water will tend to diffuse out into the water, or that a shoe will tend to wear out and develop holes, but that the reverse processes simply don't happen.

In fact what we observe is the tendency for everything to achieve its most *probable* condition, and the concepts of order and probability are closely linked. Consider a simple example.

A box is divided into a square array of 25 spaces, 5 rows by 5 columns. We throw 10 balls into the box; any ball can fall into any space with equal probability. From combinatorial analysis, the number of possible arrangements is

$$\frac{25 \times 24 \times \cdots \times 17 \times 16}{10!} = 3,268,760$$

and so the probability of any single arrangement is the reciprocal of this number, or about 3×10^{-7}. Now consider the highly ordered arrangements described by "two neighboring columns of five filled spaces." Only four arrangements satisfy this rule:

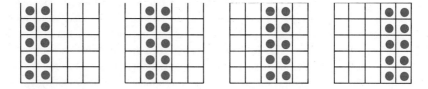

The probability that 10 balls thrown at random will fall into this pattern is only about 1.2×10^{-6}, whereas they are very likely to fall into a pattern that looks completely random to us. Similarly, a fresh deck of playing cards is highly ordered and highly improbable, for it is only one of 52! possible sequences and the probability of finding the original order in a random shuffle is about 10^{-68}.

It is important to realize at the outset that every other arrangement of balls in the box or sequence of playing cards has the same low probability as the ordered ones we have described, but they are not interesting to us and do not seem remarkable when they occur by chance. The point is that the few arrangements that are remarkable are improbable.

The probabilities for various states of ordinary physical systems can also be calculated. Every molecule of a gas in a closed container can be described by six numbers—three coordinates and three components of momentum along those coordinates. We construct an imaginary, six-dimensional *phase space* consisting of little boxes that are just big enough to hold one molecule each. In a real gas, the distribution of points in these boxes is random; particles are found everywhere in the container, moving in all directions with an enormous range of velocities, subject only to the restrictions imposed by the temperature and internal energy of the gas. But now consider the physical meaning of the six-dimensional equivalent of the ordered arrangement of balls in the box. If all of the points in phase space were in a limited set of "rows" and "columns," this would mean that the particles were all in one small region of the container, all moving in the same direction with the same momentum. This is a highly improbable condition; no one seriously expects ever to find themselves in a room in which all of the air suddenly moves over against one wall and freezes there, leaving a vacuum in the rest of the room. By appropriate mathematical treatment, the probability of such a highly ordered arrangement could be calculated, but it would be small to the point of vanishing.

Probability is clearly an appropriate measure of order, and the thermodynamic probability of a physical system can be determined. For any given internal energy, statistical analysis tells us that n_1 molecules of the system have energy E_1, n_2 have energy E_2, and so on, in a characteristic distribution. However, we cannot tell which molecule has which energy, just as we cannot tell one ball from another when they are thrown into the box. We therefore have two levels of description: the *macrostate* description is the measured value that tells how many molecules are in each state and the *microstate* description tells which molecule is in which state. Together they give a measure of probability for the system. We assume that all microstates are equally probable; each corresponds to one distribution of the particles in phase space, and one random distribution is no more likely than another. But each macrostate corresponds to a certain set of microstates, and its probability is proportional to the number it contains. We define the thermodynamic probability of any macrostate as the number of microstates corresponding to it, divided by the total number of microstates possible. For example, there are many arrangements of particles in phase space that correspond to molecules moving randomly in a container, but very few that correspond to all the molecules being packed against one wall, and the first arrangement is clearly more probable than the second.

exercise 3-3 The macrostate of a pair of dice is their total in a given throw. The microstate description of them tells which die shows what number. On this basis, calculate the probabilities of the macrostates "two," "seven," and "twelve."

We now define a new function, the **entropy** of a physical system, by

$$S = k \ln P \qquad (3\text{-}14)$$

where P is the thermodynamic probability of the system and $k = 1.38 \times 10^{-23}$ joules/deg is Boltzmann's constant. Since k is just R, the gas constant, divided by Avogadro's number, we use R for molar quantities. In practice, we generally deal with differences in entropy, for which

$$\Delta S = k \ln \left(\frac{P_2}{P_1} \right) \qquad (3\text{-}15)$$

Notice that the dimensions of entropy are energy per temperature; S is usually measured in calories per degree or entropy units.

Entropy is clearly suitable for measuring the order of a system. If the thermodynamic probability is high, the system is disordered and S is high. If the probability of a system is low, it is more highly ordered and its entropy is low.

There are equivalent definitions of entropy using other functions, but the connection between them all may be difficult to illustrate. One of the most useful definitions says that the entropy change during any process at constant temperature is the heat exchanged in the process divided by the temperature:

$$\Delta S = \frac{\Delta Q}{T} \qquad\qquad (3\text{-}16)$$

Notice that this equation can be rewritten to give another expression for heat:
$\Delta Q = T \,\Delta S$.

3-7 FREE ENERGY AND EQUILIBRIUM

We can now return to the problem raised in Section 3-5: Why are there reactions that
do not proceed in the apparently natural exothermic direction? The answer is easy
now. The enthalpy of a system measures its tendency toward a state of minimum
energy. But the entropy must also be considered now, because it measures the
tendency of every system toward a state of maximum probability; the second law of
thermodynamics states that every system in isolation will tend to achieve its condition
of maximum entropy.

To determine the point of equilibrium for a system at constant temperature and
pressure, we take both of these tendencies into account by defining a new function G,
the (Gibbs) free energy:

$$G = H - TS$$

For a finite change with p and T constant,

$$\Delta G = \Delta H - T \,\Delta S \qquad\qquad (3\text{-}17)$$

It then follows that at equilibrium under these conditions, the enthalpy will tend toward
a minimum, the entropy toward a maximum, and the free energy toward a minimum.
In some cases, the free-energy change will depend more strongly on entropic factors
and in others it will depend more strongly upon change in enthalpy, but it always takes
both factors into account and so it is the most generally useful criterion for
equilibrium. We can now distinguish **exergonic** processes, where $\Delta G < 0$, which
are spontaneous and entail a decrease in free energy, from **endergonic** processes,
where $\Delta G > 0$, which are nonspontaneous. The second law forbids an endergonic
process in isolation; the free energy of a system can increase only if it is driven by
some other reaction whose free energy undergoes an equal or greater decrease, so
the total free-energy change is negative.

It is not hard to show that the decrease in free energy is equal to the maximum
amount of useful work that a system can perform. Suppose the system can do work
$\Delta W'$ in addition to $p\,dV$ work. Its internal energy will change by $\Delta E = \Delta Q - p\,\Delta V - \Delta W'$;
equation 3-17 can be expanded as $\Delta G = \Delta E + p\,\Delta V - T\,\Delta S$, but $T\,\Delta S = \Delta Q$.
Combining these two equations, we obtain $\Delta W' = -\Delta G$. In essence the change in
enthalpy measures the total energy change during any process, but the entropy
term $T\,\Delta S$ is the useless heat energy that must be wasted during the process and
the remainder ΔG is the amount that can actually be used. In this form, the second law
is a statement of the well-known fact that no engine can be 100% efficient.

Since we are concerned with chemical processes, it is important to have an
expression for the free-energy change in a chemical reaction. We first define a set

of standard states as 1 atm pressure for all gases and 1 m concentration for all solutions. We let the standard free energy G^0 of all *elements* in their most stable form at 25°C be zero. Then the standard free energy of a *compound* is the change in free energy necessary to form the compound from its elements in their standard states; these values are known and tabulated for many compounds. For any chemical reaction, we define ΔG^0 as

$$\Delta G^0 = G^0_{products} - G^0_{reactants} \qquad (3\text{-}18)$$

We often require an addition for biological purposes. The standard state for a solution of hydrogen ions is also 1 m, but the pH of this solution is zero, which is not very realistic for most biotic processes that typically operate closer to neutrality. We therefore define a set of $\Delta G'$ values for reactions in which all components are in their standard states, except that pH $= 7$.

With these definitions, it is relatively easy to show that

$$\Delta G^0 = -RT \ln K \qquad (3\text{-}19)$$

where K is the equilibrium constant of the reaction. In general, particularly in biotic systems, all of the components are not in their equilibrium concentrations; the actual free-energy change is then different from the standard free-energy change. If the reaction is $aA + bB \rightleftarrows cC + dD$, the free energy is

$$\Delta G = \Delta G^0 - RT \ln \frac{[C]^c[D]^d}{[A]^a[B]^b} \qquad (3\text{-}20)$$

exercise 3-4 At 25°C and 1 atm pressure, G^0 for solid glucose ($C_6H_{12}O_6$), gaseous CO_2, and liquid water are -215.80, -94.45, and -56.69 kcal/mole, respectively. Calculate the standard free energy of combustion of glucose with oxygen to yield CO_2 and water.

exercise 3-5 Two molecules of the amino acid glycine dimerize to make the dipeptide glycylglycine, with the following free energies of formation in aqueous solution:

$$2 \text{ glycine} \rightleftarrows \text{glycylglycine} + \quad H_2O$$
$$(-89.26) \qquad (-115.63) \qquad (-56.20)$$

(a) Calculate ΔG^0 for the reaction. (Note: We always consider the concentration of the water in which the compounds are dissolved to be 1M.)
(b) Calculate the equilibrium constant for the formation of glycylglycine.
(c) Suppose the actual concentrations of the dipeptide and the water of formation were 1 M and the free glycine were only 10^{-2} M. What would the actual free-energy change be?

As Exercise 3-5 illustrates, ΔG may be very different from ΔG^0 if the concentrations of products or reactants are very far from their equilibrium values. Since it is ΔG that actually determines whether or not a reaction will occur, it is important to determine the actual free energy and not some arbitrary standard.

C. Metabolism

Most of the subjects we will discuss in this book technically come under the heading of metabolism, because we are largely concerned with all of the chemical processes that occur in cells and cell complexes. This is the first in a series of views of the overall process; in later chapters we will focus on the details with increasing powers of resolution.

We have already seen how organisms solve their basic *kinetic* problem; to make their reactions go fast enough, they have a battery of enzymes that act as catalysts. It is instructive to think of organisms as having two other metabolic problems. The first is a **biosynthetic** problem. The structures of which cells are built are vastly different from most of the materials in the nonbiological environment; how do they transform these materials into their own characteristic materials? The second problem is **energetic.** Many—if not most—of the reactions required to build cell materials are endergonic; how do cells obtain the energy necessary to drive these reactions, and how do they employ this energy? To answer both questions, it is helpful to see how our current ideas about metabolism have grown up; we will then attempt to give the most modern answers to these two questions.

3-8 BACKGROUND

The history of metabolic research is largely a history of research on *respiration*. At its simplest, the word means breathing, and since it is one of the chief characteristics associated with life, a number of prominent scientists became involved with the problem toward the end of the eighteenth century. In 1755, Joseph Black discovered that a special gas which he called "fixed air" is given off by respiring animals and during combustion; today, of course, we call this gas carbon dioxide. Joseph Priestley, in 1771, studied respiration and combustion by keeping small animals and burning candles in tightly closed jars; naturally, the candles soon went out and the animals died. In the prechemical idiom of his time, Priestley said that both the candles and the plants were "phlogisticating" the air. He then made the remarkable observation that a sprig of mint had the opposite effect; it "dephlogisticated" the air so an animal would survive longer and a candle would burn again. Priestley also produced oxygen, which supported the combustion of a candle very well, by heating mercuric oxide with the focused rays of the sun, but to him this was just dephlogisticated air, for he could not interpret his experiments in any other way nor understand their full significance.

Priestley's contemporary Antoine Lavoisier was one of the principal opponents of the phlogiston theory and a founder of modern, quantitative chemistry. Lavoisier proved that only a portion of the air, which he named oxygen, is combined during either combustion or respiration, and that in both cases carbon dioxide is produced. He and Pierre de Laplace collaborated to perform the first quantitative measurements of the heat produced during these processes. They built ice-filled chambers in which this heat could be trapped and measured in terms of the amount of ice that was melted. They measured the heat produced during combustion and the amount of carbon dioxide resulting and thereby obtained the heat of formation of carbon dioxide. Then they measured the carbon dioxide produced by an animal during respiration and the attendant heat. Finding the heat of formation of CO_2 to be approximately the same in both cases, they concluded that respiration and combustion are fundamentally the same. According to Lavoisier's view, carbon is carried by the blood to the lungs, where it is slowly combined with oxygen, and the heat of the body is produced in this combustion. This crude view of respiration was gradually replaced by a more realistic conception. The site of respiration was pushed back gradually from the lungs to the blood and finally to the tissues of the body. However, Lavoisier was still essentially correct in his view of the chemical process.

Lavoisier's discoveries influenced the Italian naturalist Lazaro Spallanzani to push this line of inquiry still further. He developed simplified techniques for measuring respiration and used them to show respiratory activity in a incredible variety of objects; most of this activity was probably due to contaminating bacteria. However, among his experiments were clear demonstrations that respiration occurs in the tissues themselves, not primarily in the blood or lungs, and that it occurs in animals that have nothing like the lungs or circulatory system of a mammal.

Another contemporary, Jan Ingenhousz, performed related experiments with plants and showed that they remove carbon dioxide from the air and replace it with oxygen under the influence of light. He said that light enables the plant to split carbon dioxide into carbon, which is "fixed" by the plant, and oxygen, which is released. Moreover, he showed that in darkness, plants produce carbon dioxide just like animals. Thus, the observations of Priestley and Ingenhousz together lay the foundations for an important concept—that there is a cycle in nature in which plants produce oxygen and remove carbon dioxide from the air, while animals (and plants in the dark) do just the opposite.

The Swiss naturalist Jean Senebier made similar observations of these phenomena and showed that it is the light itself that is necessary and not merely heat from the sun. Soon afterward, Nicholas de Saussure demonstrated that water is fixed in the plant along with carbon dioxide. With the development of chemistry in the nineteenth century, it gradually became possible to describe the process in plants as a **photosynthesis** of starch under the influence of light, which could be written:

$$6H_2O + 6CO_2 \rightarrow C_6H_{12}O_6 + 6O_2$$

Respiration could be described as the opposite process, in which starch or some other carbohydrate is burned with oxygen to produce CO_2:

$$C_6H_{12}O_6 + 6O_2 \rightarrow 6H_2O + 6CO_2$$

So respiration and photosynthesis could be described as two primary, directly opposite processes that create a cycle in nature, and this was a simple, satisfying way to look at the world.

It is probably a good thing that science is basically conservative of old theories, especially simple ones, but the concept of photosynthesis as a simple, light-driven reduction of CO_2 should not have persisted as long as it did. During the 1880s, Sergei Winogradsky found some sulfur bacteria that could fix CO_2 in the dark using sulfide as their sole energy source. Shortly thereafter, Theodor Engelmann studied some of the purple sulfur bacteria and found that they photosynthesize but do not evolve oxygen. The real change in viewpoint came only after some experiments by C. B. van Niel, beginning about 1931. van Niel confirmed that the purple sulfur bacteria are dependent on light and do not produce oxygen; he found that, instead, they oxidize reduced sulfur compounds such as sulfide to elemental sulfur and sulfate, according to the equation

$$12H_2S + 6CO_2 \rightarrow 12S + C_6H_{12}O_6 + 6H_2O$$

van Niel then realized that adding six molecules of water to each side of the equation for green plant photosynthesis would give an analogous expression:

$$12H_2O + 6CO_2 \rightarrow 6O_2 + C_6H_{12}O_6 + 6H_2O$$

He therefore proposed that there can be many different photosyntheses, all dependent upon the oxidation of some hydrogen donor H_2A according to the equation

$$12H_2A + 6CO_2 \rightarrow 12A + C_6H_{12}O_6 + 6H_2O$$

In this formulation, A and H_2A can be many different compounds, including organic molecules. Or there might be no "A"; the hydrogen donor could be hydrogen itself, which leaves no byproduct.

van Niel postulated that the source of energy in all photosynthesis is the splitting of water into an [H] part that is used to reduce CO_2 and an [OH] part that is eliminated by combination with H_2A. In fact, the literal splitting of water does not seem to be a part of photosynthesis, except in the case of green plants, but at least van Niel turned thought in the field away from the classical view of the nineteenth century and toward a more modern viewpoint.

Meanwhile, investigations related to respiration were continuing. Much of this research centered about the phenomenon of fermentation, particularly alcoholic fermentation, which is not surprising when one considers the importance of alcohol to man. Lavoisier was really the first to show that during alcoholic fermentation, sugar is separated into CO_2 and alcohol, but he ignored the fermenter itself, the yeast. Even though Charles Cagniard-Latour had been able to prove to the satisfaction of his biological colleagues that yeast is really an organism, he was ridiculed by the most prominent chemists of the day, who continued to believe that fermentation was a purely chemical process. The question was finally settled by the great Louis Pasteur, who showed that fermentation is associated with the anaerobic (without air) growth of yeasts, while aerobically (with air) they respire like most other cells. Eduard Buchner accidentally discovered fermentation *in vitro* (literally, "in glass," as contrasted with *in vivo,* meaning in whole cells) when he tried to preserve some yeast extracts with sugar and found that the sugar was fermented without cell growth. Buchner thus opened the door to investigations of the chemical processes occurring in cells during respiration and fermentation. A milestone in this work was the discovery by Harden and Young, in 1905, that fermentation of glucose to alcohol is enhanced by phosphate, followed by the isolation of fructose-1,6-diphosphate from their preparations. Other sugars with phosphate groups were soon shown to be intermediates in fermentation.

At the same time, research on the mechanism of muscle contraction revealed that glycogen, a polyglucose food reserve stored in muscle, is broken down anaerobically to lactic acid. This process of **glycolysis** soon began to emerge as a kind of fermentation similar to alcoholic fermentation, and the same sugars with phosphate groups were shown to be intermediates in the process. These phosphates are now known to be among the most important cellular components from an energetic point of view, and we must consider them in some detail.

3-9 ENERGY COUPLING AND TRANSFER POTENTIALS

It should be obvious that organisms have a lower entropy than their surroundings. Their monomers are CHNOPS compounds that do not form spontaneously under present terrestrial conditions, and the polymers made from them have lower entropies than an equivalent set of free monomers, for they are constrained between two well-defined neighbors in a sequence and are no longer free to move about in all directions as they would be in a solution. Many of the steps in the synthesis of these compounds are endergonic, and there must therefore be some mechanism for driving these reactions in the right direction.

In general, endergonic reactions are driven by **coupling** them to exergonic reactions that have a greater change in free energy, so the ΔG of the two reactions taken together is still negative. Two reactions can be coupled if they share a common molecular species; a simple example from inorganic chemistry is the system

$$AgCl \rightleftharpoons Ag^+ + Cl^-$$
$$Ag^+ + NH_3 \rightleftharpoons AgNH_3^+$$

Silver chloride is extremely insoluble and has a very low dissociation constant, but it dissolves in water immediately if ammonia is added. The second reaction is so strongly exergonic that it drives the first reaction, in spite of the fact that the first one has a very small equilibrium constant. It is more instructive to write these reactions:

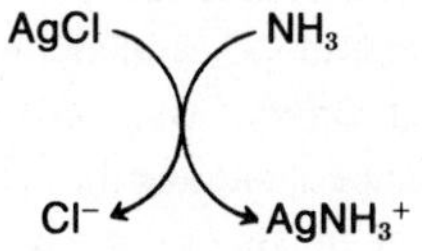

$$\begin{array}{ccc} AgCl & & NH_3 \\ & \diagdown \diagup & \\ & \diagup \diagdown & \\ Cl^- & & AgNH_3^+ \end{array}$$

to show that the Ag^+ ion is being transferred from the chloride to the ammonia. This is a kind of **group transfer reaction,** the more general form of which is

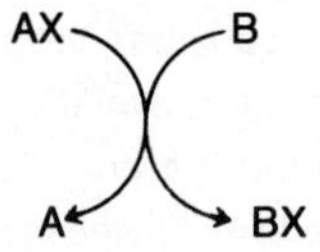

$$\begin{array}{ccc} AX & & B \\ & \diagdown \diagup & \\ & \diagup \diagdown & \\ A & & BX \end{array}$$

where X is the transferred group. Group transfer reactions are used in biotic systems to couple exergonic and endergonic reactions, but there is one class that stands out above all others in importance; this is where X is a **phosphoryl group,**

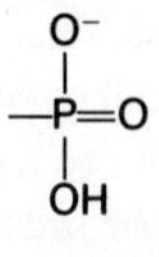

$$\begin{array}{c} O^- \\ | \\ -P{=}O \\ | \\ OH \end{array}$$

which we will often abbreviate $\enclose{circle}{P}$. The transfer of a phosphoryl group is called

phosphorylation, and its importance lies in the fact that there are several compounds that can donate a phosphoryl group with such a large change in free energy that they can drive many other reactions. We will consider only the most important of these now, to avoid obscuring the woods by trees.

To measure the ability of any compound to donate or accept a phosphoryl group, we define the **phosphoryl transfer potential,** ΔG_p^0, for a phosphorylated compound as the ΔG^0 of the reaction in which water is the acceptor:

$$Y\text{---}PO_3H + H_2O \rightarrow YH + H_2PO_4^-$$

or

$$Y\text{---}\textcircled{P} + H_2O \rightarrow YH + P_i$$

where P_i is the standard abbreviation for inorganic orthophosphate. It should be easy to see that a compound with a higher phosphoryl transfer potential (more negative ΔG_p^0) will transfer its phosphoryl group exergonically to any compound with a lower potential.

The donor in phosphorylation might be almost anything, but the most interesting ones are the nucleoside di- and triphosphates, particularly the adenosine phosphates. Adenine bonded to ribose is the nucleoside adenosine; adenosine bonded to a phosphate is adenosine 5'-phosphate (AMP) and AMP can accept two additional phosphoryl groups to make adenosine 5'-diphosphate (ADP) and adenosine 5'-triphosphate (ATP). ATP (and less often ADP) serves as a kind of universal *energy-exchange unit* in biotic systems because it is used to drive so many reactions.

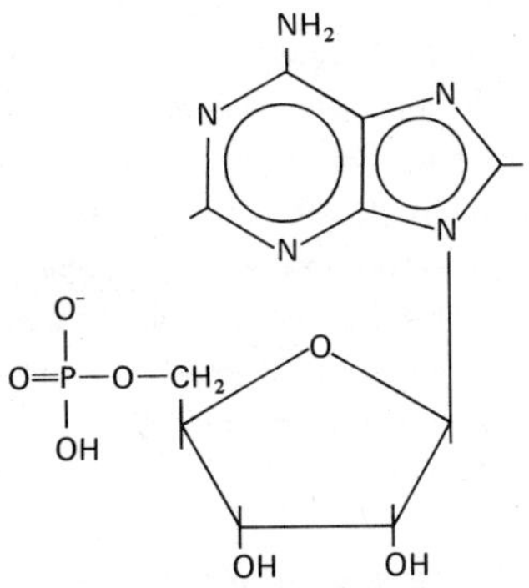

adenosine 5'-phosphate (AMP)

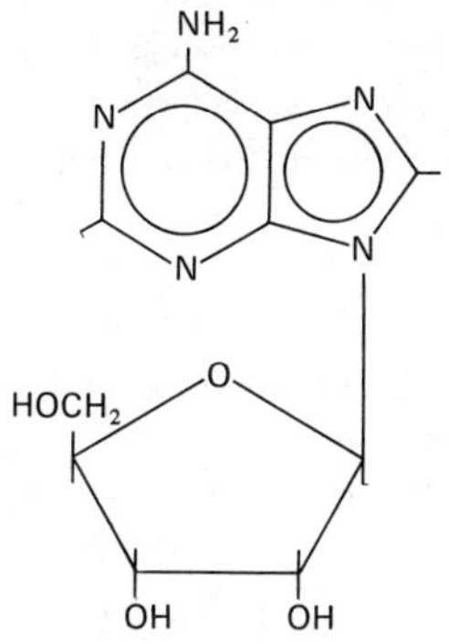

adenosine

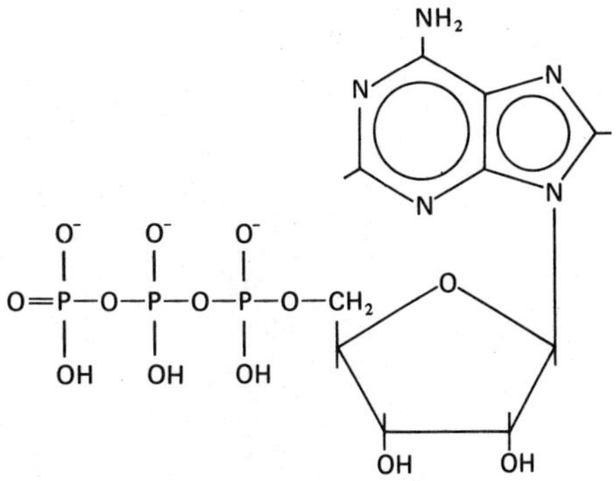

adenosine 5'-triphosphate (ATP)

Reasonable values for the transfer potentials of these compounds are

$$ATP^{4-} + H_2O \rightarrow ADP^{3-} + P_i + H^+ \qquad \Delta G_p^0 = -8.9 \text{ kcal/mole}$$
$$ADP^{3-} + H_2O \rightarrow AMP^{2-} + P_i + H^+ \qquad \Delta G_p^0 = -9.5 \text{ kcal/mole}$$

In cells, the concentrations of these compounds are much less than 1 m, so the free-energy changes are probably closer to -12 kcal/mole.

The significance of compounds with a high transfer potential was first recognized by Fritz Lipmann in about 1941. We refer to systems that generate ATP and related compounds as Lipmann systems. Lipmann suggested that some anhydrides of phosphoric acid, particularly ATP, contain "high-energy bonds" between the phosphate and the rest of the molecule that can hold enough energy to drive many other reactions. Herman Kalckar called these compounds "energy rich." However, these phrases are used here in quotes to indicate that they should not be used. There are no localized "high-energy bonds" in these compounds, and they are not "energy rich" in any ordinary chemical sense. Their standard free energies of hydrolysis are somewhat above average, although not as much as was once thought. The more realistic transfer-potential terminology of Irving Klotz is preferable and we will use it here. The really important feature of these compounds is that they are used so universally.

3-10 SOURCES OF FREE ENERGY

Now that we know how important ATP and related compounds are, we must find out where they come from; that is, how does a Lipmann system operate? Lavoisier and his contemporaries thought that respiration is simply a slow burning of foods very much like combustion; certainly, if you heat sugar with oxygen it will burn nicely and liberate a lot of heat, but there is no way to recover useful, high-potential compounds from this process. In our modern view of respiration, we still place the major emphasis on oxidation, but now we know that oxidation can take other forms than a simple combination with oxygen. Remember that in its most general sense oxidation means the removal of hydrogen ions or electrons from a compound, while reduction is the addition of hydrogen atoms or electrons. Since the two processes must always go together, we speak of them jointly as **oxidoreduction.** With this in mind, we can state the following general principle.

All free energy in biotic systems is derived from the transfer of electrons from a high-potential source to a low-potential sink. As electrons fall through this potential drop, a part of their free energy is removed in the form of ATP and related compounds.

In Chapter 2 we distinguished chemotrophy and phototrophy as the two principal modes of energy metabolism. In Table 3-1 these modes are subdivided on the basis of the following criteria. Chemotrophs obtain high-potential electrons by oxidizing reduced compounds. Phototrophs perform a photosynthesis in which the energy of light is used to raise electrons to a high potential, and they use an **accessory hydrogen (electron) donor** as a source of those electrons; this is the H_2A of van Niel's equations.

In both cases, the electrons are allowed to fall through an **electron-transport (ET)** system in which ATP is made; we will come back to this point in a moment.

TABLE 3-1
a classification of metabolic patterns

	ENERGY SOURCE	
ELECTRON DONOR	LIGHT	REDUCED COMPOUNDS
Inorganic	Photolithotrophy	Chemolithotrophy
Organic	Photoorganotrophy	Chemorganotrophy*

*Electron acceptor: organic; fermentation; oxygen: aerobic respiration; inorganic (not oxygen), anaerobic respiration.

From this point of view, van Niel's formulation of photosynthesis really means that enough light must be absorbed by a pigment system to create a potential that is strong enough to separate water into an [H] and an [OH] component. In fact, the splitting of water into its elements, $2H_2O \rightarrow 2H_2 + O_2$, can be written as two half-reactions,

$$2H_2 \rightarrow 4H^+ + 4e^-$$
$$O_2 \rightarrow 2O^= + 4e^-$$

The difference between these two reactions is about 1.2 eV. The light absorbed most strongly by chlorophyll, $\lambda = 680$ nm, has an energy of about 1.8 eV/einstein, so 1 einstein should be able to remove 1 mole of electrons from water if it is used efficiently. To evolve 1 molecule of oxygen then requires 4 electrons, so at least 4 quanta of light are necessary. This should give you some idea of the amount of energy involved in the photosynthetic process.

On the other hand, chemosynthesis (respiration) with glucose as an energy source and oxygen as an electron acceptor is essentially the reverse process, in which electrons or hydrogen ions are removed from the glucose and combined with oxygen to yield water. In fact, during this process 12 pairs of electrons or hydrogens are removed from each molecule of glucose. If you look at the old formulation for respiration that we wrote above, you will see that there are only six pairs of hydrogen ions. It was Fritz Lipmann who pointed out where the other six pairs come from; they come from water. During respiration, the only way to add oxygen is to add water and then remove the two hydrogens; these added hydrogens are then passed down the electron transport system to yield ATP. So the correct formulation for respiration is just the reverse of van Niel's formulation for photosynthesis; six extra water molecules must be added for each glucose, and these are split into an oxygen part that remains with the glucose to make CO_2 and a hydrogen part that is combined with oxygen to make water. We have come full circle, and in the most modern sense respiration of carbohydrates can again be considered just the opposite of photosynthesis.

We have been talking rather glibly about electron-transport systems, and we will continue to talk about them throughout this book. We will discuss their structure in more detail in Chapter 16. For now, it is sufficient to say that they are highly organized complexes of proteins and other materials that can be oxidized and reduced rapidly; a chain of these compounds can carry electrons just like a bucket brigade. They are generally built into specific membranes in cells, and the whole system makes a little specialized machine that is capable of turning the energy of the electrons into the chemical energy of ATP, in just the same way that a flashlight can turn the chemical energy of a dry cell into light energy or a dynamo can turn the mechanical energy of falling water into electrical energy. All of these devices that convert one form of energy into another are called **transducers.** The cell also uses some transducers to get its energy. In procaryotes, the electron-transport systems are built into the limiting membranes themselves. In eucaryotes, the photosynthetic apparatus is built into the chloroplasts and the chemosynthetic apparatus is built into the mitochondria. Even if it is a little dissatisfying, you will simply have to accept these facts for now; we are not ready yet to explain how these transducers work. After all, you used a flashlight for many years before you learned about filaments, dry cells, and electrons.

3-11 THERMODYNAMICS OF ELECTRON TRANSPORT

Oxidoreductions, like all other chemical reactions, can occur spontaneously only if they are exergonic. To measure the tendency of a compound to donate or accept electrons, we define a **reducing potential** or **electromotive force (emf)**, $\mathscr{E}$, that can be related to free energy. Consider a substance that goes from its reduced form (red) to its oxidized form (ox) with the transfer of z electrons:

$$\text{red} \rightleftarrows \text{ox} + z e^-$$

If $z = 1$, the oxidation of 1 mole of the substance requires the transfer of 6.023×10^{23} electrons, each of which carries a charge of 1.062×10^{-19} coulombs, so the total charge transferred is 96,500 coulombs or 1 **faraday,** denoted by $\mathscr{F}$. If 1 coulomb of charge is transferred through a potential drop of 1 V, then 1 joule of work is performed. Hence the free-energy change in transferring z moles of electrons through a potential of $\mathscr{E}$ volts is

$$\Delta G = -z\mathscr{F}\mathscr{E} \tag{3-21}$$

If the standard free energy of the above half-reaction is ΔG^0, the actual change is

$$\Delta G = \Delta G^0 - RT \ln \frac{[\text{ox}]}{[\text{red}]} \tag{3-22}$$

Combining equations 3-21 and 3-22,

$$\mathscr{E} = \frac{\Delta G^0}{z\mathscr{F}} - \frac{RT}{z\mathscr{F}} \ln \frac{[\text{ox}]}{[\text{red}]}$$

$$= \mathscr{E}^0 - \frac{RT}{z\mathscr{F}} \ln \frac{[\text{ox}]}{[\text{red}]} \tag{3-23}$$

where $\mathscr{E}^0 = \Delta G^0/z\mathscr{E}$ is the standard emf of the oxidized/reduced couple.

To measure the emf of an oxidation-reduction couple, a half-cell is made as shown in Fig. 3-8 with a neutral electrode, generally platinum, in a solution of the reactants. The other half-cell is standardly made of a platinum electrode with bubbling hydrogen gas, for which the half-reaction is

$$0.5\ H_2 \rightleftarrows H^+ + e^-$$

The standard emf of this reaction is *defined* as zero and all other emfs are related to it. The flow of electrons through the wire from one half-cell to the other depends on the emf of the half-reaction relative to that of the hydrogen electrode. If the half-reaction tends to go to the right (oxidation) more strongly than the hydrogen half-reaction, then electrons will flow *toward* the hydrogen electrode and the measured emf will be *negative*. If the half-reaction tends to go more strongly toward the left (reduction), then electrons will flow *from* the hydrogen electrode and the measured emf will be *positive*. Therefore, an oxidation-reduction couple will reduce any other couple with an algebraically more positive emf and will oxidize any with an algebraically more negative emf—electrons will flow in the positive direction on an emf scale.

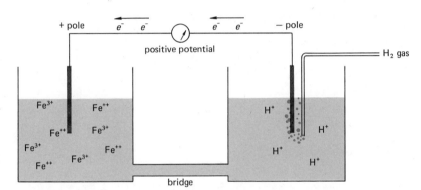

FIGURE 3-8

A coupled pair of electro-chemical half-cells. The H_2/Pt cell on the right is the reference; the Fe^{++}/Fe^{3+} cell on the left has a relatively positive potential because electrons flow toward it. The bridge between the cells allows diffusion but not rapid mixing.

exercise 3-6 If $R = 8.315$ joules/deg and $T = 300°K$ calculate:

$$\frac{RT}{\mathscr{F}} = \underline{\hspace{1cm}} \qquad \frac{2.303RT}{\mathscr{F}} = \underline{\hspace{1cm}}$$

The dissociation constant for the hydrogen half-reaction is

$$K = \frac{[H^+][e^-]}{[H_2]^{1/2}}$$

and so from equation 3-23 its emf in a solution of any hydrogen ion concentration is

$$\mathscr{E} = \mathscr{E}^0 - \frac{RT}{\mathscr{F}} \ln \frac{[H^+]}{[H_2]^{1/2}}$$

$$= 0\ - \frac{RT}{\mathscr{F}} \ln\ [H^+] - \frac{RT}{2\mathscr{F}} \ln\ [H_2]$$

and at 300°K,

$$\mathscr{E} = 0.03 \log \frac{1}{[H_2]} - 0.06\ \text{pH} \qquad (3\text{-}24)$$

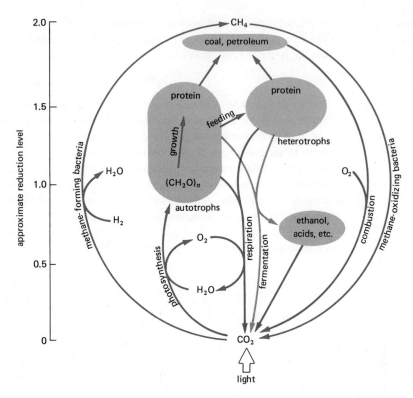

FIGURE 3-9
The carbon cycle.

which gives the dependence of emf on pH. Since the standard emf is defined in the same way as the standard free energy, for a solution in which $[H^+] = 1$ m and pH $= 0$, we again prefer to use a scale of $\mathscr{E}_0'$ values measured at pH 7. From equation 3-24, the emf of the hydrogen electrode at pH 7 is -0.420 V, and all other values are shifted accordingly. It is again important to remember that $\mathscr{E}$ values, and not $\mathscr{E}^0$, values determine whether a given oxidoreduction will occur, so it is important to consider the concentrations of the components in a system.

3-12 METABOLIC PATHWAYS

In Chapter 1 we saw that the continual flow of mass through the biosphere involves a series of food chains. Each ecosystem consists of autotrophs that reduce the CO_2 of the atmosphere to make organic compounds and heterotrophs that oxidize these compounds back to CO_2. Thus the ecosystem as a whole operates on a **carbon cycle** like the one outlined in Fig. 3-9.

We are now in a position to see generally how a single organism operates as a part of this cycle. The organism grows (or proliferates, if that is the more appropriate description) by transforming its carbon source into all of the other organic compounds of its body. Every functioning cell must carry out many sequences of chemical reactions called **metabolic pathways** to convert one compound into another, chiefly by adding and removing small chemical groups. The molecules that move through these reaction sequences are called **metabolites.**

Every organism must take its primary carbon source (along with its primary sources of nitrogen, phosphorus, etc.) and convert it through many different metabolic pathways into molecules of monomer size that are not available in the environment or cannot get across the cell membrane. It then polymerizes these monomers into polymers. It is convenient to divide metabolism into four phases to account for these operations, at least in heterotrophs (Fig. 3-10). First, there are **catabolic** reactions in which the numerous carbon compounds available in the environment are shuffled and reorganized and where a certain amount of oxidation occurs to supply energy for the rest of metabolism. Second, there are **anabolic** or **biosynthetic** reactions, which lead to all of the compounds of monomer size, including amino acids and nucleotides. Third, there is a basic set of central reactions called the **amphibolic** pathways, since they are partially catabolic and partially anabolic; these are primarily cyclical, and because of their importance we will spend considerable time trying to understand these cycles. Fourth, there are all of the polymerization reactions in which the monomers are made into polymers.

FIGURE 3-10

The general pattern of metabolism. The central cycle of reactions receives small molecules such as sugars and amino acids and transforms them into other monomers, which are then polymerized. The cycle also conserves free energy in a usable form.

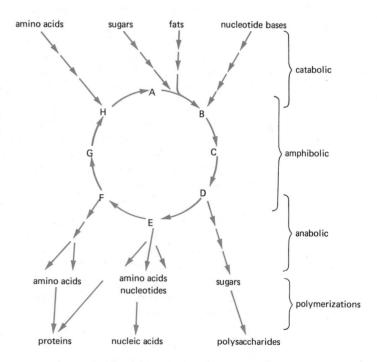

There are two primary additions that are necessary to complete this picture. In autotrophs, there are some special reactions in which CO_2 is "fixed" or converted into organic compounds. In heterotrophs, there are generally some initial reactions that must precede catabolism, because the food of heterotrophs generally consists of dead or partially decomposed organisms. Since these organisms consist primarily of polymers, and the polymers themselves cannot get across cell membranes, an initial phase of **digestion** outside the cell is necessary. In animals, digestion takes place primarily in a specialized intestine. There are a few plants, such as the pitcher plant and Venus's-flytrap, that make special cavities for digestion of insects and other small animals. In both cases, special enzymes are secreted into the digestive cavities that split the macromolecules of food, such as proteins and polysaccharides, into their monomers. Only the monomers are picked up by the organism's cells and digestion is almost entirely an extracellular process. In fact, you should realize that your entire alimentary canal, running from your mouth through the esophagus, stomach, intestines, and anus, is entirely open to the outside and is really just a long extension of the outside that runs through your body. Enzymes that digest your food are secreted into parts of this canal by your salivary glands, pancreas, and other glands; only the monomers created by their action are picked up by the walls of the intestine and passed into your blood. Similarly, bacteria and molds that grow on solid food secrete enzymes that digest this food and they pick up the monomers through their cell membranes.

One major aspect of biosynthesis involves changes in the oxidation state of compounds. We define the **reduction level** of an organic compound as the ratio of the number of oxygen atoms required to oxidize the whole compound to the number required to oxidize the carbon atoms alone. If a compound contains c atoms of carbon, h atoms of hydrogen, n atoms of nitrogen, and o atoms of oxygen, its reduction level is $(2c + 2n + 0.5h - o)/2c$. You should satisfy yourself that the reduction level can have a maximum value of 2 for CH_4, and a minimum of 0 for CO_2. A general carbohydrate $(CH_2O)_n$ has a reduction level of 1, while protein is in the range of 1.6. Therefore, biosynthesis must involve a considerable reduction, as well as the addition and removal of chemical groups. An autotroph must reduce CO_2 from 0 to about 1.6, while a heterotroph growing on glucose must reduce this sugar from 1 to about 1.6. As an adjunct to ATP formation, every Lipmann system creates a set of ubiquitous reducing agents that are used in biosynthesis.

In considering metabolism as a whole, we can see one very important reason why enzymes must exist. Even if every reaction in a metabolic pathway occurred spontaneously at a high enough rate to support growth, all of these rates would probably be different. Enzymes are needed chiefly as *regulatory devices,* to make all of these rates coordinate with one another. And there is still a broader sense in which they must be able to regulate.

Every enzyme or set of enzymes working together is a kind of *amplifier.* Like the electronic amplifiers that are now so familiar, an enzyme responds to an *input signal*—its substrate—and in a characteristic way, creates an *output signal*—its product. But an amplifier would be useless if it were not responsive to some kind of control mechanism that monitors either its input or its output and regulates the level of its activity accordingly. Figure 3-11 shows the two major classes of these control systems, which are called **feedahead** and **feedback;** either type may be **positive** or **negative,** by making the amplifier increase or reduce its activity, respectively. When we dissect the metabolic pathways, we will find many examples of these control mechanisms.

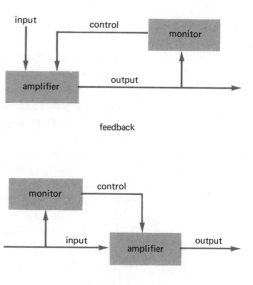

FIGURE 3-11

Control mechanisms. In a feedback control system, a monitor samples the output and sends a signal back to the amplifier to tell it to increase or decrease its activity to maintain a constant output level. In a feedahead control system, the monitor samples the input and sends a signal ahead to tell the amplifier to increase or decrease its output.

D. Information

We are now ready to return to the main line of thought developed in Chapter 2: During the course of its reproduction an organism transmits a description to the next generation of organisms which tells them exactly how to grow and metabolize, so that, with rare exceptions, each generation looks like and operates like the previous one. In other words, each organism *communicates* with its offspring by transmitting some kind of message; in ordinary English idiom, we would say that it was transmitting *information*.

Communication is primarily a process by which we limit a number of possibilities to a small range or to a single possibility. The "Twenty Questions" game is a good model: you must guess what object someone is thinking about by asking no more than 20 questions, each of which can be answered "yes" or "no." The most efficient procedure is to ask questions that divide the universe as much as possible into equal halves. (Is it in the western hemisphere? Is it in North America? Is it west of the Mississippi?) Each of these well-chosen questions narrows the range of possibilities by about a factor of 2 and we say that they all yield the same amount of information. The proper measure of information is therefore logarithmic, so any *n*-fold decrease in uncertainty is as good as any other. If there are x possibilities, information should be defined as $I = k \ln x$, but a priori they are all equally likely, so the probability that any one will be realized is $P = 1/x$ and we define information by

$$I = -k \ln P \qquad (3\text{-}25)$$

If P_1 is the probability of some condition before making a measurement or asking a question about it and P_2 is the probability afterwards, then the information gained is

$$\Delta I = -k \ln \frac{P_2}{P_1} \tag{3-26}$$

The theory of information was first developed carefully for computer systems that use a two-valued logic (switch open or switch closed) that is equivalent to the yes/no logic of the "Twenty Questions" game. We set $k = 1$, so we can ignore it, and use a scale of $\log_2$ instead of natural logarithms. The specification of one symbol on a tape or one switch in a circuit then provides one unit of information, called a binary digit or bit.

The relationship between information and entropy should now be obvious. If we let k equal Boltzmann's constant in equation 3-25, then information and entropy are just the negatives of one another. This relationship was first established in 1929 by Leo Szilard, and it has been developed in great detail by Leon Brillouin.

In fact, it is useful to have a function that is equal to Boltzmann's constant times information, and since this is the negative of entropy, it is called **negentropy**, N. According to Brillouin's development of the theory, every change in negentropy ΔN corresponds to a change in free energy $\Delta G = T \Delta N$, just as every change in entropy ΔS corresponds to an amount of heat $\Delta Q = T \Delta S$. This relationship simply means that it requires an expenditure of free energy—useful work—to specify some condition. For example, suppose a cell is making a polypeptide out of two amino acids, in equal amounts but with an apparently random sequence; the specification of one amino acid at a given position requires one bit of information, corresponding to a negentropy of $1.38 \times 10^{-23} \times 0.693$ joules/deg $= 0.96 \times 10^{-23}$ joules/deg. At a temperature of about $300°K$, the cell must therefore perform about 3×10^{-21} joules of work just to put each amino acid into its correct position, aside from the energy required to make the peptide bonds between residues. This is not very much energy, but it may be significant in the economy of the cell. You should recognize that on a molar basis, all of these values must be multiplied by Avogadro's number, and the energy changes may then be quite significant.

It will be convenient from now on to use the verb **inform** in a more technical way, meaning "to give information." David Hawkins, who is responsible for reviving the word in this sense, points out that this is the original meaning—literally, to give form to something, to give it shape, to specify its structure out of many possible structures. With these thoughts in mind, we will ask in Chapter 4 just how an organism can specify its own structure and that of its descendents.

3-13 SUMMARY

Perhaps the best way to summarize the rather wide range of topics we have discussed in this chapter is to consider a simple example; let's consider an ecosystem consisting of just you and a maple tree. The maple tree collects a certain amount of light energy in its leaves; its chloroplasts convert this energy into a stream of electrons, and they use this electrical current to reduce CO_2 to sugar. The sugar travels down into the rest of the tree, and with a little care you can tap some of it out for your food. The sugar you eat is carried by your blood to all of the cells in your body; each cell picks up some sugar and carries it into the mitochondria, where another series of reactions create an electrical current by oxidizing the sugar back to CO_2. The CO_2 can go back to the maple tree for another cycle, which will continue as long as there is sufficient sunlight.

You and the maple tree are not really very different. Both of you create minute electrical currents in specialized cell structures (chloroplasts and mitochondria) that can make ATP. Both of you use this ATP to drive all of the endergonic reactions of biosynthesis, for you must both manufacture amino acids, nucleotides, and so on, and polymerize them into compounds that are essentially the same in all cells. You both have similar sets of enzymes for catalyzing all of these reactions. The maple tree has some special enzymes for making CO_2 into glucose, and you are unable to make certain amino acids and vitamins because your ancestors lost a few enzymes during the course of evolution. But the maple tree cannot grow hair or make adrenaline. In essential things, though, your metabolic processes are very much alike. Remember that the principal reasons you need these enzymes are (1) to make many reactions proceed fast enough and (2) to provide points where metabolism can be coordinated and regulated.

Albert Szent-Györgyi suggests that this whole system may be understood most easily as an *electron cycle,* which he draws like this:

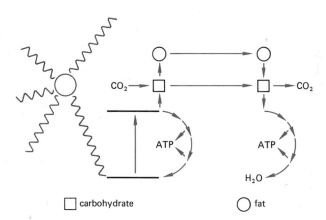

The electron is kicked up to a high energy level by a photon of light and is allowed to fall back to a lower level in the course of metabolism—either in the phototroph that starts the cycle or in some chemotroph that eats it. This is a useful picture, particularly when it is superimposed on the carbon cycle. It will help you to remember that everything we will study is an integral part of these cycles of mass and energy.

SUPPLEMENTARY EXERCISES

exercise 3-7   Determine values of ΔG^0 corresponding to K values of 10^4, 10^2, 1, 10^{-2}, and 10^{-4}, at $T = 300°K$.

exercise 3-8   Assume that in a cell the concentrations of ATP and ADP are about the same, so they cancel each other out in a fraction, but that inorganic phosphate is 10^{-4} M. What is the value of ΔG for ATP hydrolysis under these conditions?

exercise 3-9   If there are 50 million telephones in the United States, how much information is required to specify one of them? What is the negentropy of this system? What is the negentropy of the largest system that can be specified by an area code of three digits and a seven-digit number?

exercise 3-10 A consequence of the second law of thermodynamics is that no process is 100% efficient; food chains are grossly inefficient. Suppose we start with green plants (which are converting no more than 2% of the incident sunlight into the bond energy of sugar). Herbivores (plant eaters) use them for food, but only at about 5% efficiency. Carnivores (meat eaters) use the herbivores for food, with about 10% efficiency, and other carnivores generally feed on them with the same efficiency.
(a) In a (ridiculously) simple food chain of grass → deer → puma → condor, how many pounds of each animal will 10 tons of grass support?
(b) If you were trying to support the world's starving people, would you encourage them to become vegetarians or beefeaters?

The concept of a metabolic pathway is so fundamental to biology that we will take some time here to examine one series of pathways in detail. The **isoprenoid family** makes a good illustration of some important mechanisms, but the pathway itself is not important enough for you to memorize. There are at least four important points to be made:

1. Cells are economical—they make many different molecules from a few basic subunits.
2. Each metabolic pathway consists of many small steps, each of which generally requires a distinct, specific enzyme.
3. Biosynthesis generally requires oxidoreductive steps.
4. Many reactions are driven by transferring phosphoryl groups.

The isoprenoid compounds are all made out of a basic five-carbon unit called **isoprene,**

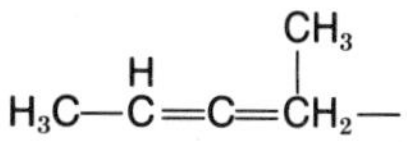

$$\underset{H_3C-\overset{H}{\underset{}{C}}=C=\overset{\overset{\displaystyle CH_3}{|}}{CH_2}-}{}$$

but the biological isoprene unit is actually a similar compound, **isopentenyl pyrophosphate (IPP):**

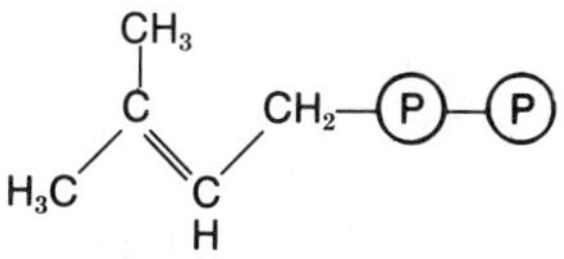

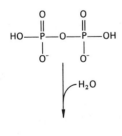 abbreviated

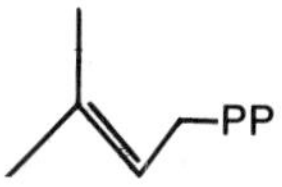

IPP is one of the many compounds that can be made out of the central amphibolic compounds and its origin is not important now. What is most important is the fact that it carries two phosphoryl groups **(pyrophosphate)** and this gives it a high enough transfer potential to carry out many reactions of the type

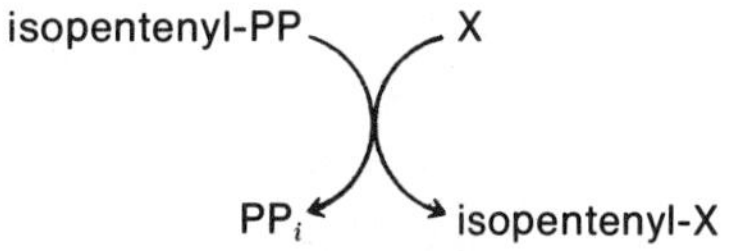

isopentenyl-PP ⟶ X

PP_i ⟶ isopentenyl-X

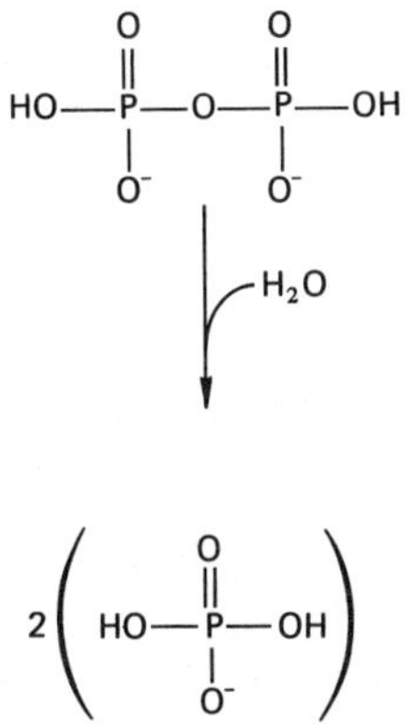

This reaction, in which free pyrophosphate (PP_i) is released, is a model for many biosyntheses; its importance cannot be overemphasized. There are enzymes called **pyrophosphatases** that cleave pyrophosphate into two orthophosphates; by removing PP_i, they make the primary transfer reaction irreversible for all practical purposes, and therefore they drive the main reaction in the direction of biosynthesis.

One of the interesting and important reactions in which IPP participates is **rubber** biosynthesis. The natural rubber produced by many plants is just a very-high molecular weight polyisoprene, made by bonding many isopentenyl units into long chains:

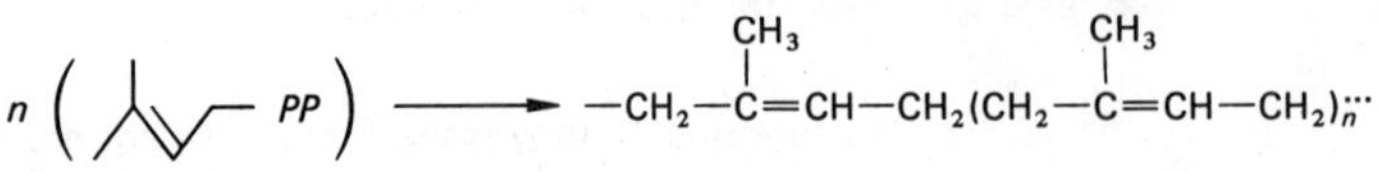

The cells where rubber is synthesized start with rather small particles and continually add these five-carbon units, one at a time in a linear sequence, to make relatively large particles; the latex bled out of these plants is a milky-white suspension of these rubber particles.

The other branch of the isoprene family leads to many important compounds. The main sequence can be summarized as follows:

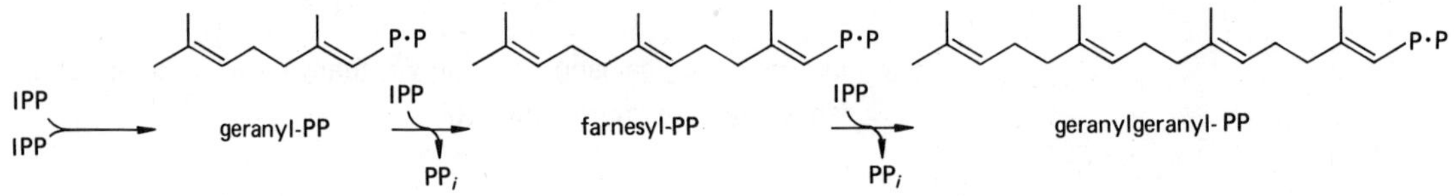

These reactions simply involve the addition of isopentenyl units. All of the resulting pyrophosphates—geranyl, farnesyl, and geranylgeranyl—can be transformed into an incredible variety of compounds called **terpenes**. These are the oils, resins, and balsams in plants that give them their characteristic odors and often make them extremely valuable commercially. Since the reactions are rather involved, we will not discuss them here.

Two molecules of farnesyl pyrophosphate can condense into squalene; squalene is the precursor of a large class of lipids called **sterols:**

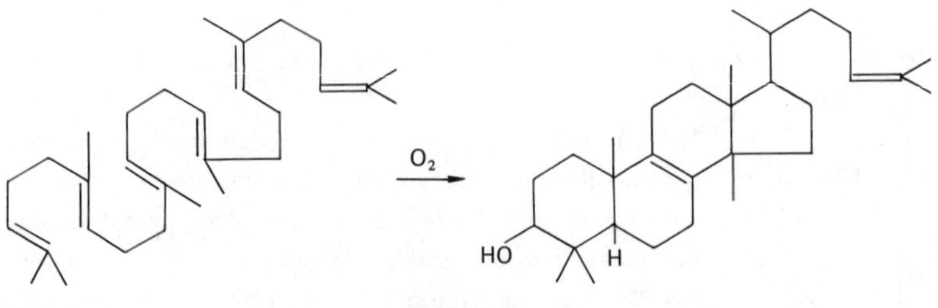

Many plant sterols are important drugs. Animal sterols are found in fats and oils, and many of them are hormones, particularly sex hormones.

Finally, there is a group of isoprenoid compounds whose biosynthesis is rather simple; these are the **carotenoids,** which are all C_{40} compounds made from eight IPP units. The basic C_{40} unit is oxidized until it is highly unsaturated and becomes strongly yellow, orange, red, purple, and so on. These pigments account for the colors of carrots, tomatoes, and other fruits and vegetables, and for the colors of many bacteria and other small organisms. They are also closely related to the visual pigments of our eyes. The group is divided into carotenes, which are simple hydrocarbons, and xanthophylls, which are oxygenated.

Carotenoid biosynthesis begins with the condensation of two molecules of geranylgeranyl-PP to make the basic compound, phytoene. Phytoene then undergoes four successive oxidations into lycopene:

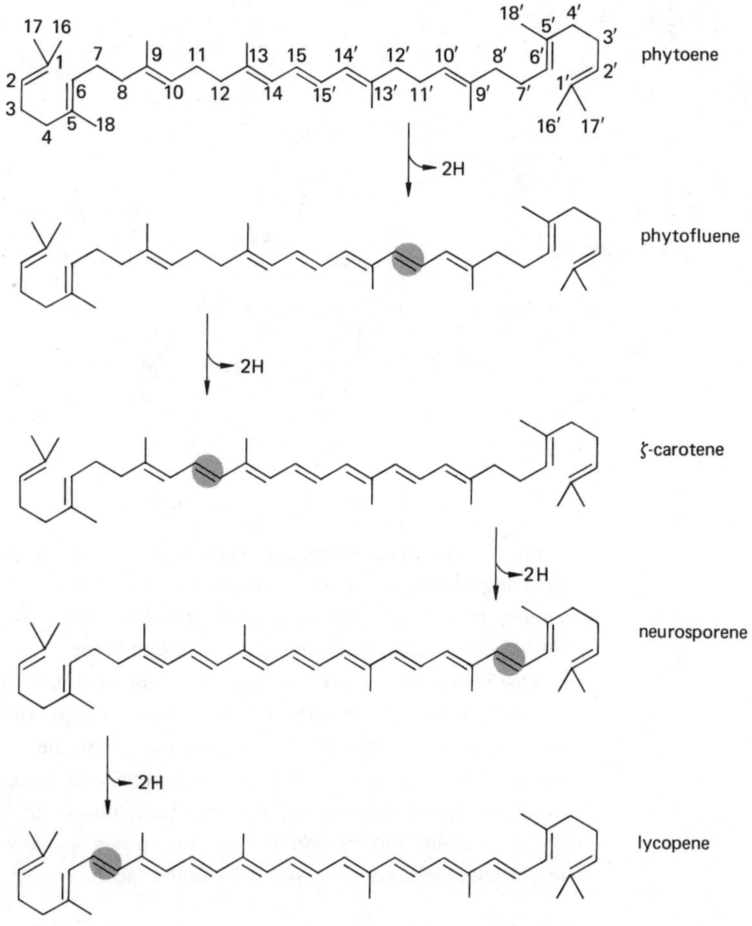

Neurosporene is the branch point leading to two other groups
of carotenoids:

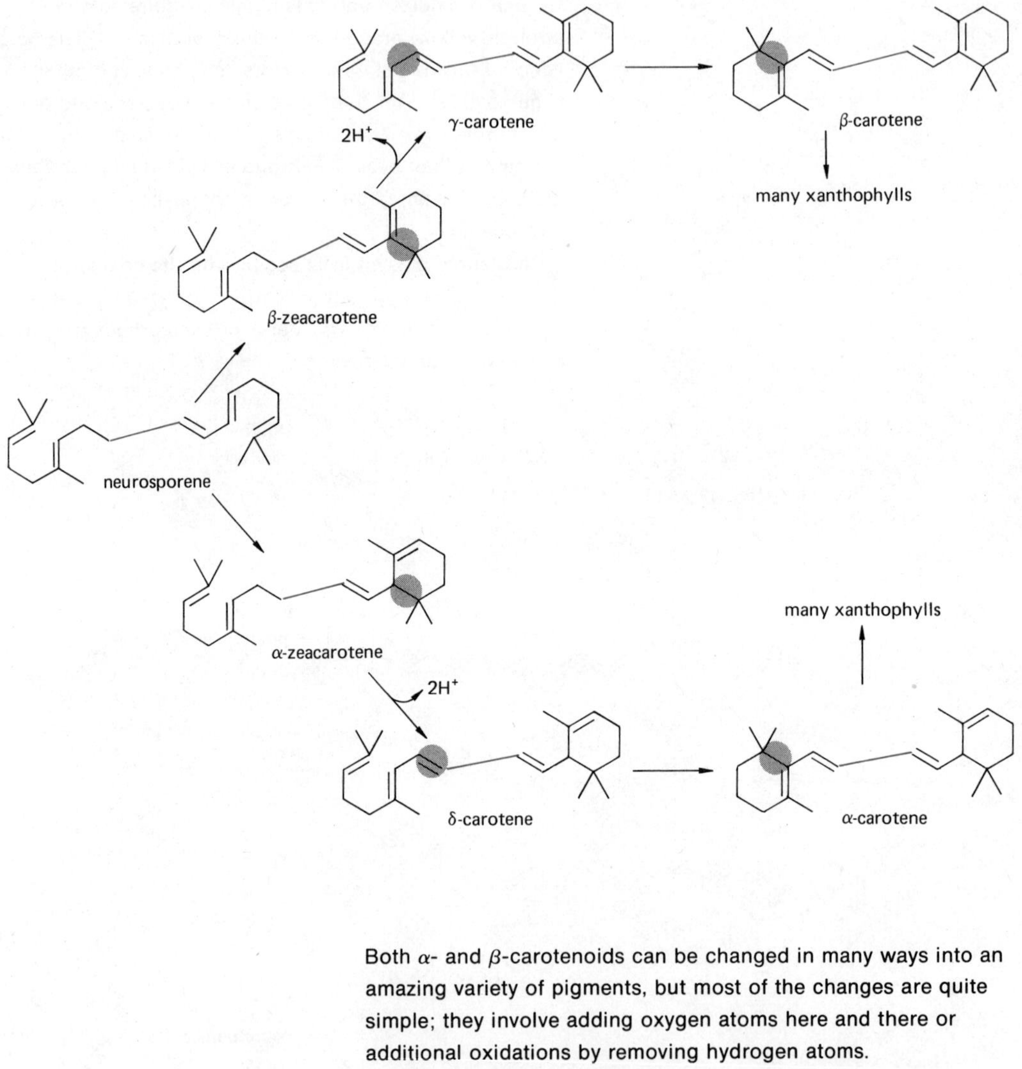

γ-carotene

2H⁺

β-zeacarotene

neurosporene

β-carotene

many xanthophylls

α-zeacarotene

2H⁺

many xanthophylls

δ-carotene

α-carotene

Both α- and β-carotenoids can be changed in many ways into an
amazing variety of pigments, but most of the changes are quite
simple; they involve adding oxygen atoms here and there or
additional oxidations by removing hydrogen atoms.

Now notice what a large number of compounds can be made
out of that simple, five-carbon building block. Large books
have been written dealing with nothing more than the
compounds made out of IPP. You will see many of these again
in later chapters. Remember, also, that biosynthesis through
pyrophosphate transfer and pyrophosphorolysis is a very
important mechanism, for you will see that again, too.

the genetic structure of biotic systems

We now have almost all of the basic ideas needed to understand how an organism operates. We have seen that cells grow and reproduce, and that during reproduction occasional mutants occur whose characteristics are slightly different from their ancestors, but which are stable enough to reproduce themselves and pass their altered characteristics on to their own progeny.

In reproducing, organisms generally behave as if they were carrying *blueprints* that specify their structure. The blueprint for a cell states, in effect, that a certain set of polymers is to be built with a certain arrangement and that certain metabolic processes are to be carried out in the system. Since the blueprint must be reproduced just like any other part of the cell, a *mutation* becomes analogous to an *error* made during copying. In fact, this is a very useful model and we will expand on it in this chapter, using slightly different terms. The model places emphasis on the *genesis* of new cells and new organisms, and therefore it is a **genetic** model. We call the hypothetical blueprint a **genome,** and since the genome specifies the structure of the organism we say that it carries **genetic information** and that it **informs** the organism. This simply means that if we could describe an organism well enough so that a very clever chemist could make one, we would have to specify an enormous amount of detailed chemical structure; somehow a similar description must reside in the genome in a form that can be transmitted to any of its daughters. We will now consider some details of biological structure to show that this model can be made chemically and biologically realistic.

4-1 PROTEINS

Chemical analysis of a virus such as phage T4 shows that it is made of roughly equal amounts of two materials: protein and nucleic acid. Both substances are linear copolymers and both make up a significant portion of all organisms and viruses; they are ubiquitous and closely related, and we shall see shortly that they are the most essential components of all biotic systems. In fact, viruses demonstrate their importance quite clearly; since they do little more than reproduce themselves at the expense of organisms and are typically made only of protein and nucleic acid, these two compounds are clearly implicated as essential elements in reproduction.

A more detailed analysis of the phage virion, or any virus particle, shows that the nucleic acid forms an internal core, with the protein forming an outer covering called a **capsid.** All of the capsid units shown in Fig. 4-1 are made of protein, and these are the pieces that give the virus its shape and allow us to

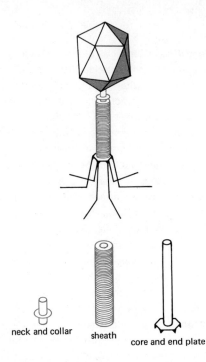

FIGURE 4-1

The capsid of phage T4 and its subunits.

neck and collar   sheath   core and end plate

tail fiber

recognize it visually. Similarly, all organisms are largely protein. Consider, for example, that all of the following components of your own body are protein: your hair, nails, and most of your epidermis (the horny, outer covering of your skin); most of the nonaqueous portion of your blood; about half the dry weight of your muscles, liver, and other internal organs; most of the tendon that links muscles to bones, the cartilage that connects adjacent bones, and the bony material itself. Bird feathers, fish and reptile scales, and mammalian hair (including wool, fur, and such strange structures as porcupine quills) are mostly protein. Perhaps most significant of all, all enzymes are proteins, and these enzymes determine how an organism metabolizes and makes all of the nonprotein components. So it is not much of an exaggeration to say that an organism is what its proteins are.

Remember that a protein—what we will call a **simple protein** from now on—consists of one or more polypeptides, each of which is a copolymer of α-amino acids. Most proteins are **conjugated;** that is, they contain some other component, a **prosthetic group,** which may be as simple as a phosphate or as complicated as a large ring molecule or an oligosaccharide. The side chain of an amino acid, which we left unspecified earlier, can also range from a hydrogen atom to a large ring. Over 100 different α-amino acids have been identified, yet, ignoring some secondary modifications, only 20 of these have been found in proteins. They are conveniently divided into five groups:

1. Hydrophilic amino acids:

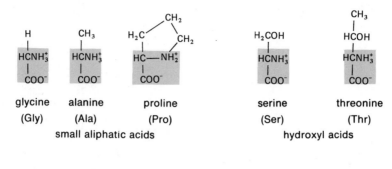

glycine   alanine   proline   serine   threonine

(Gly)   (Ala)   (Pro)   (Ser)   (Thr)

small aliphatic acids   hydroxyl acids

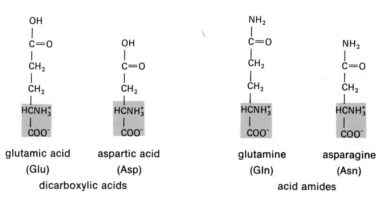

glutamic acid   aspartic acid   glutamine   asparagine

(Glu)   (Asp)   (Gln)   (Asn)

dicarboxylic acids   acid amides

2. The sulfhydryl amino acid cysteine (Cys):

$$
\begin{array}{c}
SH \\
| \\
CH_2 \\
| \\
HCNH_3^+ \\
| \\
COO^-
\end{array}
$$

3. Hydrophobic, aliphatic amino acids:

valine   isoleucine   leucine   methionine

(Val)   (Ilu)   (Leu)   (Met)

4. Basic amino acids:

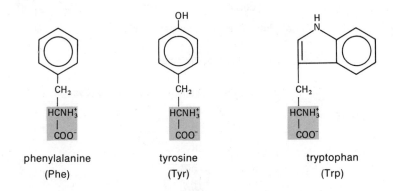

lysine	arginine	histidine
(Lys)	(Arg)	(His)

5. Aromatic amino acids:

phenylalanine (Phe)   tyrosine (Tyr)   tryptophan (Trp)

Notice the wide range of reactive groups represented by these 20 compounds, including hydroxyl, carboxyl, amino, amide, and several hydrocarbon groups. The various combinations of these groups make an enormous variety of proteins with very different properties.

If a pure protein sample is hydrolyzed carefully, the amino acids in the resulting mixture can be separated from one another chromatographically, identified, and quantitated. Generally, a sample is applied to a large piece of filter paper and run with one solvent at a time; it is then removed, dried, and run in another solvent at right angles. The distance that a compound moves relative to the distance the solvent front moves is called its R_f; by two-dimensional chromatography each amino acid is characterized by two different R_fs and it can be identified on this basis. The amino acid spots on the chromatogram can be visualized by treating them with ninhydrin (1,2,3-triketohydroindane), with which they form blue complexes:

The amount of each acid is determined by measuring the color spectrophotometrically. This procedure gives the relative amount of each amino acid in a pure protein sample. The absolute number of residues per molecule of the protein must be calculated by adding up the weights of the separate amino acids in the right proportion to yield a weight for the whole protein that agrees with physical measurements, taking account of the fact that there must be an integral number of residues of each amino acid.

Using a terminology invented by Kai Linderstrøm-Lang, we call the sequence of amino acids in a polypeptide the **primary structure.** This is written conventionally starting with the residue having a free α-amino group (the **amino-terminal** or **N-terminal** end) toward the residue having a free α-carboxyl group (the **carboxy-** or **C-terminal** end). Formally, the primary sequence would have to be a long name such as L-valyl-L-asparaginyl-L-seryl-L-seryl- . . . -L-arginyl-L-alanine, but this nomenclature would be very cumbersome for molecules of a few hundred residues. Instead, we omit the L designation, since all amino acids in proteins have this configuration, and use the standard three-letter abbreviation given above for each acid, so the above peptide becomes simply Val-Asp-Ser-Ser- . . . -Arg-Ala. Sometimes cysteine residues, whose sulfhydryl groups participate in special bonds, are emphasized as CyS or CySH. Also, if glutamine and glutamate or asparagine and aspartate have not been clearly distinguished, the uncertain residue is represented by Glx or Asx, respectively.

Once the elementary composition and molecular weight of a protein are known, the most important problem is to determine its primary structure. Most of the basic techniques for doing this were developed in the late 1940s and early 1950s by Frederick Sanger and his associates; sequence determinations are now becoming rather routine, but this is only because of the long years of hard work put in by this group and some others. It is instructive to examine briefly the methods they developed in determining the structure of *insulin,* a small protein hormone.

One of the first problems in this work is to determine the number of polypeptide chains per protein molecule. The first approach is to titrate the number of α-amino groups, for there should be only one of these per polypeptide. Sanger found he could titer them with 1-fluoro-2,4-dinitrobenzene (FDNB) which reacts with free amino groups to form a bond that is not susceptible to acid hydrolysis:

The resulting dinitrophenyl (DNP) amino acid can be separated chromatographically and identified. Of course, the ϵ-amino groups of lysine and arginine also react with FDNB, but ϵ-DNP acids can be distinguished from α-DNP acids.

Sanger obtained 1 mole of DNP-glycine and 1 mole of DNP-phenylalanine per mole of protein, indicating that there are two different chains with these N-terminal residues. He also showed that the chains are bonded to one another through cysteine residues forming disulfide bonds, the most common covalent bonds in proteins other than peptide bonds:

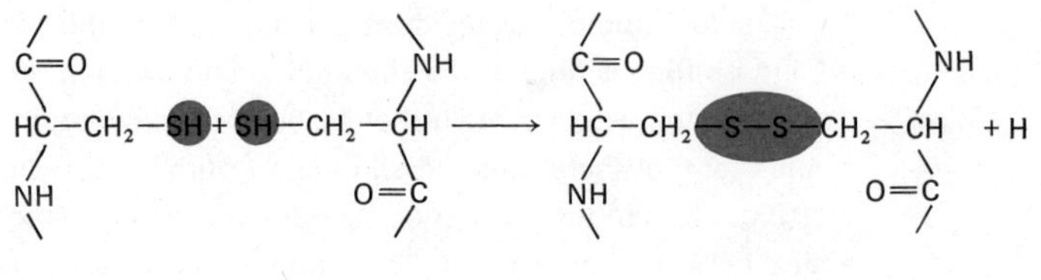

The residue on the right side of this equation is the amino acid **cystine,** but this is a secondary structure formed after the primary sequence of the polypeptide is established, so we do not count it among the 20 amino acids listed above. Disulfide bonds can be broken by treating the protein with performic acid to oxidize all of the cysteines to cysteic acid:

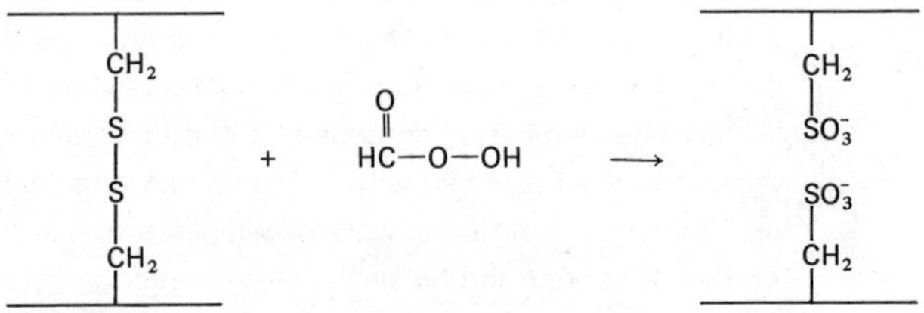

Sanger separated the two chains with performate oxidation and went on to the main task of determining their sequences. The basic principle of this work is simple: The polypeptide is broken systematically into a number of short, overlapping peptides whose sequences can be established easily and then the original sequence is established by fitting these together where they overlap. A good example of this work is Sanger and E. O. P. Thompson's analysis of the glycine chain of insulin.

First, they obtained a series of short peptides by incomplete acid hydrolysis of the protein, and separated these chromatographically. The sequence of a dipeptide can be established easily; if it contains only arginine and glycine and if FDNB labeling yields only DNP-arginine, its sequence is clearly Arg-Gly. Longer peptides must be further hydrolyzed and their parts subjected to the same analysis. Figure 4-2 shows how a series of peptides whose sequences are established in this way can be fitted

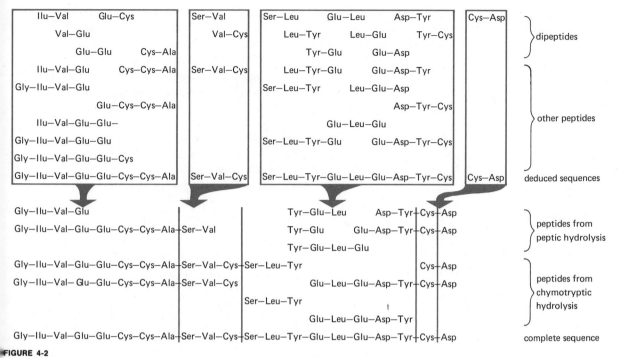

FIGURE 4-2

Deduction of the primary structure of the insulin glycine chain. The upper sequences are deduced from acid-hydrolysis peptides, the lower from peptides obtained by enzymatic hydrolysis.

together to yield three sequences of moderate length. To fit these three into a proper order, a second method must be used.

There are several hydrolytic enzymes called peptidases which are easily obtained from domestic animals and are commercially available. They are quite specific for certain bonds, so digestion of a protein sample with one of them produces a set of peptides with characteristic terminal residues:

Trypsin digests on the carboxyl side of lysine and arginine, leaving peptides with C-terminal lysyl and arginyl residues.

Pepsin attacks on the amino side of tyrosyl and phenylalanyl residues, leaving peptides with these N-termini.

Chymotrypsin attacks primarily on the carboxyl side of tyrosyl and phenylalanyl residues, leaving these as C-termini. There are no substrates for trypsin in the insulin glycine chain, but pepsin and chymotrypsin produce very useful peptides, and the lower half of Fig. 4-2 shows how these could be used to fit the whole sequence together at last.

Figure 4-3 shows what the whole insulin molecule looks like. It really shows only primary sequences, not the actual conformation of residues, but the shape of the

Ala
|
Lys
|
Pro
|
Thr
|
Tyr
|
Phe
|
Phe
|
Gly
|
Arg
|
Glu
|
Asn   Gly
|
Cy—S———S—Cy
|
Tyr
|
Asn   Val
| |
Glu   Leu
| |
Leu   Tyr
| |
Gln   Leu
| |
Tyr   Ala
| |
Leu   Glu
| |
Ser   Val
| |
Cy—S   Leu
| |
Val   His
| |
Ser   Ser
| |
Ala   Gly
| |
Cy—S———S—Cy
|
Cy—S—┘ Leu
| |
Gln   His
| |
Glu   Gln
| |
Val   Asn
| |
Ilu   Val
| |
Gly   Phe

FIGURE 4-3
*The covalent structure
of insulin.*

molecule as a whole is clearly a function of its primary structure and it must be restricted enormously just by the few disulfide bonds within it. For the sake of orientation, it would be well to think of protein molecules as having generally one of two shapes: *fibrous* or *globular.* Think of the basic polypeptide molecule as a rope. Then a fibrous protein, perhaps like the tail fibers of phage T4, is a set of ropes twisted into a braid, while a globular protein is a heavy knot of rope. This mental picture should help you to think about protein structure.

exercise 4-1 A peptide containing 17 amino acids is subjected to tryptic digestion, which yields the dipeptide Met-Arg, the tripeptide Glu-Ilu-Thr, the tetrapeptide Gly-Cys-Asp-Lys, and a longer peptide. Another sample of the whole heptadecapeptide is then digested with acid, and the following are found:

Arg-Gly	Pro-Ser	Pro-Ser-Thr
Glu-Ilu	Asp-Gly	Gly-Lys-Met
Thr-Gly	Ilu-Thr	Thr-Gly-Lys
Pro-Pro	Gly-Pro	
Asp-Lys	Lys-Glu	

The N-terminal residue of the whole peptide is Asp. What is its sequence?

4-2 DEOXYRIBONUCLEIC ACID

The other half of the phage is its nucleic acid. When this is hydrolyzed carefully, four different nucleotides are released. Remember the terminology used for these compounds: A nitrogenous base bonded to a sugar is a nucleo*side* and a nucleoside bound to a phosphoryl group is a nucleo*tide.* Nucleic acids are named for their sugar, and only two sugars have been identified in them:

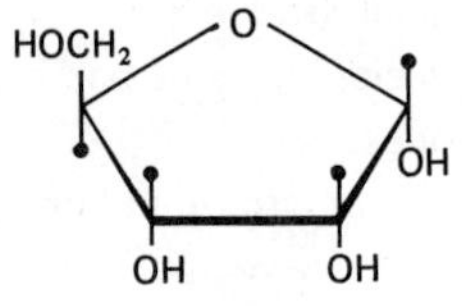

ribose 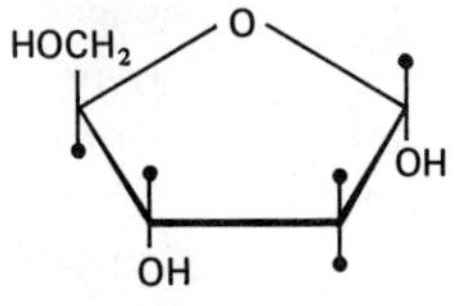   deoxyribose

The nucleic acids made from these are, respectively, **ribonucleic acid (RNA)** and **deoxyribonucleic acid (DNA);** the monomers of the latter actually should be called deoxyribonucleotides and deoxyribonucleosides. (In principle, there is no obvious reason why a mixed polymer of deoxyribonucleotides and ribonucleotides should not

deoxyribonucleic acid

exist, but no one has ever found one.) Every cell contains both DNA and RNA; every virus contains either DNA or RNA, but never both. Phages such as T4 contain only DNA.

When the phage DNA is hydrolyzed, four different nitrogenous bases are found; two of them are purines,

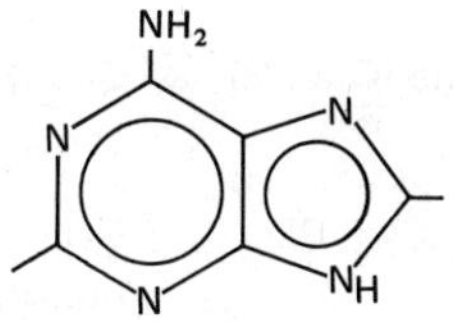

adenine

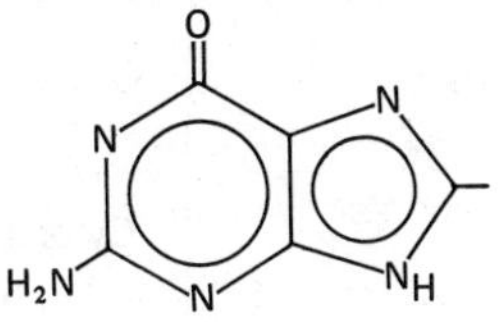

guanine

and two are pyrimidines,

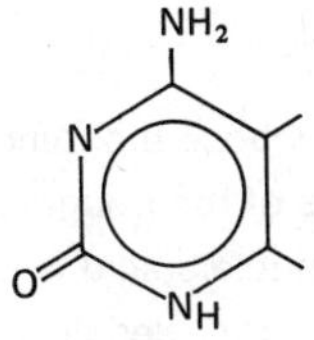

cytosine

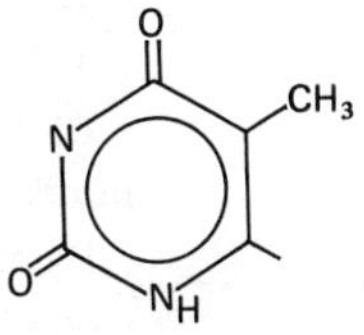

thymine

Table 4-1 gives the names of the nucleosides and nucleotides of these bases, along with those of two others,

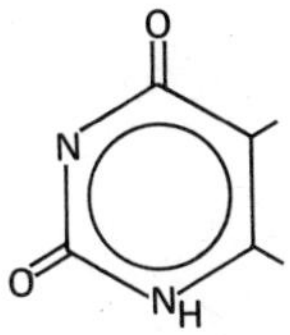

uracil

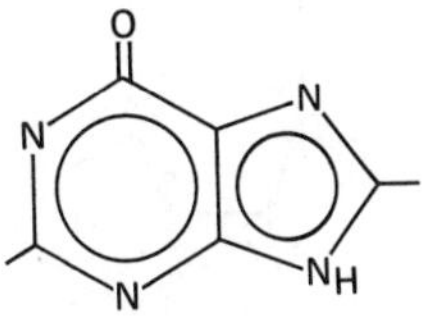

hypoxanthine

that are found frequently in RNA. (Many other bases have been found that have additional hydroxyl groups, methyl groups, etc.)

TABLE 4-1

nomenclature of bases and nucleotides

BASE	NUCLEOSIDE	NUCLEOTIDE
Adenine (Ade)	Adenosine (Ado, A)	Adenylic acid or adenosine 5'-phosphate (AMP, Ado-5'-P, pA)
Guanine (Gua)	Guanosine (Guo, G)	Guanylic acid or guanosine 5'-phosphate (GMP, Guo-5'-P, pG)
Cytosine (Cyt)	Cytidine (Cyd, C)	Cytidylic acid or cytidine 5'-phosphate (CMP, Cyd-5'-P, pC)
Thymine (Thy)	Thymidine (Thd, T)	Thymidylic acid or thymidine 5'-phosphate (TMP, Thd-5'-P, pT)
Hypoxanthine (Hyp)	Inosine (Ino, I)	Inosinic acid or inosine 5'-phosphate (IMP, Ino-5'-P, pI)
Uracil (Ura)	Uridine (Urd, U)	Uridylic acid or uridine 5'-phosphate (UMP, Urd-5'-P, pU)

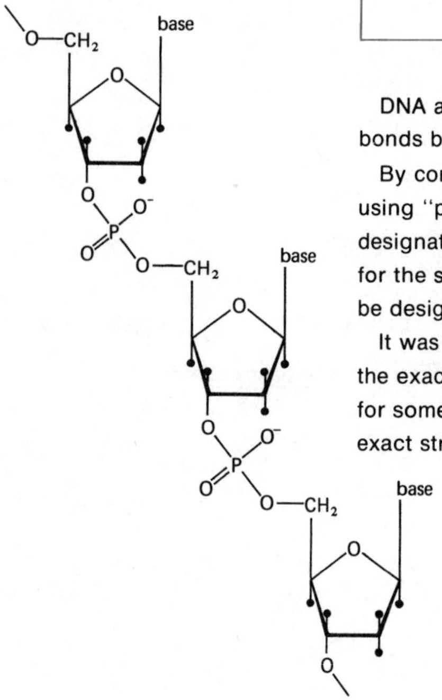

DNA and RNA are polynucleotides whose monomers are joined by phosphodiester bonds between the 3' and 5' carbons of their sugars, as shown here.

By convention, the sequence of a nucleic acid is written left to right from 5' to 3', using "p" for the phosphates and "d" for "deoxy-" in DNA. Homopolymers can be designated by abbreviations such as poly A, for the sequence pApApA . . . , or poly dA for the sequence pdApdApdA . . . ; a mixed copolymer of dA and dT, for example, can be designated poly d(AT).

It was not hard to determine that DNA consists of such polynucleotide strands, but the exact conformation of the strands within the native molecule remained a mystery for some time. And there was good reason to believe that an understanding of the exact structure of DNA was essential for biology.

4-3 DISCOVERY OF THE GENETIC MATERIAL

In 1928, Frederick Griffith performed an intriguing experiment. He was working with two strains of the bacterium *Pneumococcus,* one that was pathogenic and could cause a fatal infection in rats and one that was nonpathogenic and harmless. He showed that an unknown material could be taken from heat-killed, pathogenic cells and added to nonpathogenic ones, rendering them pathogenic. Griffith said that he had transformed the harmless cells to the pathogenic state. The transformation was not merely a passing state, since the newly transformed cells produced pathogenic progeny. This mysterious result was not understood until 1944, when the experiment was repeated by Oswald Avery, Colin MacLeod, and Maclin McCarty, who demonstrated that the transforming agent was DNA. This remarkable finding was the first indication that DNA might be the material that carries genetic information in biotic systems, but it did not really impress biologists at the time.

A definitive demonstration of the biotic role of DNA came from experiments performed in 1952 by A. D. Hershey and Martha Chase. In these experiments, they took advantage of the fact that DNA contains phosphorus but no sulfur, while protein contains sulfur but no phosphorus. Phage T4 could be grown in a medium containing ^{32}P, which would be incorporated as a label into their nucleic acid, or they could be grown in a medium containing ^{35}S, which would be incorporated into their proteins. The Hershey-Chase experiment is diagramed in Fig. 4-4. It can be seen that in a phage

^{32}P-labeled phage...   and ^{35}S-labeled phage...

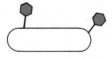
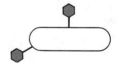

...are used to infect separate cultures of bacteria. The infected cells are then agitated in a small, high-speed kitchen blender. This does little damage to the cells, but it strips off the phage coats (capsids) attached to their surfaces. The two infected cultures are then spun in a centrifuge at speeds high enough to sediment the cells into a pellet. The stripped phage capsids remain in the supernatant.

Less than 10% of the ^{32}P
is released by the blender.

Over 80% of the ^{35}S
is released by the blender.

The infected cells are then allowed to lyse.

FIGURE 4-4

*A composite version of the
Hershey–Chase experiment.*

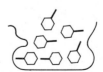

Over 30% of the ^{32}P can be found   Less than 1% of the ^{35}S can be found

infection, only the DNA gets into the cell and only DNA is transmitted to the next generation of phage. *DNA obviously must carry genetic information,* but it was not until 1953 that it became clear just how it does so.

4-4 THE WATSON-CRICK MODEL

In 1953, James D. Watson and Francis H. C. Crick undertook an analysis of DNA structure in the x-ray crystallographic laboratories at Cambridge University. We shall see later than x-ray diffraction is a powerful technique for determining molecular structure, and Watson and Crick had new diffraction data on DNA from the laboratory of Maurice Wilkins and Rosalind Franklin. Watson and Crick knew what role DNA must play in biotic systems, and this limited their speculations about its structure. They also knew of some interesting results that had been obtained by Erwin Chargaff, which indicated that in any DNA preparation there are roughly equimolar amounts of adenine and thymine and of guanine and cytosine; in other words, A = T and G = C, approximately. On the basis of all these data, they proposed the model shown in

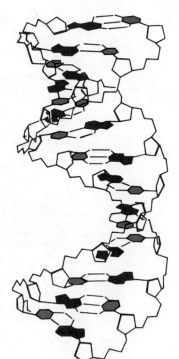

FIGURE 4-5
The Watson–Crick model of DNA structure. Deoxyribose molecules in the backbone are pentagons; pyrimidines are solid hexagons, and purines are in color.

Fig. 4-5. DNA, they said, is a helical molecule consisting of two polynucleotide strands wound around each other. The strands are antiparallel; that is, looking along the helix axis, one strand runs 5′ to 3′ and the other runs 3′ to 5′. The sugar-phosphate backbone is on the outside and the bases are inside. The molecule is held together by hydrogen bonding between the bases of the two strands, but there are restrictions on

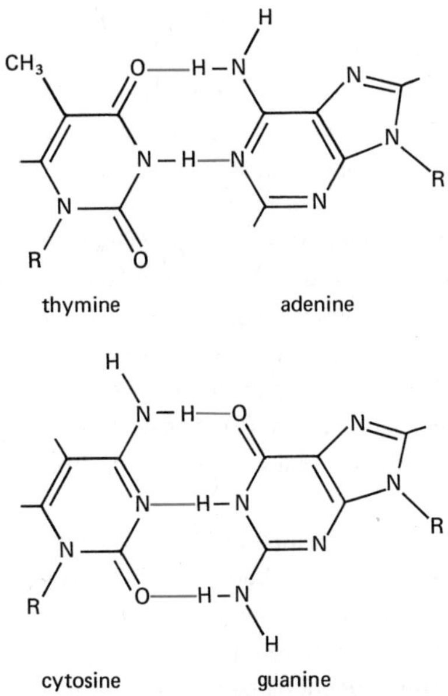

thymine   adenine

FIGURE 4-6
Hydrogen bonding between bases in the Watson–Crick model.

cytosine   guanine

which bases can pair with one another. First, one member of the base pair must be a purine and the other must be a pyrimidine—two purines are too long and two pyrimidines are too short. Second, if the purine is adenine, the pyrimidine must be thymidine; and if the purine is guanine, the pyrimidine must be cytosine. These are the combinations that fit together best, as shown in Fig. 4-6, and these pairs account nicely for the Chargaff data. The molecule therefore consists of a long sequence of A-T and G-C pairs, which lie parallel to one another, 3.4 Å apart, and perpendicular to the helix axis. There are 10 base pairs per turn of the helix, so one turn is 34 Å long. The sequence of base pairs is not specified and there are no obvious restrictions on it.

Now, if the genome of an organism is made of DNA, we can see at once what genetic information means. Every DNA molecule is a linear sequence of four base pairs: A-T, T-A, G-C, and C-G. (Or, if you please, each strand is a sequence of four bases; the principle is the same no matter how you think about it.) Since any of the four units can be put into any position in a DNA molecule, there are an enormous number of possible base sequences and information is required to specify any one of them. In fact, since $\log_2 4 = 2$, there are two bits of information per base pair. Furthermore, the four base pairs can be used just like the letters of an alphabet or the dots, dashes, and spaces of Morse code. Given the proper code book, any message could be written in a DNA molecule, and a molecule of sufficient length could contain all of the information in an encyclopedia—or a complete description of an entire organism.

One of the reasons Watson and Crick were so confident of their model is that it contains another feature that is biologically essential. There must be a way to make copies of the genome so that each new organism produced by reproduction can have its own copy. If we were really dealing with a blueprint, we would probably copy it by some kind of photographic process. That is, we would make a *negative* of the blueprint and from it print a new *positive* identical to the original. The positive and negative are *complementary* to one another, for wherever one is dark, the other is light. Similarly, a statue and a mold are complementary to each other; if you wanted an exact copy of a plaster statue, you would make a mold around it, strip off the mold, and fill it with plaster to make a new statue. In this case, wherever the statue has a bump, the mold has a complementary cavity. In both the photographic process and the molding and casting process, we are making *replicas* of the original object; we call the process **replication.**

Watson and Crick observed that adenine and thymine are complementary to each other, as are guanine and cytosine; the two bases in each pair fit together through hydrogen bonding and one base of the pair specifies the other. Therefore, DNA can easily be replicated as shown in Fig. 4-7. The two original strands separate and new nucleotides from the cell are brought together into complementary positions and polymerized to make two new strands. Since each strand specifies its complement exactly, the result is two double-stranded molecules that are identical to the original one.

Incidentally, notice how energetically simple the synthesis of DNA really is. Remember that ribonucleotide triphosphates such as ATP are the main carriers of useful free energy in cells. Cells also contain the corresponding *deoxy*ribonucleotide triphosphates (dATP, dGTP, dCTP, and dTTP) which have equally high transfer potentials and are the precursors of DNA. The phosphodiester bonds between them are formed one by one with the release of inorganic pyrophosphate. This is the same general reaction we discussed in the Picture Essay following Chapter 3 for the transfer of isoprene units from IPP.

Finally, we can begin to see what a *mutation* is. Grossly, it is any *error* in replication that is stable enough to be passed on in future replications. It is the equivalent of the kind of accidental change that occurs in human communication when messages are transmitted. Figure 4-8 shows a few examples of errors in English, classified by the *form* of the error. We also classify errors according to their *meaning* as **nonsense** or **mis-sense;** the latter is intelligible, but conveys the wrong message.

Mutations in DNA molecules may be similar; nucleotides can be added or deleted, and pieces of DNA can be removed and turned around. A more interesting change, which we will discuss in some detail in Chapter 5, may involve the exchange of one base pair for another; we will see that there are rather simple chemical mechanisms whereby an A-T pair might be substituted for a G-C pair, or vice versa, during the course of replication.

exercise 4-2 Quickly write down the complements of:
(a) A-T-C-C-G-A-C-T-G-C-T-A-A-C-G-T;
(b) C-C-G-A-C-T-G-C-C-A-T-G-G-C-A-A;
(c) A-T-A-T-G-C-C-G-T-A-C-C-T-G-A-T.

exercise 4-3 Try to draw A-C and G-T as hydrogen-bonded pairs. Notice where the hindrances to bonding lie.

exercise 4-4 Two intertwined helices are called *plecto*nemic if they must be separated by unwinding and *paranemic* if they can be separated by simply pulling them apart sideways. Is DNA plectonemic or paranemic? What does

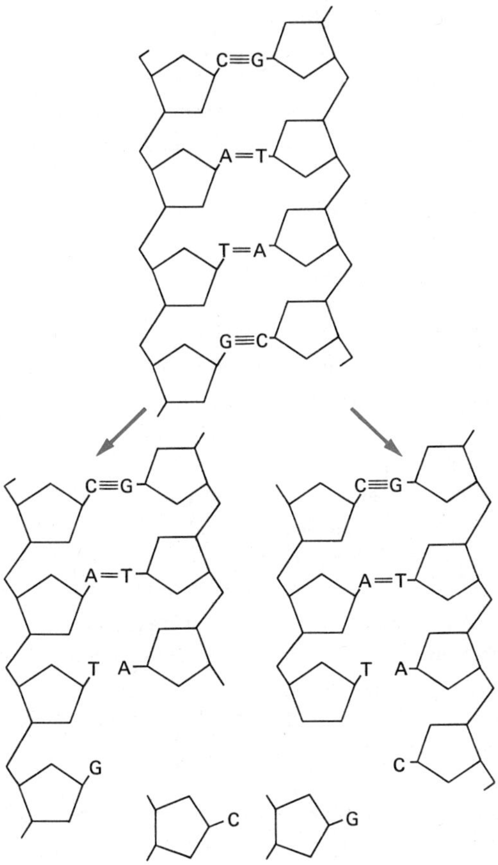

FIGURE 4-7
General method of DNA replication according to Watson and Crick's hypothesis.

DUPLICATION

"This unnecessary repetition, said
"This unnecessary repetition, said Johnson, "is what makes school work so boring for the good students."

SUBSTITUTION

As the happy couple rushed from the church, they were given the traditional shower of mice.

TRANSLOCATION

Merriweather said he landed the boat being tossed around in the fish while trying to steady a rough, white-capped ocean.

NONSENSE

The President announced today that an additional 50,000 troops would be wuletaoinshrdluetaoins be withdrawn from Viet Nam by the end of the year.

INSERTION

As Graham came on stage, he was greeted by a standing ovulation from the audience.

FIGURE 4-8
Some mutations in English.

this imply about the process of DNA replication? (Hint: Chargaff once said that DNA replication would require a special enzyme that he called an "unscrewase.")

4-5 THE MESELSON–STAHL EXPERIMENT

In the replication process shown in Fig. 4-7, notice that each original strand remains intact, but separates from its complement and acquires a new complement; this mode of replication is called **semiconservative.** This certainly seems like a natural way for DNA to replicate, but it is not the only possible way. For example, one could imagine a **conservative** mode of replication in which the two original strands separate temporarily and form their complements, but then the two old strands go back together and the two new strands form a new molecule. The most naive interpretation of the Watson-Crick model would predict the semiconservative mode, and it should be possible to devise an experiment to test this prediction. A very elegant experiment was performed by Matthew Meselson and Franklin Stahl; it depends upon an important technique they developed with Jerome Vinograd for separating molecules of different densities.

If a dense solution of a salt such as cesium chloride (CsCl) is spun rapidly in an ultracentrifuge for several hours, it will form a kind of density gradient; Cs^+ and Cl^- ions will be driven toward the bottom of the centrifuge tube but will tend to diffuse back upward and will finally come to an equilibrium distribution. If any macromolecules, such as DNA, are added to the solution, they will eventually come to rest at a point where their bouyant density is equal to that of the solution; molecules of different densities will go to different points in the gradient. They can be separated in this way and their densities and molecular weights can also be calculated if certain characteristics of the gradient are known.

Meselson and Stahl grew bacteria in a medium containing the heavy isotope ^{15}N so that all of their DNA was dense. They then transferred the cells to a medium containing normal ^{14}N, allowed them to go through two doublings, and took samples at various times. Figure 4-9 shows what one would expect if the Watson-Crick model is correct and if replication is semiconservative. Figure 4-10 shows what Meselson and Stahl actually observed. The DNA extracted before any growth has occurred in light medium is all dense. After one generation, it is all half-dense. After two generations, half of it is half-dense and half is completely light. This demonstration of semiconservative replication strengthened the Watson-Crick hypothesis materially.

exercise 4-5 If DNA were replicated conservatively, what would Meselson and Stahl have observed after one generation; after two?

exercise 4-6 Fill in the following molecular-weight data: deoxyribose _____
$HPO_4 =$ _____

adenine _____ guanine _____ cytosine _____ thymine _____
dAdo _____ dGuo _____ dCyd _____ dThd _____
dAMP _____ dGMP _____ dCMP _____ dTMP _____

an A-T nucleotide pair _____ a G-C nucleotide pair _____
Since the two nucleotide pairs differ in weight by only 1 dalton, you can easily write the approximate weight of a random sequence DNA molecule 100 pairs long: _____

exercise 4-7 Suppose the DNA molecule whose weight you have just calculated were made in a medium where every nitrogen atom was ^{15}N. Calculate (a) its weight and (b) the percentage increase in density over the molecule made from ^{14}N.

4-6 INFORMATION TRANSFER

Now let's think again about the events in phage infection. A phage particle, consisting of protein and nucleic acid, attaches itself to a bacterium. The Hershey-Chase experiment demonstrated that only the DNA enters the bacterium and that the protein

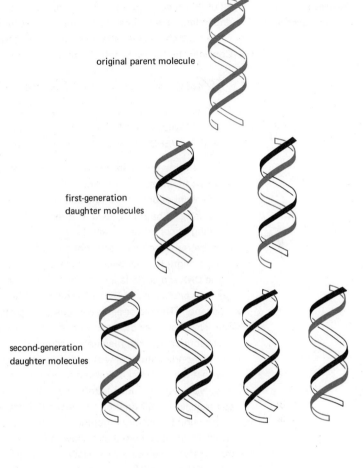

original parent molecule

first-generation
daughter molecules

second-generation
daughter molecules

FIGURE 4-9

Semiconservative replication of DNA according to the Watson–Crick model. A parental molecule that is heavily labeled (color strands) is allowed to go through two rounds of replication in a light medium so the new DNA that is formed is unlabeled (black strands). The first-generation molecules should have one light and one dense strand and should all be half-dense. In the second generation, half of the molecules should be half-dense and half should be all light.

FIGURE 4-10

The results of the Meselson–Stahl experiment. During the first generation in light medium, the DNA moves from the dense to the half-dense position. During the second generation, half of it remains half-dense and the rest becomes light.

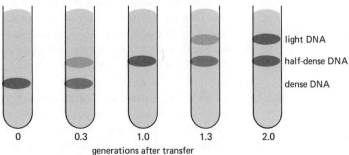

light DNA

half-dense DNA

dense DNA

0   0.3   1.0   1.3   2.0

generations after transfer

can be stripped off the cells without interfering with infection. After ½ hour, 200 new phage particles emerge from the cell, and every one of them is made of the same kind of DNA and protein as the infecting particle. We interpreted this experiment to mean that the genome is made of DNA. Since we know generally how DNA replicates, there is no mystery in the fact that much more DNA comes out and that the new molecules have the same sequence as the original one. However, we have yet to explain where all of this specific protein—the complex phage capsid protein—comes from. We can understand this process only if we say that *the genome directs the synthesis of protein.*

We need to be clear about some words used to describe the transfer of information. Consider the following four forms of a single message:

Mary. Will arrive Saturday. I love you. John

—— •— •—• —•—— •—— •• •—•• •—•• •— •—• •—• •• ••••— •

MARY. WILL ARRIVE SATURDAY. I LOVE YOU. JOHN.

MARIA. WERDE SAMSTAG ANKOMMEN. ICH LIEBE DICH. JOHANN.

The first three forms are related by **transcription** from one set of symbols directly to another; the messages are linear sequences and they are **collinear** with one another, for the symbol orders are retained. Transcription has nothing to do with meaning; even gibberish can be transcribed. But the fourth form is derived by **translation;** here meaning becomes important, for a meaningful group of symbols in one language must be replaced by a corresponding group in the other. Moreover, corresponding groups might be collinear with one another, but they do not have to be.

The DNA molecule is a message written in a language of nucleotides. It is easy to see how one strand of DNA can transfer its information to another strand during replication. The old strand serves as a **template** for the new one; that is, the old strand consists of a series of bases with specific hydrogen-bonding capacities, and these bases influence a complementary series of bases on free nucleotides to line up so they can be bonded into a new strand. One series specifies the other in exactly the same way that the shape of a drawing template constrains the movement of your pencil when you use it to make letters and curves. However, it is not easy to see how the DNA can transfer its information to a protein, since a protein (polypeptide) carries its information in the very different molecular form of a series of amino acids. The protein structure must be coded somehow in the DNA structure, and there must be some machinery in the cell for translating from one to the other.

It is not hard to place some severe restrictions on the translation mechanism. First, since DNA and protein are both linear copolymers, it is reasonable to assume that they are collinear; in other words, it must be possible to read along the DNA sequence and translate directly into a polypeptide sequence, if only you have the right dictionary. This principle was first stated formally by Francis Crick, in essentially the following form:

Crick's law: A sequence of nucleotides in DNA informs a collinear sequence of amino acids in protein, but there is no reverse information transfer from protein to DNA. (The reason for adding the second half of this statement will become clear as we go along.)

Given this principle, it is clear that when we try to understand how the primary structure of some particular protein is specified, we do not have to consider all of the DNA in the genome. Every organism can make many different proteins, each of which is a distinctive polypeptide with a specific length, generally of about 200 to 500 amino acids. The genome must therefore consist of many distinct regions, which we will call **genes,** *each of which informs a single polypeptide.* The word gene was introduced into biology at the turn of the century, and since then it has been used in so many ways that the matter has sometimes become very confused. It is almost impossible to eliminate the word from our language now, so the best we can do is to define it in the most modern way. This is the culmination of a long series of changes in our conception of genetic specificity, and it comes most directly from the work of George Beadle and Edward L. Tatum, who performed an important series of experiments which led them to propose that a gene is *a unit that specifies a single enzyme.* The one gene-one enzyme hypothesis was a milestone in the development of modern biology, and it stimulated many other people to perform useful experiments. However, most of Beadle and Tatum's intellectual descendents now prefer the definition given above, since most proteins are combinations of polypeptides and there are many proteins that are not enzymes.

It is not hard to imagine silly and complicated ways in which a sequence of DNA nucleotides could inform a series of amino acids, but there is really only one straightforward and obvious way. The DNA must somehow serve as a template for the protein, in much the same way that one strand serves as a template for the other. However, we must consider an experimental fact before elaborating on this idea: Protein is not made on the DNA directly, but rather on special protein factories found elsewhere in the cell. Therefore, there must be a special device for getting the information from the DNA to these factories. About 1941, Torbjorn Caspersson and Jean Brachet suggested independently that the intermediate is RNA; we now know that they were correct. Remember that RNA differs from DNA chiefly in its sugar; it also contains uracil in place of thymine, but uracil can hydrogen bond to adenine just as thymine does. Using the Watson-Crick rules of hydrogen bonding, with U substituting for T, a strand of RNA can be synthesized along a complementary strand of DNA by an enzyme called an RNA polymerase. The transfer of information from DNA to RNA at this point is called *transcription.* The RNA molecule then goes to the sites of protein synthesis, where it is *translated* into protein. We can imagine that each gene is transcribed into a separate RNA molecule; the problem of translation now is to specify how this RNA can serve as a template.

Since a polypeptide is a series of amino acids and we are assuming that this series corresponds to a collinear series of nucleotides, each gene must be a series of units, which we call **codons,** each of which informs *a single amino acid.* However, there are 20 different amino acids and only four nucleotides. Clearly, a single nucleotide cannot inform an amino acid; nor can two, since there are only $4^2 = 16$ pairs of nucleotides. Three nucleotides taken together, however, can easily inform an amino acid, since there are $4^3 = 64$ possible triplets. In Chapter 10 we shall examine the experimental evidence that a codon is a set of three nucleotides. At present, we must think of a general way to make every amino acid line up specifically with a codon.

Crick went on to point out an important difficulty in the concept of RNA as a template for protein. Nucleotide bases can recognize one another by hydrogen bonding, but they cannot recognize amino acids in the same way. Furthermore, even if there were an obvious way to make every amino acid bond directly to a specific triplet of RNA bases, they would be too far apart to form peptide bonds. Crick therefore proposed that there are **adaptor** molecules that can bond to the template RNA on the one hand and carry amino acids on their other ends so they bring these acids into proper alignment for peptide synthesis. These adaptors are now known to be smaller RNA molecules of a different type; they are covalently bonded to amino acids,

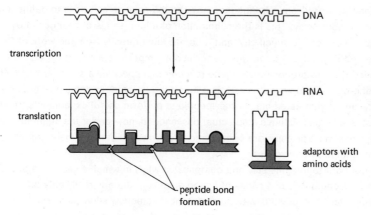

transcription

translation

DNA

RNA

adaptors with
amino acids

peptide bond
formation

FIGURE 4-11

A template model for the transfer of information from DNA to RNA to protein. The adaptors line up a distinctive amino acid with each section of the RNA message. The RNA, in turn, is complementary to one strand of DNA.

and they pair up with the template RNA by means of Watson–Crick hydrogen bonding. The whole scheme is shown in outline in Fig. 4-11.

Shortly after the Watson–Crick model was published, George Gamow wrote the first in a series of clever attempts to define the DNA-protein (or RNA-protein) code. Since only 20 of the 64 triplets are really needed, one can imagine several different kinds of dictionaries for translating from DNA to protein. One might have a **nondegenerate code,** in which only 20 of the triplets make **sense** and the others are **nonsense;** this might mean that there are only adaptor molecules for 20 triplets. However, if all 64 triplets make sense, then the code is **degenerate** and all codons that specify the same amino acid are **synonyms.** The code has now been broken experimentally, and we know that only a few of the 64 triplets are nonsense. This means that there are three types of mutations from an informational point of view. Some mutations merely change one codon into a synonym; these might never be detected because they do not produce changes in protein structure. Others are *mis-sense* because they lead to the substitution of one amino acid for another; these will generally be seen. A few mutations are *nonsense* because they produce nonsense triplets that do not specify any amino acid.

Two points ought to be stressed. First, the DNA informs only the primary structure of the protein. We believe that this is sufficient and that a polypeptide will fold up spontaneously into the right conformation simply because that is one of its most stable states. (However, experiments are still being performed to confirm this point.) Second, the flow of information is unidirectional. The RNA translation mechanism is irreversible and there appears to be no way for information in the primary structure of a protein to be translated into a DNA sequence. Modifications induced in protein structure by environmental conditions cannot be transformed into modifications of DNA structure, and therefore acquired characteristics cannot be inherited. For example, treating a bacterium with a strong reagent that changes the structure of its proteins has no effect on the genome that specifies those proteins. This also means that evolution proceeds by the Darwinian mechanism we have described — the selection of random mutations that cause useful changes in proteins — and not by the opposite mechanism.

exercise 4-8   The following is a hypothetical DNA molecule:

ATTCCGGTACGTTCGGAAATC
TAAGGCCATGCAAGCCTTTAG

Divide it into codons, reading from left to right.

exercise 4-9   Assume that RNA is transcribed only from the lower strand of this molecule; write the sequence of this RNA and divide it into codons.

exercise 4-10   Here is a part of the real dictionary for translating RNA into protein:

CCG = Pro   CGU = Arg
GAA = Glu   GUA = Val
UCG = Ser   AUU = AUC = Ilu

Write the sequence of the short peptide informed by the RNA.

exercise 4-11   In the fifth codon from the left, a mutation occurs that changes the T-A pair into a C-G pair. Write down the mutated DNA codon, the corresponding RNA codon, and tell what change will occur in the peptide. What kind of a mutation is this?

exercise 4-12   The Martians have only 12 different amino acids in their proteins, although their DNA is just like ours. What is the most likely size of a codon in the Martian system?

4-7   THE GENOMES OF CELLS

The genome of a phage is the piece of DNA that it injects during infection, but where is the genome of an uninfected cell? This question can be answered by using several different approaches, the simplest of which is to locate the bulk of the cellular DNA.

Bacteria or cells of higher organisms may be fixed, cut into thin sections, and subjected to a mild acid hydrolysis, which removes purines from their DNA, leaving aldehyde groups. When the slices are treated with the dye basic fuchsin (Shiff's aldehyde reagent), these aldehydes combine with the dye and make a purple stain, the so-called **Feulgen reaction** for DNA. Feulgen staining reveals that the DNA of eucaryotic cells is in the *nucleus* and that in bacteria it resides in the *nuclear bodies.* The dark material that we called chromatin earlier is largely DNA. The nucleus, or its procaryotic equivalent, is therefore the functional, as well as the topographical, heart of the cell.

Feulgen staining of eucaryotic cells has always shown positive spots here and there which have been largely ignored. Genetic evidence, however, clearly indicates that some extranuclear structures contain genetic information. When the DNA extracted from eucaryotic cells is spun in a CsCl gradient, it generally forms one main peak and one or two minor peaks with different densities. It can now be shown that some of these minor peaks consist of DNA from the mitochondria and chloroplasts, at least, and possibly from other cell components. While it is not clear just what information these DNA molecules carry, and it is certain that the bulk of the genetic information is carried by nuclear genes, these minor components must also be included in the genome.

Many experiments have been performed to illustrate the importance of the nucleus, but some are more elegant than others. The interesting green algae *Acetabularia* is a single, large cell with an umbrella-like cap and a long *rhizoid* containing the nucleus. The several species of *Acetabularia* have different cap shapes; J. Hämmerling performed the experiment shown in Fig. 4-12 by grafting sections of rhizoids from different species together and letting them regenerate new caps. The cap was always characteristic of the species that contributed the nucleus, showing clearly that the information for the cell morphology resides there.

Daniel Mazia performed experiments with large amebas, whose nuclei can be removed by microsurgical techniques. An enucleated cell continues to metabolize for weeks, but it cannot regenerate a nucleus or proliferate, and it eventually dies. However, a small cell created by cutting most of the cytosome away from a nucleus eventually grows into a normal cell again. Howard I. Adler has obtained similar results with bacteria; he has isolated a strain of *E. coli* that regularly produces "minicells" about a tenth the size of normal cells and without any DNA. These cells maintain themselves for some time and metabolize normally, but they cannot grow and proliferate.

4-8 ORGANIZATION OF THE GENOME

DNA molecules are so long and fragile that they are easily broken by shearing forces during ordinary operations such as pipetting, so molecules normally isolated from cells weigh only a fraction of a percent of the total DNA mass. It is not particularly informative to measure the masses and lengths of these molecules because they do not represent intact cell structures. However, there are now techniques for breaking cells gently at an air-water interface so their DNA spreads out quite thoroughly with little or no damage. When bacteria are broken in this way, all of their DNA appears to be in the form of long, thin filaments, visible with an electron microscope, which are just about as long as the DNA in a single nuclear body ought to be. (In fact, the molecule is probably circular, with no ends, but this is a matter we will discuss later in more detail.) This long molecule, which apparently contains the whole genome, is called the bacterial **genophore.** We use the same term for the single, long nucleic acid molecule that can be isolated from a phage particle by similar techniques.

While a bacterium contains on the order of 10^{-14} g of DNA, typical cells of higher organisms have several picograms per nucleus—a factor of 100 difference. All of this DNA is not in a single, long molecule; a cell probably cannot handle such a long molecule. This DNA is usually seen as an apparently structureless chromatin, but just prior to cell division it starts to condense into visible threads, which get shorter and fatter, until they can be seen easily with a light microscope. Each of these pieces of DNA, with associated material, is a **chromosome;** the number of chromosomes per nucleus varies widely from species to species, with most typical values in the range of about 10–50, but the chromosome set of any particular species is extremely constant and highly characteristic. This is another major difference between procaryotes and eucaryotes: the single genophore per nuclear body of the former contrasted with the complement of chromosomes in the nuclei of the latter.

DNA really is an acid; its phosphate groups carry strong negative charges that bind counterions such as Mg^{++} and NH_4^+. In cells, most of the neutralization of their charge

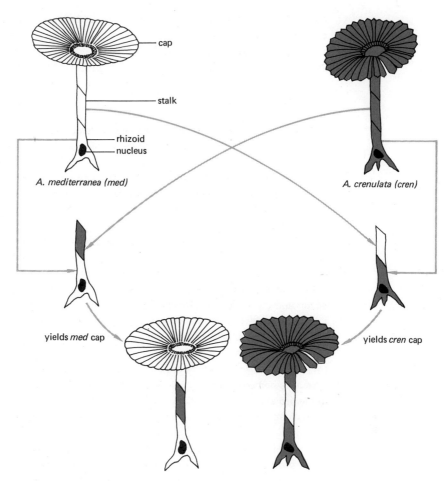

FIGURE 4-12

Hämmerling's experiments with Acetabularia. *The nucleus always determines the cap form. If two different rhizoids are grafted together, caps with intermediate shapes can be generated.*

is accomplished by strong polycationic molecules. Bacteria (and many viruses) contain large amounts of polyamines such as putrescine and spermidine on their nucleic acids.

$$H_2N—(CH_2)_4—NH_2$$
putrescine

$$H_2N—(CH_2)_3—NH—(CH_2)_4—NH_2$$
spermidine

The chromosomes of most eucaryotes contain basic proteins called **histones** that weigh about 10,000–20,000 daltons. There are many species of histones, varying from a fraction I type containing 24–29% lysine and only 1% arginine to a fraction IV type that contains about 12% arginine and only 9% lysine. Chromosomes contain large amounts of histone bound to their DNA, a complex sometimes called nucleohistone, but apparently not by covalent linkages, since the protein can be removed by solutions of high ionic strength. In the mature sperm of fish and other animals, histones are replaced by small (5000 daltons) polypeptides called protamines that are almost entirely arginine.

4-9   ORGANISMS AND VIRUSES

It should now be quite clear what the essential features of an organism are; it remains only to state them more explicitly, and the following summarizes the main points:

An organism is a structure that can obtain energy and raw materials from its environment and transform them so that it grows, reproduces itself, and occasionally mutates.

The most important single feature of an organism is its genome, which has two major functions:

1. Its **autocatalytic** function is to *replicate;* during replication, occasional errors can occur, and these are *mutations* in the strict sense.
2. Its **heterocatalytic** function is to direct the synthesis of the rest of the organism— that is, to inform specific RNA and protein molecules.

The difference between an organism and a virus should also be clear. A virus particle is composed of protein and nucleic acid; the nucleic acid informs the structure of the protein capsid as well as other proteins necessary for replication and multiplication. However, the virus *lacks* at least three things:

1. a Lipmann system for making ATP and reducing agents;
2. a protein-synthesizing system for translating the genome; and
3. a complete set of enzymes for making all of the monomers of the virus polymers.

Lacking these, the virus is dependent upon intact cells that can supply them. The virus is essentially a little genome that moves from cell to cell, using the cell's apparatus to make more of itself. It is like a robot that walks into the Ford plant with a set of blueprints under its arm and forces everyone to stop making cars and start making more robots.

All the information necessary for producing new virus particles exists in the viral genome; the virus is reproduced entirely from the genome. However, a cell is a complex, functioning system that must continue to function in order to maintain itself. There is no known way to make a new cell except to enlarge an existing one and cut it in half. The cell, in other words, could not exist unless a part of it already existed. A major part of its structural information is probably in the form of non-DNA components, so the cell is reproduced from more than its genome alone.

At first glance, these statements appear to be paradoxical; how can it be that cells exist only on condition that previous cells already exist? The solution of the apparent paradox lies in the nature of the system. Through Darwinian evolution, the cells of each generation can become slightly different from those of the last generation, because mutated genomes express their new information in existing, functioning cytosomes. When carried back for many generations, there is no paradox, and the only problem is to explain how the first functioning cells evolved out of simpler structures; this is not a question that we can think about yet.

A sophist now comes along and laughs at our definition of an organism. "That mule over there," he says, "is not an organism by your definition because he can't reproduce; and I'm not an organism either, because I certainly can't reproduce unless I can get some girl to cooperate with me." He has pointed out the problem of extending this definition to organisms that reproduce **sexually,** where two individuals combine copies of their genomes to make a new individual. In fact, this is a very interesting point, but let's try to understand it by considering some other organisms, like robins.

Within a certain age range, any male robin and any female can mate and produce young robins. As in any biotic system, there is a constant flux through the population of breeding birds, as some individuals enter it by reaching maturity and some leave it as they grow old, or sick, or die; but the *breeding population* remains, and its past, present, and future members can be identified. Now, it is not merely one male and one female who are potentially capable of reproducing, but any male paired with any female and with different pairs formed at different times. In fact, the adult robins themselves are relatively unimportant to this picture; it is sufficient to imagine a large population of sperm contributed by all the males collectively and a similar large population of eggs from all the females. A new robin is made by drawing one sperm and one egg at random and combining them. In other words, the population really consists of a **genome pool** to which all individuals contribute at some point and from which all new individuals are formed. If we use reproduction as a definition of an organism, then it may indeed be difficult to strictly define a sexual organism, but it is very easy to define the set of all individuals that contribute to the genome pool; we call this population a **species**—the set of all individuals that are actually or potentially capable of interbreeding with one another.

On the other hand, it is very easy to define an *asexual organism,* like a bacterium, but it is very difficult to define an *asexual species.* If we grow a clone of bacteria long enough, mutations will accumulate in the population and many different types of bacteria will diverge from one another like the branches of a tree. But how can we draw lines across these branches and say that one type of cell is a different species from another? We can only do so arbitrarily, and in fact we do draw rather arbitrary lines between different groups of asexual organisms just for convenience.

In summary, we define asexual organisms and sexual species rather easily, chiefly using the criterion of reproduction, but we define their counterparts—asexual species and sexual organisms—more or less by analogy. Remember that we are still not trying to define "life"; the sophist and his mule are "alive" by any ordinary criterion, and no one denies this. If he wishes to take the isolationist view that he alone is an organism, it may comfort him to learn that a few cells taken from his body can be cultured and grown like any mass of bacteria, so every cell is, in principle, capable of reproduction and he is just a clone of these cells. However, it is more enlightening for him to realize his position in the biotic world as a part of a whole species. He may even find that such a realization may serve as a guide to certain moral and philosophical issues.

4-10 TWO MECHANISMS OF EVOLUTION

The details of evolution are really understood by carefully analyzing the behavior of genes in populations, a task that we are not ready for. However, it is important to understand some aspects of evolution in a general way at this time.

We pictured reproduction in a sexual population as a random picking of sperm and eggs out of pots, and we assumed that every individual contributes equally to the pots. This is generally not true, for the essence of evolution is variability among the individuals in a population, so some will contribute more sperm or eggs than others because they are healthier, longer lived, and generally better adapted. For example, suppose that most individuals in the population carry a gene *A* and a very few carry a different gene *a*, but the latter are somewhat healthier than the former. Quantitatively, suppose that the better health of an *a* individual makes him contribute just a few percent more sperm or eggs than an *A* individual. Then when sperm or eggs are randomly withdrawn from a pot in the next generation, the probability may be 0.55 of getting one with gene *a* and only 0.45 of getting an *A*. Therefore, there will be more *a* individuals each generation, and eventually the population will become almost entirely *a*. While sexuality makes things somewhat more complex than this, it is still easy to see how selection will operate in this situation.

An important variation occurs when the population is small. Random variations that are damped out when the sample size is large may become significant as the number of sperm and eggs withdrawn from the pot decreases. Then even if there is some gene in the population that has a selective advantage, in a small sample there is a good probability that most of the individuals withdrawn from the mating pool will have the disadvantageous gene. The rapid change in gene frequencies in small populations is called **genetic drift.** This is one of the factors that makes it harder for a population to survive when it becomes small, and one of the reasons that is is so hard to save species that are on their way to extinction. The time to save grizzly bears, passenger pigeons, and whooping cranes is before their numbers get dangerously low.

Now let us consider a fairly large population of sexual organisms with a wide range where individuals can move rather freely; a population of birds is a good example. Because of the size of this population, genetic drift is not a significant factor in its dynamics. Furthermore, any gene mutation that might be a significant advantage in one part of the range may not be an advantage in another part, so the population as a whole is not likely to change very much as the result of new mutations. It will probably remain quite uniform. Suppose, however, that a part of the population becomes isolated from the rest by a geographic barrier; it may come to live on the other side of a mountain range, on an island, or merely on the other side of a different type of vegetation—a broad, grassy plain, for example, when the birds' normal habitat is a forest. Now the isolated population has a chance to change considerably; it may be small enough for genetic drift to take effect. It may live in a rather different environment where certain mutations will have a distinct advantage compared with the environment occupied by the original population.

Because of such factors, the isolated individuals may come to have slightly different coloration; they may behave a little differently; they may mate at different times from the main population; they may fall into a slightly different niche, so they tend to feed in a different part of the forest; and the structures of their sperm and eggs may become different enough from those of the main population so that fertilization of one by the other is not generally successful. Such changes can occur within a period of a few thousand years if the population is quite isolated.

Now what happens if the original isolating barrier is removed so the two populations can come together again? The small population will have gone its own way and acquired some of these isolating mechanisms during its separation. When the two populations come together again, even though they may be very similar, they will not mate (Fig. 4-13). Now we can say that they are separate species—separate mating populations.

A rather uniform species with a wide range

is cut in two by a geographic barrier. The isolated population undergoes a separate evolution so that

when the barrier is removed and the two populations come together again, there is no interbreeding.

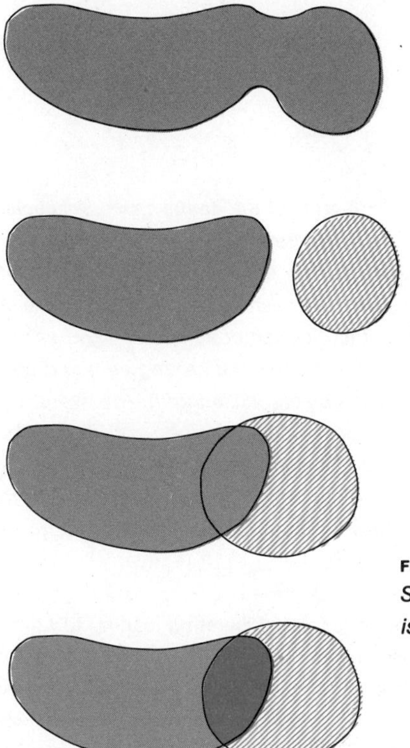

FIGURE 4-13
Speciation through geographic isolation.

Alternatively, the isolated population may not have changed enough, so some hybridization occurs upon contact with the main population.

This is the general picture of **speciation**—species formation—developed primarily by Ernst Mayr from his studies on birds and from studies by many other students of evolution. While there are many other important processes that contribute to species

Larus argentatus

Larus glaucoides

Larus fuscus

cachinnans type

FIGURE 4-14

Speciation in gulls of the genus Larus. *A species of gull was probably widely distributed around the North Pole before the Pleistocene or Ice Age, perhaps a half million years ago. During the Ice Age, populations were isolated in North America, Siberia, the Caspian-Mediterranean region, and elsewhere. After the Ice Age, perhaps 10,000–15,000 years ago, the Siberian population spread eastward across North America and the Atlantic* (L. argentatus, *the herring gull) and westward across Asia and Europe* (L. fuscus, *the black-backed gull). The two populations met in Europe, where they behave like good species and do not interbreed. Meanwhile, the American birds had become a distinct species* (L. glaucoides, *the glaucous-winged gull). The yellow-legged gulls* (cachinnans *races) of the Mediterranean and southern Russia are still an enigma; they are apparently more isolated from* argentatus *than from* fuscus.

formation, geographic isolation leading to an independent gene pool is perhaps the most significant.

Figure 4-13 shows that the two populations may not have become completely separated during their isolation, so they will hybridize with one another when they meet again. This simply means that speciation was not complete. Many examples are known; in North America, there are common, ground-feeding woodpeckers called flickers. The eastern yellow-shafted flicker has yellow underparts and the western red-shafted flicker has red feathers. They are largely separated by the great plains, but where they meet hybrids occur with orange feathers. This is an excellent example of an intermediate stage in evolution, though it may make nomenclature hard. Some people would consider the two flickers merely geographical variants ("subspecies") of a single species; others would call them two different species. It is most instructive to acknowledge them for what they are—an interesting stage in evolution—and call them two **semispecies** of one **superspecies.** Other interesting examples of this type of speciation include rings of populations (Fig. 4-14) that spread out in a circular path and gradually get to be so different that the last population will not mate with the first one where they meet, even though each population will mate with its neighbor in the other direction. This again presents a problem in nomenclature, but it is an excellent example of evolution through geographic isolation.

In this way, the evolutionary tree acquires its branches. A single population is restricted to one niche and represents one solution to the problem of survival, but the process of speciation can split it into two or more populations that can go their own ways, find new niches, and perhaps become very different in structure. The great variety of organisms we observe today is the result of continual divergences of this kind for over a billion years.

4-11 MOLECULAR INDICATIONS OF EVOLUTION

Since the result of a mutation is some change in the primary structure of a protein, and evolution consists of the accumulation of different combinations of mutations, the course of evolution can be followed very well by comparing amino acid sequences of similar proteins in different organisms. The determination of primary structure has become such a routine task that sequence studies are being pursued in many laboratories. The *Atlas of Protein Sequence and Structure,* in which many of the sequence data have been collected, has a doubling time of about 1 year. Margaret O. Dayhoff and Richard V. Eck, who compile this *Atlas,* have performed some very interesting computer analyses that have revealed some features of evolution.

If a certain protein occurs in two or more species with some modifications, we say that the various forms of it are *homologous* to one another. (If two proteins with the same function, such as catalysis of a single enzymatic reaction, were found to have very different structures, we would say that they were *analogous* to one another.) During the evolution of one homologue into another bits of the protein may be removed by deletions, or extra bits may be added, particularly at the ends, but it should be possible to align the homologues so that many positions remain identical and the homology is maximized; we can then talk about homologous positions within the amino acid sequences, and these may not be identical to the numbered positions assigned to each amino acid by simply counting from the N-terminal end.

We must recognize at the outset that the sequences of amino acids we observe are the result of both mutation and selection, and that the latter is by far the most important. It is possible for a codon to mutate into several others that code for different amino acids, but most of the replacements are probably *unacceptable* because they change the characteristics of the protein too much. Natural selection allows only certain accepted substitutions, and for the most part amino acids that replace one another are very similar. As you would expect, serine and threonine are frequent replacements for one another. The branch-chain amino acids valine, isoleucine, and leucine, as well as methionine, can also replace each other frequently. Similarly, the basic amino acids are often exchanged with one another and the large, aromatic amino acids phenylalanine, tyrosine, and tryptophan are often exchanged. Accepted substitutions outside of these groups become very rare in proportion to the differences between the amino acids; an unusual substitution is probably acceptable only in noncritical regions of the protein, where the side chain does not matter much.

A computer can be programmed to compare homologous sequences and pick out those that are most nearly identical. Since we now know the genetic code (to be discussed in Chapter 10) and can predict the most likely patterns of mutation, and since we know which amino acids are most likely to be acceptable replacements for one another, the computer can be programmed to print out the sequences of hypothetical ancestors of living species. As a very trivial example, if one species has an amino acid sequence -ABCDEG- and another has the sequence -AHCDEF-, there was probably a common ancestor with the sequence -ABCDEF-. With this technique, family trees can be drawn that show relationships between different species; so far, these trees merely confirm known relationships, but in principle they could clarify many mysteries about patterns of evolution.

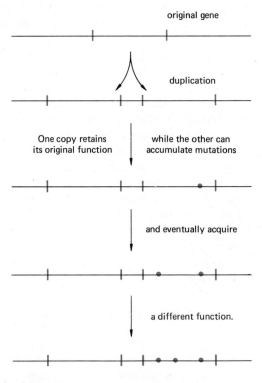

FIGURE 4-15
Diversification through gene duplication.

Comparisons of several proteins have revealed an important general pattern of evolution. One major strategy of evolution can be described as "duplicate, then diversify." The most important source of diversity comes from *duplications* of portions of the genome; once there are two copies of a single gene in the genome, they can undergo independent evolutions (Fig. 4-15). There will be a very strong pressure to keep one copy performing its original function, but the second can diverge greatly and it may eventually come to produce a homologous protein with a very different function.

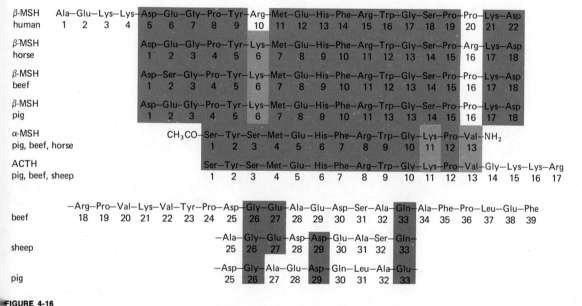

FIGURE 4-16

Primary structures of some mammalian polypeptide hormones. The melanocyte-stimulating hormones (MSH) induce color changes in pigmented cells. Adrenocortico-tropic hormone (ACTH) affects a number of processes.

Figure 4-16 shows some mammalian peptide hormones aligned to maximize homologies. It is clear that they are closely related, in spite of their different functions, and that they have a common ancestor. Notice, also, how similar are the sequences of a single hormone in related species.

elementary genetic analysis

5

Now we can begin to develop the program of *genetic analysis* that we will employ wherever possible throughout this book. The most generally useful — and most singularly biological — method of analyzing biotic systems takes advantage of their definitive characteristics: reproduction, inheritance, and mutation. We first used this approach in Chapter 2, where we saw how the isolation of certain mutants with different inheritable characteristics provided some clues to understanding the process of inheritance. This method works very well throughout biology; it is a truism that a problem is half solved if you can find a mutation that affects the system you want to study. Once found, a series of mutations can be employed in a series of operations that yield a self-consistent picture that has never failed to agree substantially with the pictures derived from more chemical or physical methods.

The elementary principles are most readily illustrated by using bacteriophage. Here we will pursue a rather idealized system of a phage such as T4 that multiplies in *E. coli*. However, keep in mind that our chief concern is the set of principles that emerges from this work, not the phage itself.

A. Basic principles

5-1 THE RATIONALE OF USING MUTATIONS

From the discussion of Chapter 4 you ought to have a kind of idealized conception of a genome. The genome is made of DNA; in a phage or a bacterium, it probably consists of one very long molecule. With an incredibly powerful microscope and more knowledge than anyone possesses at present, we could read along this molecule and translate the DNA structure into protein structures. Notice: structures, in the plural, because every real genome must have many different functions and therefore must inform many different RNA and protein molecules; and so the genome must be divided into distinct functional regions that inform distinct molecules. These are the regions that we called genes, so if we could read along the DNA we would be able to see points that mark the beginnings and ends of these genes. We don't really expect to see anything unusual at these points, and there may simply be distinctive codons or groups of codons that mean "start" and "stop."

The genome of a phage is a long piece of DNA that carries all of the information for phage multiplication. There may be genes that inform enzymes that help in DNA replication and RNA transcription. There may be other genes that inform proteins that interfere with the normal functions of the host cell, so the phage can take over the cell's biosynthetic capabilities. Other genes may inform proteins that are structural units of the phage capsid, such as head units, tail units, and tail fibers. We can imagine all of these genes, and this makes a very reasonable picture. But, in fact, we could imagine these units and discuss their relationships to one another until doomsday and yet have no more assurance that we are talking about reality. In fact, we do not have the powerful microscope of our imagination; even if one were theoretically possible—which is highly debatable right now—we would not necessarily be able to make good use of it, because just determining the sequence of bases in a DNA molecule does not really tell us how the molecule functions. However, at this moment we actually have a very powerful method of genetic analysis that yields a great deal of information about the structure and function of the genome, simply by comparing two different genomes with one another.

The only way we really know a gene exists is that we occasionally find a mutant in which the gene's function is missing. If all cats were black, we would not be inclined to think about their color; but as soon as we saw the first white cat, we would recognize that the cat must have a system for making its black fur and that in this case the system is defective. Most of the mutants we use are defective in some way, simply because normal organisms are so highly organized that most changes that occur randomly are likely to reduce their organization. However, we might also some day see a yellow cat, and then it would become clear that the system for making fur color can be changed drastically without making it particularly defective. In any case, we have to begin by establishing a standard type to which all mutants can be compared; since this is usually the type most often found in the wild, it is more often called the **wild type.** So our wild-type cat is black. We could establish a wild-type human, too; I have just decided that he is 5' 10'' tall, has black hair, brown eyes, brown skin, and is incapable of synthesizing 10 different amino acids. The decision was obviously arbitrary; I could just as well have decided that he is 6' 0'' tall, has blond hair, blue eyes, white skin, and is incapable of synthesizing eight different amino acids. Obviously, we don't make any such standard types in dealing with humans (although we distinguish mutants for clear-cut, inheritable characteristics such as genetic diseases). However, in dealing with experimental organisms we do make arbitrary decisions and establish a wild type.

We may begin with a simple (but experimentally unrealistic) example using phage. Suppose my wild-type phage has a head that looks hexagonal and a long, straight tail. I have no reason to give this phage type any particular designation until I find a mutant of it; and the first mutant found has a round head. Now we can make labels that are based either on the mutant type or the wild type, but in this case we will use the former; we call the mutant a *round* mutant and abbreviate it *rnd*. The wild-type characteristic must now be named for comparison, so we call it *rnd*+, the plus sign being the standard designation of wild type. The name given to the wild type does not imply that it has suddenly changed—merely that we have become aware of one feature that gives it a hexagonal head, so we can name that feature. We are not even saying that *rnd*+ is the name of a gene, because there might be many genes that have to cooperate in making a hexagonal head.

Suppose we now find a second mutant that has a strange *corkscrew* tail. Its head is still hexagonal, so the mutation obviously affects only the tail structure and we must give it a different name; we call it *crk* and the wild type is therefore *crk*+. Now, having two separate and distinct mutants, we can begin to compare one with the other and

perform some experiments to find out how their mutations relate to one another. To complete the process of naming, we can write down a formula for each one, telling which mutations it does and does not have; this formula is called its **genotype.** The round-headed phage has the genotype *rnd crk⁺* and the phage with the strange tail has the genotype *rnd⁺ crk.*

The discovery of these two mutants has opened a new door. At first we could conceive of the genome only as a featureless line. We could imagine arbitrary places that mark the ends of genes:

but we had no way of knowing anything more. Now we know that there is at least one part of the genome responsible for a hexagonal head, and we know that a mutation there can produce round-headed phage, whose genomes must look like this:

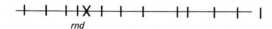
rnd

We also know that there is at least one region of the genome that makes the phage have a straight tail, and that a mutation there can produce phage with a corkscrew tail; the genomes of these phage must look like this:

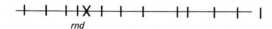
crk

But notice that we are still *imagining* where these mutations must lie; now we shall see that by performing simple experiments we can actually determine where the mutations are relative to one another. The general method is called recombinational analysis.

5-2 RECOMBINATION AND MAPPING

The basic idea of recombinational analysis is so simple that it can be said in one sentence. If two mutations are located very close to one another, they will tend to remain together; and if they are far apart, they will tend to separate from one another. Therefore, by putting two genomes together so they can interact with one another and following the mutations they carry through all of their interactions, we can make an estimate of the distance between the various mutations and thereby draw a **genetic map** of the genome. The map is simply an attempt to draw a realistic picture of the genome. Mapping is accomplished by allowing two genomes that carry different mutations to interact in the same cell, so they have an opportunity to **recombine** and make new genomes with different combinations of mutations. When we put two different genomes together in this way, we perform a **cross** or **mating.** We have learned from many years of experience that this type of experiment permits us to draw consistent maps; it should be emphasized that mapping is an internally consistent procedure that is independent of any other type of experiment.

Before we try to become more formal (and more realistic) we can fix this idea by seeing how a cross can be performed between the *rnd* and *crk* mutants we have been discussing. A phage cross is relatively easy to perform. We simply infect a culture of bacteria with these two phages at such a high multiplicity that almost every cell is infected with at least one phage of each type. The two types of phage that are used to initiate the infection are called the **parents** (or **parental types**). We have already learned, principally from the Hershey–Chase experiment, that during an infection the phage inject their genomes into the bacteria and that there is then a period of DNA replication, followed by the synthesis of new phage particles. It is during this period that interactions between the parental genomes can presumably occur. The only requirement for making a new phage is that an entire copy of a genome be wrapped up with some capsid protein, regardless of what mutations that genome might carry.

Since we are still in the realm of fantasy, we can draw two very different models for the structure of the phage genome; these two models will have very different genetic consequences. We have been drawing the genome as a single, long structure. However, the genome might actually consist of two or more distinct units (presumably all pieces of DNA), each of which is called a **genophore.** Now, consider these two models.

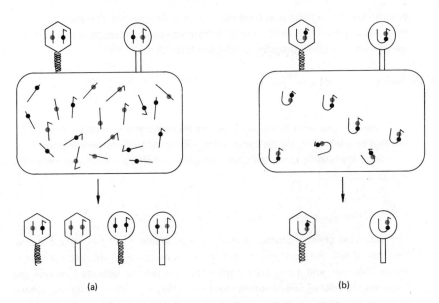

(a)   (b)

FIGURE 5-1
Recombination between two phage strains carrying different markers. (a) With two genophores, recombination occurs by random selection; the progeny consists of four different types with equal frequencies. (b) With one strong genophore bearing very close markers, there is no recombination; the progeny consists only of the two original types.

Model I (Fig. 5-1a): The genome consists of two genophores, which we represent as a straight line and a hooked line. Suppose that the mutation *rnd* is some change at a point on the straight line and that the mutation *crk* is a mutation at some point on the hooked line. Consider a bacterium infected with one phage of each type. The genomes are injected and, after a period of DNA replication, the cell will be filled with four types of genophores which are suitably marked by mutations (or the lack of mutations): *crk, crk+, rnd,* and *rnd+*. In making new phage particles, it is only necessary to pick one genophore that is straight and one that is hooked, but the assembly mechanism is blind to the mutations on these genophores. The straight genophore must be marked *rnd* or *rnd+*, and the probability of picking either of these is $\frac{1}{2}$. The hooked genophore must be marked *crk* or *crk+*, and again the probability of picking either of these is $\frac{1}{2}$. Therefore, phage with four different genotypes can emerge from this cross, and the probability of any one of them is $\frac{1}{2} \times \frac{1}{2} = \frac{1}{4}$. About half of the new phage will have the parental genotypes, *rnd+ crk* and *rnd crk+*; the other half will have the **recombinant** genotypes *rnd+ crk+* and *rnd crk,* which are the wild type and the double mutant, respectively.

Model II (Fig. 5-1b): The genome consists of a single genophore; the *rnd* and *crk* mutations are changes at two distinct points on this structure. However, they are located so close together that it is practically impossible to separate them. Therefore after the genophores have been injected and have replicated many times, there are still only the two types in the cell. When new phage particles are made, they all incorporate one genophore or the other and so all of the new phage are of the parental type; there are no recombinants.

Now we define the **frequency of recombination** R for any pair of mutations as the fraction of recombinants that emerge from a cross between them. In Model I, R is clearly equal to $\frac{1}{2}$, because half of the new phage are recombinant; in Model II, $R = 0$ because there are no recombinants. We describe the results of a cross in terms of the **linkage** between the two mutations, and the map we finally draw is actually a **linkage map.** Linkage is a measure of the probability that two mutations will remain together when they are introduced into a cross together. It should be obvious that if two mutations are on separate genophores, so there is no linkage between them, they will behave like independent objects and the frequency of recombination between them will be $\frac{1}{2}$. And if the two mutations are so tightly linked that nothing can separate them, the frequency of recombination will be 0.

In fact, when crosses are performed the frequency of recombination generally falls somewhere between 0 and $\frac{1}{2}$. We take the experimentally determined value of R to be a measure of the physical distance between the mutations, and we draw a map using numbers that are proportional to R. We have to justify this procedure, and we will do so by stating our basic assumptions, defining some terms as carefully as possible, and then turning to some real experimental systems.

5-3 AXIOMS AND DEFINITIONS

We have really been making a number of unstated assumptions. At the risk of being a little too formal, we will list some of these assumptions here. We picture a genome—or any part of one—as a long string of elements, *abcdefghijklmn.* . . . You may think of these as being nucleotides of DNA, if you wish, but their chemical structure is really irrelevant at this time. The point is that the standard, wild-type genome will have just this sequence of elements, and any genome that differs at even a single point will generally be some kind of mutant.

axiom 1   *All individuals of a given species (strain, type, variety, etc.) have genomes with substantially the same arrangement of elements.*

We have to make some such assumption at the outset to make any sense out of our operations. However, the methods we will develop are so powerful that we can detect individuals whose genomes differ substantially from the norm and find out what the differences are.

definition 1   A **point mutation** is the smallest possible alteration of a genome that is detectable as a mutation.

We may represent a point mutation as a change in one element, such as $f \rightarrow f'$. Greater changes are obviously possible; we might have $fg \rightarrow f'g'$. We must also admit the possibility that any of these mutations might be a *deletion* of the elements, rather than a simple change in their structure, but in this case we will still retain the prime marks to remind ourselves what elements are supposed to be there.

definition 2   A **genetic unit** is an operationally definable portion of the genome, such as a codon or gene.

definition 3   A genetic map is a picture showing the positions of all mutations and units in the genome.

definition 4   Every genetic unit is *defined* by the set of mutations that fall within it. A mutation used to define and map some unit is a **marker.**

definition 5   The position assigned to a point mutation on the genetic map is its **site;** a site is a point with no dimensions.

axiom 2   *No point mutation maps at two sites.*

definition 6   The position assigned to a genetic unit on the map is its **locus.** A locus covers a finite number of sites.

Finally, we make the hypothesis that a genome consists of a linear set of elements—that is, a one-dimensional string—so that the map is linear. This will simplify our illustrations, but the assumption is justified because we know that genomes are made of nucleic acids, which are linear polymers. This kind of analysis was begun about 1913 by A. H. Sturtevant with the fruit fly *Drosophila,* and it was reasonable for him to make a similar assumption because he was fairly sure that the genome of this organism consists of its chromosomes, which are visibly linear. The strict evidence for linearity comes from genetic work itself; while the hypothesis is capable of being refuted, it has never failed a strict experimental test.

5-4   SOME REAL MUTATIONS

In Chapter 2, we discussed briefly host range (*h*) mutants which can grow on bacteria that are resistant to wild-type phage. These can be used in mapping experiments. There are also a number of mutants whose plaques have an abnormal form, and these can be used very easily because recombinants make plaques that

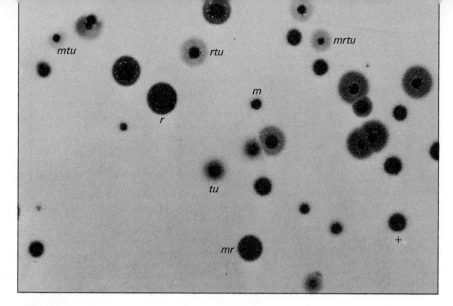

FIGURE 5-2
A variety of plaques formed by phage T4. (Courtesy of Jette M. Foss.)

combine the abnormalities of the parents. A *minute* mutant (*m*) makes small plaques; a *turbid* mutant (*tu*) makes turbid or cloudy plaques because not all of the bacteria in the plaque region are lysed; and a *rapid lysis* (*r*) mutant makes plaques that are larger than normal, with very sharp edges. Figure 5-2 shows the plaques made by several of these mutants and their recombinants; an experimenter can easily learn to recognize each type of plaque, so by merely counting plaques he can determine how many parental and how many recombinant phage are emerging from even the most complicated crosses with these mutants.

These mutants are valuable for pure mapping exercises, but in fact we want eventually to use the mutants for learning about the normal biosynthetic processes that occur during a phage infection, and we will want to find similar mutants for bacteria and other organisms so we can explore their normal processes. Most of the mutations that occur are probably *lethals,* which make such gross deficiencies that the virus (or organism) simply cannot survive. However, it occured to Norman Horowitz that the most useful kind of mutation might be a **conditional lethal,** which is only deleterious under certain prescribed **restrictive** conditions but not under some other, **permissive** conditions. Working with the common bread mold *Neurospora,* Horowitz found a number of **temperature sensitive** (*ts*) mutants (sometimes called thermolabile, *TL*) that will only grow at low temperatures, such as 30°C, but not at high temperatures, such as 42°C. In principle, a *ts* mutation could occur anywhere, for it must simply cause the substitution of one amino acid for another in a protein so that it is unstable at higher temperatures. Following this lead, Robert S. Edgar was extraordinarily successful in isolating *ts* mutants of phage T4, while in the same laboratory at the California Institute of Technology, Richard Epstein selected **amber** (*am*) mutants that will multiply in *E. coli* strain CR63 but not in strain B. (We will find out later what the difference is.) Many other people have used similar methods to find conditional lethal mutants of other viruses and organisms.

The great advantage of using these mutants is that they can be handled just like ordinary phage under permissive conditions. For example, amber mutants can be grown in CR63 and crosses can be performed in the same strain so all the amber mutants can be mapped relative to one another and to certain classical mutations. The mutants can also be grown in strain B; in general, they will not produce any whole phage under these conditions, but the infected bacteria can be studied by electron microscopy and by chemical methods to learn what processes are blocked by the mutation. Through these studies, many genes have been identified and their functions have been determined. We will study some of the most interesting results of these experiments in Chapter 20.

exercise 5-1 The little creatures shown here are groodies; they are useful (but unfortunately fictional) little organisms that are pure genetic tools. We will use them in parallel with phage to illustrate the basic concepts of this chapter. A wild-type groody has a short neck, a fat body, a long tail, and flagella. Mutants are known that have a long neck (*lng*) or a thin body (*thn*), and other mutants are tailless (*tls*) or have no flagella (*flg*⁻). Groodies can mate with each other (but they are so shy we don't know how) and produce recombinants. Just for practice, write down the genotypes of the following groodies.

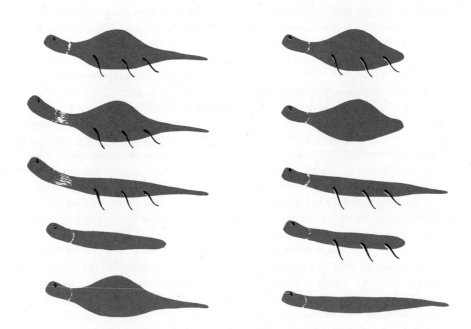

exercise 5-2 Let's perform one cross with these mutants. We mate a *tls* mutant with a *lng* mutant and pick a thousand of their progeny randomly. We obtain the following numbers:

230 264

254 257

What do you conclude about the location of these two mutations?
Test this hypothesis by calculating χ^2 (see Appendix 2).

5-5 CROSSING OVER

To explain values of R that fall between 0 and $\frac{1}{2}$, we have to assume that when two similar or identical genophores lie in the same cell, there are opportunities for them to interact quite specifically so that part of one can become attached to part of the other and therefore recombination can occur between markers that are on the same genophore. The events that produce these recombinations are exchanges or **crossovers.**

Since a genome is a kind of recording tape that contains genetic information, it is useful to use a recording-tape model. I have two tape recordings of Brahms' Fourth Symphony that are identical, except that one has a defect in the first movement and the other in the fourth movement. There are two obvious ways to get a perfect ("wild-type") copy. First, I could cut the tapes at some point in the middle and splice the two good halves together; this is break-reunion recombination. I could also get a third tape and copy the good halves of the other two tapes onto it; this is copy-choice recombination. The same two possibilities exist in genetic recombination, since genophores are nucleic acids that can break and be rejoined (presumably using appropriate enzymes) and can also replicate, so one might replicate part of a molecule off of one strand and part off of another. Because it is conceptually simpler—and probably more realistic—we will imagine that genophores actually recombine by breakage and reunion. We picture a crossover as an exchange of material, as in Fig. 5-3, and we will often use the stylized drawing shown in the figure to indicate that some information is coming from one genophore and some from the other.

We assume that crossovers occur at random and that within an infected cell there are many opportunities for genophores to come together and exchange material with one another. Suppose that d is some appropriate measure of the distance between two markers on a genophore; in fact, the average number of crossovers between the markers should be proportional to d, so the average number of crossovers between the markers is a very appropriate measure of distance. At first glance, one is tempted to say that R should also be proportional to distance because a crossover produces a recombinant. However, if two crossovers occur between the markers, they will cancel

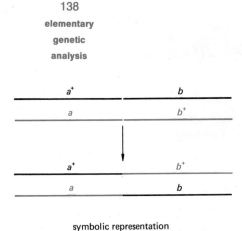

FIGURE 5-3

*Recombination by exchange of segments.
The event yielding an a⁺b⁺ recombinant is
shown symbolically.*

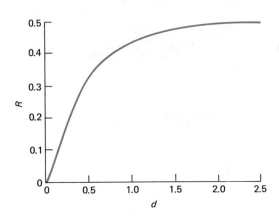

FIGURE 5-4

*A graph of the Haldane equation, showing
recombination values corresponding to values
of d.*

each other; in fact, only odd numbers of crossovers produce observable recombination. If crossovers are Poisson (randomly) distributed among pairs of genophores, and if d is taken to be the average number that occur between the markers, then R should equal the sum of the odd terms of the Poisson distribution:

$$R = e^{-d} + \frac{d^3 e^{-d}}{3!} + \frac{d^5 e^{-d}}{5!} + \cdots$$

$$= e^{-d}\left(1 + \frac{d^3}{3!} + \frac{d^5}{5!} + \cdots\right)$$

$$= e^{-d}\left(\frac{e^d - e^{-d}}{2}\right)$$

$$= \tfrac{1}{2}(1 - e^{-2d}) \tag{5-1}$$

Equation 5-1 was first derived by J. B. S. Haldane and is often called the Haldane equation; it is graphed in Fig. 5-4. Notice that when d is small, $R = d$, and for moderate values of d the curve does not depart much from proportionality, but as d becomes larger, R approaches 0.5. This means that if two markers are quite close together, the frequency of recombination between them is a good measure of their distance, but as they get farther and farther apart, recombination data will tend to indicate that they are not linked, even though they are. Therefore, we cannot depend upon recombination frequencies that begin to approach 0.5, because they are not critical enough.

The distance d used above is an arbitrary measure; from now on we will employ a **map distance** defined by the frequency of recombination, where 1% recombination ($R = 0.01$) is 1 map unit. To construct a map, we have to use markers that are close enough to each other that frequencies of recombination between them are small.

The most general procedure is to derive a *mapping function* that takes a variety of factors into account and relates R values to map units over a very wide range.

5-6 LINEARITY AND ADDITIVITY

We can now perform a test of the hypothesis of linearity, for if we use markers that are close enough together the map distances between them, as determined by re-combination frequencies, should be additive. We use three phage mutants carrying markers *a, b,* and *c.* First, we perform two crosses:

CROSS 1: $ab^+ \times a^+b$		CROSS 2: $ac^+ \times a^+c$	
ab^+	497	ac^+	420
a^+b	443	a^+c	440
a^+b^+	44	a^+c^+	81
ab	36	ac	59
total	1000	total	1000

From the definition of R, the frequencies of recombination in these two crosses are $R_{ab} = (44 + 36)/1000 = 0.080$ and $R_{ac} = (81 + 59)/1000 = 0.140$. Therefore, *a* and *b* are 8 map units apart and *a* and *c* are 14 map units apart. But from these data alone the markers might be on a three-dimensional object with the relationships shown in Fig. 5-5. If the map is really linear, the markers must fall on some diameter through the

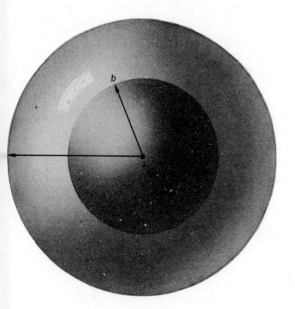

FIGURE 5-5
On the basis of the two crosses used to establish distances between a *and* b *and between* a *and* c, *the three markers might lie on a three-dimensional structure like this.*

sphere and the distance between *b* and *c* must be either $8 + 14 = 22$ map units or $14 - 8 = 6$ map units. We therefore perform the decisive cross $bc^+ \times b^+c$, with the results:

$bc^+ \times b^+c$	
bc^+	473
b^+c	467
b^+c^+	35
bc	25
total	1000

$R_{bc} = (35 + 25)/1000 = 0.06$, so b and c are 6 map units apart, and the order of markers is a _____ b _____ c. The third cross might have shown 22% recombination between b and c, so the map would be b _____ a _____ c. However, when this experiment is performed the results are always consistent with one of these two sequences; even though the data might have shown nonlinearity—and thus constitute a critical test of the hypothesis—they never show enough departure from additivity to make us abandon the hypothesis. Of course, no data ever show perfect additivity, and occasionally a system yields recombination data that force us to look for special mechanisms. In Section 5-8 we will examine some very critical evidence for linearity of the phage genome.

We can now go one step further:

axiom 3   *Linkage is transitive; if a is linked to b and b is linked to c, then a is linked to c.*

definition 7   A set of markers that are all linked to one another constitute a **linkage group.**

Thus, even though two markers may be so far apart that they recombine with a frequency of about 50%, linkage between them can be demonstrated by showing that they are linked to a common marker. Our conception of linkage groups and of the map as a whole is subject to constant revision just for this reason. The first linkage map of a phage (Fig. 5-6) was derived by A. D. Hershey and Raquel Rotman for T2, which is very similar to T4. There are three linkage groups; a marker on any one of them gives about 15% recombination with markers on any of the others, and there is no way to orient the three groups from these data. Some rough map distances can be assigned on group II. Notice the cluster of r mutations there that one is tempted to call a gene.

Some years later, the map had been filled in with other markers but it was still drawn with three linkage groups. This was disturbing because physical measurements indicated that the T4 genome is a single, long piece of DNA. George Streisinger and Victor Bruce thought they detected linkage between the three groups and performed

FIGURE 5-6
Hershey and Rotman's first map of mutations in phage T2, showing three linkage groups. There is about 15% recombination between mutations in different groups and the three cannot be ordered any further on the basis of these data.

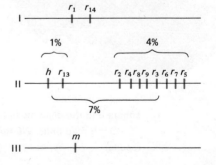

more delicate tests which showed that, in fact, all of the markers can be placed in one linear sequence. The same thing has happened many times in genetic analyses as additional data are accumulated. In fact, Streisinger's data actually indicated that the markers are not only linked in a single group, but that the end markers are linked to each other so the map is a circle. The circularity was shown later by more critical experiments with additional markers.

5-7 THREE-FACTOR CROSSES

We can perform more critical and revealing tests by using more than two markers in each cross. Crosses with three markers readily give information about the *sequence* of markers and reveal other interesting features of recombination.

theorem 1 *In a genetic cross with three linked markers, the marker in the middle will recombine with the lowest frequency.*

Let us perform a cross between two phage strains marked $a^+b^+c^+$ and abc. It is convenient to number the sites of these markers, assuming them to map in the order given:

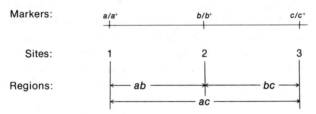

The genotypes that may emerge from this cross are

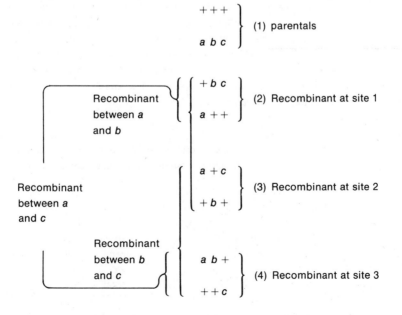

Notice, first, that the genotypes in class 3 are simultaneously recombinant in both regions *ab* and *bc* — they are the result of a *double recombination*. This double-recombinant class should be considerably less frequent than either of the single-recombinant classes, simply because the probability of two recombinations is much less than that of one. So if we find that this class is the least frequent, we are forced to place marker *b* in the middle, regardless of any previous assumptions. This is one of the most important principles of genetic mapping and we will apply it again and again. Notice, incidentally, that the sites are ordered arbitrarily from left to right; an absolute configuration could be given only in relation to a fourth marker outside this region.

It should be clear from the above tabulation that R_{ac} is not simply the sum of R_{ab} + R_{bc}. R_{ac} is the fraction of genotypes in classes 2 and 4; if you ignore marker *b* in this cross, you can see that R_{ac} is the frequency we would measure in a cross involving markers *a* and *c* alone. In counting genotypes to calculate R_{ab} and R_{bc}, we count those in class 3 — in fact, we count them twice, and so we must subtract them twice in calculating R_{ac}:

$$R_{ac} = R_{ab} + R_{bc} - 2R_{ab}R_{bc} \qquad (5\text{-}2)$$

This equation covers the ideal case in which a crossover in one region has no effect on a crossover in an adjacent region, so their probabilities are independent. However, we are dealing with physical structures that depart from the ideal, and the frequency of double recombinants often departs significantly from the expected value. Therefore we define a parameter *S*, the **coefficient of coincidence,** as the ratio of *actual* frequency of double recombinants to the *expected* frequency. Equation 5-2 should then be rewritten

$$R_{ac} = R_{ab} + R_{bc} - 2SR_{ab}R_{bc} \qquad (5\text{-}3)$$

In eucaryotes, particularly, *S* values less than 1 are frequently observed; this is called **interference.** It indicates that crossovers in one region of the genome tend to suppress crossing over in an adjacent region. On the other hand, particularly in phage, *S* is often greater than 1. Quixotically, this is called **negative interference;** *S* values often become very large for markers that are close together, and this is an indication that events are occurring during recombination for which our simplistic model cannot account. We will consider these when it is time to examine the physical basis of recombination (Chapter 14).

exercise 5-3 The following crosses are performed with phage carrying the markers *r*, *tu*, *h*, and *m*. Calculate map distances and draw the best map you can.

$m^+h \times mh^+$		$r^+h \times rh^+$		$r^+tu \times rtu^+$	
m^+h	206	r^+h	211	r^+tu	223
mh^+	236	rh^+	219	rtu^+	255
m^+h^+	20	r^+h^+	16	r^+tu^+	22
mh	29	rh	17	rtu	37
total	491	total	463	total	537

$h^+tu \times htu^+$		$m^+tu \times mtu^+$	
h^+tu	156	m^+tu	201
htu^+	214	mtu^+	231
h^+tu^+	14	m^+tu^+	10
htu	10	mtu	13
total	394	total	455

exercise 5-4 Considering the measured distances between h and tu and between tu and m, what is the expected distance between h and m? What is the coefficient of coincidence in this region?

exercise 5-5 A cross is performed between phage marked $a\,b\,c$ and $+++$. One thousand plaques that are counted have the following genotypes:

$a\,b\,c$	498
$+++$	482
$+\,b\,c$	7
$a\,++$	4
$a\,+c$	3
$+\,b\,+$	2
$a\,b\,+$	1
$++c$	1

(a) Arrange the markers in their proper order.

(b) Calculate the map distances between them.

(c) Calculate the coefficient of coincidence in this region.

exercise 5-6 A wild-type groody is crossed with the triple mutant $thn\ tls\ flg^-$. A thousand baby groodies are obtained with the following numbers of each type; on this basis, map the three markers.

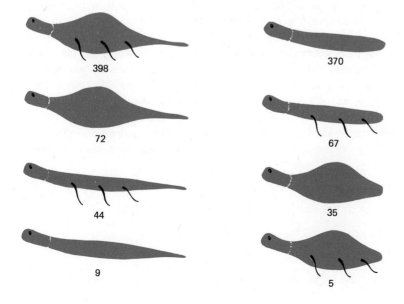

398

370

72

67

44

35

9

5

B. Genetic fine structure

5-8 FINE STRUCTURE ANALYSIS: TOPOLOGY

We noted above that mapping experiments generally indicate that the genome is linear, since distances established on the basis of frequencies of recombination are approximately additive. However, positive or negative interference always makes distances slightly nonadditive, and sometimes grossly so, so none of these experiments present an unambiguous proof of linearity, much as our knowledge of nucleic acid structure makes us want to believe in it. We are now in a position to understand a very severe test of the linearity hypothesis that was performed by Seymour Benzer with the T4 *r*II mutants. Not only does Benzer's analysis provide proof of the hypothesis, but it also allows us to explore a region of the genome down to its finest details.

Benzer made the important discovery that the *r*II mutants are a kind of conditional lethal; they will multiply in *E. coli* strain B, but not in strain K. This allowed him to perform many phage crosses in a short time. He infected a culture of strain B with a high multiplicity of two different *r*II mutants; in these cells, the phage multiply and produce recombinants if recombination is inherently possible. He then plated the progeny phage on K and looked for plaques, which could be produced only by wild-type recombinants emerging from the cross. The test asks one simple question: Can the two mutants recombine to make the wild type or can't they? It is so sensitive that Benzer could detect one recombinant in 10^6, or 10^{-4}%!

The test of linearity requires a special class of *r*II mutants that appear to be *deletions;* there are several reasons for thinking that this is what they are. Most mutants back-mutate or revert occasionally to wild type at rates comparable to other mutation rates, about 10^{-6}–10^{-8} per phage. These are presumably point mutations resulting from a minor change in the genome. However, occasional mutants are found that never seem to revert—that is, no revertants have ever been found. When these are crossed to a series of point mutations, they fail to recombine with several mutations that map at different sites, indicating that their position on the map must be represented as a line covering several sites. Therefore, they cannot be point mutations, such as $f \rightarrow f'$. They are also not likely to be double mutants due to independent events, such as $f \rightarrow f'$, $k \rightarrow k'$, for these would be rare and the mutations in question occur too frequently. However, a single mutational event could remove several elements in a row, to make a deletion, and such a mutation would cover several sites and would not revert with any measurable frequency. In fact, Benzer and Masayasu Nomura later showed that when two point mutations are put on either side of a presumed deletion, their frequency of recombination is reduced, indicating that a piece of the genome is really missing.

Every deletion could be represented by a line covering the series of elements that are missing, so a representative set would be

$$abc\underline{defghijk}lmnopqr\ldots$$
$$abcde\underline{fghijkl}mnopqr\ldots$$
$$abcdefghij\underline{k}lmnopqr\ldots$$
$$abcdefghi\underline{jklm}nopqr\ldots$$

and so on. Now, if two deletions overlap at all—even in a single element—they cannot recombine to restore the wild-type genome. (Of course, they can recombine to make a different mutation, but this will not be detectable in the *r*II system, which is why it is so valuable.) If they do not overlap—even if they are immediately adjacent to one another—then they should be able to recombine to produce a wild-type genome, and this recombination should be detectable by Benzer's sensitive test.

The four deletions shown above are in *dictionary order,* because they are alphabetized according to the elements that are missing. If the *r*II mutations are really part of a linear structure, it should be possible to arrange them in an analogous sequence:

If no wild-type recombinants come from a cross of deletions 1 × 2 or 2 × 3, but a cross of 1 × 3 does yield recombinants, the deletions must map in the pattern:

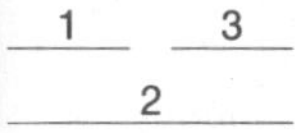

    1   3
      2

Any deletion in this series will give no wild-type recombinants with itself and with some deletions that are close to it, but it will recombine with deletions that are far enough to the left or to the right so that there are no overlaps between them. We test such a series by crossing them with one another in all possible combinations, representing the results in the form of a matrix (Fig. 5-7) in which 1 means that wild-type recombinants are formed and 0 means that they are not formed. The matrix shown

	184	215	221	250	347	455	459	506	749	761	782	852	882	A103	B139	C4	C33	C51	H23
184	0	0	1	0	1	0	1	1	1	1	0	1	1	1	1	1	0	0	0
215	0	0	1	1	1	1	1	1	1	1	1	1	1	1	1	1	1	1	0
221	1	1	0	1	0	1	0	0	0	0	0	0	0	0	0	0	0	1	0
250	0	1	1	0	1	1	1	1	1	1	1	1	1	1	1	1	0	0	0
347	1	1	0	1	0	1	1	1	1	1	0	1	1	1	1	1	1	1	0
455	0	1	1	1	1	0	1	1	1	1	1	1	1	1	1	1	1	1	0
459	1	1	0	1	1	1	0	1	0	0	0	0	1	1	1	0	1	1	0
506	1	1	0	1	1	1	1	0	1	1	0	1	1	1	1	1	1	1	0
749	1	1	0	1	1	1	0	1	0	0	0	0	1	1	1	0	1	1	0
761	1	1	0	1	1	1	0	1	0	0	0	0	1	1	1	0	1	1	0
782	0	1	0	1	0	1	0	0	0	0	0	0	0	0	0	0	0	1	0
852	1	1	0	1	1	1	0	1	0	0	0	0	1	1	1	0	1	1	0
882	1	1	0	1	1	1	1	1	1	1	0	1	0	1	1	0	1	1	0
A103	1	1	0	1	1	1	1	1	1	1	0	1	1	0	0	1	1	1	0
B139	1	1	0	1	1	1	1	1	1	1	0	1	1	0	0	1	1	1	0
C 4	1	1	0	1	1	1	0	1	0	0	0	0	1	1	1	0	1	1	0
C 33	0	1	0	0	1	1	1	1	1	1	0	1	1	1	1	1	0	1	0
C 51	0	1	1	0	1	1	1	1	1	1	1	1	1	1	1	1	0	0	0
H 23	0	0	0	0	0	0	0	0	0	0	0	0	0	0	0	0	0	0	0

FIGURE 5-7

An arbitrarily drawn matrix showing the results of crosses between 19 nonrevertible mutants in all combinations. A 0 indicates an overlap between the mutants and a 1 indicates no overlap.

in Fig. 5-7 is just a jumble because the deletions are arranged in an arbitrary order. Figure 5-8 shows how a map can be made from the data in this matrix so that the deletions appear in dictionary order. When another matrix is made with the deletions in this order (Fig. 5-9), an interesting pattern appears. All of the diagonal elements are

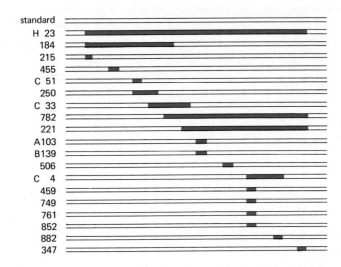

FIGURE 5-8
The 19 mutations can be arranged in this dictionary order. The length of each deletion and the precise location of its ends cannot be determined so this is an approximation which is consistent with the data.

FIGURE 5-9
The matrix of Fig. 5-7 has been redrawn with the mutants arranged so there are now continuous blocks of zeros and ones.

	H 23	184	215	455	C 51	250	C 33	782	221	A 103	B 139	506	C 4	459	749	761	852	882	347
H 23	0	0	0	0	0	0	0	0	0	0	0	0	0	0	0	0	0	0	0
184	0	0	0	0	0	0	0	0	1	1	1	1	1	1	1	1	1	1	1
215	0	0	0	1	1	1	1	1	1	1	1	1	1	1	1	1	1	1	1
455	0	0	1	0	1	1	1	1	1	1	1	1	1	1	1	1	1	1	1
C 51	0	0	1	1	0	0	1	1	1	1	1	1	1	1	1	1	1	1	1
250	0	0	1	1	0	0	0	1	1	1	1	1	1	1	1	1	1	1	1
C 33	0	0	1	1	1	0	0	0	0	1	1	1	1	1	1	1	1	1	1
782	0	0	1	1	1	1	0	0	0	0	0	0	0	0	0	0	0	0	0
221	0	1	1	1	1	1	0	0	0	0	0	0	0	0	0	0	0	0	0
A103	0	1	1	1	1	1	1	0	0	0	0	1	1	1	1	1	1	1	1
B139	0	1	1	1	1	1	1	0	0	0	0	1	1	1	1	1	1	1	1
506	0	1	1	1	1	1	1	0	0	1	1	0	1	1	1	1	1	1	1
C 4	0	1	1	1	1	1	1	0	0	1	1	1	0	0	0	0	0	0	1
459	0	1	1	1	1	1	1	0	0	1	1	1	0	0	0	0	0	1	1
749	0	1	1	1	1	1	1	0	0	1	1	1	0	0	0	0	0	1	1
761	0	1	1	1	1	1	1	0	0	1	1	1	0	0	0	0	0	1	1
852	0	1	1	1	1	1	1	0	0	1	1	1	0	0	0	0	0	1	1
882	0	1	1	1	1	1	1	0	0	1	1	1	0	1	1	1	1	0	1
347	0	1	1	1	1	1	1	0	0	1	1	1	1	1	1	1	1	1	0

zeros, because no deletion recombines with itself, and most of the near-diagonal elements are zeros, because of overlaps between neighboring deletions. But as soon as the first one appears in any row or colum, all of the other elements in the same row or column are also ones. When the deletions are properly arranged, this feature of the matrix reflects the linearity of the map.

This can be shown more clearly by considering an imaginary two-dimensional genome with four overlapping deletions:

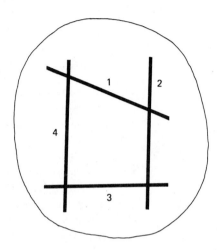

As an exercise, convince yourself that the matrix representing all possible crosses with these deletions cannot be arranged so that the zeros fall only on and near the diagonal.

Now, it is true that the 19 deletions shown above do not constitute a great test of the hypothesis of linearity, but Benzer obtained the same result with a set of 145 *r*II deletions! If these were not a part of a simple, linear structure, there is a very small chance that they could be arranged in a sequence compatible with linearity and give a matrix with the proper arrangement of zeros and ones. Thus, a very critical genetic test is perfectly consistent with the assumption that the genome is a DNA molecule or a linear sequence of DNA molecules.

exercise 5-7   Suppose a genome is a branched structure in which the following deletions occur:

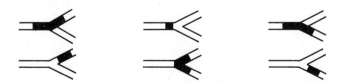

Draw a matrix for recombination between these deletions. Can it be transformed into one compatible with a linear structure?

exercise 5-8   Draw a map of the deletions represented by the following matrix. You might find it easiest to rearrange the matrix first.

	19	36	54	72	304	711
19	0	1	1	0	1	1
36	1	0	0	1	1	0
54	1	0	0	0	0	0
72	0	1	0	0	0	1
304	1	1	0	0	0	1
711	1	0	0	1	1	0

5-9 COMPLEMENTATION

The deletion mapping experiments support our concept of the genome as a long, linear structure, presumably a piece of DNA. We can now ask about the function of the genome: Can it be divided into units that correspond to the hypothetical genes? If these genes exist, then they must be defined and mapped by mutations that fall within them (from Definition 4). We will first see how mutations can be assigned to functional units by using the criterion of **complementation.**

We defined a gene as a unit of the genome that specifies the structure of a single polypeptide. A complementation test, in the simplest case, determines whether two mutations affect the same or different genes when defined in this way. If they fall in the same gene, then they affect the same polypeptide; if they fall in different genes, they affect different polypeptides, and in this case it might be possible to combine them so that each one supplies a function that the other cannot supply. Two mutations that mutually correct one another are said to *complement* each other. The basic idea is not too different from that used in repairing machinery with spare parts. If you have two cars with broken crankshafts, you can't get either one to run, but if one has a broken crankshaft and the other a broken carburetor, you can combine their good parts to make a functional car.

The concept of the **phenotype** becomes useful here, although it is sometimes hard to define precisely. If the genotype is the information present in the genome, the phenotype is the visible expression of that information. For example, there are several different *tu* mutants of phage T4. They have different genotypes because they carry different mutations that map at different sites, but they all have the same phenotype — a turbid plaque morphology. There are also many *r*II mutants, but they all have the same phenotype, consisting of a distinctive, large plaque and the inability to multiply in strain K. If two mutants can complement each other, their genotypes are not affected and the mutations stay just as they are, but we recognize them as complementing because they show the wild-type phenotype when combined.

When cells of strain K are infected with an *r*II mutant, no phage are produced. If the cells are infected simultaneously with both an *r* and an *r*⁺, both types of phage are produced. This means that the wild-type phage can supply the necessary functions for the multiplication of the *r* phage as well as itself. Now if the cells are infected with two different *r*II mutants, sometimes they make phage and sometimes they don't; if they do, it is because the two phage mutants have complemented each other. Figure 5-10 shows that two mutants can be put into a host cell in two positions, either *cis* — on the same genophore — or *trans* — on different genophores. The double mutant, with the mutations in cis position, will never multiply by itself but will always be complemented by the wild type. In the trans position, the two mutants will complement one another if their mutations fall into different genes, but not if they are in the same gene.

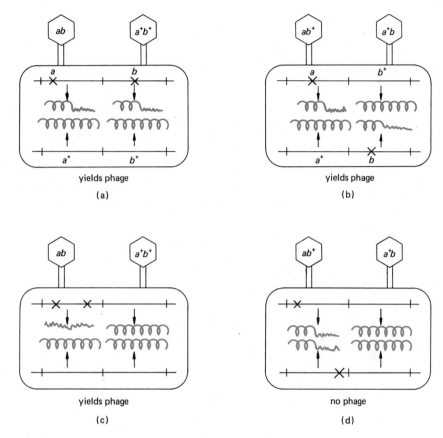

yields phage

(a)

yields phage

(b)

yields phage

(c)

no phage

(d)

FIGURE 5-10

The cis-trans (complementation) test. Genomes are in black and × marks a mutation. The gene products (proteins) are colored helices. There is no complementation when the two mutations are in the same gene and are in the trans position (d). In all other cases, one genome yields a nondefective protein to correct the other's deficiency.

Benzer performed complementation tests with a large number of *rII* mutants in different pairwise combinations. The mutants all fell into two sets, called complementation groups A and B. Any mutant in group A complements any mutant in group B, but no mutant complements any other in its group. On this basis, one is tempted to say that the *rII* region consists of an A gene and a B gene. Exhibiting a bit more logic and caution than most other mortals, Benzer did not go quite so far. Since the two groups are defined on the basis of a cis-trans test, he called the units *cistrons*. We cannot defend the idea that a cistron and a gene are identical without going at least one step further—finding out exactly where all of these mutations map.

DISPLAY 5-1

Benzer's map of the rII region of phage T4. Each square represents one mutation.

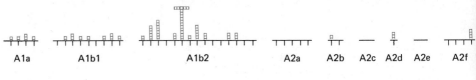

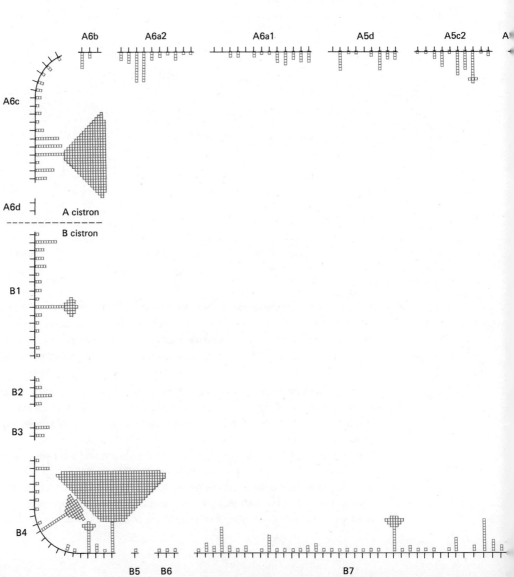

5-10 FINE STRUCTURE ANALYSIS: TOPOGRAPHY

By using a very clever system of deletion mapping, Benzer could easily assign
every rII point mutation to a specific site and in this way build up a very detailed
picture of the rII region. The procedure is shown in Fig. 5-11. Every pair of overlapping,
nonidentical deletions defines a segment of the genome, because a point mutation at a
site within that segment will recombine with one of the deletions but not the other.

A2g   A2h1   A2h2   A2h3   A3a-d   A3e   A3f   A3g

A5b   A5a   A4g   A4f   A4e   A4d   A4c   A4b   A4a   A3i   A3h

Any new point mutation can be mapped easily by attemping to cross it successively
with shorter and shorter deletions, until it is localized in a short segment. It can then
be crossed with other point mutations mapping in the same segment, to determine if
they are at different sites, and all of the sites within the segment can be mapped by
means of standard crosses.

Two important facts emerge from this work. First, there are hundreds of mutable
sites within the rII region (Dis. 5-1). Moreover, they do not all mutate at the same rate;
there are two "hotspots" that mutate very frequently, many sites that have been found
only once, and a range of intermediate sites at which two or more separate mutations
have been found. There are undoubtedly many other sites at which mutations never
have been found. The reasons for this inhomogeneity are still unknown.

Second, all of the rIIA mutations map in the left part of the region and all of the
rIIB mutations map in the right part. With this information, we can argue more
strongly that the A and B cistrons should be identified with genes, for corresponding
to each complementation group there is a long, continuous region that contains many
mutable sites. This is precisely what one would expect to find if a gene is a segment
of the DNA genome that specifies the structure of a polypeptide, and if a change in a
single base pair within the region can appear as a mutation.

Benzer performed many pairwise crosses with the point mutations that mapped
closest to one another. The smallest nonzero frequency of recombination that he found
was 0.01%, whereas he could have detected a much lower frequency with his methods.
Some simple arithmetic can translate this result into physical terms. Stahl and Edgar

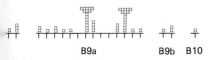

B9a   B9b   B10

have developed a mapping function that allows them to relate physical distances on the T4 genophore to map units. By performing crosses with *ts* and *am* mutants that are relatively close together, the length of the map can be estimated; its total is somewhere between 2000 and 3000 map units—let's say 2500 for convenience. The phage genophore consists of about 2×10^5 nucleotide pairs, or somewhat less than 100 nucleotide pairs per map unit. Since one map unit is 1% recombination, 0.01% recombination is about what one would expect for recombination between neighboring base pairs. In Benzer's terminology, a recon, the elementary unit of recombination, is defined as the smallest unit that can be exchanged between genomes. This analysis means that a recon is probably a single nucleotide pair.

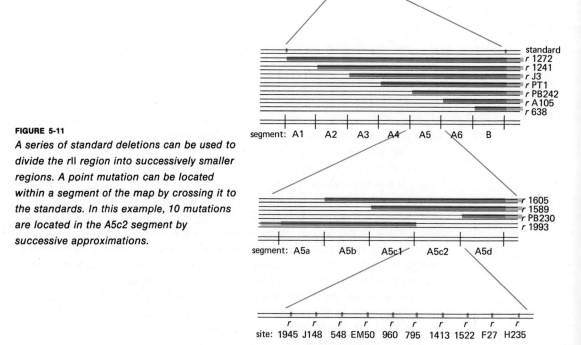

FIGURE 5-11
A series of standard deletions can be used to divide the rII region into successively smaller regions. A point mutation can be located within a segment of the map by crossing it to the standards. In this example, 10 mutations are located in the A5c2 segment by successive approximations.

In summary, Benzer's analysis of the *r*II region indicates that the genome is just what one would expect it to be—a structure divisible into functional units (genes) where mutations can occur, apparently at random, though not uniformly, and within which recombination can occur without restriction, even between neighboring nucleotides.

exercise 5-9   Place the following point mutations within the segments established by the deletion mapping of Exercise 5-8.

MUTATION	36	711	54	304	72	19
1	0	0	0	1	1	1
2	1	1	0	0	0	1
3	0	0	1	1	1	1
4	1	1	1	1	0	0
5	0	0	1	1	1	1
6	1	1	1	1	0	1
7	1	1	0	0	0	1
8	0	0	1	1	1	1
9	1	1	0	1	1	1
10	1	1	1	1	0	0
11	0	1	1	1	1	1
12	1	0	0	1	1	1
13	1	1	0	0	1	1

Be careful; some of these point mutations may force you to modify your deletion map.

C. Mutation

Genetic biology is obviously very dependent upon appropriate mutations. So far we have talked about mutations primarily with analogies and symbols; it is time to describe them in more realistic terms.

5-11 BASE-SHIFT MUTATIONS

In proposing their model for DNA, Watson and Crick pointed out that errors could easily occur in replication because of temporary shifts in the hydrogen atoms of the nucleotide bases, since base-pair specificity depends upon hydrogen bonding. Therefore, it is not strictly true that only A-T and G-C pairs are possible. Each of the bases can exist in a rare tautomeric form, with its hydrogen atoms in different positions, and in these states they can form the "wrong" hydrogen bonds; four base pairs formed in this way are shown in Fig. 5-12. Any of these errors during replication could lead to a mutation.

The mutations that might conceivably occur through the substitution of one base for another are

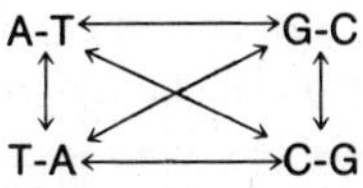

The two horizontal exchanges, in which a purine is replaced by a purine and a pyrimidine by a pyrimidine, are called **transitions.** The other changes are **transversions.**

To study these exchanges more effectively, Ernst Freese and others have used

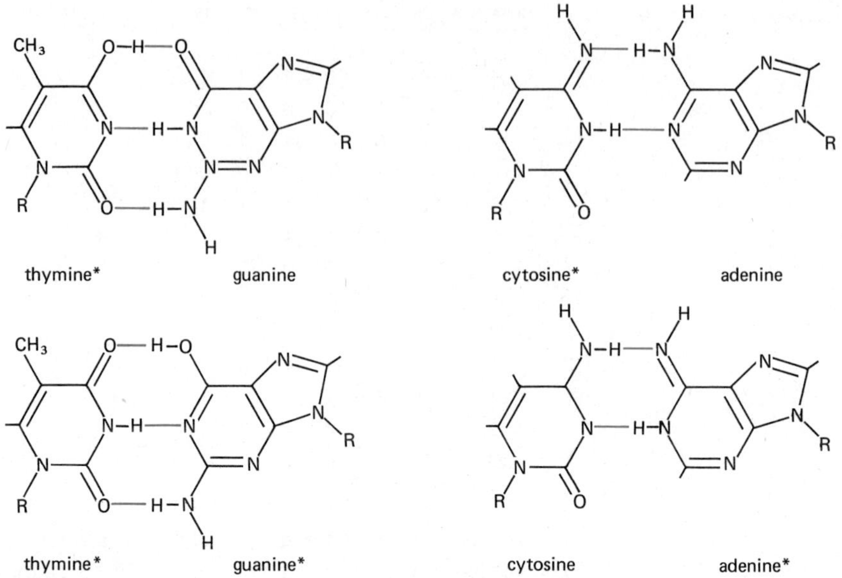

FIGURE 5-12

Some possible "wrong" base pairs permitted by tautomeric shifts of certain hydrogen atoms. In each pair, the base that is in a rare tautomeric state is indicated by an asterisk.

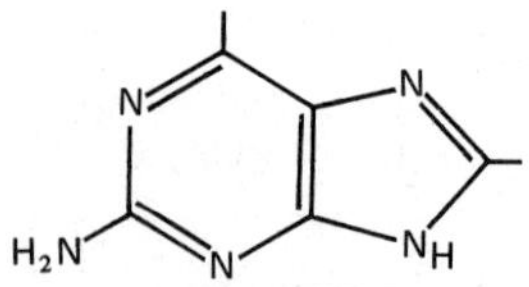

2-aminopurine

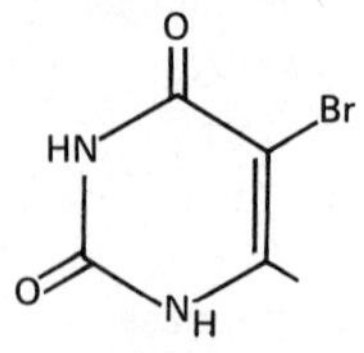

5-bromouracil

mutagens, principally analogues of the natural bases, that might be expected to shift into their other tautomeric configurations more often and thereby induce a higher frequency of exchanges. The most useful have been 2-aminopurine (2-AP) and 5-bromouracil (5-BU) or its nucleoside 5-bromodeoxyuridine (5-BdU), which are analogues of adenine and thymine, respectively. 5-BU normally pairs with adenine, just as thymine does, but it is more likely to shift into a state where it pairs with guanine. Similarly, 2-AP usually pairs with thymine, but it has a greater tendency to shift into the state where it pairs well with cytosine. Figure 5-13 shows how these changes can cause transitions from A-T to G-C pairs. Notice that in both cases the analogues are incorporated in their common states and subsequently cause a mutation during a later replication. Since the mutagen causes a mutation after its incorporation, it could be washed out of the growth medium and still have its effect; this is called clean-growth mutation.

On the other hand, the mutagens could be incorporated from the growth medium in their rare states, and as Fig. 5-14 shows, this would induce transitions from G-C to A-T; these are called *dirty-growth* mutations. Betty Terzaghi, Franklin Stahl, and George Streisinger have shown that 5-BU can induce both transitions in phage T4. We should therefore expect to find any mutation induced by either 5-BU or 2-AP to be revertible by either mutagen. Seymour Benzer and Ernst Freese have examined the topography of *rll* mutations induced by 5-BU; they find the distribution of these mutations to be quite different from the spectrum of spontaneous mutations, and in particular, the two "hot spots" are not sensitive to 5-BU. However, the distribution of 5-BU and 2-AP mutations is very similar, and it does appear that mutations induced by one are generally revertible by the other.

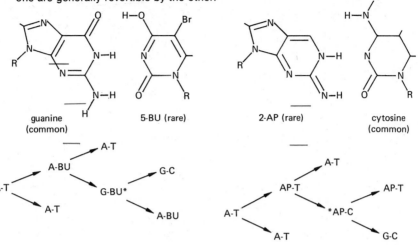

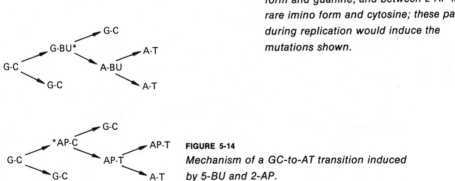

FIGURE 5-13

Base pairing between 5-BU in its rare enol form and guanine, and between 2-AP in its rare imino form and cytosine; these pairings during replication would induce the mutations shown.

FIGURE 5-14

Mechanism of a GC-to-AT transition induced by 5-BU and 2-AP.

Freese, Ekkehard Bautz, and Elisabeth Bautz-Freese have studied the mutagenic action of hydroxylamine, NH_2OH. They find that it reacts with cytosine very strongly, but only weakly with thymine, and not at all with purines. From this, they argue that it is most likely to induce one-way transitions from G-C to A-T. They have shown that it will revert rII mutations induced by 2-AP, as it should. However, other data make it likely that NH_2OH can also induce the reverse transition.

Nitrous acid, HNO_2, has been used to induce mutations in free phage particles. Since their DNA is not replicating at the moment of mutagenic action, the acid must alter the bases in a more stable way, and it apparently acts by deamination. As Fig. 5-15 shows, it converts cytosine to uracil, adenine to hypoxanthine, and guanine to xanthine.

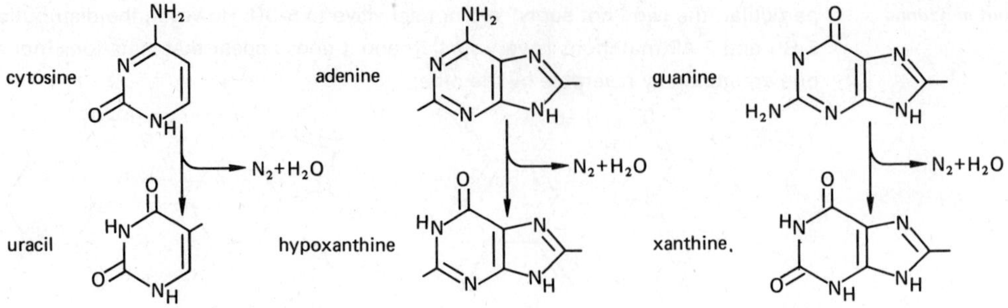

FIGURE 5-15
Alterations of nucleotide bases induced by nitrous acid.

The first change will produce a G-C to A-T transition and the second will produce an A-T to G-C transition, but the third is probably not mutagenic, as xanthine still bonds strongly to cytosine. Nitrous acid should therefore induce both transitions; Bautz-Freese and Freese have confirmed that it will revert mutations induced by 5-BU and 2-AP and that, in general, the three mutagens have similar rII spectra.

5-12 OTHER MUTAGENIC PATTERNS

Two other mutagens that affect nonreplicating phage DNA are ethyl methane sulfonate [EMS: $CH_3SO_3CH_2CH_3$] and ethyl ethane sulfonate [EES: $CH_3CH_2SO_3CH_2CH_3$]. These compounds apparently ethylate the 7 position of purines, leading to hydrolysis of the purine-deoxyribose bond, so they are called *depurinating agents*. Their effect is to leave a gap in the DNA, and since there are no obvious restrictions on the nucleotide that could be inserted opposite such a gap, depurination could result in a transition, a transversion, or no mutation at all. The mutational pattern of EMS and EES is certainly very different from that of other mutagens. No other plausible mechanism for transversions has ever been suggested.

Of course, there are many other possible types of mutation besides transitions and transversions, but it is rather hard to describe very plausible mechanisms for the greater alterations of DNA that might occur, such as deletions and inversions. Hard radiation, such as x rays, can create breaks in DNA, and these could presumably lead to gross damage if they were not repaired properly. Nitrous acid also seems to induce deletions, but its mode of action is unknown.

5-13 A LAST WORD ON TERMINOLOGY

There was a time when life was very simple. Geneticists observed that people could have either blue eyes or brown eyes, blond hair or black hair; that peas could be either yellow or green, smooth or wrinkled; and that flies could have either red eyes or white eyes, long wings or stunted wings. In each case, the characteristic behaved as if it were determined by a single gene which had two different forms;

they called the alternative forms of a gene **alleles.** Genes were supposed to be
arranged like beads on a string, and everyone had his own personal charm bracelet.
Mine would have a bead for brown hair and another bead for brown eyes, while my
neighbor would have different alleles on his, giving him blond hair and blue eyes. Or
perhaps there were several alternatives (brown eyes, blue eyes, hazel eyes, green
eyes, etc.) and one could say that there were **multiple alleles** of a given gene, meaning
simply that you had more choice in selecting beads for one position of the bracelet.

Life is no longer that simple, but the word allele is often still used as if it were, to
the dismay and confusion of beginning students. A gene is now a string of nucleotides
and the phenotypes we work with are less often gross features such as eye color than
quite simple characteristics such as the activity or inactivity of an enzyme. (My fruit
fly has white eyes because he lacks one enzyme for making a red pigment, but we
don't often get to that level of sophistication.) Suppose we select a series of rII
mutants and, by complementation tests, sort out just those in the rIIA gene. (We'll
assume that the rII genes inform an enzyme, although we don't know for sure.) By
any classical definition, any of these mutants is an allele of the wild type rII⁺, but what
about their relation to one another? If we establish a map with the mutations, we find
that any pair of them are either at the same site or at different sites; Herschel Roman
has suggested that if they are at the same site they should be called *homoalleles* of
one another and if they are at different sites, they should be called *heteroalleles.*
However, it is generally unnecessary to use such a complicated terminology. If we
perform a cross of r1267 × r319, we shall know exactly what the situation is simply
by looking at the results of the experiment, for we will either get recombinants or not.
The designation of alleles at this level is generally unnecessary.

When people started to do extensive mapping experiments with fruit flies and mice,
they began to find groups of mutations that appeared to be alleles of one another,
because they all mapped in essentially the same region, but on the other hand
appeared *not* to be alleles because a very low frequency of recombination between
them could be detected. They called these series of *pseudoalleles;* now we would
simply say that they had found mutations at different sites that were very close to one
another, probably in the same gene. Such results were puzzling simply because they
did not separate the functional conception of a gene ("something that determines a
characteristic") from the mutational conception ("something that can change into a
different form") and the recombinational conception ("something that can recombine
with another piece like itself"). They conceived of genes as alternative beads, and they
did not easily admit the possibility that a bead could itself be split up. Now that we
can resolve nucleotide pairs with the high-power microscope of recombination in
phage, the concept of pseudoalleles should be allowed to die.

We can sometimes see alternative states of genes very clearly because amino acid
sequences for specific proteins are known. One of the best examples is in human
hemoglobins, which consist of two types of chains designated α (141 amino acids)
and β (146 amino acids), whose sequences are known completely for humans and
many other animals. Occasionally someone comes into a clinic with a medical problem
that is traced to an anemia, and when his blood is sampled a different hemoglobin is
found. Over 100 cases have now been traced to simple amino acid substitutions due

to mutations in the genes for the α and β chains; some of these are shown in Fig. 5-16. The first one ever identified was the hemoglobin S found in people with **sickle-cell anemia,** in which the red blood cells take on a strange sickle shape; people with this hemoglobin often get into serious medical difficulties in later life. The mutation is common in Negro populations because hemoglobin S provides some resistance to malaria, clearly an advantage in many parts of Africa.

We might want to say that some of these mutations are alleles of one another. For example, in the β chain at position 6 there are mutations from Glu to Val (hemoglobin S) and from Glu to Lys (hemoglobins X and C). However, since it takes three bases to code for one amino acid, the mutations might be in neighboring nucleotides within the same codon, so they would not be strict homoalleles. Nevertheless, in comparing amino acid sequences of related proteins, it is useful to talk about amino acid alleles at a given position.

FIGURE 5-16

Some mutations of the β chain of human hemoglobin. At left is part of the standard amino acid sequence, numbered from the N-terminal end. Most of the mutations are designated by the locality where the aberrant hemoglobin was discovered. Notice that sickle-cell hemoglobin (S) has a change from Glu to Val at position 6.

1	Val	
2	His→Tyr	Tokuchi
⋮		
6	Glu→↗Val	S
	Glu→Lys	X, C
7	Glu→↗Lys	C Georgetown, Siriraj
	Glu→Gly	G
8	Lys	
9	Ser→Cys	Porto Alegre
⋮		
14	Leu→Arg	Sogst
15	Trp	
16	Gly→↗Arg	D Bushman
	Gly→↘Asp	J Baltimore, Trinidad, Ireland, New Haven
⋮		
22	Glu→Lys	G Saskatoon
23	Val→ – – –	Freiburg
⋮		
26	Glu→Lys	E
27	Ala	
28	Leu→Pro	Genova
⋮		
42	Phe→Ser	Hammersmith
43	Glu→Ala	G Galveston, Port Arthur, Texas
⋮		
46	Gly→Glu	K Ibadan
47	Asp→Asn	G Copenhagen
⋮		
56	Gly→Asp	J Bangkok, Meinung, Korat
⋮		
61	Lys→Asn	Hikari
62	Ala	
63	His→↗Tyr	M Saskatoon, Emory, Kurume, Chicago, Hamburg
	His→↘Arg	Zurich
⋮		
67	Val→↗Glu	M Milwaukee 1
	Val→↘Ala	Sydney

The reason for elaborating on this point is that the word allele continues to come up in discussions of genetics. We will use it as little as possible because there is usually a more precise alternative, but it is still often necessary to say that one genophore is carrying a mutation and the other is carrying the wild-type allele of that mutation.

In summary, we will use these terms:

Genome is the general conceptual term: the total structure that carries information from one generation to the next.

Genophore is the physical structure (such as a DNA molecule) that carries this information.

Genotype is a description of the information in any particular genome—generally a list of all the relevant alleles, where alternatives must be contrasted with one another.

Genetic map is the picture we draw of the genome on the basis of standard mapping experiments.

PART 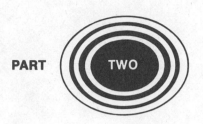 TWO

6

molecular
structure

Every biological object is made of condensed phases, either solid or liquid or both. In contrast to a gas, a condensed phase has a fairly regular structure which is a function of the amount of bonding between its atoms. In a liquid, bondings are confined to local, structured regions which are surrounded by more disordered areas. In a crystalline solid, the atoms are bonded to one another in a very extensive array; a perfect mineral crystal too large to lift exhibits a large-scale order that depends upon microscale regularities in its interatomic bondings.

The extreme order of most biotic structures is obvious. We see this all the way from the objects we see every day down to the minute objects revealed by electron microscopy. In this chapter we will discuss the physical and biological factors that create such regularities and look more carefully at the major types of molecules of which regular biological structures are made.

A. Biomolecular architecture

6-1 PRINCIPLES OF DESIGN

About 1956, Crick and Watson pointed out some interesting features of the small, spherical viruses that cause many plant diseases, as well as human diseases such as polio and common colds. Each virion of a typical virus in this class contains about 3.5 million daltons of protein and a genome of about 2 million daltons of a single-stranded RNA. Because the average nucleotide weighs about 3.3 times the average amino acid and it takes three nucleotides to inform each acid, 2×10^6 daltons of RNA can only inform about 2×10^5 daltons of protein. This is considerably less than the weight of a virus capsid alone, and the genome probably must inform several enzymes associated with virus multiplication that never become part of the capsid. Crick and Watson therefore argued that the virus capsid must be made out of many small, identical subunits; if the genome had to specify only one subunit of, say, 20,000 daltons, it would still have enough space to specify several other proteins and the whole capsid could be made out of about 180 such units.

Now if many identical subunits associate to make a large structure, they must interact through similar or identical bonds. The resulting structure must then have a certain element of *symmetry*—it must be regular—and its symmetry can be determined through appropriate optical techniques, particularly x-ray diffraction. Crick and Watson specified certain theoretical limitations on biological symmetry; within a few years, their basic predictions for small viruses were confirmed in several cases.

The same arguments advanced by Crick and Watson for small

viruses have been extended by Donald L. D. Caspar and Aaron Klug. We are now in a position to state some basic principles of biomolecular architecture on the basis of some firm physical and biological principles.

The major biological principle is Darwinian selection. Everything we know about the mechanism of evolution and all that we see of specific cases of evolution lead us to believe in what George Gaylord Simpson calls the "opportunism" of evolution. The fundamental variations in a biological population are caused by mutations, which occur with enough frequency to be fairly common in a natural population. Mutations that lead to any improvement in an organism tend to be selected, so that over a number of generations a population tends to become better adapted. The population thus has an opportunity to experiment with novel structures over enormously long periods of time. In Simpson's words,

> . . . over and over again in the study of the history of life it appears that what can happen does happen. There is little suggestion that what occurs *must* occur, that it was fated or that it follows some fixed plan, except simply as the expansion of life follows the opportunities that are presented.

Clearly, the implication is that if it is advantageous for anything to happen, it will happen, given enough time. Therefore, if a structure, a chemical mechanism, or a type of organism survives it must be well designed by Darwinian mechanisms for the function it serves; if it were not well designed, it would have been eliminated or replaced by one that was fundamentally better. Biotic designs are always being improved upon, because evolution is a continuing process and because environmental demands are always changing, but anything that survives for a significant time must be reasonably efficient and free of extraneous features that serve no real function. We will talk frequently about the design of a biological system with this concept in mind.

The primary physical limitation on design is thermodynamic. The functional form of a structure should be a relatively stable form whose free energy is a minimum relative to that of similar forms. This is not to say that an enzyme or some other protein functions only in a state of absolutely minimum free energy; in fact, Daniel Mazia has made the interesting suggestion that a functioning structure must be in a slightly energized condition and that when it settles into an absolutely stable state, it is literally dead. But at least the object should assume a functional structure more or less spontaneously, without a lot of additional energy and information from special enzymes. For example, we have already indicated that when a polypeptide of a given primary structure is put together it will fold up spontaneously into its proper shape and all of the bonds that hold it in this shape will form spontaneously. We believe this to be quite generally true for other molecules and for different degrees of molecular complexity.

These evolutionary and thermodynamic principles imply two other biological limitations on design. First, they imply that the genome should specify a minimum amount of structure, so it can be as small and efficient as possible. Second, the organism should have a maximum of control over the quality and quantity of its components. Caspar and Klug now argue that all of these considerations are satisfied by two fundamental design principles: subassembly and self-assembly.

The **subassembly principle** states that any large molecular structure should be built out of a number of small molecules that are similar or identical. We have already seen that polymers are built out of monomers; we will soon see that the polymers are typically assembled into larger and larger complexes until a structure of the required size is built. The **self-assembly principle** states that these subunits should assemble themselves into a functional structure by spontaneously forming the correct bonds between themselves and assuming a minimum energy conformation.

If biotic structures are designed according to these principles, then the genome must specify only the structure of the subunits; as we saw in the case of the small viruses, this is very economical of information. If the subunits assemble themselves, the genome does not have to carry information for special assembly enzymes whose only function is to make other structures. If there are several subassembly stages, control can be exerted at each stage, so there are several points where the quantity of material flowing into the assembly can be regulated. Furthermore, if a defective subunit is made it will probably have the wrong shape and will simply be unable to bond correctly with other subunits; this kind of regime therefore contains an internal error-correcting mechanism that automatically discards defects without a special mechanism.

The elimination of errors is very important. Suppose that a large object were to be made out of n monomers in a one-stage process, without intervening subassembly stages, and that an average of k errors per monomer are made, where k is much less than 1. Then the probability that the structure is free of errors is e^{-nk} and the probability that it contains at least one wrong monomer increases rapidly as n increases. If one wrong monomer is inserted, the structure will probably be defective and the other $n - 1$ monomers will also be wasted. It is much more efficient to make the entire structure out of m small subunits of n/m monomers each, so the probability of an error in the subunit is much less and fewer monomers will be wasted if an error does occur.

exercise 6-1   A structure is made out of 10 proteins, each of which contains 100 amino acids. The probability of inserting a wrong amino acid at any point is 10^{-4}. (a) What is the probability of making a defective protein subunit? (b) If defective subunits are rejected, what is the probability of making a defective structure? (c) If the structure were made out of 1000 amino acids in one polymer, what would be the probability of making a defective structure?

6-2 SYMMETRY

As we have seen, these principles of biological design imply that biotic structures should be symmetrical, at least at the macromolecular level. An object is symmetrical if it shows some regular geometric relationship between its elements. More precisely, it is symmetrical if some set of operations can be performed on it that will superimpose one part of it precisely on another.

The subunits of a symmetrical object are not generally symmetrical themselves; the simplest whole unit from which the entire object can be made is the **asymmetric unit.**

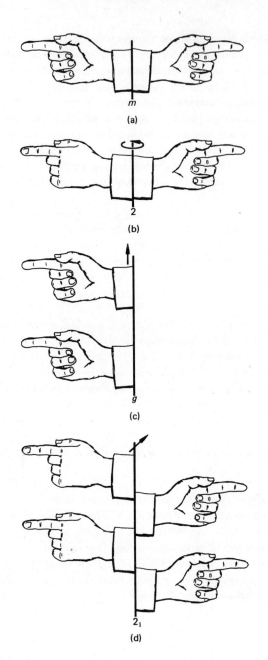

FIGURE 6-1

Symmetry elements. (a) Mirror reflection. (b) Dyad rotation.
(c) Translation. (d) Translation with dyad rotation.

There are three types of imaginary operations that can be performed with this unit to form the whole structure.

1. The object can be inverted through a mirror (**mirror reflection,** denoted by *m*). This operation, for example, will convert your left hand into your right hand; your two hands together are said to have *bilateral symmetry* (Fig. 6-1a).

2. The object can be rotated by 360°/*n* around an axis of *n*-fold **rotational symmetry** (Fig. 6-1b). If *n* = 2, the object is superimposed on itself by 180° rotation around a dyad axis. If *n* = 3, the axis is a triad; similarly, tetrad (*n* = 4), pentad (*n* = 5), hexad (*n* = 6), nonad (*n* = 9), and other symmetry axes have been found.

3. The object can be **translated** along a glide axis, *g* (Fig. 6-1c). For example, the bricks in a wall are asymmetric units that can be superimposed on one another by translation along a vector at 45° to the lines of the wall.

Of course, any symmetrical object can and usually does combine several of these symmetry elements (Fig. 6-1d). If you examine a perfect cube you can easily find several planes of mirror symmetry as well as triad and tetrad axes of rotational symmetry.

While mirror inversions are important elements in inorganic crystals, they cannot be elements of biological molecules, for they would convert L-amino acids into D-amino acids, D-sugars into L-sugars, and so on. Therefore, all molecular symmetries in biotic systems are combinations of rotations and translations. This imposes severe restrictions on the forms of biological objects. Long ago, crystallographers catalogued all of the possible combinations of symmetry elements and thus defined 230 crystal classes; only a few of these ever appear in biological molecules.

Roy Markham devised a clever way to determine the value of *n* for objects seen by electron microscopy that seem to have *n*-fold rotational symmetries. The first operation is the hardest—to find the exact center of the object in a photograph; this may require some experimentation. Once the center has been found, you stick a pin in it and then rotate the picture successively by 360°/*n*, taking a picture of it each time to make a multiple exposure on another piece of film. This is repeated for different values of *n*, say, 5, 6, and 7. If the object really has hexad symmetry, then all the pictures taken when the photograph is rotated successively by 60° will reinforce one another and make a sharper picture, while those rotated by 72° (*n* = 5) and 51.4° (*n* = 7) will be more blurred than the original.

This is a very limited procedure, though. We are generally interested in molecules that cannot be seen by even the finest electron microscopes and that have more complex symmetries. To resolve points that are very close together, we need very short-wavelength radiation, which means x radiation. x-Ray diffraction studies of macromolecules were started by W. T. Astbury and his group in the 1930s, and after a long period of relatively meager results, the x-ray crystallographers have come up with the most exciting and detailed pictures available of biological structure.

6-3 X-RAY DIFFRACTION

Since x rays are simply very high-energy (short-wave) photons, they interact with the electrical fields of atoms much like lower-energy photons and are affected in much the

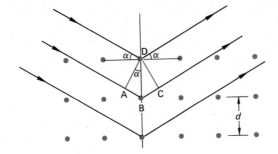

FIGURE 6-2
A lattice of spacing d *being irradiated by an x-ray beam.*

same ways. An ordinary light lens looks at the diffraction pattern of light coming from an object and translates the pattern back into one that is familiar to our eyes. However, there are no lenses for focusing x rays in the same way, so we have to look at the diffraction pattern directly and translate it into a picture with a computer.

Figure 6-2 shows a two-dimensional lattice of atoms with spacing d vertically. A beam of x rays with wavelength λ is aimed at it at $\alpha°$ to the horizontal axis; the x rays are scattered by the atoms of the lattice at an equal and opposite angle and a photographic plate is placed where the diffracted x-ray beam can be detected. Since d is very small compared with the distance to the plate, x rays coming to the plate from different atoms travel essentially parallel paths. In general, the angles and phases of the diffracted x rays will bear very complex relations to one another; they will interfere with each other and nothing will be detected on the plate. However, there is at least one value of α for which the diffracted rays will be in phase, so they will add to one another and produce a spot of high intensity on the film. Compare the ray that strikes point D with the ray that strikes point B; the latter must traverse the extra distance ABC as compared to the former. The condition that the two rays remain in phase is that there are an integral number of wavelengths, $n\lambda$, in this extra distance. Clearly AB = AC = $d \sin \alpha$; therefore, $n\lambda = 2d \sin \alpha$. This is the Bragg relationship; it relates distances in the lattice to the direction and wavelength of the x-ray beam and allows us to calculate the former from the latter. This analysis can be extended to two and three dimensions with analogous results.

The analysis of a crystal containing only one kind of atom is relatively simple. Now consider a crystal of a more complicated molecule, such as a protein. Each asymmetric unit in the crystal is a single molecule; within the crystal structure these may be related to one another by rotations, but it is always possible to pick out a block, the **unit cell,** containing one or more asymmetric units and having just the right dimensions and angles so that the entire crystal can be made by translation of this block alone. The position of the unit cell is entirely arbitrary, but it is generally drawn to contain whole molecules as much as possible. The two-dimensional analogue is shown in Fig. 6-3.

The dimensions of the unit cell define the repeating distances throughout the crystal, for if there is a carbon atom at coordinates (x, y, z) in one unit cell there will be an identical atom at the same coordinates in every other unit cell. These carbon atoms form a lattice, and there will be a similar lattice for every other atom

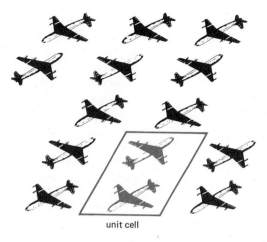

FIGURE 6-3
*A symmetrical pattern with
an arbitrary unit cell superimposed.*

unit cell

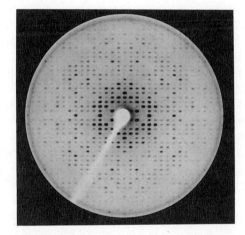

FIGURE 6-4
*x-Ray diffraction pattern from a globular protein
(horse heart cytochrome c).
(Courtesy of Richard Dickerson.)*

in the molecule. An x-ray beam aimed at the crystal will detect each of these lattices; it will also detect more complex regularities in the crystal due to combinations of the elementary lattices. Each of these regularities causes at least one reflection that can be picked up as a spot on photographic film, as shown in Fig. 6-4. The angles by which the reflected beams that make these spots diverge from the axis of the x-ray beam measure the spacings between the atoms in each lattice. The intensities of the spots measure the electron densities of the diffracting atoms.

Another way of looking at the crystal lattice may be helpful. Since each point of the lattice consists of an atom with a certain electron density, one can imagine waves of electron density running through the crystal in all directions. Each of these waves can be represented by some combination of sine and cosine functions, in general called a Fourier function, just as a musical tone can be represented by a sine wave or a combination of sine waves. The entire crystal is the sum of all these Fourier functions, just as the sound produced by an orchestra is the sum of all the tones produced by its separate instruments. The x-ray beam picks out these waves of electron density and produces a single spot on the film for each one, just as one can put down a musical note for every tone produced by the orchestra. The regular pattern of spots is a *reciprocal lattice,* so called because distances within it are the reciprocals of distances in the real crystal lattice; spots close to the center of the film represent regularities in the crystal with long spacings, while those closer to the edge represent those with short spacings. The intensity of a spot is proportional to the square of the electron density of the atoms that produce the wave.

A good musician can listen to the sounds produced by each instrument and transcribe them into a score; if he is very good, he can look at the score and hear the corresponding music. The reciprocal lattice consists of the crystal analyzed into Fourier functions, and anyone with some mathematical training and a good calculator can perform the analysis. However, no x-ray crystallographer is a good enough "musician" to look at the reciprocal lattice and transform it directly into a picture of the crystal lattice. He can measure the intensities and the angles of the beams that make the photograph, but there is no direct way to obtain the *phase* of each wave relative to all of the others. This is where his ingenuity comes in.

It occurred to Max Perutz that one could get a measure of phase relationships by introducing into each unit cell of the crystal one or two atoms that are so electron dense that they change the whole reciprocal lattice. Then a comparison of crystals with and without the extra atoms would yield the needed phase information. This is called the *isomorphous replacement* method, since the substituted protein must have a unit cell with precisely the same size and shape as the unsubstituted one. It is not easy to obtain proper crystals, but several people have solved their problems by attaching mercury atoms to free sulfhydryl groups or other atoms to free amino groups, for example.

A Fourier synthesis from the diffraction patterns is primarily a lot of hard work after the phase angles have been established, and now this is done by computers. No one tries to establish the whole crystal structure at one time. Remember that as you move away from the center of the film you look at finer and finer structural details, and this is the way structures are established in practice. For example, the structure of myoglobin was first established with a resolution of 6 Å by analyzing 400 reflections. Once the general shape of the molecule was established, 9600 reflections were used to give 2-Å resolution, and finally 25,000 reflections were used to give 1.4-Å resolution. As one focuses on the crystal, certain groups and atoms with characteristic shapes appear, and by making intelligent guesses about details at each step, one can save a lot of time and effort.

6-4 DIFFRACTION FROM A FIBER

The globular proteins are obtained in crystalline form by artificial methods; they do not exist as crystals in cells. However, there are many proteins that form fairly regular, quasicrystalline structures in cells; these produce x-ray diffraction pictures that are simpler than those obtained from pure crystals. They are easier to interpret because they contain less information; but the information they give is still very useful. The molecules in these quasicrystalline structures all have the same general orientation, but they may be displaced from one another by rotations and by nonintegral translations along the fiber axis. The pattern from such a fiber is essentially what one would obtain from a pure crystal that was rotated and moved slightly up and down.

A typical fiber diffraction pattern, such as Fig. 6-5, shows a small number of reflections. The fiber is oriented vertically in the x-ray beam; the blurring of each spot

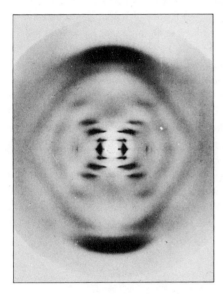

FIGURE 6-5

x-Ray diffraction pattern of the lithium salt of DNA in the B form. Note the cross pattern in the center of the photograph and the very large, black meridional reflections which are on the tenth layer line at 3.4 Å; the off-meridional reflections near the center are on the first few layer lines, and you can use them to establish the spacing of these lines. (Photograph by Maurice Wilkins, courtesy of Richard Dickerson.)

in an arc shape is due to slight disorientations of the molecules from the fiber axis. Reflections above and below the center are **meridional,** and they reflect regularities along the fiber axis. Reflections to the side of center are **equatorial** and they reflect regularities across the fiber axis. There is typically a diagonal cross pattern around the center of the film reflecting diagonal regularities. A series of regular, horizontal lines of reflections, the *layer lines,* can usually be picked out; these may be understood by remembering that each reflection represents a regular sine wave of electron density and that, like any other sine wave, it can have harmonics whose frequencies are integral multiples of its own. If there is a regularity in structure in the protein with a spacing of a Å, there will be a reflection with a reciprocal spacing of $1/a$ Å on the film. There will also be weak reflections at reciprocal spacings of $2/a$, $3/a$, $4/a$, . . . , due to the harmonics of the fundamental density wave. These are the first, second, third, fourth, and so on *orders* of the fundamental spacing in the protein, and their positions on the film are the first, second, third, fourth, and so on layer lines.

The higher-order harmonics will give very weak reflections. However, if there is an additional regularity in the fiber whose fundamental wave has the same frequency as one of these harmonics, the two will add to one another and give a very strong reflection. For example, the DNA helix contains nucleotide pairs separated by 3.5 Å along the fiber axis and the helix repeats every 10 bases, to make an additional regularity at 35 Å. The 35-Å spacing will produce a reflection at $1/35 = 0.0276$ Å^{-1}, as well as harmonics, including one at $10/35 = 0.276$ Å^{-1}. But the 3.5-Å spacing will also produce a reflection at $1/3.5 = 0.276$ Å^{-1}, and the two will add to produce a very strong reflection on the tenth layer line. This was an important piece of information that led Watson and Crick to the correct model for the DNA helix.

There is only one way to make a linear structure out of subunits: repeated translation of the subunit along one axis, with or without rotation. In every known case, there is at least a two-fold rotation in the basic polypeptide unit of a protein, so all of them form *helices*. In principle, every fibrous protein helix can be described by two numbers giving the distance along the fiber that every residue must be translated and the number of degrees it must be rotated around the axis to bring it into the position of the next residue; and the designation right- or left-hand must be given, because the helix can be turned in either direction. The solution of fibrous protein structure is primarily a matter of finding the correct helix, and there is now a well-developed theory of diffraction from helices to help in this work. In practice, one generally builds a model and then modifies it suitably by trial and error to make it conform with the x-ray diffraction data. We will discuss this after a brief review of some facts about chemical bonds.

6-5 CHEMICAL BONDS

Before going on, we will summarize briefly some points about chemical bonding that are important in thinking about protein structures. By this time, you should be quite familiar with covalent bonds, the principal type of bond encountered in organic molecules, in which atoms are held together at close distances and with relatively high energies by combining their atomic orbitals into molecular orbitals and sharing electrons. These bonds determine the fundamental structures of organic molecules; but much weaker bonds become important in determining the conformations of macromolecules like proteins that might take on a variety of shapes after their covalent bonds have all been made. That is, these weak bonds severely restrict the number of degrees of freedom that very long polymers might enjoy. They are also extremely significant in interactions between macromolecules.

Ionic bonds, which are so important in inorganic crystals, arise from the attraction between oppositely charged ions. Each ion is considered a point charge of z electrostatic units, and the energy of interaction between ions a and b at a distance r from one another is

$$E_{ab} = \frac{z_a z_b}{Dr}$$

where D is the dielectric constant of the medium between the two charges. The latter may not be easy to evaluate; while D is about 80 for water when measured in bulk, it is probably much smaller for charges that are very close together and may be separated by very few atoms of water.

Electrostatic interactions are certainly important in biological materials; charged, polar groups within proteins and between proteins attract one another and also interact with small organic molecules and ions in solution nearby. These electrostatic interactions also occur between the charged organic groups and water molecules; for this reason, they are considered part of a class of **hydrophilic** interactions, and given a choice of local environments, we expect ionic groups to assume positions in which they can interact with water in preference to interacting with uncharged organic groups.

Finally, there are a number of interactions under the heading of *London dispersion forces.* At any given moment, the electrons around an atom are distributed so the atom as a whole has a dipole moment. Furthermore, when an ion approaches an atom it causes a slight displacement of electrons in the latter and gives it an induced dipole moment. All of the possible interactions of this type—charge-dipole, dipole-dipole, dipole-induced dipole, and so on—are relatively weak, and the forces of attraction between atoms that they generate decrease with large powers of the distances between the interacting groups, as about r^{-6}. However, these interactions are sufficient to create the so-called **van der Waals forces** between atoms; the atoms approach each other up to the point where their electron clouds begin to overlap and the London attractive forces are balanced by corresponding repulsive forces. Each atom therefore has a characteristic van der Waals radius, which is the closest distance to which another atom will approach because of these forces. These radii are always greater than covalent bond distances.

van der Waals forces come under the heading of **hydrophobic** bonds. Uncharged organic groups tend to associate with one another much more strongly than they associate with the polar molecules of water and therefore, given any significant degree of freedom, they tend to form regions from which water is excluded. Their preferences for aqueous or nonaqueous environments can be measured by the energy and entropy changes that occur in passing from one to the other; these values have been measured for certain hydrocarbon groups (methane, ethane, benzene, etc.) and the results show a very interesting pattern. The enthalpy ΔH for the transition from an organic to an aqueous environment is always zero or less than zero; that is, this transition is *energetically* favorable. However, ΔS is also negative, and so the sum $\Delta G = \Delta H - T\,\Delta S$ at temperatures about 300°K is always positive because of the unfavorable *entropy* change. This may be interpreted to mean that nonpolar organic groups tend to associate with one another because their intrusion into an aqueous environment produces entropically unfavorable disruptions of the water structure. Notice that this also means that hydrophobic interactions tend to be more stable as the temperature increases.

B. Proteins

It was clear from Astbury's early work that all of the fibrous proteins have a great deal in common with one another, and he divided them into two major groups (Table 6-1). The **k-m-e-f group** is named for the type proteins **keratin, myosin, epidermin,** and

TABLE 6-1

representative fibrous proteins

k-m-e-f group
α SUBGROUP
1. *Keratins:* hair, wool, fur, horn, hooves, porcupine quill.
2. *Myosins:* major muscle protein.
3. *Epidermin:* skin of mammals, amphibians, and some fishes.
4. *Fibrinogen:* inert blood protein that is converted to *fibrin,* which makes blood clots.
β SUBGROUP
1. *Keratins:* feathers, beaks, claws, scales of birds and reptiles.
2. *Actin:* secondary muscle protein.
3. *Fibroin:* protein of natural silk.
COLLAGEN GROUP
1. *Collagen:* main protein of cartilage, tendon, bone, and skin.
2. *Vitrosin:* vitreous humor of the eye.
3. *Ichthyocol:* skin, tendon, swim bladder of fish.
4. *Spongin:* skeletal fibers of some sponges.
5. *Secreted collagens:* anchor threads, egg cases, cuticles of insects, molluscs, and other animals.
ELASTIN GROUP
1. *Elastin:* an elastic protein of tendons, etc., that gives them resiliency.

fibrinogen, and is divided into α and β subgroups which are very similar. Some of these proteins can be changed from the α to the β form by processes such as wetting, stretching, and steaming. We will examine these proteins to help understand basic protein structures.

The **collagens** include a number of closely related proteins from different animals, including a variety of excreted substances with which some animals protect themselves, attach themselves to rocks to avoid being washed away, and perform other functions. We will discuss these in Chapter 19 in relation to processes of assembly.

6-6   THE α-PROTEINS

The principal x-ray reflections from the k-m-e-f proteins, such as α-keratin, are strong equatorial reflections at 9.8 Å and strong meridional reflections at 1.5 and 5.1 Å. The object of the early work on these proteins was to explain this pattern by means of a satisfying model. Linus Pauling, Robert B. Corey, and R. R. Branson laid down some requirements for the model on the basis of their studies of artificial polypeptides:

1. The atoms around a peptide bond can be arranged in either a cis or trans configuration; the latter is more stable. Figure 6-6 shows the structure of a trans polypeptide chain; in a protein, these distances and bond angles must be preserved with little distortion.

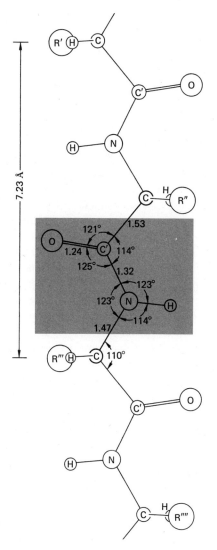

FIGURE 6-6

The conformation of a trans polypeptide chain, with the peptide bond region in color.

2. Because of resonance, the atoms in the peptide bond must all lie in a plane.

3. The structure should allow a maximum number of hydrogen bonds between the —NH and —C=O groups of the backbone.

Considering these requirements, Pauling and Corey proposed an **α-helix** structure to explain the regularities of the α pattern; the helix can be either right- or left handed, and the latter is shown in Fig. 6-7. The appropriate hydrogen bonds are formed between imino and carbonyl groups of every third residue. There is a distance of 1.5 Å between residues along the helix axis, and since there are 3.6 residues per turn, the distance along the axis between corresponding points is $3.6 \times 1.5 = 5.4$ Å. These numbers are in good agreement with values from the diffraction pattern. The slight discrepancies can be accounted for by a suitable modification of the α-helix shown in Fig. 6-8. The whole helix is bent very slightly into a **coiled-coil** form, and several of these helices can be put together to make a larger fibril.

More recently, A. M. Liquori found that the same type of structure occurs in hydrocarbon polymers where hydrogen bonds are clearly impossible. Therefore, he suspected that the hydrogen bonds in the α-helix are less significant than van der Waals interactions in establishing the conformation of the protein, and he has been able to demonstrate that this is true.

Liquori has studied homopolypeptides such as polyglycine, poly-L-alanine, and poly-L-valine; because of their regular structure, the conformation of the whole polypeptide will be determined once the conformation of any residue is determined. He therefore places all of the known restrictions on the conformation of a residue, allows free rotation around all of the unrestricted bonds, and uses the known energies of the van der Waals interactions to determine the conformation of minimum energy. For poly-L-alanine, the conformation is the α-helix that allows almost perfect hydrogen bonding of the Pauling–Corey type, yet the hydrogen bonds merely add a little more stability to an already stable structure. On the other hand, polypeptides such as poly-L-valine, threonine, and isoleucine, with their bulky side chains, show energy minima for somewhat distorted α helices, and in fact these residues in real proteins tend to disrupt α-helical structures. Liquori also finds that polyglycine and poly-L-proline do not form α-helices; their conformations agree very well with structures proposed for collagen, of which they are the major components.

In summary, hydrogen bonding, which was one of the primary considerations in early work, is now being shown to be of secondary importance in proteins, with van der Waals contacts supplying most of the stability.

Just to get an idea of size relationships and the higher levels of organization in proteins, Fig. 6-9 shows how the keratin fibers are organized into a strand of hair or wool. There is still considerable uncertainty about the conformations of α-proteins at the protofibril-microfibril level, and the details at this level are supposed to be indicative rather than definite.

6-7 THE β-PROTEINS

The β-proteins have equatorial reflections at 9.8 and 4.65 Å and a meridional reflection at 3.33 Å. The β pattern is satisfied by the parallel and antiparallel **pleated-sheet** structures, also proposed by Pauling and Corey (Fig. 6-10). The peptide bond planes are tilted alternately up and down, as shown in Fig. 6-11. The side chains point upward and downward from this backbone and neighboring peptides are joined through one or the other pattern of hydrogen bonding.

Silk fibroin differs from other k-m-e-f proteins because of its strange amino-acid

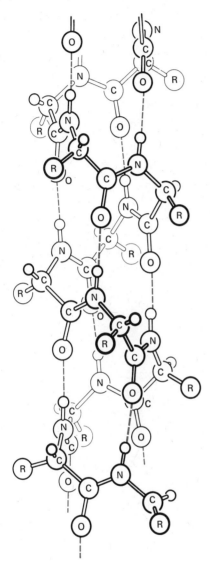

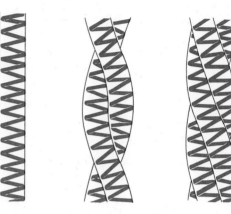

FIGURE 6-8

Coiled-coil forms of the α-helix, in which the helices are given a slight twist and wrapped around each other.

FIGURE 6-7

The α-helix.

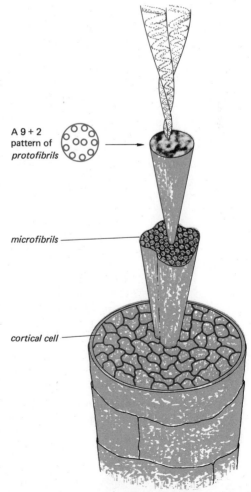

A 9 + 2 pattern of *protofibrils*

microfibrils

cortical cell

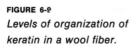

FIGURE 6-9

Levels of organization of keratin in a wool fiber.

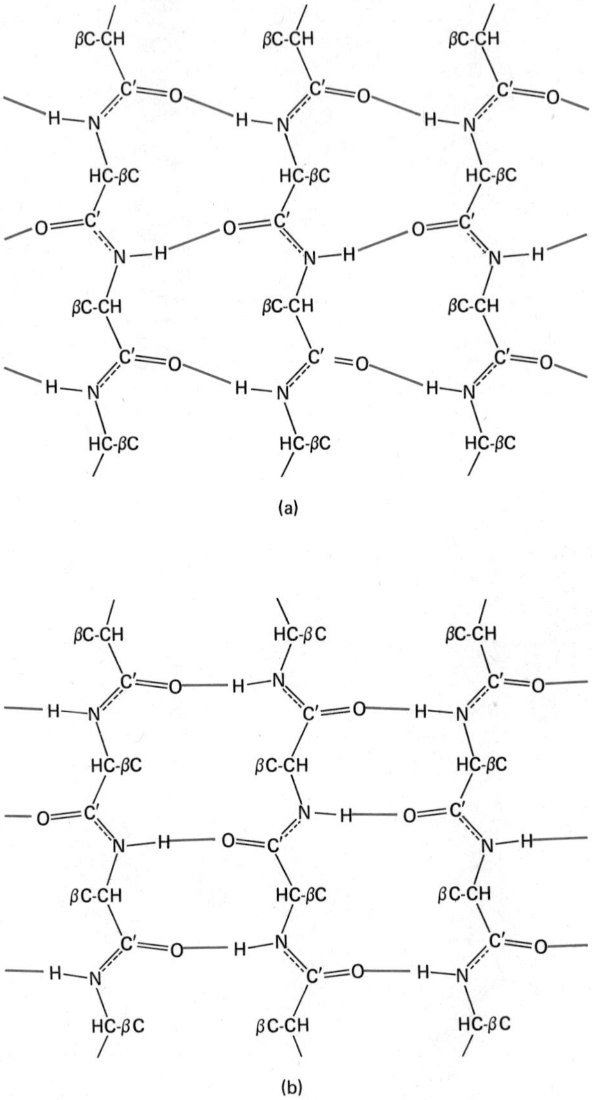

FIGURE 6-10

Hydrogen bonding in the pleated-sheet structures.
(a) *Parallel.* (b) *Antiparallel.*

composition. More than 80% of the residues are glycine, alanine, and serine, and most of the rest are tryptophan. The more regular segments of the molecule can be accounted for by a pleated-sheet structure in which every other residue is glycine; there are no satisfactory models for the regions that contain all of the tryptophan residues.

All of this work on protein structure illustrates a point about scientific inquiry that is not emphasized enough. This endeavor is motivated as much by aesthetic considerations as any artistic work. The best models tend to be beautiful and elegant, even though they must be judged primarily on their ability to explain and predict.

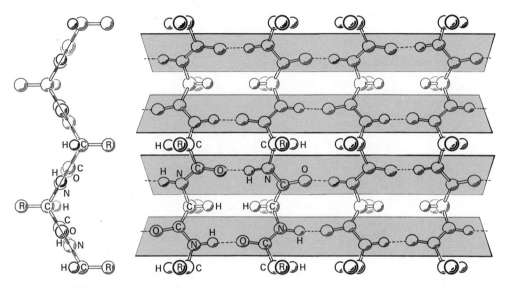

FIGURE 6-11

Perspective views of a pleated sheet. The tilted planes are drawn through the peptide bonds, and the R groups extend above and below the backbone.

The classification schemes for elementary particles that have been invented by physicists are based on considerations of symmetry and simplicity. The Watson–Crick model for DNA is among the most elegant of ideas, and again it is based on a concept of symmetry. Work on the fibrous proteins has also been motivated by a desire to find simple, beautiful, and symmetrical structures; it happens that the protein models are not only pleasing to the eye but to the sense of simplicity as well. By means of two or three structures, we understand a great diversity of apparently unrelated materials from many sources, and we find that at the molecular level they are all essentially alike.

6-8 DIGRESSION: THE TETRAPYRROLES

Even with modern techniques, an enormous amount of time and effort is required to determine a protein structure, and so the most interesting proteins have been selected first. Among these are some proteins that carry oxygen and other gases in animals, the **myoglobins** and **hemoglobins.** These are conjugated proteins whose prosthetic group is a tetrapyrrole molecule.

Most of the **tetrapyrroles** are groups of four pyrrole rings joined by methene (—CH=) bridges. The two major members of the group are **chlorin** and **porphin**. A **porphyrin** is a porphin ring with a specific set of side chains on positions 1 through 8. There are many possible porphyrins, but they are chiefly of interest in pure chemistry; all of the porphyrins normally found in biotic systems are related to the so-called etioporphyrin III class, which have methyl groups on positions 1, 3, 5, and 8 and ethyl groups on the other four.

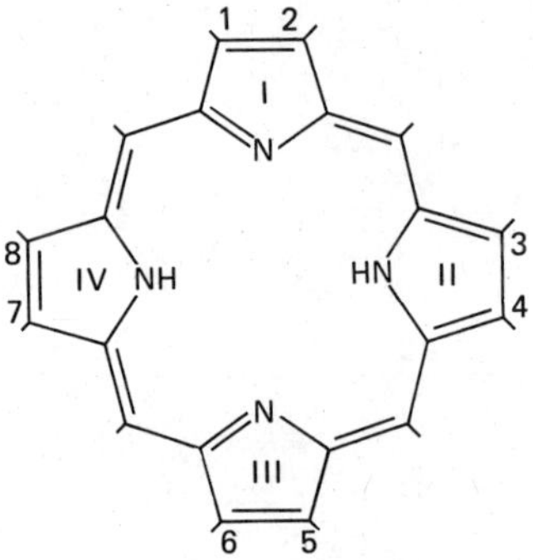

porphin

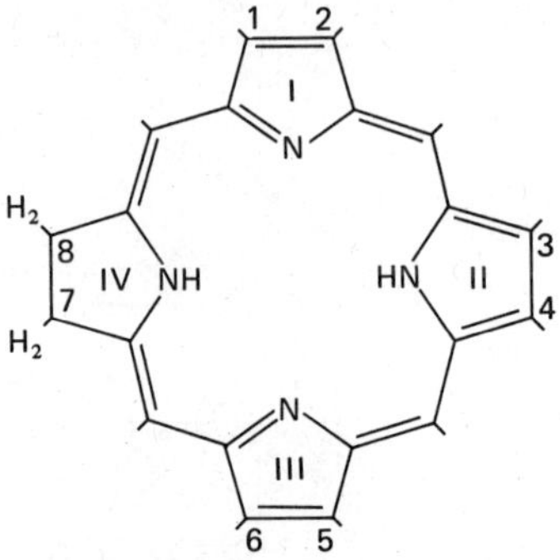

chlorin

The tetrapyrroles are important because they can chelate a metal ion in the center of the ring between the four nitrogen atoms. Almost any metal ion can be bound in this way, but only four are found in nature. A copper porphyrin, turacin, forms a red component in the feathers of some birds. A cobalt complex, cobalamin, is the nucleus of vitamin B_{12}. The magnesium complexes of chlorin are the all-important *chlorophylls*. But the most widespread are the iron porphyrins or **hemes,** which are the prosthetic groups of the hemoproteins.

Hemes are named for the porphyrin from which they are derived; for example, most hemes in nature are iron chelates of protoporphyrin IX and are called protohemes. If the oxidation state of the iron is known, the heme may be called ferroheme [for Fe (II)], ferriheme [for Fe (III)], or ferrylheme [for the rare Fe (IV)]. An Fe (II) atom can accept 12 electrons, which may be donated by six coordinating groups (*ligands*). In a porphyrin, four of the ligands are the ring nitrogens that lie in a plane around the iron; the other two may be groups on opposite sides of the plane, and these groups largely determine what the heme does. In a **hemochrome** complex, both of these groups are nitrogen atoms, and in a protein these nitrogens are in the imidazole rings of histidines, so the heme is buried in the middle of the protein somewhere. Alternatively, only the fifth ligand is a histidine nitrogen and the sixth position is

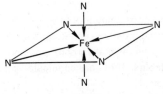

hemochrome

"open" so it can be occupied by many different groups. The hemoglobins and myoglobins are the latter type; they can carry oxygen, CO_2, CO, water, and other small ligands in their sixth position and this makes them useful in transporting gases from outside the body through the blood to the tissues and back out again. The degree of binding to a hemoglobin or myoglobin heme is apparently primarily a function of the partial pressure of the gas, and there are simple equilibria for competitive binding of different gases. Thus, oxygen is primarily bound at the outside, where its tension is high, and released in the tissues where its tension is low, while CO_2 produced in the tissues does just the opposite.

Iron (III) can also bind six ligands, but when it is coordinated in this way it still has a positive charge and can form an additional ionic bond to a negative group. Hemoproteins that contain ferriheme are enzymes called *catalases and peroxidases* which catalyze oxidations by peroxide (H_2O_2).

In the hemoproteins called **cytochromes,** the heme group is a hemochrome that undergoes reversible oxidation and reduction from the Fe (II) to the Fe (III) states. These are very rapidly oxidized and reduced and a string of them can transport electrons somewhat like a bucket brigade. The cytochromes are the essential components of all electron-transport systems. They are always built into membranes in association with specific lipids.

We will now consider the one hemoprotein whose structure is known more completely than any other, myoglobin.

6-9 MYOGLOBIN STRUCTURE

The structure of whale myoglobin has been determined by John Kendrew and his colleagues; it serves as a model for contemporary thought about the structure of globular proteins in general. Figure 6-12 shows the general form of the molecule with the side chains omitted for clarity, although the positions of virtually all of the atoms are now known.

There are several straight sections, lettered A through H, that appear to be helical. To see their shape more clearly Kendrew asked his computer to print out the electron density along a cylinder directed through one of these sections (Fig. 6-13). The agreement with an α-helix could hardly be better; the dimensions are almost exactly those proposed by Pauling and Corey.

There are nonhelical regions between the α-helical stretches. Four of these bends contain the four prolines of myoglobin; since proline has a slightly different form from other amino acids, it produces a bend in any polypeptide backbone and cannot fit into an α-helix. At other bends, various residues are used in special ways to provide stability. In two places, the hydroxyl groups of serine and threonine form hydrogen bonds with nearby amino groups of the backbone. A similar bond is formed in one place by a lysine bonded to a carboxyl group and in another by a tyrosine bonded to an amino group.

Hydrogen bonds of this kind are clearly important in stabilizing the structure of the molecule, but again, they are not as important as had once been thought. Kendrew has asked his computer to print out lists of all possible hydrogen bond contacts and all possible van der Waals contacts; the latter outweigh the former by far and are clearly the most important factors in stability. One would expect all the nonpolar side chains

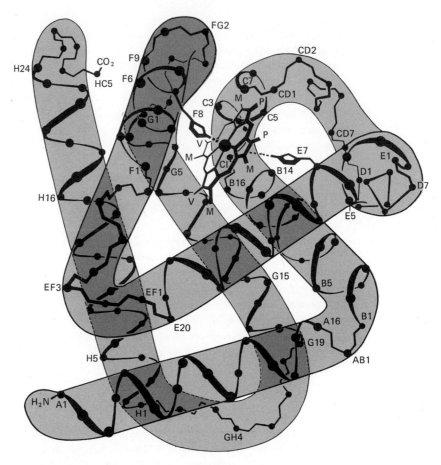

FIGURE 6-12

The general structure of myoglobin. Each helical portion is given a letter and residues are numbered. Nonhelical regions are designated by the letters of the regions they join. (*Courtesy of Richard Dickerson.*)

FIGURE 6-13

Electron density through a helical region of myoglobin, with the cylinder of the helix cut open and spread out flat. The outline of a theoretical α-helix is superimposed. (*Courtesy of Richard Dickerson.*)

of amino acids to be in a hydrophobic region inside and all the polar side chains to be outside; in fact, they tend to be arranged in this way, although not as strongly as one might have predicted. Seventy percent of the polar groups face outward. Forty-five percent of the nonpolar groups are buried, including all of the large groups; the ones that are not buried are mostly small groups, such as the methyl group of alanine.

The center of interest in the molecule is its porphyrin ring; the ring itself has side chains that bond with the protein and it must be a major factor in holding the protein in shape (because of its size). Its fifth ligand position is occupied by a nitrogen of histidine residue F8 and some distance away there is another histidine (E7) that may be able to occupy the sixth position when the protein is in a different conformation. However, in this picture the sixth position is occupied by water. The theory of coordination bonding suggests that the bonding of various ligands in the sixth position may be strongly influenced by conformational changes that alter the positions of other ligands around the iron atom.

We will discuss the matter of conformational changes in Chapter 15 in much more detail, but it is well to start thinking of globular protein molecules as loose structures that can have several different conformations. Furthermore, consider the fact that the surface of the protein is really bristling with at least 20 different kinds of side chains, not counting their modifications and the other molecules that can be added to them. The protein therefore presents opportunities for a wide variety of specific interactions with small molecules and with the side chains of other proteins. It should be easy to see, in a very general way, how a protein can act as an enzyme and how it can perform a variety of other tasks. A protein can easily be designed to interact specifically with any other molecule and this interaction can take the form of binding the other molecule and changing it chemically. However, at the same time the protein may be changed by this interaction and may be forced into a very different shape. The possibilities of these interactions are endless; we will not try to expand on them now, but you should be aware of them.

At the same time, you should be aware that the interactions may be between two protein molecules. Two or more macromolecules can be designed so that their side chains just fit together in a specific way to form a larger structure. Now we can see what self-assembly really means at the molecular level; if two molecules are made from the right sequence of amino acids, they will fold up spontaneously in the right way and they will then associate in the right way, without any further information. Again, this is too great a subject to discuss here and we will delay it for Chapter 15.

The description of the way separate polypeptide chains associate to make larger protein units has generally been called quaternary structure. It may seem strange that we have used primary structure (amino-acid sequence) and quaternary structure, but have ignored secondary and tertiary structure. The trouble is that these terms are falling into disuse. Secondary structure was defined as that due to interactions between the atoms of the polypeptide backbone, which were assumed to be responsible for patterns such as the α-helix and pleated sheets. Tertiary structure was defined as that due to interactions between the side chains of the amino acids, which were assumed to be responsible for the folding of the helices into the larger patterns visible in Fig. 6-11. However, from the work of Liquori, Kendrew, and others it has now become apparent that these distinctions are very artificial. The helical structure of the polypeptide is clearly influenced as much by the side-chain atoms as by those of the backbone, and the stabilization of the myoglobin molecule in its peculiar shape is due to proline residues and to interactions between atoms of the backbone and the side chains, as well as all of the van der Waals contacts throughout the molecule. As so often happens, experimental facts may make theoretical distinctions meaningless.

C. Lipids

6-10   NEUTRAL FATS

Most of the lipids are built around the carboxylic or fatty acids. The important saturated fatty acids have an even number of carbons and fall in the range from about C_{12} to C_{18}. These acids are listed in Table 6-2 for reference. Table 6-3 lists the most important

TABLE 6-2
some natural saturated fatty acids

SYSTEMATIC NAME	COMMON NAME	FORMULA
	acetic	CH_3COOH
	propionic	CH_3CH_2COOH
	n-butyric	$CH_3(CH_2)_2COOH$
n-hexanoic	caproic	$CH_3(CH_2)_4COOH$
n-octanoic	caprylic	$CH_3(CH_2)_6COOH$
n-nonanoic	pelargonic	$CH_3(CH_2)_7COOH$
n-decanoic	capric	$CH_3(CH_2)_8COOH$
n-dodecanoic	lauric	$CH_3(CH_2)_{10}COOH$
n-tetradecanoic	myristic	$CH_3(CH_2)_{12}COOH$
n-hexadecanoic	palmitic	$CH_3(CH_2)_{14}COOH$
n-octadecanoic	stearic	$CH_3(CH_2)_{16}COOH$
n-eicosanoic	arachidic	$CH_3(CH_2)_{18}COOH$
n-docosanoic	behenic	$CH_3(CH_2)_{20}COOH$
n-tetracosanoic	lignoceric	$CH_3(CH_2)_{22}COOH$
n-hexacosanoic	cerotic	$CH_3(CH_2)_{24}COOH$

TABLE 6-3
some natural unsaturated fatty acids

CARBON ATOMS	SYSTEMATIC NAME	COMMON NAME	FORMULA
16	$\Delta^{9:10}$-hexadecenoic	palmitoleic	$CH_3(CH_2)_5CH\!=\!CH(CH_2)_7COOH$
18	$\Delta^{9:10}$-octadecenoic	oleic	$CH_3(CH_2)_7CH\!=\!CH(CH_2)_7COOH$
18	$\Delta^{11:12}$-octadecenoic	vaccenic	$CH_3(CH_2)_5CH\!=\!CH(CH_2)_9COOH$
18	$\Delta^{9:10,12:13}$-octadecadienoic	linoleic	$CH_3(CH_2)_3(CH_2CH\!=\!CH)_2(CH_2)_7COOH$
18	$\Delta^{9:10,12:13,15:16}$-octadecatrienoic	linolenic	$CH_3(CH_2CH\!=\!CH)_3(CH_2)_7COOH$
18	$\Delta^{9:10,11:12,13:14}$-octadecatrienoic	eleostearic	$CH_3(CH_2)_3(CH\!=\!CH)_3(CH_2)_7COOH$
20	$\Delta^{5:6,8:9,11:12,14:15}$-eicosatetraenoic	arachidonic	$CH_3(CH_2)_3(CH_2CH\!=\!CH)_4(CH_2)_3COOH$
24	$\Delta^{15:16}$-tetracosenoic	nervonic	$CH_3(CH_2)_7CH\!=\!CH(CH_2)_{13}COOH$

unsaturated fatty acids. Other acids with hydroxyl groups, methyl groups, and even cyclic regions have been found.

The fatty acids are stored primarily as the depot fats of animals, the oils of seeds, and so on, in the form of neutral fats or triglycerides, which are esters of glycerol with three acids:

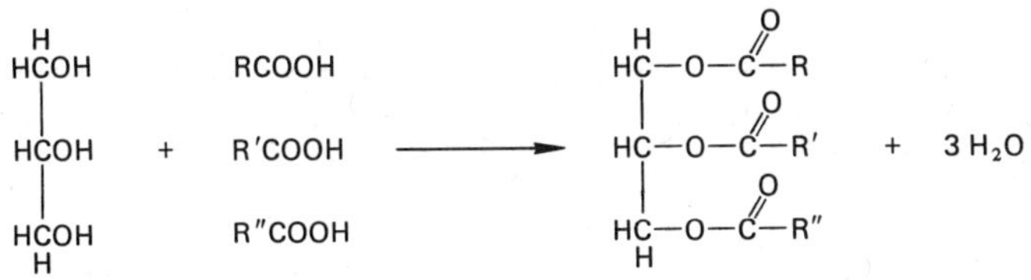

glycerol   fatty acids   triglyceride

where R, R′, and R″ are fatty-acid radicals that may or may not be identical. A molecule made of three oleic acids is called triolein, one of three palmitic acids is tripalmitin, and so on. A mixed molecule may be designated, for example, β-stearo,α,α'-diolein, meaning that it has a stearate residue on the middle carbon and oleate residues on the two outer carbons.

There is growing evidence that triglyceride structures occur regularly in which ether bonds (—C—O—CH$_2$—R) are substituted for ester bonds. Such compounds have been detected here and there and given special names, but they are now showing up so regularly that they are becoming commonplace and we will assume that any glyceride linkage may be either an ether or ester. An ether bond would have to be formed from a fatty aldehyde, and these must also be normal constituents of the cells that make glycerides.

The neutral fat deposits not only form food stores as an insurance against hard times (hibernation, overwintering, the early stages of seed germination, etc.) but are also protective agents. Masses of fat are shock absorbers that protect against sudden blows and provide insulation against the cold; the blubber of whales and other marine animals is primarily for insulation. The *waxes* are neutral fats that are also used for protection (as coverings over the leaves and petals of plants) and for structure (as in bee hives); their general formula is R—CO—O—CH$_2$—R′, where R and R′ may be very long fatty-acid residues, up to C$_{36}$.

6-11 PHOSPHOLIPIDS

The more important and interesting compounds of fatty acids contain phosphates and are called **phospholipids.** There are two main series, as summarized in Table 6-4. The

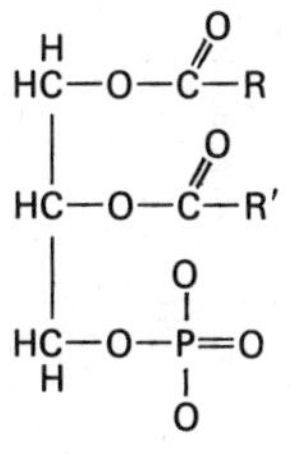

α-phosphatidic acid 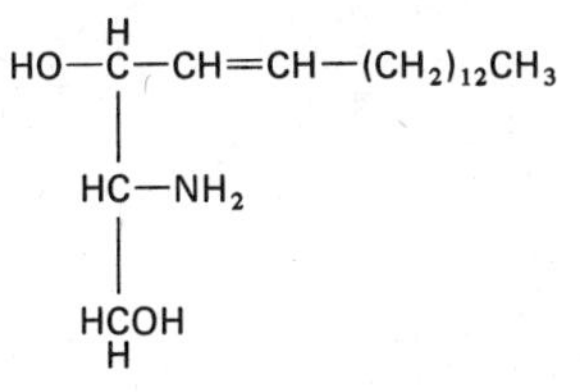   sphingenine 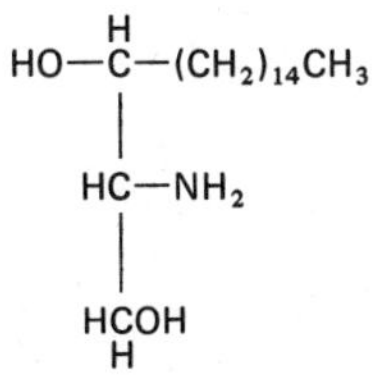   sphinganine

TABLE 6-4

structures of the simple phospholipids

Class / X =	Choline	β-Ethanolamine	L-Serine	Glycerol	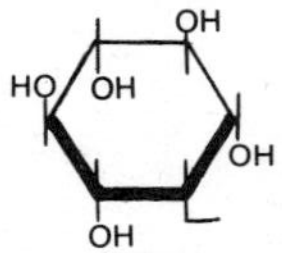 myo-Inositol
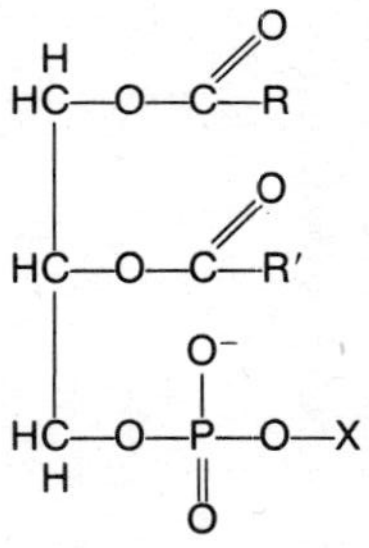 Phosphoglycerides	glycerophosphoryl-choline (GPC) or phosphatidyl-choline	glycerophosphoryl-ethanolamine (GPE) or phosphatidyl-ethanolamine	glycerophosphoryl-serine (GPS) or phosphatidyl-serine	glycerophosphoryl-glycerol (GPG) or phosphatidyl-glycerol	glycerophosphoryl-inositol (GPI) or phosphatidyl-inositol
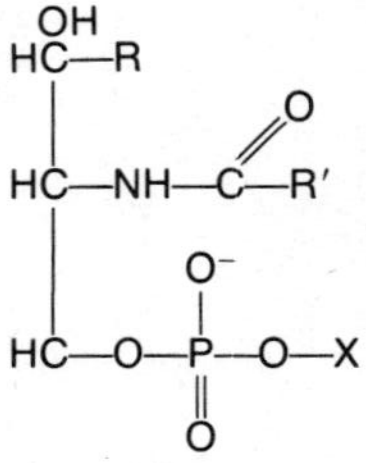 Phosphosphingosides	ceramide phosphoryl-choline	ceramide phosphoryl-ethanolamine	ceramide phosphoryl-serine	ceramide phosphoryl-glycerol	ceramide phosphoryl-inositol

Choline: $-(CH_2)_2-\overset{+}{N}(CH_3)_3$

β-Ethanolamine: $-CH_2CH_2NH_3$

L-Serine: $-CH_2CH(COO^-)NH_3$

Glycerol: $-O-CH_2-HOCH-HOCH_2$

phosphoglycerides are all considered derivatives of α-phosphatidic acid, in which the first two positions are occupied by fatty acids in ester linkage (or fatty aldehydes in ether linkage) just as in the neutral fats. The **phosphosphingosides** are built on serine instead of glycerol. Serine bonded to certain hydrocarbons forms a **sphingosine,** of which there are several kinds. A sphingosine with another hydrocarbon peptide bonded to C-2 is a **ceramide.** Any of the phosphorylated bases found in the phosphoglycerides may then be bonded to the C-1 position.

One of the larger lipids frequently found is a triglycerol phosphatide, **cardiolipin,** which is about as close as the lipids ever come to forming polymers. Lipids do not form large, covalently bonded structures like proteins and polysaccharides. Their importance lies in the extensive, hydrophobically bonded membranes that they form. We will discuss these in more detail in Chapter 7 in relationship to membrane structure, but it is important to see how they can be expected to associate with one another.

$$
\begin{array}{lll}
R{-}CO{-}OCH_2 & \quad\quad \overset{\displaystyle O}{\overset{\displaystyle \|}{C}}{-}O{-}\overset{}{P}{-}O{-}CH_2 & \\[4pt]
| & \quad\quad | \quad\;\; \underset{\displaystyle O^-}{|} \quad\; | & \\[4pt]
R'{-}CO{-}OCH & \quad\quad HCOH & HCO{-}CO{-}R''' \\[4pt]
| \quad\quad \overset{\displaystyle O}{\overset{\displaystyle \|}{}} & \quad\quad | & | \\[4pt]
H_2C{-}O{-}\overset{}{P}{-}O{-}CH_2 & H_2CO{-}CO{-}R'' \\[4pt]
\quad\quad\;\; \underset{\displaystyle O^-}{|} &
\end{array}
$$

Choline, ethanolamine, and serine are strongly charged molecules in neutral solution. The first two carry strong positive charges, and serine is amphoteric, with a basic amino group and an acid carboxyl group that are both charged. Moreover, all of the phospholipids have very acidic phosphate groups that are negatively charged. All of these strongly polar groups are concentrated at one end of the phospholipid molecule, so this end is very hydrophilic and tends to form ionic bonds with water and with polar groups of proteins and other molecules. However, the rest of the molecule consists of long hydrocarbon chains that are hydrophobic and tend to associate with one another through van der Waals forces. The phospholipids may be considered *amphiphilic* because of this dual affinity, one end for water and the other for other hydrocarbons. We therefore expect phospholipid molecules to form molecular associations in water that are rather similar to globular proteins, with their polar groups in association with the water and their hydrocarbons turned inward to form regions from which water is excluded. This is generally what we do find, and in Chapter 7 we will evaluate the importance of these associations for membrane structure.

D. Carbohydrates

6-12 MONOSACCHARIDES

All of the basic sugars may be derived conceptually from either *glyceraldehyde* or *dihydroxyacetone.* The designation D or L is set by comparison with glyceraldehyde; in the standard drawing, the hydroxyl group on the asymmetric carbon atom farthest from the carbonyl group is drawn to the right for the D series and to the left for the L series.

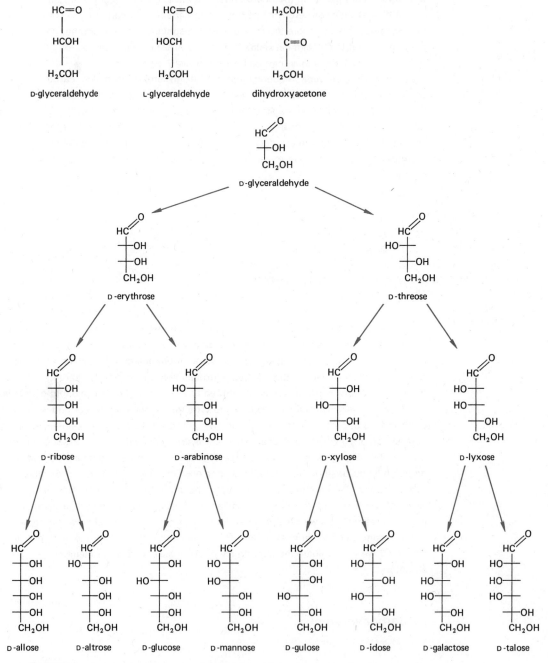

FIGURE 6-14

The D-aldose family. The L-aldoses are the mirror images of these.

Then, starting with D-glyceraldehyde we mentally add one HCOH group at a time next to the carbonyl group; since this group is asymmetric, the hydroxyl group can point to the right or the left. Therefore, there are two different D-*tetroses* (with four carbons), four D-*pentoses* (with five carbons), eight D-*hexoses* (with six carbons), and so on; these are shown in Fig. 6-14. All of the sugars generated in this way are called **aldoses.** We can do exactly the same thing by starting conceptually with dihydroxyacetone; this produces a corresponding family of **ketoses.**

An aldehyde may react with an alcohol to form a *hemiacetal:*

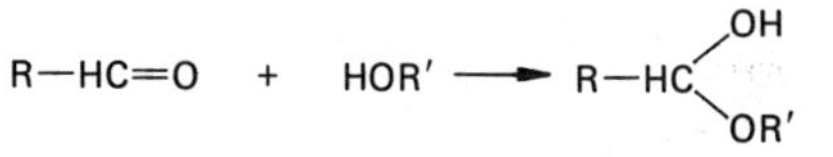

$$R-HC\!=\!O \quad + \quad HOR' \longrightarrow R-HC\overset{\displaystyle OH}{\underset{\displaystyle OR'}{}}$$

which may combine with another alcohol to form a (full) *acetal:*

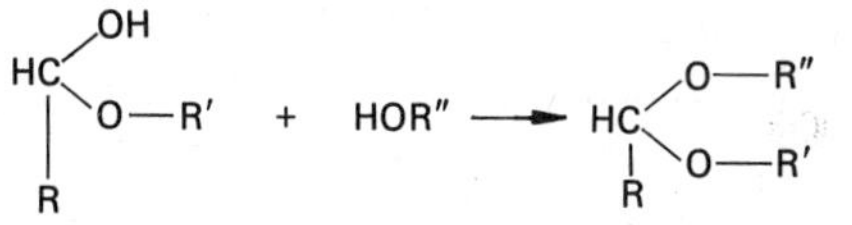

$$HC\overset{\displaystyle OH}{\underset{\displaystyle O-R'}{\underset{\displaystyle |}{}}} \quad + \quad HOR'' \longrightarrow HC\overset{\displaystyle O-R''}{\underset{\displaystyle O-R'}{\underset{\displaystyle |}{}}}$$

The aldehyde group of a pentose or hexose is in a position to form an internal hemiacetal with one of its own hydroxyl groups to make a ring structure:

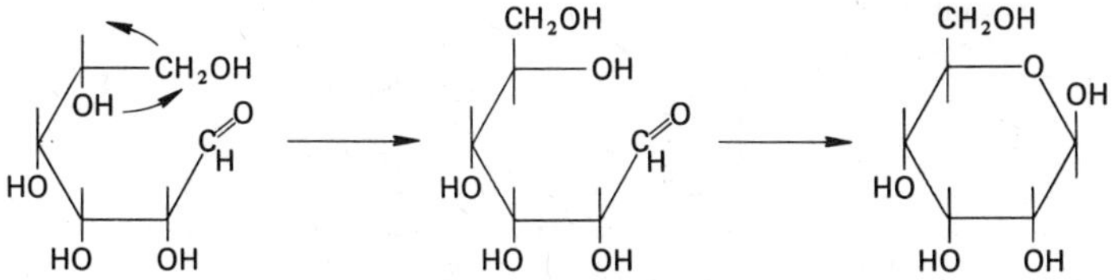

These pentagonal and hexagonal rings are named **furanoses** and **pyranoses,** respectively, after the rings furan and pyran. These are not aromatic rings, so they are not planar; the most stable configuration for a pyranose is the "chair" form shown here. The bonds that can be formed to this ring point either up and down (*axial* bonds, shown by solid lines) or to the side (*equatorial* bonds, shown by dotted lines). All of these bonds point either upward (β bonds) or downward (α bonds) and it is easier to draw the molecules as hexagons and pentagons with substituents above or below.

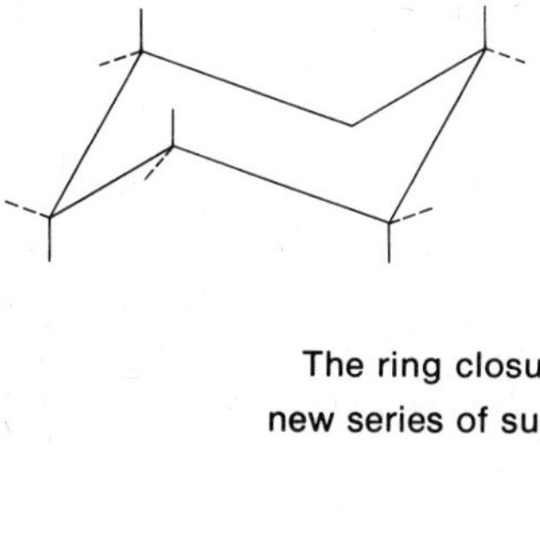

The ring closure produces an additional asymmetric center at C-1 and therefore two new series of sugars designated α and β. Carbon C-1 is then called the **anomeric**

TABLE 6-5

*simple and derived monosaccharides**

Hexoses	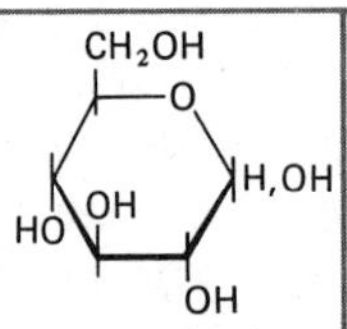 D-glucose (Glc)	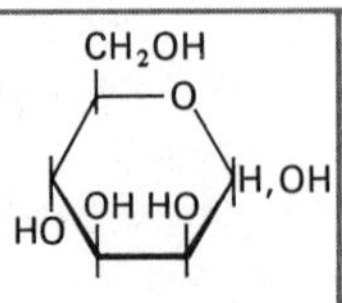 D-mannose (Man)	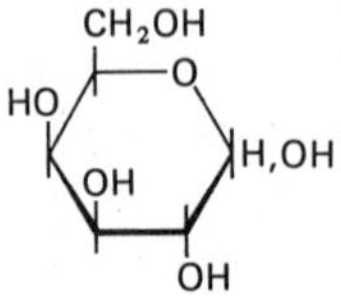 D-galactose (Gal)	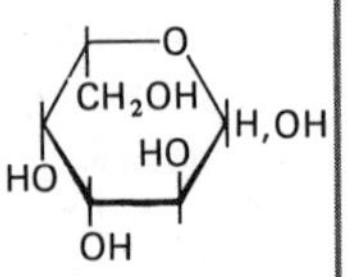 L-galactose (L-Gal)	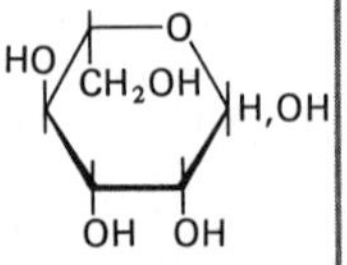 L-mannose (L-Man)
Hexosamines	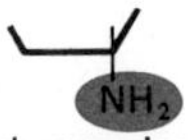 D-glucosamine (GlcN)	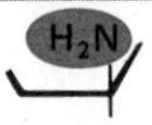 D-mannosamine (ManN)	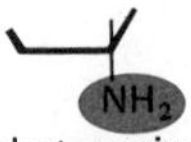 D-galactosamine (GalN)		
N-acetyl hexosamines	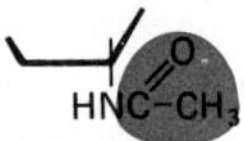 N-acetylglucosamine (GNAc)		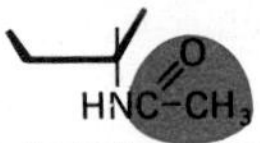 N-acetylgalactosamine (GalNAc)		
Uronic acids	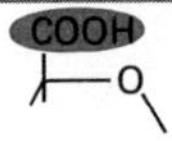 D-glucuronic acid (GUA) 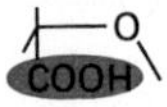 / L-iduronic acid (IdUA)	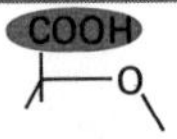 D-mannuronic acid (MUA) 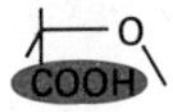 / L-guluronic acid (GulUA)	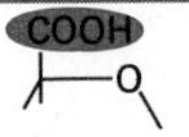 D-galacturonic acid (GalUA)		
Pentoses	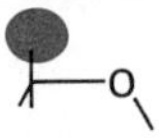 D-xylose (Xyl)		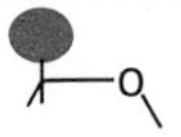 L-arabinose (L-Ara)	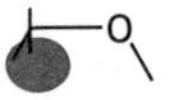 D-arabinose (Ara)	
6-Deoxyhexoses	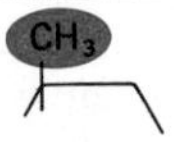 D-quinovose (Quin)		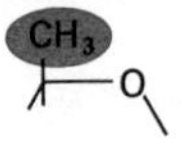 D-fucose (Fuc)	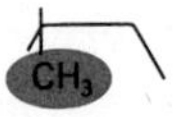 L-fucose (L-Fuc)	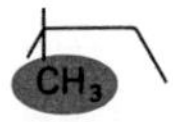 L-rhamnose (L-Rha)
3, 6-Dideoxy hexoses	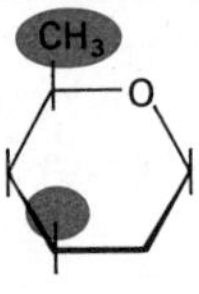 paratose (Para)	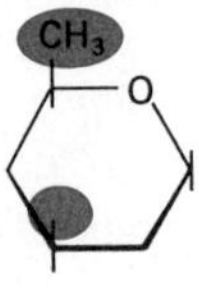 tyvelose (Tyv)	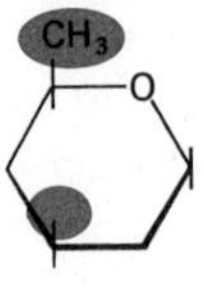 abequose (Abe)	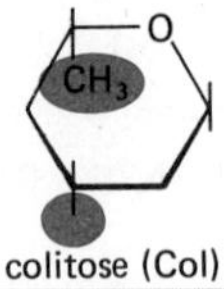 colitose (Col)	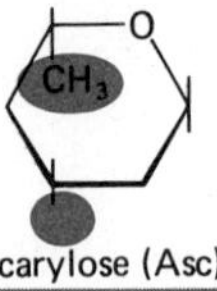 ascarylose (Asc)

*All sugars in a column may be derived conceptually from the hexose in the first row by altering only the part of the ring shown. Notice the standard abbreviations for most of these compounds.

carbon; two sugars that differ only in their configuration there are **anomers.** In the D series, the hydroxyl group at C-1 is trans to the CH_2OH group in the α form and cis to it in the β form, but in a monosaccharide in solution, the ring can open and close continually and so the two anomers exist as an equilibrium mixture along with a small amount of the open aldehyde form. We will generally denote the structure of this equilibrium mixture by $|$ H, OH at C-1.

One cannot transform directly from the straight-chain structures to the rings with two-dimensional drawings, so to avoid errors we will follow some rules. We will usually draw the ring with O at the back and C-1 to the right, and in this case the —CH_2OH group is up in D-sugars and down in L-sugars. In α-D-glucose, the hydroxyls are alternately up and down, and this will serve as a reference.

Most hexoses assume a pyranose form; the furanoses have never been isolated in solution. While we ought to include the ring designation in every name (as α-D-glucopyranose), we will generally omit it and assume the pyranose form with one exception: the substituted forms of fructose are always furanoses.

The simple sugars can be modified in several ways to make derived sugars that are very important in polysaccharides; the most important of these are summarized in Table 6-5. Notice that most of them are related to D-glucose, D-galactose, and D-mannose. The most common modifications are

1. Removal of an oxygen atom to make a *deoxy* sugar.
2. Oxidation at C-6 from CH_2OH to COOH to make a *uronic acid*.
3. Replacement of a hydroxyl group by an amino group.
4. Substitution of the amino group by either a sulfate or an acetyl group.

6-13 GLYCOSIDES AND GLYCANS

Just as a hemiacetal can combine with an alcohol to make an acetal, a sugar that has formed an internal hemiacetal with itself can combine with an alcohol to make an acetal analogue called a **glycoside.** The alcohol residue is an **aglycone** and the bond between it and the sugar is a **glycosidic bond.** Methyl groups, phosphates, and other small molecules are common aglycones; but the second alcohol can just as well be another monosaccharide and the resulting glycoside is a disaccharide. In general, the second sugar will be able to form another such bond, so oligo- and polysaccharides can be built.

Once the anomeric carbon of a sugar is engaged in a glycosidic bond, the ring is no longer free to open and close; the bond is designated α or β according to the configuration at this position. The sugar is given the suffix "-syl" or "-side" and the bond is written as a directed linkage from the anomeric carbon to the carbon of the aglycone. For example, glucosyl-α(1 $\rightarrow$ 4)-galactose has a bond from C-1 of α-glucose to C-4 of galactose. In a few disaccharides, such as trehalose and sucrose, the glycosidic bond is between two anomeric carbons.

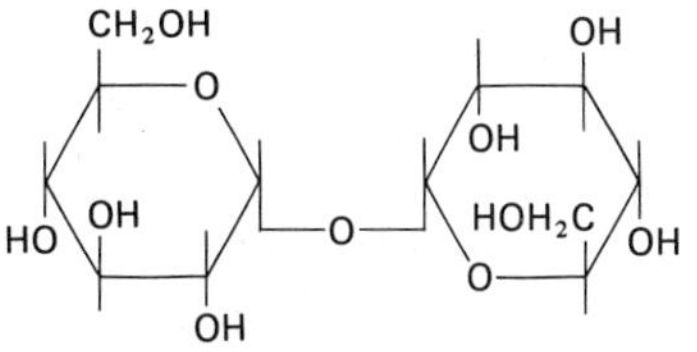

α, α'-trehalose

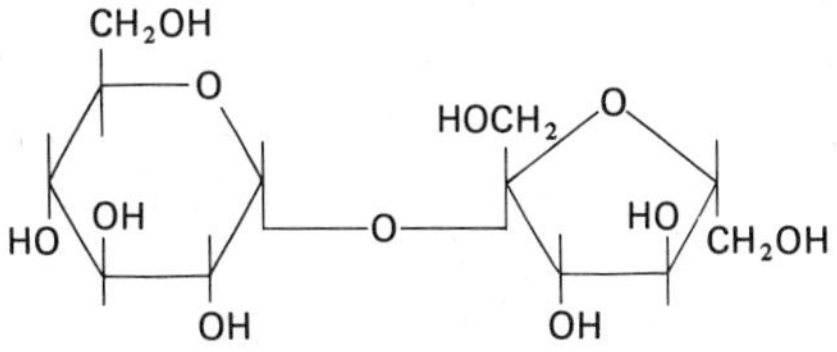

sucrose

There are many different oligosaccharides; almost every plant seems to have some unique di- or trisaccharide, although sucrose is the most common and important one. Those that you may encounter now and then are

maltose	glucosyl-$\alpha(1 \rightarrow 4)$-glucose
cellobiose	glucosyl-$\beta(1 \rightarrow 4)$-glucose
gentiobiose	glucosyl-$\beta(1 \rightarrow 6)$-glucose
isomaltose	glucosyl-$\alpha(1 \rightarrow 6)$-glucose
lactose	galactosyl-$\beta(1 \rightarrow 4)$-glucose

The polysaccharides are the chief structural materials and reserve foodstuffs of most plants, and there is an enormous variety of them. They are named after their monosaccharides by substituting "-an" for "-ose," so a glucose polymer is a *glucan*. "Glycose" is a general word for any monosaccharide, so *glycan* means any polysaccharide. The major groups are as follows:

1. *Homoglycans.* Linear or branched polymers of a single neutral sugar, including glucans, galactans, mannans, fructans, arabinans, and xylans. This group includes the starches and other reserve food materials as well as some polymers of plant cell walls.

2. *Heteroglycans.* Complex polymers containing several sugars, including uronic acids. This group includes plant gums, mucilages, and components of wood.

3. *Glycuronans.* Polyuronic acids, which are parts of the cell walls of plants and algae.

4. *Mucopolysaccharides.* Polymers of amino sugars and uronic acids, which are important lubricants and structural components, primarily in animals.

The polysaccharides play two main roles: as food reserves and as structural materials. The two types differ primarily in the bonds between monomers and between chains, not in the type of monomer. For example, Fig. 6-15 shows the structures of the starch **amylose,** an $\alpha(1 \rightarrow 4)$-glucan, and **cellulose,** a $\beta(1 \rightarrow 4)$-glucan. The starch is a food reserve; it forms an open helix that can be penetrated easily by water molecules. (A well-known test for starch—a blue color in combination with iodine— depends upon iodine molecules entering the intrahelix space where their optical properties change.) The cellulose, however, is a structural material, the chief component of wood. Cellulose molecules are straight chains that align themselves in regular, crystalline arrays with strong hydrogen bonding to make fibers. Water does not penetrate these complexes easily and they are very strong.

Many of the glycans are branched polymers. **Amylopectin** is another starch made of $\alpha(1 \rightarrow 4)$-glucan chains like amylose with about 4% of $1 \rightarrow 6$ linkages (Fig. 6-16). Typical starch deposits are about one-fourth amylose and three-fourths amylopectin and these can be seen as visible granules in cells. Animal cells contain granules of **glycogen,** which is almost identical to amylopectin except that about 9% of its bonds are $1 \rightarrow 6$, so it is more highly branched. There are many other food reserve compounds in other organisms.

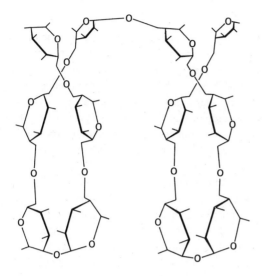

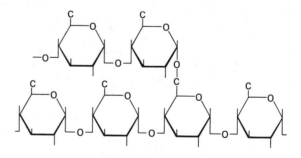

FIGURE 6-15
*The helical structure of amylose
compared with the straight chains
of cellulose.*

FIGURE 6-16
*Structure of amylopectin at a branch
point, with a schematic indication of the
general form of the molecule.*

Aside from the nucleic acids, which are primarily informational macromolecules, biological structures and activities can be understood largely in terms of three types of molecules: protein, lipids, and polysaccharides. The proteins are often fibers, either with an α-helix or a β-sheet structure; otherwise they are irregular knots, although even these often contain a lot of helix. The lipids are essentially long hydrocarbons with a charged, hydrophilic end (generally containing phosphate). The polysaccharides are long sugar chains that can hydrogen bond into a variety of tough structures. Biological systems have selected this small group of compounds out of all the possible organic structures that clever chemists can create.

We feel justified in assuming that when some particular molecule is used to perform a job in a biotic system, there is a very good reason for that selection; we assume that the selection has been made through a long evolution. It is therefore rational to talk about the *purpose* of a structure within this framework, and this way of thought is often an important clue in research. The question, "How would I do this if I were designing the system?" is often a very intelligent one. With a little biological and chemical insight—of the sort we are trying to provide in this book—you can often guess at a reasonable design, and this may suggest the proper experiment. What we earlier called principles of "biomolecular architecture" are just that. They are as significant as any principles that guide a human architect, and it is important to keep them in mind.

SUPPLEMENTARY EXERCISES

exercise 6-2 The amino acid composition of a strain of *E. coli* given here may be considered fairly typical for cells. Since molecular weight × mole fraction = weight fraction, you can easily fill in the last column:

AMINO ACID	MOLECULAR WEIGHT	MOLE FRACTION	WEIGHT FRACTION
Ala	89.1	0.128	
Arg	174.2	0.054	
Asp	133.1	0.100	
Cys	121.1	0.017	
Glu	147.1	0.105	
Gly	75.1	0.079	
His	155.2	0.010	
Ilu	131.2	0.047	
Leu	131.2	0.080	
Lys	146.2	0.070	
Met	149.2	0.034	
Phe	165.2	0.033	
Pro	115.1	0.046	
Ser	105.1	0.062	

AMINO ACID	MOLECULAR WEIGHT	MOLE FRACTION	WEIGHT FRACTION
Thr	119.1	0.048	
Trp	204.2	0.010	
Tyr	181.2	0.021	
Val	117.1	0.055	
Total		1.000	

The total of the weight fractions is the average molecular weight of an amino acid in *E. coli;* now subtract one molecule of water to get the weight of the average amino acid residue: _____.

exercise 6-3 How many amino acid residues are there in the average protein with a molecular weight of (a) 30,000 daltons and (b) 50,000 daltons?

exercise 6-4 If three nucleotides code one amino acid, the average nucleotide residue has a weight of 330 daltons, and the distance between nucleotides averages 3.33 Å, calculate the molecular weights and lengths of RNA molecules which code for proteins with molecular weights of (a) 30,000 daltons and (b) 50,000 daltons.

exercise 6-5 If the smallest RNA virus has a single-stranded genome that codes for two proteins of 50,000 daltons each and one protein of 30,000 daltons, what is the length of its genome?

7

cells: general features and some techniques

The viruses are useful tools for understanding some of the fundamental genetic properties of biotic systems simply because they are little more than genomes that infect cells and multiply; they served well to illustrate some genetic principles because we could ignore the cells themselves, by and large. Now we must begin to think about the cells, with their complex apparatus for producing energy, making a variety of monomers for polymers, and synthesizing proteins and other polymers from them. We will begin by examining cell structure in a little more detail, with some emphasis on the optical techniques that are used here. As with many other subjects that are peripheral to biology, microscopy is primarily an art that must be learned through experience; we will merely touch on some points that are necessary for understanding what microscopes can really show us about cells.

A. Microscopy

7-1 RESOLUTION AND CONTRAST

It is relatively easy to construct a system of lenses to magnify an object many times, but there is little point in doing so if other physical limitations prevent the object's being seen clearly. In fact, there is a major limitation imposed by light diffraction, so that while one can magnify almost without limit, diffraction determines what can be *resolved*.

The image of a point source of light is not a point itself, but a peak of intensity surrounded by rings of lesser intensity (Fig. 7-1). The diffraction rings of two points will interfere with one another if the points are too close, and this sets a physical limitation on the distance to which two points can approach one another and still be resolved. They are said to be resolved if the intensity of their peaks is at least twice the intensity at the point midway between them (Fig. 7-2). This limitation is a function of the lens used to observe the points and the medium through which the light travels.

If there is a medium of refractive index n between a point light source and a lens, and if the half-angle of the cone of light is α (Fig. 7-3), the **numerical aperature** of the lens (NA) is defined by $NA = n \sin \alpha$. If the wavelength of the light is λ, the resolution d of the lens is $d = 0.61\lambda/NA$. The resolution of a microscope that uses a series of lenses with different numerical aperatures is calculated from the average of the separate NA's.

197

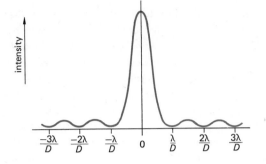

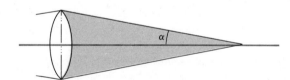

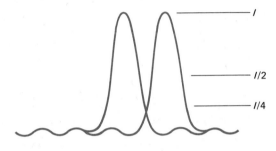

FIGURE 7-1
The intensity of light of wavelength λ from a point source of diameter D is a major peak surrounded by diffraction rings.

FIGURE 7-2
Two point sources at the minimum separation at which they can be resolved. The sum of their intensities at the midpoint is just half of their maximum intensities.

FIGURE 7-3
Numerical aperature of a lens.

exercise 7-1   A lens 2.0 mm in diameter has a focal length of 13.8 mm. Calculate (a) its *NA* in air and (b) its resolution for light with a wavelength of 500 nm.

There are obviously only a few ways of increasing the resolution of a lens system. The refractive index of the medium can be raised by substituting a clear oil, for which $n \sim 1.5$, for the air between the object and the objective lens. The angle α can be increased within limits by using higher-power lenses, with shorter focal lengths. The wavelength of the light can be decreased slightly. The resolution with green light ($\lambda = 550$ nm) in a good compound microscope is about 0.2 μm; this can be decreased to about 0.1 μm by using ultraviolet light as short as $\lambda = 250$ nm. An ultraviolet microscope requires quartz lenses, but it is difficult to use light with a shorter wavelength because the optical path must be evacuated.

The other principal microscopic limitation is *contrast*, for it does little good to resolve two points if they do not contrast enough with their surroundings to be seen. In the simplest microscopic techniques, contrast is achieved by staining with agents that have more affinity for some kinds of cellular material than others. These techniques have been discussed in Chapter 1. It should be pointed out that in ultraviolet microscopy, contrast arises because the light is absorbed strongly by certain materials in the cell, so artificial stains are not necessary.

7-2 INTERFERENCE MICROSCOPY

Any object whose refractive index differs from that of the surrounding medium will cause a phase shift in the light passing through it relative to the light passing through the medium. This would make every such object contrast very well with its surroundings if our eyes or our photographic plates could perceive phase differences, but these receptors can only see differences in intensity; therefore, phase differences have to be transformed into intensity differences, and there are several types of microscopes classed as *interference* instruments that make this possible. They all depend upon the principle that two rays of coherent light (i.e., rays that originate from the same source) of the same wavelength will interfere with one another if they are out of phase and add to one another if they are in phase (Fig. 7-4).

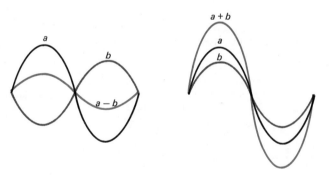

FIGURE 7-4

Summation of the intensities of two light rays, a and b. *At left, the rays are out of phase by 180° and their sum* (a − b) *is a wave of lower intensity than either. At right, they are in phase and their sum* (a + b) *is a wave of higher intensity than either.*

The simplest interference instrument is the **phase microscope.** As Fig. 7-5 shows, the light source is shielded by an annular opening; light passes through the opening and through the specimen and is focused on the back focal plane of the objective lens. As it passes through the specimen, it is broken into a diffracted and an undiffracted beam; the former spreads out across the back focal plane of the objective, while the latter is concentrated in the narrow annular image of the light source. This undiffracted beam is then focused on a silvered annulus in a plate at the front focal plane of the objective, where its intensity is cut down to a level comparable to the intensity of the diffracted beam. The two beams are then brought back together at the image plane.

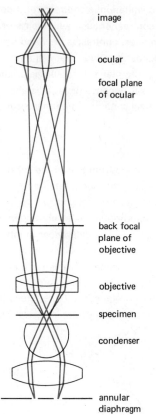

image

ocular

focal plane
of ocular

back focal
plane of
objective

objective

specimen

condenser

annular
diaphragm

FIGURE 7-5

Optical system of a phase microscope. Undiffracted rays are solid lines, diffracted rays are dashed.

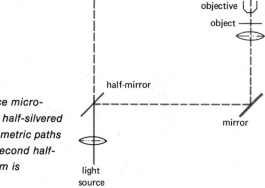

image

mirror   objective

blank   half-mirror

objective

object

half-mirror

mirror

light
source

FIGURE 7-6

Schematic optics of an interference microscope. The light is split at the first half-silvered mirror and passes along equal geometric paths but different optical paths to the second half-silvered mirror where a single beam is reformed.

The waveform of the undiffracted beam is represented by $s = a \cos \omega t$ and the waveform of the diffracted beam is represented by $s = a \cos (\omega t + \phi)$, where ϕ is the phase-angle difference between the two and $\omega = 2\pi/\lambda$. Each beam will travel a different optical path, which is the product of the geometrical path times the refractive index of the medium. The difference is given by $\Delta = -\lambda\phi/2$. The beams generally arrive at the image plane about $\frac{1}{4}$ to $\frac{1}{2}$ wavelength out of phase, and when superimposed they show either a dark object on a lighter field or a light object on a darker field, with excellent contrast in a good, well-adjusted instrument.

Even finer definition is achieved by an **interference microscope** in which the two beams are separated mechanically before entering the specimen, rather than being separated by the specimen. Figure 7-6 shows schematically the arrangement in the

Dyson interference microscope. A half-silvered plate is introduced below the specimen and another one above it, to split the light into a beam running through the specimen and another beam through the background nearby. If there were no specimen, the two beams would arrive together in phase at the image plane, but a specimen in one path will bring that beam out of phase with the other. There will be maximum contrast between them when they are out of phase by an integral number of half-wavelengths; in practice, the microscope is arranged so the phase difference can be altered and measured. This allows one to measure the *retardation* Φ of one beam relative to the other:

$$\Phi = (n_s - n_0)t \tag{7-1}$$

where t is the thickness of the specimen, n_s is its refractive index, and n_0 is the refractive index of the medium.

It is relatively easy to measure n_s by using sets of oils with different refractive indices. When the specimen is mounted in an oil of the same refractive index, it becomes invisible except for diffraction at its edges. The retardation Φ is determined by the microscope, and since n_0 is also known, the thickness of a specimen can be measured. This is an important application of the microscope, but measurements of the mass of a specimen are even more significant. Since the area of a specimen can be measured and t can be obtained from equation 6-1, the mass can be obtained if the density of the specimen is known. This, in turn, can be obtained from the refractive index.

The refractive index of pure water is 1.3330, that of a 1% solution of the protein egg white albumin is 1.3348, and that of a 2% solution of albumin is 1.3366. Every 1% increase in protein concentration is thus accompanied by a **specific refraction increment** of 0.0018, defined by

$$\delta = \frac{(n_s - n_0)}{C} \tag{7-2}$$

where C is in grams per 100 ml of solution (weight percent). Similar measurements for other proteins give values of δ between 0.00170 and 0.00195, with 0.0018 being a good average for a protein mixture. If the mass of a cell can be approximated by the mass of its protein, we can proceed as follows. From equations 6-1 and 6-2,

$$\Phi = \delta t C \tag{7-3}$$

Since C is measured in grams per 100 ml, we multiply by 100 to put everything in terms of grams per milliliter (g/ml). But milliliters, a volume, can be replaced by area $\times$ thickness, At, and we can rewrite equation 6-3 as

$$\Phi = \frac{100 \delta g t}{At}$$

or

$$g = \frac{\Phi A}{100 \delta} \tag{7-4}$$

The mass of an object can therefore be obtained from optical data. If the specimen is nonuniform, its mass can be calculated by integrating local masses over the entire area. Data obtained in this way are in excellent agreement with data from nonoptical sources.

exercise 7-2 Human red blood cells are circular discs about 7.5 μm in diameter. They contain essentially one protein (hemoglobin) for which $\delta = 0.00193$. Calculate the mass of a red blood cell if 550-nm light is retarded by $\frac{1}{4}$ wavelength in passing through it.

7-3 POLARIZING MICROSCOPY

Most biological materials are *anisotropic,* which means that their structure is different in different planes. This can be seen in differences in the indices of refraction of these materials to polarized light passing through them in different planes. If a beam of polarized light passes through an anisotropic material, it is split (Fig. 7-7) into an *ordinary beam* that continues as before, in the same plane of polarization, seeing an index of refraction n_0, and an *extraordinary beam* vibrating at right angles, seeing an index of refraction n_e. The two beams will travel different optical paths, and, as before, the difference at the image plane will be $\Delta = (n_e - n_0)t$. The quantity $(n_e - n_0)$ is called the **birefringence** of the specimen.

Birefringence can be either positive or negative; its value is characteristic of many materials and can be an important indication of the composition of a cellular structure. Birefringence can also have its origin in several different factors. *Intrinsic* birefringence reflects directional differences in basic molecular structure; for example, the amino and carbonyl groups of a polypeptide may all extend in one direction while the principal axis of the backbone is at right angles. *Form* or *textural birefringence* reflects similar orientation of molecules, as when many protein molecules line up in parallel. Strain birefringence arises from deformations of materials that might otherwise be isotropic, due to tension and pressure. In theory, one can distinguish different sources of birefringence through a careful, quantitative analysis of a specimen, but we will not go into this analysis here.

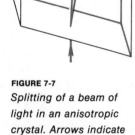

FIGURE 7-7
Splitting of a beam of light in an anisotropic crystal. Arrows indicate the plane of vibration of each beam.

7-4 ELECTRON MICROSCOPY

Electron microscopy is truly an art in itself; we can hardly begin to consider its techniques in these few pages, but it is useful to look briefly at some of the methods used in preparing specimens for electron microscopy, particularly at methods for bringing out contrast in the specimens, so we have a general idea of what one is really looking at in viewing an electron micrograph.

Electron beams are sensitive to differences in the electron density of specimens, but the differences in density between the various CHNOPS atoms in biological materials are trivial as far as a beam of electrons is concerned. The beam cannot distinguish between proteins, lipids, and nucleic acids because they are all made of essentially the same materials. Contrast is only achieved by using heavy metals and depositing them on or in the specimen so they occur only at selected points. There are basically three ways to do this.

Shadow casting was one of the first techniques to be developed. As Fig. 7-8 shows, a heavy metal such as platinum or palladium is evaporated from a filament so it is deposited over the specimen. The beam of atoms is equivalent to a light beam in creating a shadow behind the specimen which stands out in contrast to the metal background. Electron micrographs of such specimens resemble aerial photographs of a landscape taken when the sun is near the horizon, and objects that rise above the surface can be seen very well. The height of a specimen can be calculated easily from the geometry of the shadow-casting apparatus.

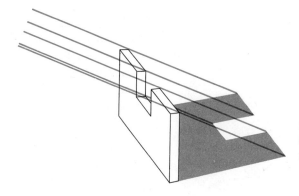

FIGURE 7-8
Shadowing a block with vaporized metal for contrast in electron microscopy.

Shadow casting has been used primarily for viewing little particles, such as viruses and discrete pieces isolated from cells, that are large enough to stand above the background and make distinctive shadows. They also lend themselves to a structural analysis through model building. If an object seen in a shadow-cast electron micrograph appears to have a recognizable shape, it is easy to build a model with the hypothetical shape and photograph it on a black background, with the only illumination coming from a small spotlight aimed at the same angle as the shadowing apparatus. This photograph should look very much like the micrograph if the object really has the shape of the model, and of course the model can be refined continuously until it gives the best possible picture.

The very new and useful technique of **freeze-fracturing** and **freeze-etching** also makes use of shadowing. Figure 7-9 shows the basic operations. One advantage of the method is that the specimen is first frozen quickly; there is no need for extensive fixation and staining that may change basic structures enormously in unknown ways. It is then struck sharply with a cold microtome blade to fracture it and make a clean break. Sometimes this face is *etched* by leaving the specimen in a vacuum for a few minutes so that water can sublime off and leave nonaqueous structures exposed. A replica of the face is then made by shadowing with carbon and a heavy metal, such as platinum, and this replica is removed and mounted for electron microscopy. This procedure has been most valuable in investigations of membrane structures, for it apparently preserves the membrane in a rather pristine condition through freezing and then splits the membrane open and exposes its components to the shadowing.

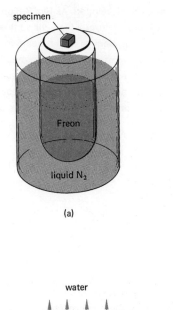

specimen

Freon

liquid N₂

(a)

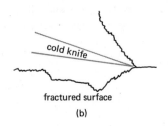

cold knife

fractured surface

(b)

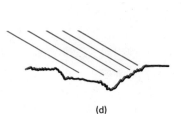

water

(c)

(d)

FIGURE 7-9

The freeze-fracture and freeze-etch technique. (a) The specimen is attached to a copper disk and frozen rapidly. (b) It is placed in a cold vacuum chamber and fractured with a sharp blow from a cold knife (−185°C). (c) If desired, the exposed surface is etched for a few minutes in the vacuum. (d) The surface is then shadowed with platinum and carbon to make a film which is examined microscopically.

The most widely used technique for biological materials is still **positive staining.** The osmium compounds that are such excellent fixatives also have a high-enough electron density to produce dark images in electron micrographs. These compounds are supplemented by a variety of others, such as potassium permanganate and lead hydroxide. Most of the electron micrographs of cells that are used in this book have been made from positively stained material.

Another new and useful technique is **negative staining.** Figure 7-10 shows how an object standing above the surface can be surrounded completely with a stain, such as a 1% phosphotungstic acid solution, so it stands out as a light object in contrast

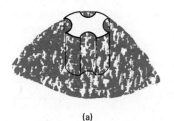

(a)

(b)

FIGURE 7-10

Negative staining of a cog-shaped piece for electron microscopy. (a) The piece is surrounded by a droplet of phosphotungstic acid. (b) Its appearance in the electron microscope.

to its dark surroundings. This method has been very successful in revealing the structures of viruses and some subcellular components.

Like any other microscopic method, electron microscopy is constantly plagued by the problem of artifacts. The most common stains and fixatives are fairly well understood now, so we know their limitations and are not likely to be fooled by the artifactual changes they produce. However, we must always be wary, particularly in looking at the finest structures, for objects that appear to be only a few nanometers in diameter may be artifacts of the interaction between the stain or fixative and the native structure. It is also important to realize that everything we see is being bombarded by a powerful electron beam and no one really knows what reactions might be occurring between molecules subjected to so much energy. In spite of these limitations, electron microscopy has revealed a cellular ultrastructure that is coherent and consistent with other observations.

B. Cell boundaries

A cell is essentially defined by its boundary, which is a membrane that separates everything inside from everything outside. One of our major tasks will be the development of a picture of membrane structure; we will begin here by considering some properties that cells would have if their membranes were little more than thin cellophane films, and gradually we will make this picture more and more realistic.

7-5 CHEMICAL POTENTIALS

Work can always be expressed as the product of a driving force times the distance through which it acts. Since free energy is a measure of a system's ability to perform work, the driving force is always some function of free energy. A few examples will illustrate this point:

1. When a mass m is lifted in a gravitational field of acceleration g, it acquires the ability to do work because of its height h; the higher it is lifted, the more work it can do. The driving force in this case, mg, is actually equal to dG/dh, at constant pressure and temperature, and the maximum work the mass can do is $mg \, \Delta h$.

2. When a gas is compressed in a cylinder of volume V, it acquires the ability to do work because of its pressure p. The driving p can be shown to be equal to dG/dV, at constant pressure and temperature, and the maximum work that the gas can do is $p \, \Delta V$.

3. When an elastic cord is stretched by a length l, it acquires the ability to do work because of its tension τ; in this case the driving force τ is equal to dG/dl, at constant tension and temperature, so the maximum work the cord can do is $\tau \, \Delta l$.

There ought to be a similar function of free energy that depends upon the concentration of a substance, c, so on analogy with the other work functions we define the **chemical potential** μ as dG/dn at constant pressure and temperature, where n

is the number of moles. The potential μ is sometimes called the partial molar free energy, and it is useful in measuring the tendency of a diffusible substance to move from a region of high concentration to one of low concentration, for the work performed in this case is $\mu \, \Delta n$. For reference, we define the standard chemical potential μ^0 as the potential of a 1-m solution; then the chemical potential at any other concentration is

$$\mu = \mu^0 + RT \ln c \qquad (7\text{-}5)$$

The energy required to transfer 1 mole of material from a point where its concentration is c_1 to a point where its concentration is c_2 is

$$\Delta G = \mu_2 - \mu_1 = RT \ln \frac{c_2}{c_1} \qquad (7\text{-}6)$$

Most of the molecules we have to deal with are charged, and so a difference in their distribution across a region produces an electrical potential ψ. But, in fact, there is no way to separate an electrical potential from a chemical potential in this case, so we must define an **electrochemical potential** $\bar{\mu}$:

$$\bar{\mu} = \mu + z\mathscr{F}\psi \qquad (7\text{-}7)$$

where z is the number of charges per molecule or ion. If there is no electrical potential, $\bar{\mu}$ is obviously equal to μ.

With these definitions in mind, we can consider some properties of solutions that can be related to some general properties of cells.

7-6 OSMOTIC PRESSURE

A 1-M glucose solution exhibits several physical properties that are different from those of pure water. It boils at 100.514°C and freezes at −1.86°C. It is easy to show that these changes in boiling and freezing points are functions of the amount of glucose in solution, so a 2-M glucose solution freezes at −3.72°C, for example. Moreover, the changes seem to be independent of the nature of the solute, since equimolar solutions of sucrose, glycerol, ethanol, etc. all have the same freezing and boiling points. The temperature differences $\Delta T = -1.86°$ and $\Delta T = 0.514°$ are called the **molar freezing-point depression** and the **molar boiling-point elevation,** respectively.

Solutions of electrolytes do not seem to fit the rule for solutions of nonelectrolytes, such as sugars. A 1-M NaCl solution freezes at −3.46°C, for example; but this is very close to twice the freezing-point depression of a glucose solution, and this value $(2 \times -1.86° = -3.72°)$ is just what we would expect if every molecule or ion in solution has the same effect as every other, regardless of their mass, because NaCl dissociates into two ions. The ratio $0.93 = 3.46/3.72$ is a correction factor which reflects the fact that the salt does not dissociate completely. Following the same line of reasoning, a 1-M solution of $CaCl_2$ should freeze at $3 \times -1.86° = -5.58°$; it actually freezes at −4.80°, requiring a correction factor of 0.86.

Freezing-point depression and boiling-point elevation are two of the **colligative properties** of solutions; these properties depend only upon the number of molecules or ions in solution, not upon their chemical nature or their mass. The changes in freezing and boiling point actually occur because the vapor pressure of the solvent is proportional to its mole fraction X; if n_s is the number of moles of a substance in a mixture that consists of n moles altogether, then $X_s = n_s/n$. As the number of solute molecules increases in a solution, the mole fraction of solvent molecules falls, and its vapor pressure falls proportionally. There is at least one other colligative property, the **osmotic pressure** of the solution, that also changes in proportion to these variables, and this has some biological importance.

Suppose we set up a 1-M glucose solution separated from pure water by a semipermeable membrane (Fig. 7-11) that allows water molecules to pass freely but retards glucose molecules. The chemical potential of water is now considerably higher

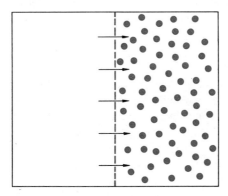

FIGURE 7-11
Osmosis of water from a pure solvent into a solution separated by a semipermeable membrane.

in the pure solvent than in the solution, so there is a diffusion of water molecules into the solution; this diffusion is called **osmosis.** Because the walls of the containers are rigid, an osmotic pressure builds up in the solution until the system finally comes to an equilibrium where the flow of water into the solution is balanced by the flow of water back into the solvent. The magnitude of this pressure is

$$\Pi = \frac{nRT}{V} \qquad (7\text{-}8)$$

where n is the number of moles of glucose and V is the volume of the solution compartment; the appropriate value of R is 0.082 liter atm/mole deg. At 273°K, the pressure exerted by a 1-M solution is 22.4 atm.

To provide a simple basis for comparing the osmotic pressures of different solutions, we define the **osmolarity** of a solution, O, as the product of its molar concentration c, the number of ions into which the solute theoretically dissociates i, and the correction coefficient, ϕ:

$$O = \phi ic \qquad (7\text{-}9)$$

Then the osmotic pressure is proportional to osmolarity:

$$\Pi = \phi icRT = ORT \qquad (7\text{-}10)$$

Just as we talk about the number of *moles* of a substance dissolved in a solution, we can talk about the number of *osmoles*. Two solutions are **isosmolar** if they contain the same number of osmoles per unit volume—in other words, if they exert the same osmotic pressure. Of two solutions with different osmotic pressures, the one with the greater pressure is **hyperosmolar** and the one with the lower pressure is **hyposmolar.**

Cells would exhibit an osmotic pressure if they were nothing but cellophane bags; however, if their contents were simply glucose solutions, they would exhibit only an **osmotic transient** when placed in water, for their osmotic pressures would rise initially and then fall to zero as glucose molecules diffused out into the solvent. This is what we actually observe with bags of glucose solutions. A bag containing a NaCl solution shows a similar osmotic transient, but the pressure falls much faster, which indicates that Na⁺ and Cl⁻ ions pass through the cellophane faster than glucose molecules. This is an important observation, and the rate at which an osmotic transient occurs has been used to characterize real cell membranes.

7-7 DONNAN EQUILIBRIA

A slightly more realistic cell model is shown in Fig. 7-12. Proteins are too large to diffuse through a cellophane membrane and they are typically charged; we begin with a bag containing a solution of negatively charged proteins balanced by an equal number of Na⁺ ions. The solution outside is an ordinary NaCl solution with equal

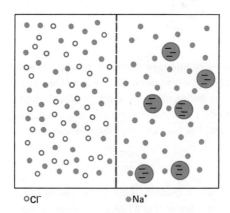

○Cl⁻   ●Na⁺

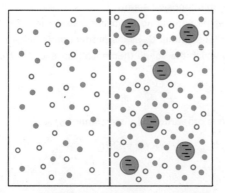

FIGURE 7-12
Donnan equilibrium. (a) A sodium-proteinate solution is separated from a NaCl solution. (b) Ions diffuse into the protein solution to make a concentration difference across the membrane at equilibrium.

numbers of Na^+ and Cl^- ions. This system eventually comes to an interesting equilibrium, and since it was first discussed by Sir Frederick Donnan, it is called a **Donnan equilibrium.**

Since the bag is freely permeable to both ions, x moles of Cl^- ions will diffuse inward accompanied by an equal number of Na^+ ions. At equilibrium, the system must satisfy the condition

$$[Na^+]_i[Cl^-]_i = [Na^+]_o[Cl^-]_o \qquad (7\text{-}11)$$

From the equilibrium concentrations shown in Fig. 7-12,

$$([Na^+]_i + x)(x) = ([Na^+]_o - x)^2$$

and solving for x, we obtain

$$x = \frac{[Na^+]_o^2}{[Na^+]_i + 2[Na^+]_o} \qquad (7\text{-}12)$$

Equation 7-12 allows us to calculate the final distribution of ions from any set of initial conditions. It is instructive to apply this equation to some representative cases.

exercise 7-3 Calculate the final concentrations for Donnan systems that begin with the following initial concentrations:

INITIAL		FINAL			
$[Na^+]_i$	$[Na^+]_o$	$[Na^+]_i + x$	$[Na^+]_o - x$	x	$[Cl^-]_o - x$
0.5	1.0				
1.0	1.0				
1.0	0.5				
0.1	1.0				

exercise 7-4 For several of the above distributions, calculate the ratios $[Na^+]_i/[Na^+]_o$ and $[Cl^-]_o/[Cl^-]_i$ at equilibrium. Do you think there will be any electrical potential across the membrane in any of these cases?

As Exercise 7-4 demonstrates, the ratios of the ions in a Donnan equilibrium are such that there can be no electrical potential. However, one of the most important characteristics of real cells is a transmembrane potential of several millivolts, which accounts, for example, for the ability of nerve cells to carry signals. It is surprisingly simple to improve our model so it develops a potential; we need only use a membrane material that is more permeable to one ion than the other. For example, if the membrane is permeable to cations but not anions, then the anions become unequally distributed at a concentration c_1 on one side and c_2 on the other side, and the more concentrated side becomes negative with respect to the other. The condition for equilibrium in this case is that the electrochemical potential shall be zero, and from equations 7-6 and 7-7 the electrical potential across the membrane will be

$$\psi = \frac{RT}{z\mathscr{F}} \ln \frac{c_1}{c_2} \qquad (7\text{-}13)$$

From your solution to Exercise 3-6, you will recognize that a tenfold difference in concentration will produce a potential of about 60 mV. The measured potentials of several animal and plant cells are actually around 50–100 mV.

It should now be clear that we can explain quite a few properties of cells by using rather simple nonbiological models. However, most of the membrane phenomena we shall study depend upon the specific biomolecular organization of real cell boundaries, so we will now go one step further and look at some of the properties of these boundaries.

7-8 TONICITY AND CELL ENVELOPES

During the latter half of the nineteenth century, Carl Nägeli, one of the most farsighted and influential biologists of his time, began investigations of cell surfaces that laid the foundations for all future work in this field. Among these early experiments were some that anyone can repeat with little more than a good microscope.

Elodea is a common water plant whose thin leaves are built of neat, rectangular cells that are in constant cyclosis. The solution bathing a bit of leaf on a microscope slide can be changed quickly just by absorbing the old solution with a paper towel, and by substituting different sucrose or NaCl solutions for one another you can see some remarkable effects. The cells look quite normal in 0.25-*M* NaCl, but as stronger salt solutions are substituted, they begin to **plasmolyze**—they shrink (Fig. 7-13) and their insides pull away from their walls. When more dilute salt solutions are used, the cells appear to swell, though their strong walls keep them from expanding much.

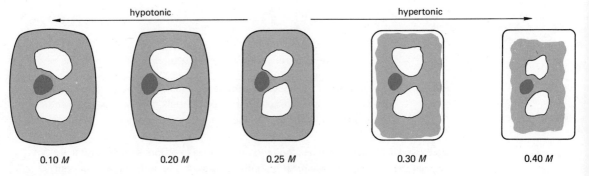

FIGURE 7-13
Behavior of a plant cell as it is changed to more hypotonic solutions (to the left) and more hypertonic solutions (to the right).

Nägeli observed these phenomena, and he gave the name *plasma membrane* to the internal envelope that can be seen during plasmolysis. These experiments indicated that the plasma membrane is selectively permeable, like the cellophane membranes we have just studied, and does not allow solutes to pass through as rapidly as water. A solution of 0.25-*M* NaCl is **isotonic** to the cells, for their shape remains normal in this concentration of salt. A more dilute solution is **hypotonic** and a more concentrated one is **hypertonic.** We avoid the terms based on the word "osmolar" in this case because we are describing the behavior of cells, not measuring osmotic pressures, and isosmolarity may be very different from isotonicity. For one thing, the osmolarity or tonicity depends upon the solute molecules that the membrane "sees," and every membrane may be different in this respect. We also find that cells do not act like perfect osmometers; that is, their volumes do not change in solutions of known osmolarity exactly as one would predict. Some of the discrepancy can be explained by excluding a certain volume of the cell from consideration, perhaps by saying that the volume occupied by the nucleus, mitochondria, and other cellular structures is different from the rest of the cell.

From 1890–1899, E. Overton employed plasmolysis to investigate the rates at which materials enter plant cells. If plant roots, which have very fine hairs in which plasmolysis and cyclosis can be seen, are suspended in a hypertonic solution, they will show an osmotic transient. The root hair cells become plasmolyzed and then return to their normal volume as the external solute enters. Continued cyclosis is an indication that the cells are not permanently injured. After many years of work, Overton stated a few general rules about permeation rates. For small, uncharged molecules, the rate of penetration is inversely proportional to molecular size; this implies that the molecules are passing through pores in the plasma membrane. Larger uncharged molecules penetrate at a rate proportional to their solubility in lipids; this is an important result, for it indicates that the membrane is built partially of lipids. Finally, charged molecules of any kind enter very slowly; this observation suggests that the pores in the membrane carry fixed charges at their boundaries that repel charged solute molecules.

This picture of a lipid membrane with charged pores is still quite valid, but we have had to modify it considerably and revise our ideas about permeation. For example, it is not quite true that charged molecules always enter slowly. Inorganic ions, in particular, are transported very actively and specifically across cell membranes, as studies using radioactive tracers have shown. Most cells concentrate K^+ ions to many times their concentration in the surrounding media, while simultaneously excreting large amounts of Na^+ and Cl^- ions. For example, a typical measurement on an animal nerve cell showed that the ratios inside/outside, in millimoles per liter, are as follows: K^+, 345/10; Na^+, 72/455; and Cl^-, 61/540.

Many cells have been shown to transport specifically sugars, amino acids, and many other substances, and bacterial mutants have been found that have lost the ability to transport certain compounds. Thus, the plasma membrane is not merely a passive barrier that excludes or admits solutes; we will develop the thesis that it carries highly specific sites that recognize various compounds and move them in or out of the cell, and that these sites are proteins that are essentially indistinguishable from enzymes. However, we will not develop this picture of membrane structure until Chapter 16, because it requires considerably more background information about the structure of biological molecules. For now, we will just go one step further and see what membranes look like with modern electron microscopy.

exercise 7-5   Red blood cells (erythrocytes) do not have strong walls and their contents are mostly the red protein hemoglobin. A few drops of an erythrocyte suspension are added to NaCl solutions ranging from

M/8 to M/13, and after a few minutes, the solutions are centrifuged. The M/8, M/9, and M/10 solutions are clear, with all of the red color in a pellet at the bottom, while the M/11, M/12, and M/13 solutions are reddish. The latter are said to show hemolysis; what has happened in the dilute solutions, and what concentration of NaCl is most nearly isotonic to the erythrocytes?

7-9 THE UNIT MEMBRANE CONCEPT

After Overton's experiments had indicated that cell membranes contain lipid and the observations of others had shown that they contain protein and probably some carbohydrates, W. J. Schmidt suggested a general arrangement for these molecules. Schmidt developed polarizing microscopy to a fine art and applied it to the **myelin sheath** that surrounds most nerves (Fig. 7-14). This sheath is built of many closely packed membranes, and it has been a favorite subject for studies of membrane structure for this reason. Schmidt could show that there is one component with the birefringence of lipid oriented radially (perpendicular to the membrane surface) and another component with the birefringence of protein oriented at right angles (parallel to the membrane surfaces). Francis O. Schmitt, Richard Bear, and their associates came to similar conclusions after doing x-ray diffraction studies on myelin sheaths and on red blood cells.

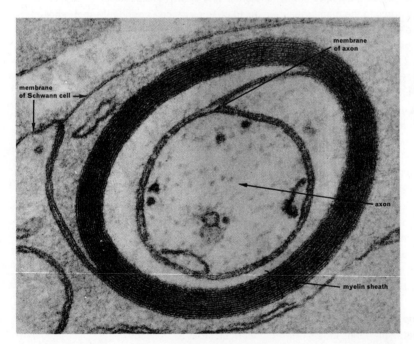

FIGURE 7-14
Cross section through a myelin sheath.
(Courtesy of J. D. Robertson.)

As we pointed out in Chapter 6, we expect lipid molecules to form associations in which the hydrocarbon chains are in a hydrophobic region from which water is excluded and the polar groups are in a hydrophilic region in association with water.

The properties of soap solutions—the K^+ and Na^+ salts of the fatty acids—have been studied extensively, and associations called mesophases are found in them that have just these properties. The most stable arrangement at soap concentrations near 0.01 M is shown in Fig. 7-15a; as the concentration increases to 50–60%, the mesophase tends to become the bimolecular layer of Fig. 7-15b. These layers can be very extensive and quite stable, and this is the lipid form generally proposed as a model for the lipid part of a cell membrane.

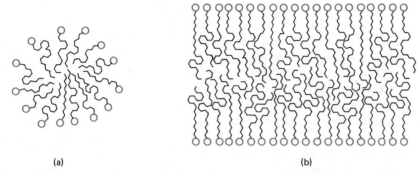

(a)   (b)

FIGURE 7-15
Cross section through typical soap mesophases; circles represent polar carboxyl groups and zigzag lines are hydrocarbons. (a) A micelle form. (b) A smectic or bimolecular-layer form.

In fact, the lipids of cell membranes are primarily phospholipids. The phospholipids can also form bimolecular layers and they may do so in membranes. This has been the chief assumption upon which pictures of membrane structure have been founded.

While the optical studies of membranes were going on, J. F. Danielli and Hugh Davson were measuring the permeability, surface tension, and electrical conductivity of membranes, and on the basis of their data and the optical results, they proposed a famous and important model in which the membrane is likened to a "fat sandwich," in which the bread is protein and the filling is lipid. The lipid is assumed to be in a bimolecular layer form, with the polar ends of the molecules in contact with the polar groups of the protein. This model agreed very well with the available data; it was modified somewhat over a period of years, but there was never any need to discard it. Furthermore, it agreed very well with the first pictures of membranes taken by electron microscopy. Humberto Fernandez-Moran and J. B. Finean made a very careful study of myelin sheath structure and found essentially the pattern shown in Fig. 7-14. The membranes appear to be sandwiches of two dark lines with a light space in the middle, and with osmic acid fixation the width of this sandwich is around 150–170 Å, depending on the amount of water in the tissue. The Davson–Danielli model did not really specify the number of bimolecular layers that exist between the proteins; if one assumes that the proteins are stained by the osmic acid and that there are two layers of lipid with varying amounts of water between the proteins, the electron micrographs agree very well with the model.

When the myelin sheath is stained with KMnO₄, an additional line appears in the middle of the light zone of each membrane. J. D. Robertson looked at the myelin more closely with KMnO₄ fixation, keeping in mind some newer information from Betty Ben Geren's work on the formation of the sheath. As Fig. 7-16 shows, the sheath is layed down on a nerve cell by a Schwann cell that wraps itself around many times, leaving just its cell membranes in close association with one another. If you look closely at this picture, you can see that the Schwann cell alternately brings two inner faces of the membrane together, then two outer faces, then two inner faces again, and so on. Robertson therefore said that the 150–170-Å unit consists of *two* membranes face to face. Osmic acid only stains the inner protein layer of the membrane, but KMnO₄ brings out the outer one as well. This indicates that the inner and outer layers of the membrane are somewhat different, and since osmic acid stains carbohydrates poorly, it suggests that the outer face contains more carbohydrate.

The dark-light-dark sandwich revealed by KMnO₄ staining is just about 75 Å wide, which is the theoretical thickness of one lipid bilayer and two coats of protein. On the basis of his work and related observations by many others, Robertson proposed to call this structure the **unit membrane** and suggested that the structure shown in Fig. 7-17 is a basic and ubiquitous membrane structure. He showed that the same trilaminar dark-light-dark pattern, with a width of about 75 Å, can be seen in many different membranes and suggested that all membranes have essentially the same structure.

Robertson's unit-membrane concept was an important development of the basic Davson–Danielli model. But in recent years it has become all too clear that this model does not jibe with the facts and that it is too simple. For example, the model is based upon the assumption that osmic acid binds primarily to the polar regions of lipids and proteins, but Edward Korn and others have shown clearly that osmic acid acts primarily on the double bonds of the hydrocarbon chains. It binds across the double bond and can bind two neighboring hydrocarbon chains together by forming bonds to four carbons simultaneously. It is also an oversimplification to say that all

FIGURE 7-16
Formation of the myelin sheath by a Schwann cell.

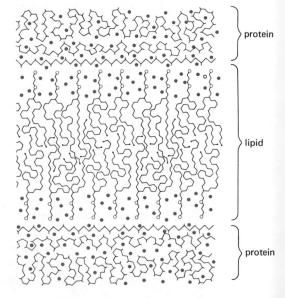

FIGURE 7-17
Postulated arrangement of lipid and protein molecules within the unit membrane. The black dots indicate regions where osmium compounds are assumed to bind. [*Courtesy of Walther Stoeckenius, from* The Interpretation of Ultra-structure *(R. J. C. Harris, ed.), 1962, reprinted by permission of Academic Press, Inc., New York.*]

FIGURE 7-18
Fracture faces of myelin sheath membranes, 100,000X. Notice how smooth and regular the membranes are, in comparison with Fig. 7-19. (Courtesy of Daniel Branton.)

membranes look alike in the electron microscope; in fact, many different widths can be seen and there are indications here and there of a finer structure that does not fit the unit-membrane concept. Perhaps the most telling evidence against the unit-membrane concept is the fact that the trilaminar pattern can be seen clearly in membranes from which most of the lipid has been extracted, which shows that basic membrane structure does not depend upon the lipids. It is interesting to note in passing that the myelin sheath, which has been such a favorite subject for study, is really quite abnormal in composition when compared to most other membranes; it is primarily lipid, with much less protein, whereas most membranes are primarily protein.

Some of the main reasons for deemphasizing the unit-membrane concept at this time have to do with the organization of cellular metabolism. It is clear that enzyme systems are associated with membranes in many cases—perhaps most cases. Following this line of thought, it is apparent that every membrane ought to be different because it is carrying different functional units. The differences can be seen most clearly by using the freeze-fracture technique. Figures 7-18 and 7-19 show

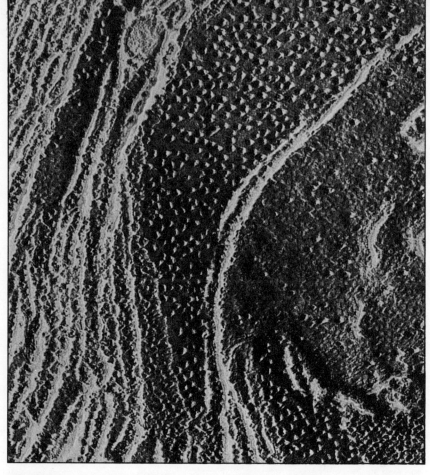

FIGURE 7-19

Fracture faces of spinach chloroplast membranes, 130,000X. Notice the abundance of globular structures, which are presumably protein complexes. (Courtesy of Daniel Branton.)

myelin sheath membranes and chloroplast membranes that have been freeze-fractured; the differences are obvious, and they are what you would expect if you consider that chloroplasts are highly organized systems of enzymes and pigments.

We have not gone through this historical build-up of a lovely membrane picture just for the purpose of letting you down at the end and leaving you with the idea that the membrane is again a mystery and we do not know how it is built. The fact that a trilaminar structure appears so regularly must mean that membranes have some kind of underlying pattern; it is simply not likely to be the unit-membrane pattern. We will be looking at many different cells and we will see many important membranous structures; you should be able to look at these pictures and recognize the existence of the trilaminar pattern wherever it appears. You should keep this pattern in mind as we develop a picture of biomolecular architecture that will include the cell membrane; but we will develop this picture against a background of some principles of molecular structure and metabolism.

There is one last general matter to be considered. Cells are membranous bags, and their membranes are largely responsible for determining what their contents are and how they operate. But why, in view of all this, are cells so small? Why aren't bacteria 10 times larger than they are? Why isn't a mouse made out of a few very large cells instead of many tiny cells? You can practically answer the question for yourself after you solve Exercise 7-6.

exercise 7-6 Calculate the surface areas, volumes, and surface/volume ratios for cubes with the following edge lengths (in arbitrary units):

EDGE	SURFACE	VOLUME	SURFACE/VOLUME
1			
2			
5			
10			
20			
50			
100			

Since surface area is a function of the square of a linear dimension and volume is a function of the cube of a linear dimension, the surface/volume ratio falls by a factor of 100 as the edge increases by a factor of 100. In a cell, metabolism takes place throughout; metabolism demands an input of materials from outside the cell and produces wastes and byproducts that must be removed to the outside. But the rate at which transport to the cell occurs is a function of the surface area. Therefore, cells must stay small enough to retain a favorable surface/volume ratio. Bacteria are a good size from this point of view; they can carry out all of the necessary exchanges with their environment simply by virtue of diffusion and specific transport processes in their membranes. But most eucaryotic cells that are very active have to develop complex infoldings of their membranes to achieve a more favorable surface area, and some of the protozoa whose cells are very large have developed specialized exchange mechanisms. When we come to organisms the size of most plants and animals, we find that special transport systems such as the circulatory system, the lungs, kidneys, and the various vessels of higher plants have been developed to facilitate exchange of materials from one part of the organism to another and from the interior of the organism to the environment.

8

the
procaryotic
cell

We are now ready to examine bacterial cells in detail. These are relatively simple cellular systems and we have enough information about them now to give an outline of their structure and behavior. We will continue to concentrate on bacteria such as *E. coli* that have been used for most of the experimental work, in the hope that a firm understanding of one system will allow us later to examine other organisms more easily.

A. Cell structure

8-1 RIBONUCLEIC ACID

After one has looked at many micrographs of cells, one of the most striking features of electron micrographs of *E. coli* and related bacteria is their apparent lack of structure. The cells have one or more nuclear bodies, a wall and membrane complex with several layers, and not much else. However, the lack of organization is only apparent; it indicates that most electron micrographs of slices through bacteria are simply not good enough to show all of the fine detail we would like to see.

In the best pictures, all of the cell outside the nuclear region appears to be filled with little black dots. Even with better stains and higher magnification, it is hard to see any structure in these dots, and the best way to investigate them and other cell components is by isolation and chemical analysis. Most cell components can be purified quite well by **differential centrifugation;** in this procedure, broken cells are subjected to a series of centrifugations at higher and higher gravitational forces to make relatively homogeneous pellets. The material sedimented by the first centrifugation consists of the largest pieces; as the gravitational force increases, smaller and smaller components are brought down. There is not much to the process in bacteria because there are so few distinct components. A low-speed centrifugation, at about 5000g, brings down a pellet with almost all of the cell wall and membrane and a major part of the DNA. The supernatant contains most of the protein, including enzymes, and practically all of the RNA.

When a bit of the supernatant is put on top of a sucrose density gradient, it separates into fractions like those shown in Fig. 8-1. Most of the protein remains near the top of the tube. However, most of the RNA appears in two or three peaks lower in the tube; these peaks also contain a lot of protein. The two peaks that always appear have sedimentation coefficients of 30S and 50S; as the Mg^{++} concentration is raised, these peaks start to disappear and a peak at 70S increases in size. This indicates that the 30S and 50S units are pieces that aggregate to make the 70S unit when there is enough Mg^{++}.

219

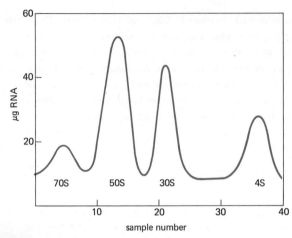

FIGURE 8-1
Distribution of RNA species in a sucrose gradient.

FIGURE 8-2
Four enlarged views of 50S ribosome units, shadowed with tungsten, with a model of the 50S unit photographed by low-angle light. The model seems to conform very well with the surface features of the ribosomes. (Courtesy of Roger G. Hart.)

The 70S particles are called **ribosomes.** They are the dots that appear in electron micrographs of cells, and when isolated and observed with very high magnification, a little of their structure can be seen (Fig. 8-2). The ribosomes isolated from bacteria are about two-thirds RNA and one-third protein; their 50S units contain an RNA molecule that weighs 1.1 million daltons and their 30S units contain an RNA half this size. The whole 70S complex weighs about 2.6 million daltons.

The first indications of ribosome function really came from eucaryotic cells, where ribosomes are attached to membranes and can be separated from the rest of the cell only as a "microsome" fraction which contains RNA, protein, and lipid. Phillip Siekevitz separated different eucell fractions and tested them for their ability to synthesize protein; his criterion was the ability to incorporate radioactive amino acids into acid-precipitable material, which is now a standard criterion of protein synthesis. He found that the microsome fraction had the highest activity by far and that the activity remains with the RNA-protein units, the ribosomes, when they are separated from the lipid membranes. This made perfectly good sense, since RNA had already been implicated in protein synthesis. It is now clear that *the ribosomes are the protein factories of all cells.*

You will notice that there is a fraction of RNA amounting to about 20% of the total that remains near the top of the sucrose gradient. We will call it "soluble RNA" for now; it is important in protein synthesis and was really discovered by investigators who were studying this process.

exercise 8-1   (a) If a ribosome weighs 2.6 million daltons, calculate its protein and RNA composition as follows:

	PERCENT	MOLECULAR WEIGHT (DALTONS)	GRAMS
Protein	37		
RNA	63		

(b) If the average protein subunit weighs 25,000 daltons, how many subunits are there per ribosome?

(c) If a cell has 10,000 ribosomes, how much ribosomal RNA does it have?

8-2   THE CELL DNA

While the bacterial DNA can be seen by spreading it out at an air-water interface, this picture is not very revealing. The genophore appears to have no ends, but is quite a tangled mass and cannot be seen very clearly. It is more enlightening to view the DNA by a process of **autoradiography** (or **radioautography**). When cells are grown with ^{3}H-thymidine, they incorporate the nucleoside into their DNA; the DNA can then be extracted and spread out gently at an interface, but rather than viewing it directly it is covered with a liquid photographic emulsion and left in the dark. Beta particles from the decaying tritium atoms cause local excitations in the emulsion, which appear as dark silver grains after the usual photographic development process. Because these β particles have such low energy, they travel an average of only 1 μm from their points of origin and the position of a silver grain is a good indication of the position of the tritium label. The density of grains is a measure of the density of tritium atoms in the underlying material, and by counting grains one can tell the difference, for example, between one strand and two strands of labeled DNA.

John Cairns has developed this technique very well and he can get quite a good picture of DNA structure and replication. He sees the bacterial genophore as a line of silver grains whose length is somewhere between 1100 and 1400 μm, which is just about right for the genome, and in his best preparations the molecule appears to be circular.

Cairns was interested primarily in the replication of the genophore, a problem which has always been difficult and is in a very unsettled state at this time. Watson and Crick outlined the essential features of replication, and no one seriously doubts that they were basically correct; but the details of the process remain very elusive. In the first place, DNA is a plectonemic helix (see Exercise 4-4) and it must be unwound as it is replicated. The viscosity of the cytosol and the speed with which replication must occur are inconsistent with any picture in which the replicating arms of the molecule whirl in ever-widening circles through the cell, but Cyrus Levinthal and L. J. Crane pointed out that the problem can be solved if the *unreplicated* part of the molecule rotates on its axis like a speedometer cable. Then only a molecule with a radius of about 10 Å must be rotated and the viscous drag must be fairly low. Levinthal and Crane's calculations indicated that this process is energetically feasible.

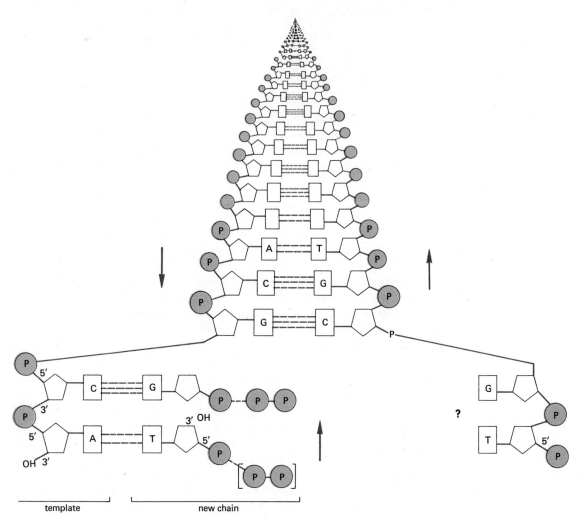

FIGURE 8-3

DNA synthesis by addition of 5'-nucleotide triphosphates.

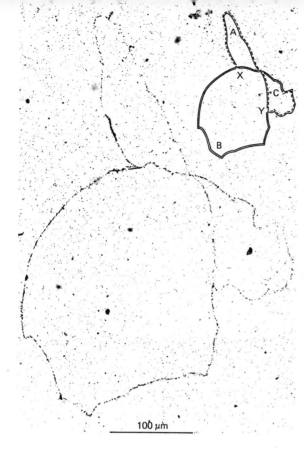

FIGURE 8-4

Autoradiograph of the E. coli *genophore. The cells were grown with ³H-thymidine for two generations and the DNA was extracted by gentle disruption of the cell wall, spread out, and exposed to photographic emulsion for 2 months. The insert shows the presumed structure of the genophore; X and Y are branch points, one of them presumably at the swivel. See Fig. 8-5 for further explanation. (Courtesy of John Cairns.)*

100 μm

If DNA is replicated in this way, then both strands are being replicated simultaneously from the same end of the molecule. But the two strands have opposite polarities, one going from 5' to 3' while its complement goes from 3' to 5'. Chemically, the addition of a nucleotide to the 5' end is different from adding it to the 3' end, and one would therefore assume that two different enzymes would be required for replication. Enzymes that appear to be part of the DNA replication system have been found—first in *E. coli* by Arthur Kornberg and then in mammalian cells by Frederick Bollum. These enzymes, which are called **DNA polymerases,** require the four nucleotide triphosphates (dATP, dGTP, dCTP, and dTTP), Mg⁺⁺ ions, and a DNA "primer." They apparently copy the "primer" and make new DNA molecules that are complementary to it, but they work only in the 5' to 3' direction; they can perform the process shown in Fig. 8-3, but it is not clear what is happening on the other arm of the molecule. The situation was summarized quite well recently by a picture of a replicating molecule with the replication point covered by a fig leaf.

Whatever is happening at the branch point of the Y, it seems clear from Cairns' pictures that the genophore is replicated along both arms more or less simultaneously, from the same direction (Fig. 8-4). The resolution of this picture is only about 1 μm, so it gives no information about details at the point of replication, but on a gross level

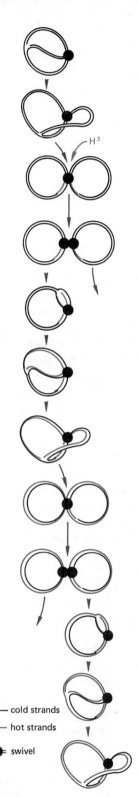

the process shown in Fig. 8-5 is clearly indicated. The only addition that must be made to the model of Levinthal and Crane is a "swivel" point that allows free rotation of the unreplicated portion; so far, there is no indication of what the swivel might be.

exercise 8-2 A bacterial cell has a dry weight of 2.5×10^{-13} g.
(a) Calculate the molecular weight of the whole cell in daltons.
(b) If its DNA is 3% of the total, calculate its

molecular weight in daltons	_____
number of nucleotide pairs	_____
length in micrometers	_____

(c) If this DNA is divided into two nuclear bodies and it takes 40 minutes to replicate each one, how fast must the DNA be rotating on its axis?

8-3 WALLS AND MEMBRANES

Although procaryotic cells are defined largely by their lack of internal membranes that define a nucleus and other regions, they still have cell membranes, which usually seem to be nothing more than simple bags. The cell membrane is recognized in electron micrographs by its trilaminar structure and does not appear to be different from membranes seen in other cells. In addition, every bacterium has some kind of a wall just outside of the membrane; the wall is very strong and is apparently required to keep the cell from bursting apart because of its high internal osmotic pressure. In fact, the only bacteria in which the wall may be reduced or absent are some that live in sea water, which is more isotonic to their cytosol. The walls of other bacteria can be removed by gentle enzymatic digestion and some other treatments, leaving only **spheroplasts** with essentially naked membranes; these are viable if kept in high concentrations of sucrose, but they burst when placed in a hypotonic medium.

In 1884, Christian Gram discovered a test that divides bacteria into two groups. A smear of cells is fixed by heating on a glass slide and then stained with the dye

FIGURE 8-5
A replication mechanism for circular DNA that will yield the pattern of Fig. 8-4 if tritium is incorporated for about two rounds of replication in the indicated pattern. Replication is assumed to start at the swivel, and one of the daughter genophores is removed at the completion of each round. (Courtesy of John Cairns.)

—— cold strands
—— hot strands
⇒●= swivel

crystal violet followed by a weak iodine solution; it is then rinsed with ethanol and stained again with a red dye such as safranin. Cells that take up the crystal violet-iodine stain appear blue and are called **gram positive,** while those that do not are stained red by the safranin and are called **gram negative.** (*E. coli* is gram negative.) The physical basis of the stain is still rather obscure, but it is related to the structure of the cell wall. Gram-positive cells have little more than a thickened outer layer around their cell membranes, while gram-negative cells have an elaborate outer layer which looks like another trilaminar membrane outside of the cell membrane. We will see in Chapter 19 that the gram-negative cell wall is much more complex chemically than the gram-positive wall.

Many bacteria are actively motile and swim around by means of flagella that spring from their membranes. These flagella are thin, flexible protein helices. Their positions are very characteristic; some species have only a few flagella at one end of the cell, while in others many flagella are attached over the whole cell surface.

The cell membrane seems to be important primarily as an organizer of cellular enzyme systems. This is a theme we will return to again and again, for this appears to be one of the chief functions of almost all membranes. In procaryotes, the most important systems built into the cell membrane are the electron-transport systems. The cell's capacity for transporting electrons and making ATP may therefore be severely limited by the amount of membrane it has, and some procaryotes have developed an elaborate system of internal membranes just to increase the area available for electron transport. Figure 8-6 presents a rogue's gallery of bacteria with different forms of electron-transport membranes; these have been selected partially on the basis of aesthetic criteria, for they are often very beautiful. Three of these bacteria are phototrophs which make extensive membranes in order to trap as much light as possible and the fourth is a chemolithotroph whose electron-transport system is not very efficient, so it compensates by devoting more of its mass to the process. There is one point to be emphasized: In every case the internal membranes are merely complicated infoldings of the cell membrane and it can be shown that they are continuous with it. Only in a few bacteria, such as the green bacteria shown in Fig. 8-6b, do a few internal vesicles pinch off as completely separate pockets. There are never internal membrane systems as complex as those in eucaryotes.

8-4 BACTERIAL CELL DIVISION

At least two processes must occur in order to divide a bacterial cell properly into two daughters: Copies of the genome must be apportioned equally to the two halves of the cell and then new cell wall and membrane must be made to separate the halves from one another. The cell membrane is obviously essential to the second process and appears to be involved in the first as well.

About 1960, several people began to notice a curious type of whorled internal membrane in some gram-positive bacteria; Phillip Fitz-James called them **mesosomes.** They appeared to be connected to the nuclear bodies in many cells, and in 1963 Antoinette Ryter showed a clear connection in *Bacillus subtilis* (Fig. 8-7). The nuclear bodies seem to be connected to the cell membrane in both gram-positive and gram-negative bacteria, although elaborate mesosomes are not always visible. Ryter

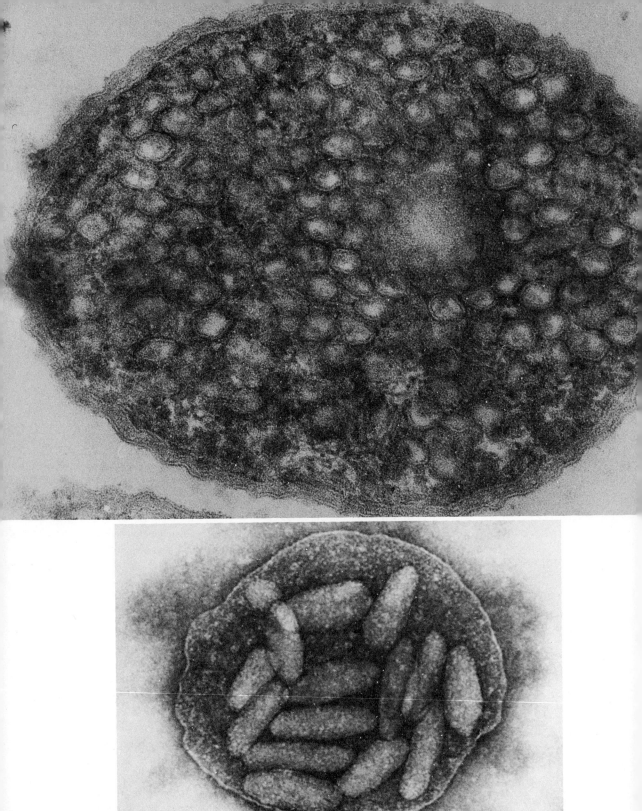

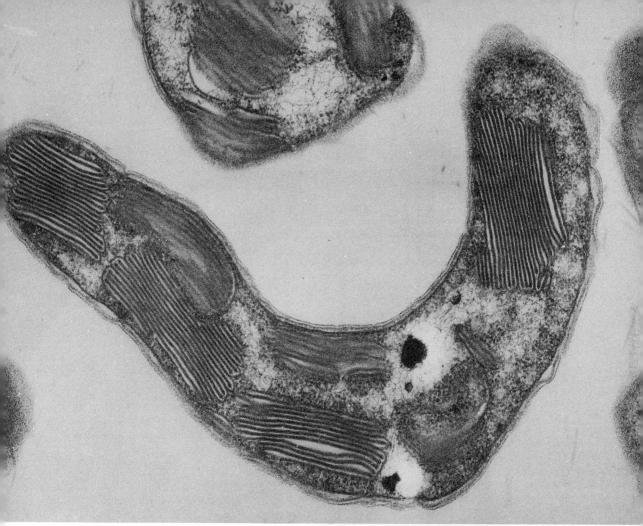

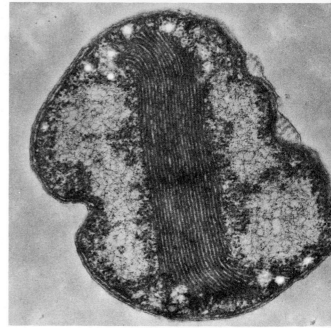

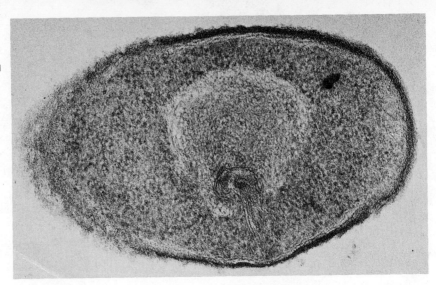

FIGURE 8-7

Above, cross section through a Bacillus subtilis *cell showing its light-colored nuclear body with its fine fibrous structure, connected to a mesosome that can be traced to the cell membrane.* [*Courtesy of Antoinette Ryter, reprinted from* Protides of Biological Fluids *vol. 15, 1967 (H. Peeters, ed.) by permission of Elsevier Publishing Co., Amsterdam.*]

FIGURE 8-8

Below, a Bacillus subtilis *protoplast that has been subjected to a hypertonic medium to force the mesosome to extrude and reveal its structure. Negative staining. (Courtesy of Antoinette Ryter, reprinted from* Comptes Rendus Acad. Sci. Paris, *vol. 265, 1967 by permission of Gauthier-Villars, Paris.)*

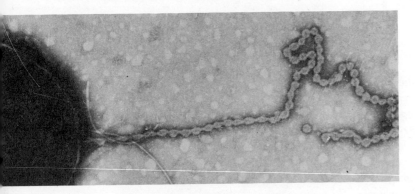

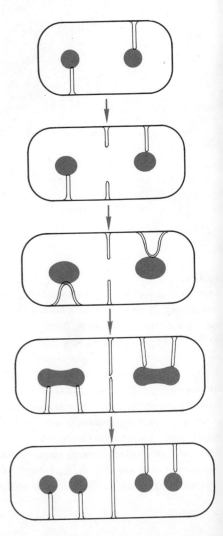

FIGURE 8-9

At right, Ryter's conception of the role of mesosomes in separating daughter nuclear bodies and in forming the septum between cells.

has shown that the mesosomes are characteristically shaped like a series of link sausages (Fig. 8-8).

By taking serial sections through dividing bacteria, Ryter has shown that the mesosomes actively separate daughter nuclear bodies (Fig. 8-9). As the nuclear body enlarges during its replication, it becomes attached to two mesosomes which separate as the cell elongates until the body is divided.

As nuclear division is going on, another set of mesosomes may be involved in forming new cross walls. There appear to be some differences between gram-positive and gram-negative bacteria in this respect. Gram-positive bacteria apparently lay down a new layer of cell wall across the middle of the old cell, while gram-negative bacteria form walls through invagination from the outside, gradually moving toward the center (Fig. 8-10). The exact role of mesosomes in these processes is still in dispute.

FIGURE 8-10

Cell division in the blue-green alga Anacystis nidulans, *which divides very much like a gram-negative bacterium through invagination of the wall into the center of the cell. Notice the several layers of wall material and the internal photosynthetic membranes parallel to the wall and membrane. (Courtesy of Mary M. Allen, reprinted with permission from* Journal of Bacteriology.)

B. Metabolism: a second view

In Chapter 3 we outlined some very general principles of cellular metabolism and presented a diagrammatic picture of the process. Now it is time to focus in a little more closely and look at some of the central metabolic processes with a little higher resolution. Since one has to break into the continuous flow of mass through the cell at an arbitrary point, we will begin in the middle.

FIGURE 8-11
*Reaction of the TCA cycle (black arrows) with the Wood-Werkman
reactions, the glyoxylate cycle (colored arrows), and the transamination reactions (gray arrows).*

8-5 THE TRICARBOXYLIC ACID CYCLE

Harold Morowitz has proved a very general theorem regarding the flow of mass
through an open system. His argument is beyond the scope of this book, but his
conclusion is critical. Morowitz states, in effect, that the flow of matter through an
open system in a steady state must always be accompanied by a cyclic process. This
is a very appealing concept for a biotic system; it is useful to conceive of a cell, as we
outlined in Section 3-12, having a main core of cyclic reactions that takes in a variety
of molecules from the environment and turns out a variety of necessary monomers
and other small molecules. In fact, this main driving wheel is the **tricarboxylic acid
cycle** (TCA cycle) discovered by Sir Hans Krebs and often called the citric acid or
Krebs cycle.

The cycle and some associated reactions are shown in Fig. 8-11; mechanically, it is
very simple. A C_4 compound, oxalacetate, is condensed with a C_2 compound, acetate,
to make a C_6 compound, citrate. The citrate is transformed through a cycle during
which CO_2 groups are removed in a couple of **decarboxylation** reactions, so the C_4
oxalacetate is regenerated. Most of the reactions are designed to yield high-potential
electrons that can be fed into an electron-transport system or used to reduce other
compounds.

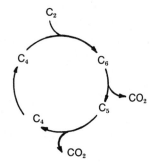

The TCA cycle is a vehicle for conserving and redistributing both mass and energy;
the two functions cannot be separated from one another. At its simplest, the cycle may
be seen as a case of the rather crude conception of respiration outlined in Section 3-8,
because it takes a C_2 compound (acetate) and oxidizes it to CO_2 and water—the latter
coming from further oxidation of the hydrogens in an electron-transport system. But
there is really much more to the cycle than that. All the organic compounds that enter
the cell must either be used directly (mostly as monomers for polymerizations) or
broken down into smaller, more usable fragments and these must be redistributed to
make whatever other molecules are required. Now as these reactions proceed, there
are many points of oxidation that lead directly or indirectly to the synthesis of ATP, so
that much of the energy of these compounds is partially conserved. But it is also
important that the organic fragments themselves be retained in a form that is directly
usable for biosynthesis. These fragments can then be donated in **group-transfer
reactions,** catalyzed by **transferase** enzymes, which are exactly like the phosphoryl-
transfer reactions in which ATP and its relatives participate. Each of these transferases
uses a specific coenzyme that has a rather high transfer potential for its particular
group; and now we can see ATP as one of this class of coenzymes that can transfer
phosphoryl groups with a high potential. Most of the other transferase coenzymes

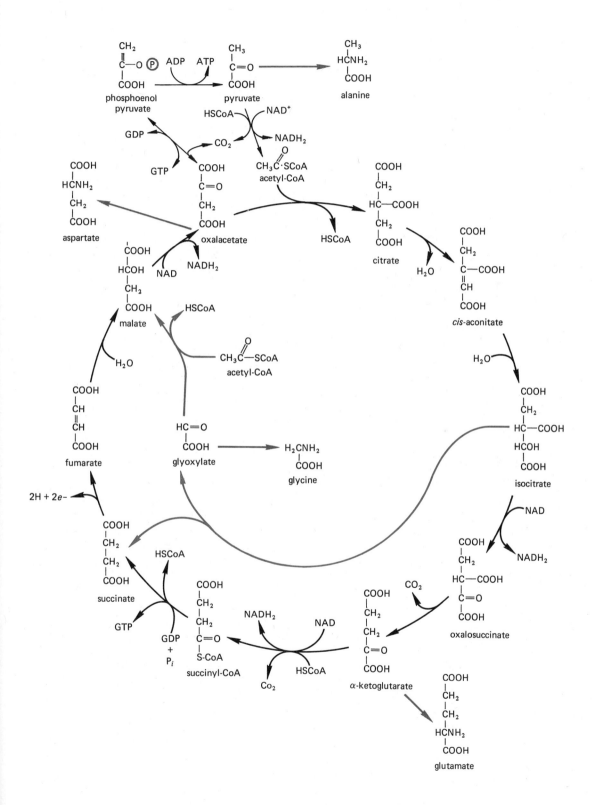

are also adenosine nucleotides, and this includes a special class of coenzymes that mediate oxidoreductions, in which the transferred group is an electron pair (generally with a hydrogen-ion pair). We will take a moment to describe these coenzymes and then come back to consider the reactions in which they participate, with emphasis on the TCA cycle and reactions close to it.

8-6 TRANSFERASES AND ADENOSINE NUCLEOTIDES

(A) METHYL TRANSFERASES

One of the most important classes of group-transfer coenzymes consists of molecules with a formal positive charge on one atom, the **onium pole.** The onium pole is almost invariably nitrogen (ammonium compounds) or sulfur (sulfonium compounds), and it is susceptible to attack by an electron pair. The primary coenzyme of methyl transfer (**methylation**) is a sulfonium compound, **S-adenosylmethionine,** which methylates an acceptor, A, with a large change in free energy:

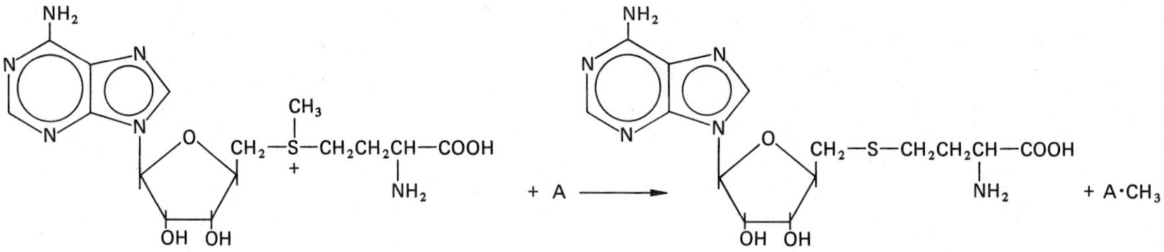

S-adenosylmethionine   S-adenosylhomocysteine

B) OXIDOREDUCTASES

Four nucleotides are the principal coenzymes of oxidoreduction. There are two basic mononucleotides, **nicotinamide mononucleotide (NMN)** and **flavin mononucleotide (FMN).** Through combination with ATP, the dinucleotides **nicotinamide-adenine dinucleotide (NAD)** and **flavin-adenine dinucleotide (FAD)** are formed:

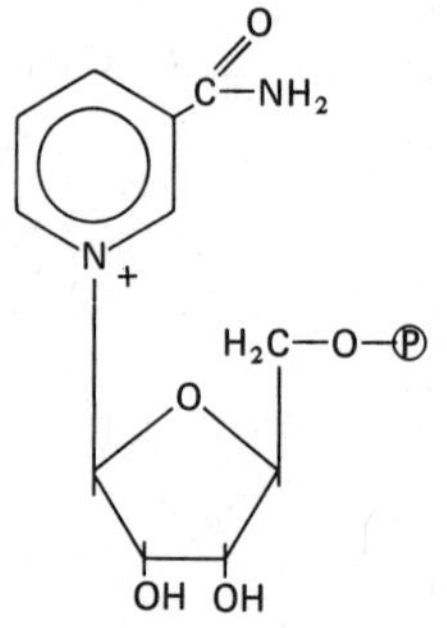

NMN 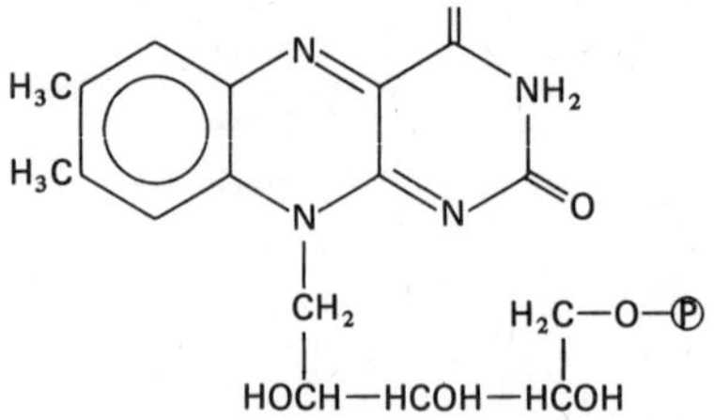   FMN

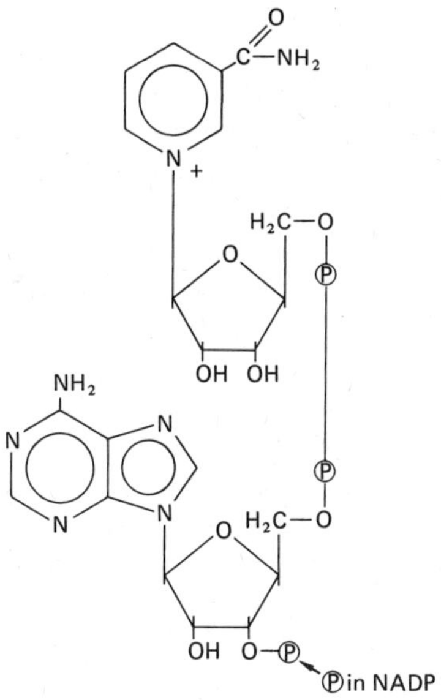

NAD

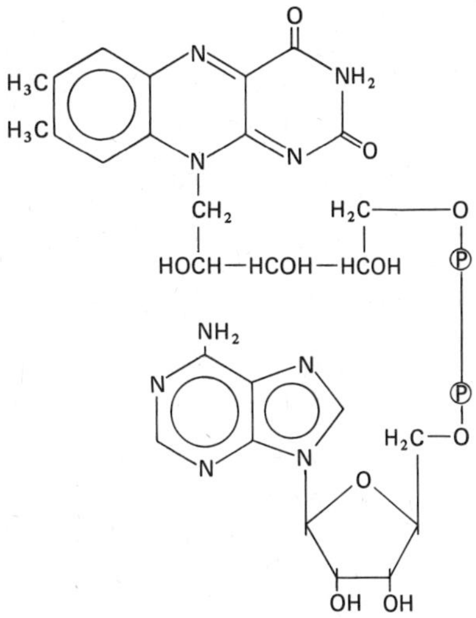

FAD

A variant of NAD, **nicotinamide-adenine dinucleotide phosphate (NADP)** has the extra phosphate group shown.

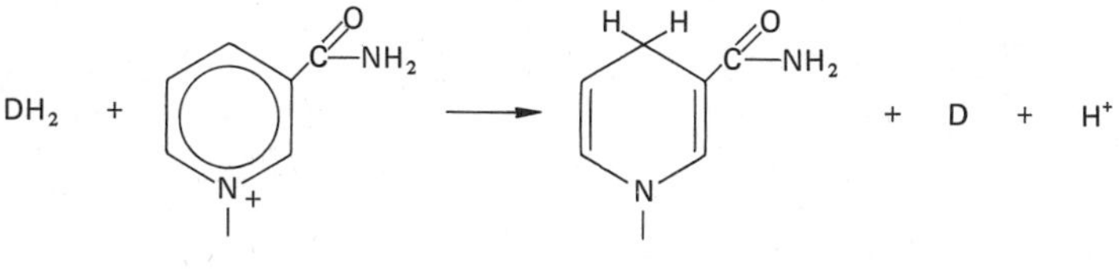

$$DH_2 \; + \; \text{[nicotinamide ring]} \longrightarrow \text{[reduced ring]} \; + \; D \; + \; H^+$$

The nicotinamide ring has an onium pole that can be reduced by a proton and an electron pair, with one proton going off into the medium. The oxidation of an organic compound effected by a nicotinamide nucleotide is typically a **dehydrogenation** in which two hydrogen atoms are removed:

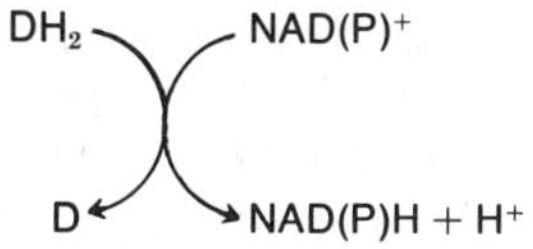

The reduced coenzymes, NADH and NADPH, can then act as reducing agents since this reaction is reversible, but NADPH is the more common reducing agent in cells and the cell must build up a supply of this coenzyme for biosynthesis. However, only NADH can reduce an electron-transport system, and so electrons from NADPH are first transferred to NAD^+ in a transhydrogenation reaction: $NADPH + NAD^+ \rightarrow NADH + NADP^+$.

The nicotinamide nucleotides are important in cell economy because they form a common pool of electrons from a large number of dehydrogenation reactions. The flavin nucleotides are less often part of a dehydrogenation from an organic compound; but they are very important as immediate acceptors of electrons from NADH. FMN and FAD are always bound covalently to proteins as prosthetic groups, and the holoenzyme is called a **flavoprotein (FP)**. It always contains one or more metal ions— Mo, Mn, Fe, or Cu—and may therefore be called a **metalloflavoprotein (MFP)**, specifically a molybdoflavoprotein, cuproflavoprotein, and so on. The oxidation of the metal seems to be a part of the overall reaction of the enzyme.

The flavin ring system may be reversibly oxidized and reduced in two consecutive one-electron transfers:

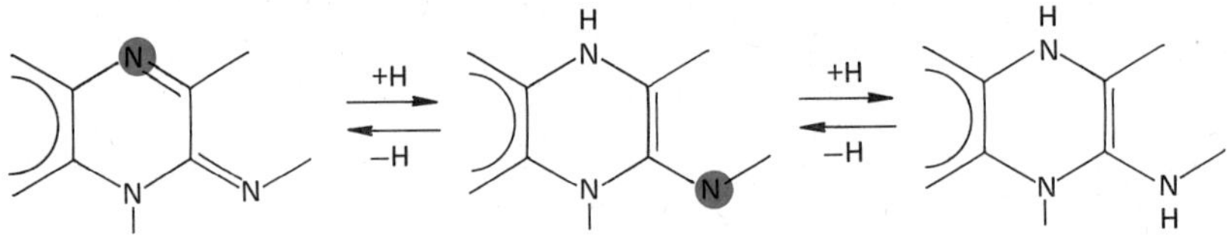

These molecules can therefore act as bridges between one-electron and two-electron processes in ET systems.

(C) ACETYL- AND ACYLTRANSFERASES

The primary coenzyme of acetyl transfer reactions is **coenzyme A (CoA: pantetheine-adenine dinucleotide phosphate)**:

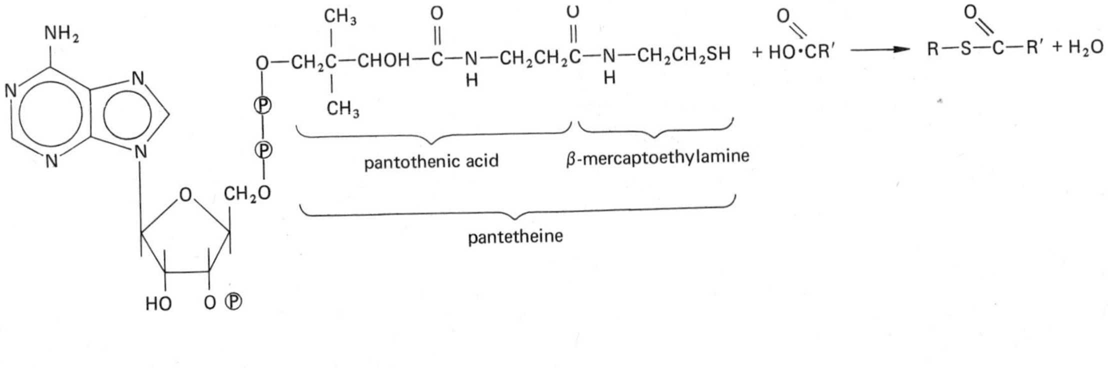

CoA forms thioesters with carboxylic acids through its sulfhydryl group; acetyl-CoA is the most important of these compounds. CoA can then transfer this group with a standard free energy of about −8 kcal/mole. We sometimes write HSCoA to indicate the sulfhydryl group that participates in the reaction.

(D) SULFATE TRANSFERASES

The transfer of a sulfate group requires the coenzyme **adenosine 5′-phosphosulfate (APS)** in its phosphorylated form, **3-phosphoadenosine-5′-phosphosulfate (PAPS)**:

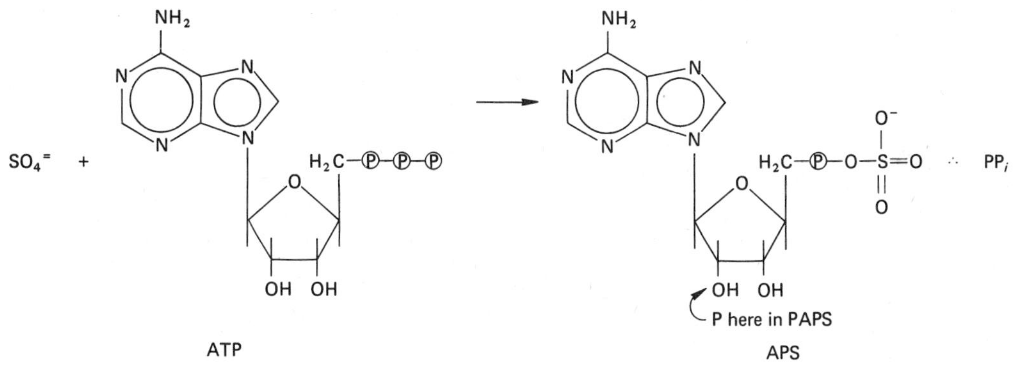

Reduction from sulfate to sulfite occurs in the coenzyme form, but the path from sulfite to more reduced forms and into the sulfhydryl groups of organic compounds remains much less certain.

8-7 REACTIONS IN AND NEAR THE CYCLE

Just as all roads lead to Rome, all reactions lead to the TCA cycle; all of the degradative pathways lead to a remarkably small number of compounds that are in or near the TCA cycle. The bulk of this material comes from organic acids through CoA derivatives, chiefly through oxidative decarboxylation. An α-keto acid is first picked up by the coenzyme **thiamine pyrophosphate (TPP)**:

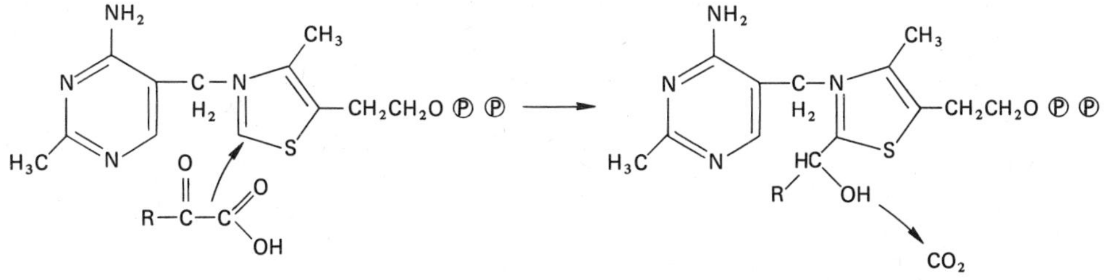

The decarboxylated residue is then transferred to **lipoic acid** (a prosthetic group bound to lysine) and thence to coenzyme A:

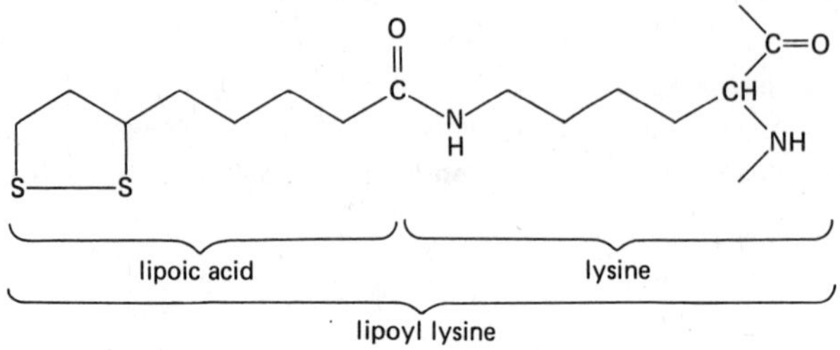

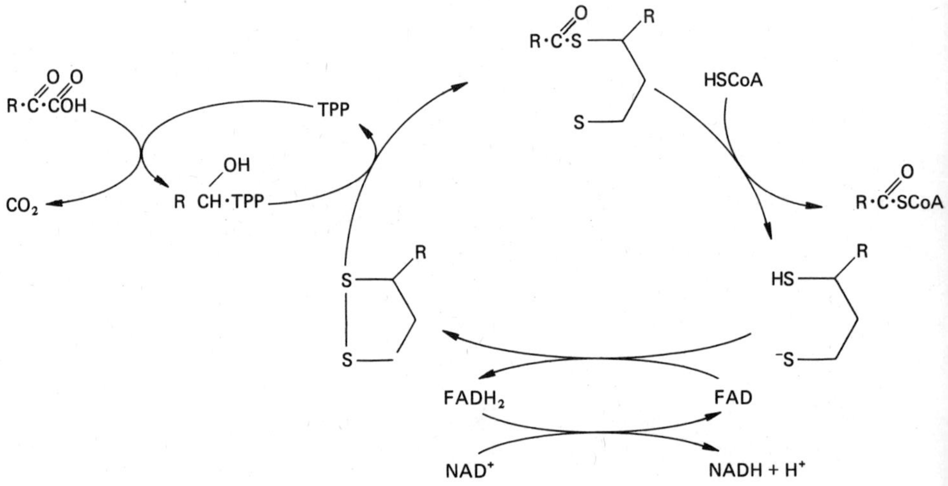

Meanwhile, the lipoic acid is oxidized for the next round of the cycle. This whole process occurs twice in the Krebs cycle—once from pyruvate to acetyl-CoA and again from α-ketoglutarate to succinyl-CoA. It will be good exercise for you to write out these two reaction sequences separately to see what is happening in each case.

The amount of material that can funnel through these two points is enormous. The oxidation of all sugars proceeds via pyruvate to acetyl-CoA and all fats are oxidized directly to acetyl-CoA. Amino acids such as glutamate, histidine, arginine, and proline are degraded to α-ketoglutarate and others, such as threonine, valine, and isoleucine, are degraded directly to succinyl-CoA.

The condensation of acetyl-CoA and oxalacetate is catalyzed by the condensing enzyme citrogenase. The citrate thus formed undergoes an interesting reaction used quite often in metabolism—**dehydration-hydration.** A molecule of water is removed, leaving a double bond across which water is added again to form the isomer isocitrate. Enzymes that catalyze these reactions are called **lyases.** The formation of malate occurs in a similar way; the fumarate formed by dehydrogenation of succinate has a double bond across which a molecule of water is added.

The dehydrogenations of the cycle are among its most important features. The isocitric and malic dehydrogenases are linked to NAD^+, which supplies electrons to the electron-transport system, and the succinic dehydrogenase is a flavoprotein that may be considered an early part of the electron-transport system itself, since it directly reduces the rest of the electron carriers built into the cell membrane. Notice also that the formation of succinate is directly coupled to the formation of GTP (sometimes ITP). Thus, there are several points where energy is removed from intermediates of the cycle and conserved directly or indirectly as nucleotide triphosphates.

The cycle as we have outlined it would work perfectly well and provide a fine supply of ATP and NAD(P)H if the cells were in a medium with monomers such as sugars and amino acids, where everything was feeding *into* the cycle and nothing had to flow *out* of it. But this hardly ever happens; intermediates of the cycle have to be drawn off into biosynthetic pathways; therefore, there must be other reactions that can provide some of these intermediates to keep the cycle turning. The first of these, discovered by H. G. Wood and C. H. Werkman, are ways of getting from C_3 compounds to oxalacetate through condensation with a CO_2. The coenzyme of CO_2 transfer is biotin (vitamin H), which is converted to carboxybiotin in a reaction coupled to ATP:

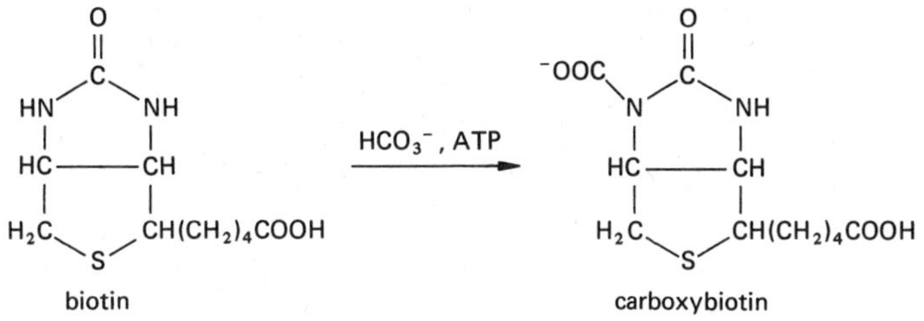

biotin   carboxybiotin

Biotin is a true prosthetic group, linked to its enzyme covalently in amide linkage to lysine.

In the first of the Wood–Werkman reactions, carboxybiotin transfers its CO_2 group directly to pyruvate:

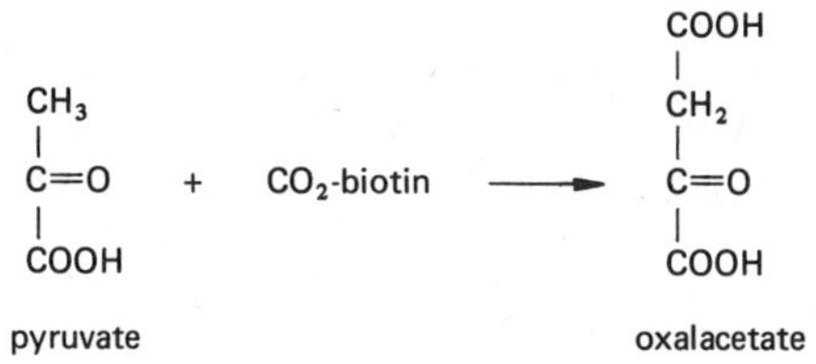

pyruvate   oxalacetate

The second is similar, but CO_2 is condensed with phosphoenolpyruvate and GTP is formed:

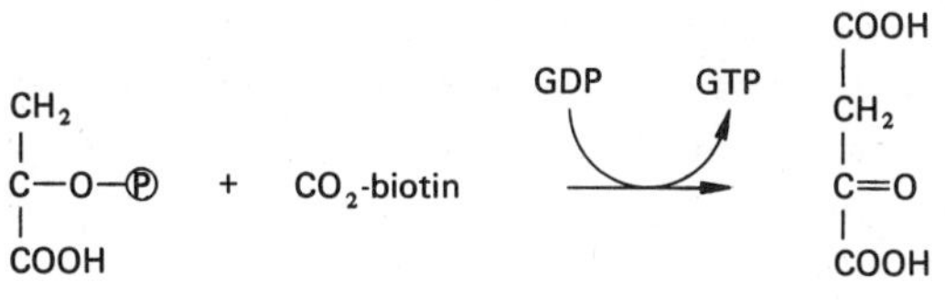

The standard abbreviation for phosphoenolpyruvate, PEP, sounds like a bad pun because the compound is a strong phosphorylating agent ($\Delta G^0 = -12$ kcal/mole) and is used in a number of reactions.

The third reaction does not involve biotin; it is a reductive synthesis of malate from pyruvate:

$$\text{pyruvate} + CO_2 + NADPH + H^+ \rightarrow \text{malate} + NADP^+$$

The Wood–Werkman reactions will supply TCA intermediates if C_3 compounds are available, but an organism may have to grow with only C_2 compounds. In this case, two other reactions can be superimposed on the TCA cycle to make a **glyoxylate cycle** that allows acetate to be incorporated at two points and yields additional TCA intermediates. The first reaction, catalyzed by isocitratase, is

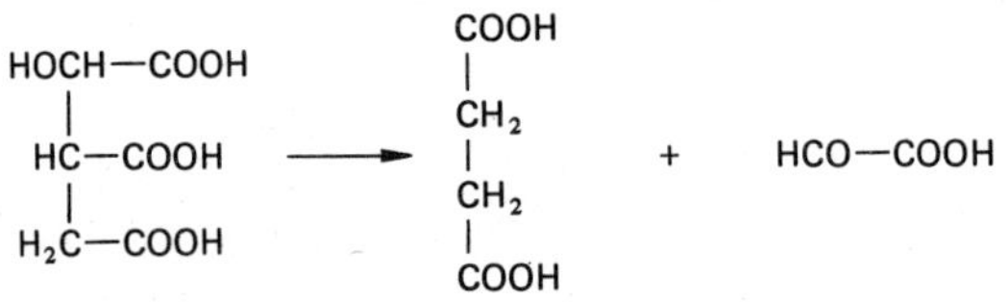

```
HOCH—COOH          COOH
   |               |
  HC—COOH   ──→   CH2       +    HCO—COOH
   |               |
 H2C—COOH         CH2
                   |
                  COOH

 isocitrate       succinate          glyoxylate
```

The second, catalyzed by malate synthetase, is

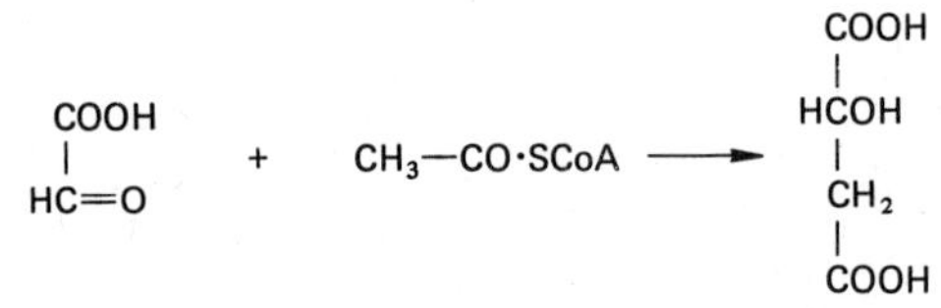

```
                                      COOH
                                      |
   COOH                              HCOH
   |      +   CH3—CO·SCoA   ──→       |
  HC=O                               CH2
                                      |
                                     COOH

 glyoxylate      acetyl-CoA          malate
```

The net effect of these reactions is to form succinate and malate out of acetates. The glyoxylate cycle is simply the substitution of these synthetic reactions for the oxidative reactions of the TCA cycle that occur between isocitrate and succinate.

8-8 BIOSYNTHESIS FROM THE CYCLE

The TCA cycle, with the addition of the Wood–Werkman and the glyoxylate reactions, can potentially generate all of the other small molecules needed in a cell, including all of the amino acids, sugars, nucleotides, and lipids. We will discuss several of these biosynthetic routes in Chapter 17. At this time let's just look at some reactions very close to the cycle itself to see generally how these biosyntheses occur.

The first point to notice is that four of the TCA intermediates shown in Fig. 8-11 are the keto acids corresponding to four important amino acids:

$$\text{glyoxylate} \leftrightarrow \text{glycine}$$
$$\text{pyruvate} \leftrightarrow \text{alanine}$$
$$\alpha\text{-ketoglutarate} \leftrightarrow \text{glutamate}$$
$$\text{oxalacetate} \leftrightarrow \text{aspartate}$$

The transformation from one to the other can occur via at least two different reactions. In the **amination-deamination** reactions, the keto acid picks up free ammonia in a reductive process:

$$\alpha\text{-ketoglutarate} + NH_4^+ + NADPH + H^+ \rightarrow \text{glutamate} + NADP^+ + H_2O$$

Reactions of this type strongly favor synthesis of the amino acid and are among the most important mechanisms for assimilation of ammonia.

The other process is **transamination,** where a keto acid and an amino acid exchange groups with one another. For example:

$$\text{glutamate} + \text{oxalacetate} \rightleftharpoons \alpha\text{-ketoglutarate} + \text{aspartate}$$
$$\text{glutamate} + \text{pyruvate} \rightleftharpoons \alpha\text{-ketoglutarate} + \text{alanine}$$

There are several different transaminases that catalyze these reactions. Depending upon the balance of other factors in the cell, they can increase and decrease the levels of each amino or keto acid and can thus favor either biosynthesis or degradation.

The coenzyme of transamination is **pyridoxal phosphate** and its relatives (the vitamin B_6 group):

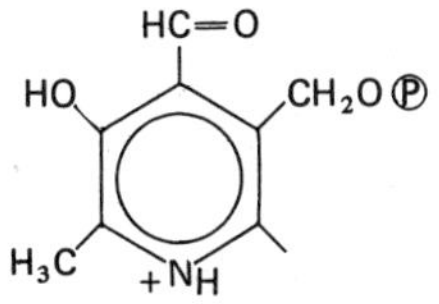

pyridoxal phosphate (PAL) 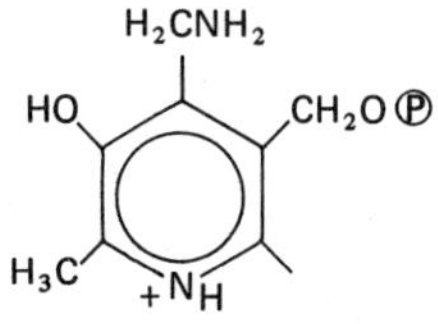   pyridoxamine phosphate (PAMP)

In transamination, the coenzyme is reversibly converted from PAL to PAMP:

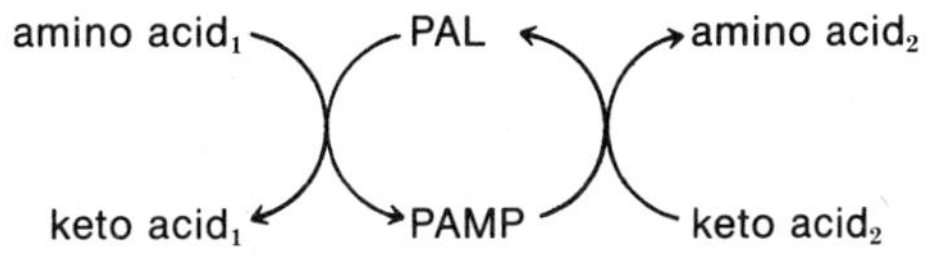

We ought to note in passing that these coenzymes are actually involved in a large number of different reactions of amino-acid metabolism in addition to transamination. The mechanisms of these reactions form an interesting chapter in the biochemistry you may study in a future course.

Aspartic acid may be converted into seven of the amino acids found in proteins and into the pyrimidine nucleotides, while glutamate is converted into three other amino acids. Several other major components of the cell arise from points on or near

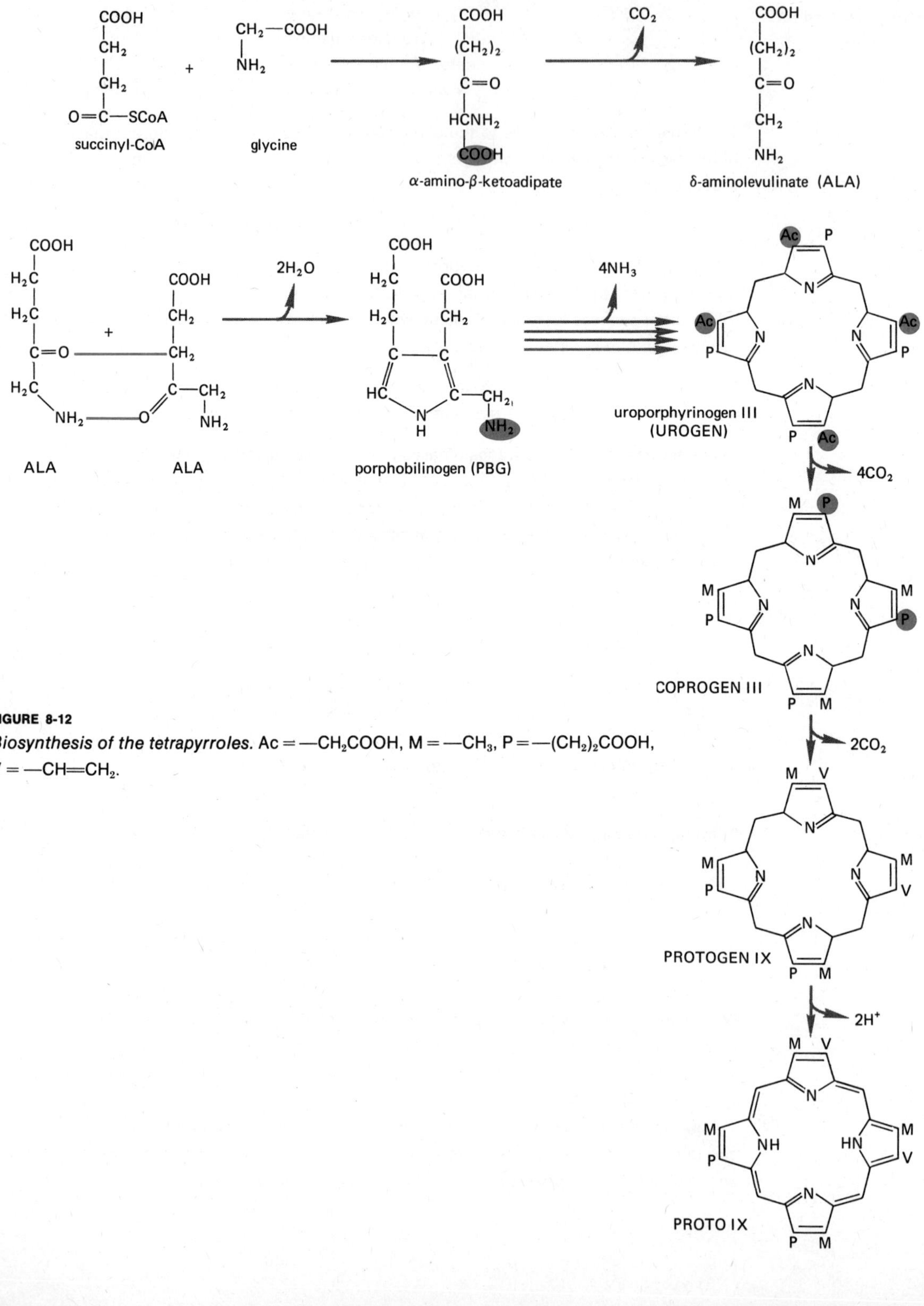

FIGURE 8-12

Biosynthesis of the tetrapyrroles. Ac = —CH$_2$COOH, M = —CH$_3$, P = —(CH$_2$)$_2$COOH, V = —CH=CH$_2$.

the TCA cycle. For example, Fig. 8-12 shows how the tetrapyrrole ring comes from succinyl-CoA; you can see that all of the tetrapyrroles have essentially the same structure because they are all made through the same mechanism. In general, you should try to keep in mind a picture of many branched pathways of biosynthesis arising from points on or near the TCA cycle.

8-9 ENERGY CONSERVATION IN THE CYCLE

Let us consider briefly the amount of energy actually removed during oxidative steps of the TCA cycle and retained in the form of ATP.

The electron-transport system is a bucket brigade of cytochromes and lipids that conserves the energy of electrons as ATP. The mechanism of this process is not important here, but the stoichiometry of it is. Electron pairs arising from dehydrogenations by flavoproteins, such as succinic dehydrogenase, are able to produce two molecules of ATP as they traverse the cytochrome system, while those coming from NADH can produce three molecules; we can use this information by itself to calculate the amount of energy conserved by reactions of the TCA cycle and by others that feed into the ET system.

exercise 8-3 Count the number of ATPs that can be made from the reactions within the TCA cycle proper. Then count the number that can be made from one molecule of pyruvate taken through acetyl-CoA and around the TCA cycle.

exercise 8-4 The oxidation of glucose proceeds intracellularly to two molecules of pyruvate and then into the TCA cycle. If glucose is burned to CO_2 and water, it liberates 686 kcal/mole. Assuming that 1 mole of ATP conserves 8 kcal, calculate the efficiency with which glucose is oxidized via just the reactions we have outlined here.

Similar calculations can be made for other carbon sources; they yield similar values and show that the pathways of the TCA cycle and related oxidative pathways are really quite efficient in conserving the bond energy of incoming organic compounds in a useful form as ATP.

With a general picture of metabolism in mind, we can now see some of the broader aspects of bacterial cell growth, metabolism, and regulation.

C. Steady states of growth

8-10 ENZYME INDUCTION AND REPRESSION

E. coli grows in the human intestine, and one of the carbon sources it can use is lactose, a major component of milk, which is an important part of the diet of young mammals. The bacteria can use lactose because they can make β-galactosidase. However, lactose is not always present in their growth medium, and it is therefore advantageous for them to make the enzyme only when its substrate is present. In fact, the cells have a very tight control over the amount of enzyme they produce; β-

galactosidase is made only when a significant amount of lactose or some other β-galactoside is present. This enzyme is therefore said to be **inducible** and the lactose or lactose-analogue is called an **inducer.**

A parallel situation exists for enzymes involved in the biosynthesis of cell monomers. *E. coli* can make all of its amino acids with biosynthetic pathways out of the TCA cycle, using the carbon in glucose plus a few inorganic salts. Because these monomers are made by intracellular enzyme systems, they are said to be *endogenous.* However, *E. coli* will also use *exogenous* amino acids that are supplied in the growth medium, and in fact they will use these in preference to endogenous metabolites. As soon as an exogenous amino acid is added to the medium, the synthesis of all the enzymes necessary for endogenous production of that acid stops. These enzymes are therefore said to be **repressible** and the added amino acid acts as a **repressor.**

Induction and repression occur very rapidly; the kinetics of the transition from one condition to another can be illustrated very well with the β-galactosidase system. However, we generally substitute an artificial inducer for the lactose; if we did not, some cells would naturally become induced slightly before others, they would start to use lactose as a carbon source, and the kinetics of growth might become very complicated. We use *gratuitous* inducers that induce but cannot be metabolized, and one of the best is isopropylthio-β-galactoside (IPTG).

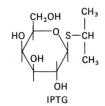

IPTG

The kinetics of induction with IPTG are very simple. If E is the amount of enzyme per bacterium and B is the number of bacteria per milliliter, then $Z = EB$ is the amount of enzyme per milliliter, which we can measure by using the assay procedure outlined in Chapter 3. We add inducer to bacteria growing exponentially; there is always a short lag of a few percent of a generation, after which enzyme synthesis proceeds at its maximum rate, which we denote by σ. The enzyme is stable, but it is diluted by cell growth at a rate αE, so the net rate of protein synthesis is

$$\frac{dE}{dt} = \sigma - \alpha E \tag{8-1}$$

It is then easy to show that the differential rate of enzyme synthesis, dZ/dB, is

$$\frac{dZ}{dB} = \frac{\sigma}{\alpha} \tag{8-2}$$

This is the rate at which the one enzyme is made with respect to the rate at which the whole cell is made. A graph of Z versus B is a straight line with slope dZ/dB (Fig. 8-13). When the bacteria are uninduced, this slope is low, but within a very short time after inducer is added it may rise by a factor of 1000, and it is this change in slope that we use as a measure of induction. The kinetics of repression are the same except that the rate of synthesis decreases sharply upon addition of the repressor.

exercise 8-5 Students who enjoy calculus might like to derive equation 8-2. Consider both equations 8-1 and 2-2.

8-11 CHARACTERISTICS OF STEADY STATES

Bacteria such as *E. coli* can be grown in many different kinds of media; under each growth condition, the cells have a particular size and composition and grow at a characteristic rate. We will now try to establish a very general conception of growth and its regulation, which may be summarized as follows. Cells must have a set of monomers for making their polymers and a number of other small molecules that are not polymerized. Either they can be given these compounds exogenously or they can be forced to make them endogenously; the more they are given, the less they will do for themselves, and the less they have to do for themselves, the faster they will grow. If they are put into a rich broth, full of amino acids, sugars, nucleosides, and perhaps some coenzymes, they will take advantage of the opportunity by growing big and fat and proliferating rapidly, while if they are placed in a simple, minimal medium, with nothing except glucose and some salts, they will have to make all of their own organic compounds themselves and they will be small and will proliferate slowly.

This is what we observe, and it is just what we expect. Evolution is primarily a race in which the winners are the organisms that can proliferate and reproduce most rapidly; we therefore expect them to have evolved regulatory mechanisms that will permit them to grow as rapidly as possible in any situation. These regulatory mechanisms are largely the controls of induction and repression of enzyme synthesis, as well as some more general controls superimposed on these. A major part of this book will be devoted to these regulatory mechanisms; we will begin here by examining the characteristics of cells under different growth conditions.

To illustrate the profound effect that addition of exogenous materials can have on the growth rate, some data are presented in Fig. 8-14. The cells are being grown in

FIGURE 8-13

Kinetics of β-galactosidase induction.

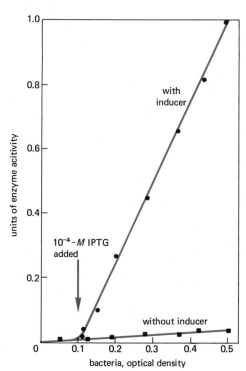

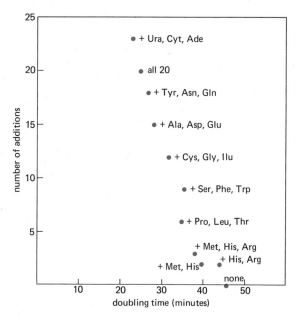

FIGURE 8-14

Charles Helmstetter's data on the effect of various additions to the growth medium on the doubling time of E. coli. Some amino acids obviously have a greater effect than others.

a simple medium that is made more and more like a broth by adding amino acids. Each addition represses a certain set of enzymes, and the fewer enzymes the cell has to make, the faster it grows.

When bacteria are left in constant conditions for several generations, they achieve a steady state of growth where their composition does not change and the amount of material entering as nutrients is just balanced by the amount of waste material and new cell mass being produced. This has also been called a state of *balanced growth*. We try to deal with populations of cells in steady states because they can be so well characterized, and the regularities one observes are really quite remarkable. Much of what we know comes from the work of Ole Maaløe and Niels Kjeldgaard and their associates over the years at the University of Copenhagen; their book (see the references at the end of the book) should be consulted for a more detailed account of this subject.

Most of the observed regularities in cell composition can be summarized by a simple rule. If m is the mass of some component per cell, then

$$\log \frac{m}{m_0} = \gamma\alpha \tag{8-3}$$

where α is the growth-rate constant and m_0 is the value of m extrapolated to $\alpha = 0$. In a surprising number of cases, the graph of log m/m_0 plotted against α is a straight line with slope γ. Some simple examples will illustrate this point; plot the data in Exercises 8-6 and 8-7 carefully, because they are intrinsically interesting and can be a valuable reference for the future.

exercise 8-6　　The data in Table 8-1 give the composition of *Aerobacter aerogenes* for different growth rates. Complete the table and plot these data on graph paper.

TABLE 8-1

composition of Aerobacter aerogenes *as a function of growth rate**

	GROWTH-RATE CONSTANT								
	0.1	0.2	0.3	0.4	0.5	0.6	0.7	0.8	0.9
Cell mass	1.84	2.00	2.16	2.40	2.67	3.05	3.55	4.30	6.60
RNA/cell	0.135	0.173	0.216	0.273	0.340	0.430	0.550	0.730	1.180
DNA/cell	0.085	0.089	0.092	0.095	0.100	0.106	0.115	0.131	0.185
Nuclei/cell	1.34	1.37	1.40	1.43	1.50	1.57	1.66	1.89	2.68
RNA/nucleus									
DNA/nucleus									
Ribosomes/cell									

*All masses are multiples of 10^{-13} g. "Nuclei" means nuclear bodies, and is an average. Calculate the number of ribosomes on the assumption that all RNA is ribosomal; this is an approximation.

exercise 8-7　　The data in Table 8-2 give the composition of *Salmonella typhimurium* as a function of growth rate. Plot these data in the same way.

TABLE 8-2

composition of Salmonella typhimurium *as a function of growth rate**

	GROWTH-RATE CONSTANT					
	0.35	0.70	1.05	1.40	1.75	2.10
Cell mass	2.2	3.2	4.6	6.5	9.4	13.5
RNA/cell	0.22	0.37	0.63	1.05	1.75	2.95
DNA/cell	0.035	0.045	0.058	0.075	0.097	0.123
Nuclei/cell	1.13	1.30	1.63	2.00	2.50	3.20
RNA/nucleus						
DNA/nucleus						
Ribosomes/cell						

*All masses are multiples of 10^{-13} g.

It is not necessary to comment further on these regularities at this time. They exist; they imply that there are some remarkable regulatory devices in these organisms for maintaining this balance. There is another meaning of balanced growth that does require some comment. If a population of bacteria is growing exponentially according to equation 2-2 and if the cells are in steady state, so their composition is not changing, then every separate component must also be increasing in mass according to equation 2-2. Thus, if P is the mass of protein per cell, then $dP/dt = \alpha P$; if D is the mass of DNA per cell, then $dD/dt = \alpha D$, and so on. The rate of protein synthesis per cell can therefore be obtained if both the growth rate and the amount of protein per cell are known. Of course, the rate of synthesis can also be determined experimentally.

Kjeldgaard and Charles Kurland measured the amounts of various cell components as a function of growth rate; they found a very interesting regularity. Over a wide range, the amount of "soluble RNA" per cell remains essentially constant. However, the amount of ribosomal RNA per cell increases dramatically with increasing growth rate. Over a wide range, *the rate of protein synthesis per ribosome is constant.* This implies that each protein factory—each ribosome—is working at full capacity at all times, and if the total rate of protein synthesis per cell is to increase, (that is, the cells are to grow faster), then the cell must make more factories.

We will now consider some of the greater changes that occur in cells during transitions from one steady state of growth to another.

8-12 TRANSITIONS BETWEEN STEADY STATES

When a cell is changed from a rich to a poor medium, or vice versa, interesting changes occur in its gross composition, which reflect the induction and repression of different enzymes and the reshuffling of its RNA components. A change from a rich medium to a poorer one, where the cells grow more slowly, is called **decelerating growth or shift-down.** A change from a condition of slow growth to one of faster growth is called **accelerating growth** or **shift-up.** Both types of shift reveal the very tight coupling between the major cell components.

Figure 8-15 shows the general pattern of synthesis during decelerating growth. There is always a short lag during which no RNA and protein synthesis occurs to any extent, apparently because the cells are suddenly starved and they must induce a number of new enzymes. During this time, there may be some breakdown of old cell polymers—a *turnover* of cellular materials—to supply the monomers required for new, essential enzymes.

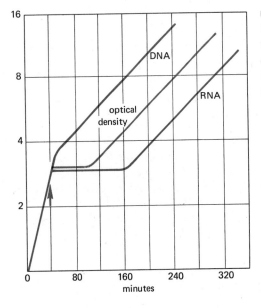

FIGURE 8-15

Pattern of biosynthesis during a shift-down experiment. Each division of the ordinate represents a doubling of the component; before the shift, the lines are drawn superimposed on each other. Optical density measures total bacterial mass.

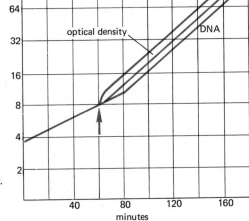

FIGURE 8-16

Pattern of biosynthesis during a shift-up experiment. The graphs are drawn as in Fig. 8-15.

After this lag, RNA and protein synthesis begin again at new rates characteristic of the new steady state. The ratios of RNA and protein are now adjusted so that the rate of protein synthesis per unit of ribosomal RNA is again constant. Let us remember that there is a much greater *variety* of enzymes being synthesized now, but that the total rate of synthesis is much lower than it was before the transition.

Just the opposite occurs during accelerating growth (Fig. 8-16). The first obvious event is a synthesis of new RNA at an accelerated rate so there are now more sites for protein synthesis and the cells can begin to grow faster. This again indicates that the existing ribosomes must be fully saturated.

The tight coupling between RNA and protein synthesis can be seen even when a

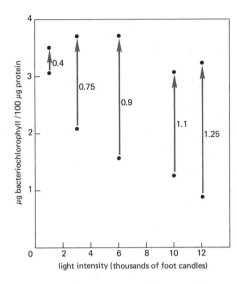

FIGURE 8-17

Steady states and transitions in Rhodopseudomonas
*spheroides. Bacteria grown at various light intensities
were shifted to 80 foot candles. Their bacteriochlorophyll contents (μg/100 μg protein) shifted to
higher values. Numbers on the arrows are rates of
bacteriochlorophyll synthesis over protein synthesis
(μg/μg phenylalanine incorporated).*

single enzyme is induced. In some strains of *E. coli*, 10–20% of the cell protein in a
fully induced population is β-galactosidase. After induction, these strains grow
somewhat more slowly, but they also make more ribosomes and thus achieve a steady
state that conforms with the patterns of growth we have seen elsewhere.

8-13 STEADY STATES IN PHOTOTROPHS

Phototrophic bacteria should be expected to have control mechanisms that are very
similar to those of chemotrophs; while they have not been studied as much, some
regularities have emerged that are clearly very similar to chemotrophic bacteria.

The growth rate of a phototroph can be changed greatly by manipulating the light
source. Cells grown in only 100 foot candles of light may have a doubling time of about
6 hours, but as the light intensity increases their growth rate increases also; it starts
to level off at higher intensities, and at 5000 foot candles the cells have a doubling time
of about 2 hours.

G. Cohen-Bazire, Roger Stanier, and William R. Sistrom have found that the
chlorophyll (actually bacteriochlorophyll) content of phototrophic bacteria is inversely
proportional to the light intensity. Cells filled with electron-transport membranes, like
those in Fig. 8-6, are characteristic of low light intensities; under these conditions,
the cells make an enormous photosynthetic apparatus to trap as much light as
possible. As the light intensity increases, the cells adjust their composition to reduce
the amount of chlorophyll. At the same time, they reduce their protein content; we now
understand this in terms of the composition of the photosynthetic apparatus, for since
the electron-transport membranes are largely protein, the protein and chlorophyll
contents must change in parallel. There is a very tight coupling between the two
components; Sistrom has shown that in *Rhodopseudomonas* chlorophyll synthesis is
dependent upon protein synthesis, so when the latter is inhibited, the former is also.
Figure 8-17 shows the results of an experiment in which cells grown at various high

light intensities were transferred to a low light intensity. Notice that their chlorophyll contents per unit of protein change from various low values to approximately the same high value and that during the transition their ratios of chlorophyll to protein synthesis are proportional to the magnitudes of the transitions.

SUPPLEMENTARY EXERCISES

exercise 8-8   Suppose we grow some heterotrophic bacteria with radioactive carbon dioxide ($^{14}CO_2$) which they incorporate into oxalacetate through a Wood–Werkman condensation with pyruvate. Which carboxyl group of oxalacetate is labeled?

exercise 8-9   Now allow this labeled oxalacetate to condense with acetyl-CoA and go into the TCA cycle.

(a) The resulting citrate is a symmetrical molecule; is it now labeled symmetrically or asymmetrically?

(b) If this citrate is converted into α-ketoglutarate, which carboxyl groups should be labeled? Remember that in solution you cannot tell one end of citrate from another.

(c) When the experiment was actually performed, the label was found in only one of the carboxyl groups of α-ketoglutarate. How can you explain this? (This result, which was considered a fatal blow to the TCA cycle as Krebs proposed it, was explained by A. G. Ogston. The citrate is *not* in solution; it is bound to enzymes, and if the binding requires a fit at three points, as shown here, the symmetrical molecule is made effectively asymmetrical and a label in one of its carboxyl groups will not be randomized.)

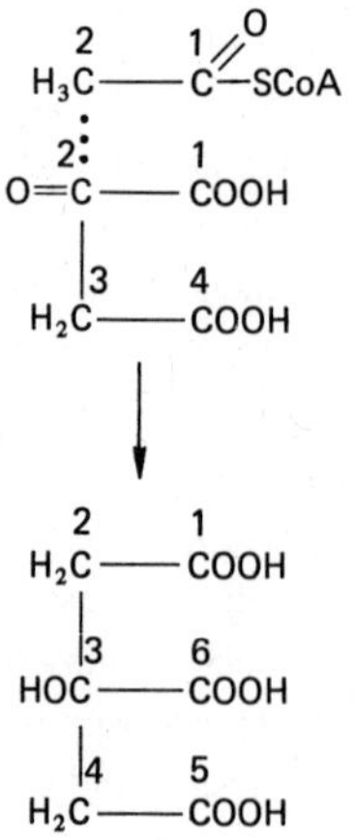

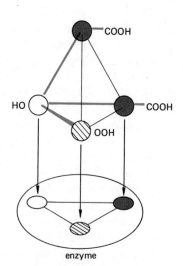

The gray carbon atoms are *not* equivalent—only one can attach to the enzyme.

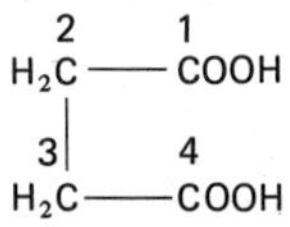

2   1
H₂C ——— COOH

3| 4
H₂C ———COOH

succinate

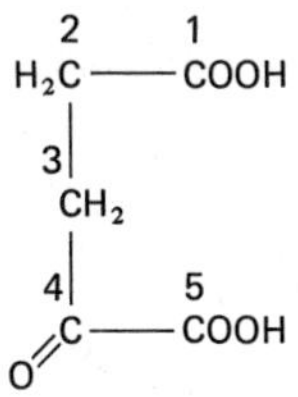

2   1
H₂C ——— COOH

3|
CH₂

4| 5
O⁄C ———COOH

α-ketoglutarate

exercise 8-10 Succinic acid is also symmetrical, but it is not bound to an enzyme and you cannot tell one of its carboxyls from the other. If we start with α-ketoglutarate that has one carboxyl group labeled, and if we define the specific activity of ^{14}C in that carboxyl group to be 1.0, then the specific activity in each carboxyl group of succinate will become 0.5. Now, let us start with unlabeled oxalacetate and use acetyl-CoA that has a specific activity of 1.0 in its carboxyl carbon only. Follow the label around and around the cycle, using radioactive acetyl-CoA with each turn, and determine the steady-state specific activity of each atom in citrate, α-ketoglutarate, and succinate. [Hint: Notice that the label never gets into the "backbone" carbons of these acids (carbons 2, 3 and 4).]

exercise 8-11 An equilibrium mixture of citrate, *cis*-aconitate, and isocitrate contains 90%, 4%, and 6% of the three acids, respectively. What are the free-energy changes in the two reactions from citrate to isocitrate?

picture essay
intracellular
pools

In trying to imagine how cells must operate and in performing experiments to probe their insides, we soon encounter the concept of a **pool** of materials as an obligatory stage of metabolism. A cell contains a set of enzymes for the biosynthesis of an amino acid like valine; when these are operating normally they must create a pool of valine molecules that are withdrawn at a certain rate for protein synthesis. The pool must have a steady-state size determined by the rates of valine synthesis and withdrawal, and the regulatory devices of the cell may be looked upon as mechanisms for keeping the valine at a certain level. Repression of the valine biosynthetic enzymes by valine is just one of these devices; we will encounter others later.

We encounter this valine pool when we try to label cells with radioactive valine by adding it to the growth medium. The cells behave as if the free amino acid is first taken up into the pool and then proceeds to protein biosynthesis. We will explore this phenomenon briefly to see what labeling reveals about the pathways of flow of materials in cells.

We add ^{14}C-valine to a bacterial culture and start to withdraw samples. Each one is divided into TCA-soluble and TCA-precipitable material; the former should contain a soluble pool of valine and the latter should be protein with incorporated valine.

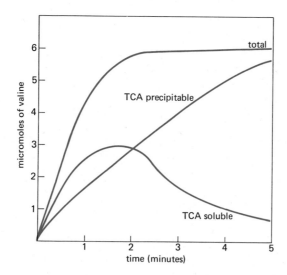

As shown here, the pool is filled more rapidly than the protein, but after about 2 minutes it starts to empty. Meanwhile, the radioactivity in protein rises steadily. The decline in radioactivity in the pool is due to exhaustion of the exogenous label from the medium; the intracellular concentration may be several hundred times the extracellular concentration.

Now the pool may be on the direct line from the outside to the protein, or it may be a side line:

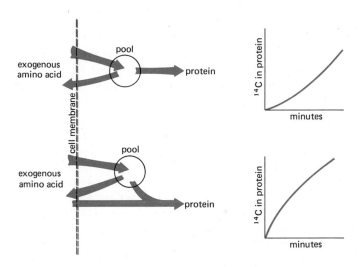

We can test this by first adding some unlabeled valine to fill the pool and then adding labeled valine. If the pool is on a side line, it will be filled up with unlabeled exogenous valine, so when radioactive valine is added it will first go into protein very rapidly, and enter the protein as labeled valine exchanges with the pool. However, if the pool is on the main line there will be a slight lag before exogenous valine gets through the pool and into protein. The graph shows that radioactivity enters protein with a lag, so the pool is on the main line:

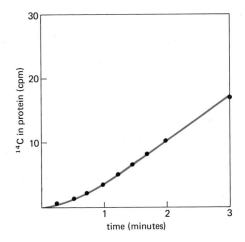

What happens if we starve the cells temporarily by taking away their glucose? The pool size doesn't decrease, but it doesn't increase, either. This suggests that the uptake of valine requires an energy source—it is an active process.

We can mistreat the cells in a variety of ways that might make valine in the pool leak out. However, if we chill them to 0°C (without freezing them) or if we take away their carbon source, the pool remains. The pool is suddenly decreased if we give the cells a brief osmotic shock, but as soon as the cells start to recover their ability to synthesize protein, they recover their pool. The pool is really quite stable.

Because valine looks like leucine and isoleucine, it might be interesting to see what happens when we put radioactive valine in the medium and then add various amounts of the other amino acids in unlabeled form. This experiment yields a spectacular result: The addition of either leucine, isoleucine, or some analogues of these compounds decreases the rate of valine uptake, but amino acids that do not have branched hydrocarbon chains have no effect. This means that there are specific sites somewhere in the cell membrane that are responsible for taking the three branched amino acids out of the medium, and these acids compete with one another for the sites. In several cases, mutants can be found that do not have the ability to pick up certain exogenous nutrients. This, and the great specificity of the sites, suggests that they are specific proteins, not very different from enzymes. We will call these sites **permeases;** the materials they are capable of picking up are **permeants,** and the whole process can be called **permeation.** Since the permeants are being carried across from the outside to the inside, many people prefer to call this **translocation.** (An older term with many emotional overtones these days is *active transport*.)

Sometimes, in order to explain the kinetics of translocation and of the movements of materials through intracellular pools, we have to assume that the cell also has "leaks" in it that allow materials to escape slowly. Quite generally, we have to assume that there is at least one step in translocation that requires the input of some free energy—particularly because the concentration of material in an intracellular pool may be many times the concentration extracellularly.

The question of the form the pools actually take inside the cells is more difficult. A material like valine may actually exist in several distinct pools in the cell, so the rate constants that we measure for translocation in or out are actually averages of several rate constants. Some of the valine may be free, some may be bound to proteins, and some may be bound in a special form at the sites of protein synthesis. Moreover, the valine pool may be exchanging rapidly with pools of other metabolites, since valine may be converted to other compounds. The only way to tell is to perform very rapid tracer experiments and try to follow the various fates of the labeled material.

On the whole, we can expect the size of any pool to be regulated at two different levels. First, by the *induction* and *repression* of enzymes, including permeases, which synthesize each metabolite, use it for other syntheses, and bring more of it into the cell. Second, by regulating the *activities* of enzymes which have already been made. The first is a long-term control, whose mechanism we shall discuss in Chapter 10; the second is a short-term control, whose mechanism we shall discuss in Chapter 15. Both are essential for the normal operation of the cell.

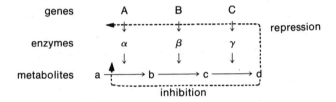

picture essay glycolysis and fermentation

Glucose and sugars that are easily converted to glucose constitute a major source of carbon and energy on earth. The pathway of glucose catabolism—**glycolysis**—is very important. It yields energy itself, and it transforms carbohydrates such as glucose into pyruvic acid. From pyruvate, some organisms use pathways into the TCA cycle, while others conduct a series of interesting and useful fermentations.

Glycolysis occurs via the Embden–Meyerhof–Parnas (EMP) pathway, which has two phases. In the first reactions, ATP molecules are invested, while in later reactions extra ATP molecules are obtained as dividends. The whole pathway is shown here:

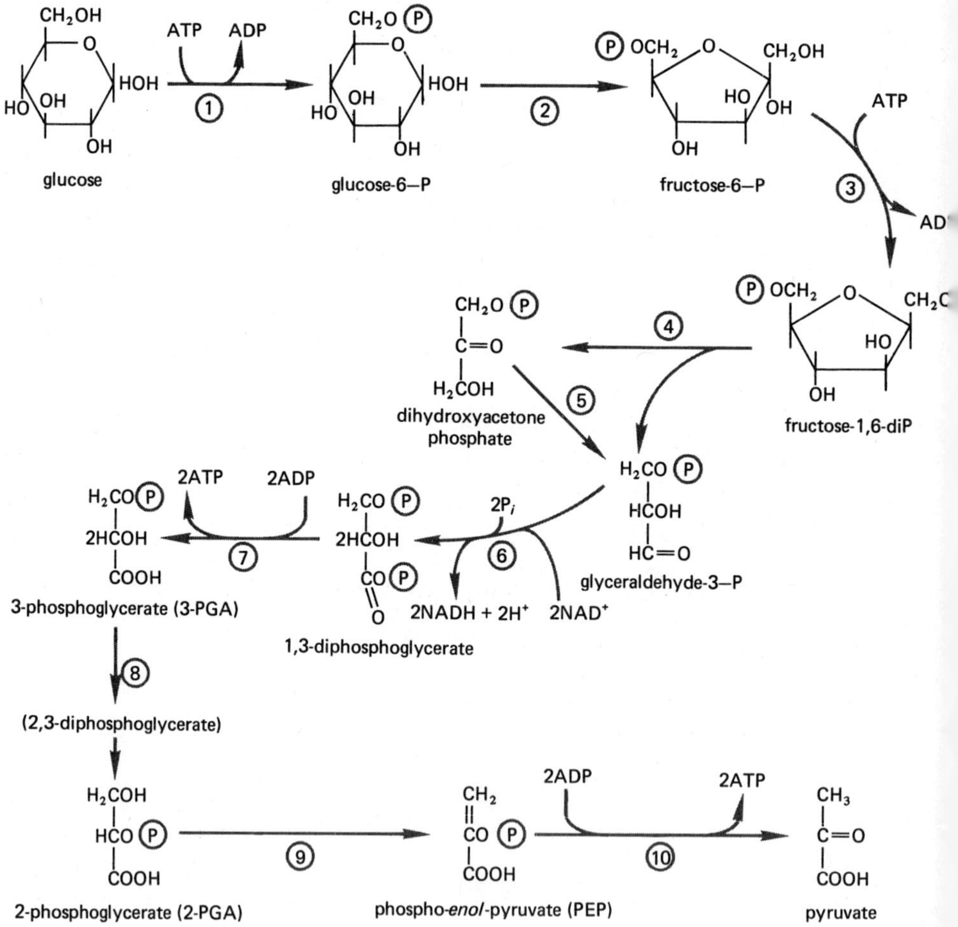

You see that two molecules of ATP used in steps 1 and 3 lead to the recovery of four molecules of ATP in steps 7 and 10.

Furthermore, two molecules of NADH are formed in step 6, and these can yield additional ATP molecules if they can reduce an appropriate electron-transport system. Since the ΔH of glucose is 686.5 kcal/mole, these reactions do not provide a very efficient conservation of energy in ATP, but they yield enough energy to support the growth of many organisms very nicely.

The first three reactions are quite simple. In step 4, the C_6 chain is split by the enzyme aldolase into two C_3 chains. This reaction is readily reversible, since the equilibrium mixture is 89% hexose and 11% trioses. (The enzyme gets its name because the reaction between an aldehyde group and a CH_2 group is called an aldol condensation.) In step 5, the acetone is isomerized to the aldehyde. From this point on, most of the reactions are energy yielding.

The oxidation of glyceraldehyde-3-phosphate is an example of a **substrate-level phosphorylation** (in contrast to the phosphorylations associated with electron-transport systems). The enzyme glyceraldehyde dehydrogenase has a tripeptide prosthetic group called **glutathione** (γ-L-glutamyl-L-cysteinyl-glycine) which reacts with glyceraldehyde-3-phosphate through its sulfhydryl group to form a bound substrate that is reduced with NAD:

```
HOOC
 |
H2NCH      SH
 |          |
(CH2)2     CH2  O      COOH
 |          |   ||      |
O=C—N —— C —— C —N—CH2
    H    H        H
```

```
H2CO Ⓟ                    H2CO Ⓟ          NAD+      NADH + H+    H2CO Ⓟ
 |                         |                                      |
 HCOH                      HCOH                                   HCOH
 |              HS-enz     |                                      |
 HC=O      ————————→       HCOH        ———————————————→           HC=O
glyceraldehyde-3—P         S-enz                                  S-enz
```

This reduced complex has a high transfer potential; its energy is conserved by substituting a phosphoryl group for the glutathione and then forming ATP directly:

```
H2CO Ⓟ          Pi                 H2CO Ⓟ       ADP       ATP     H2CO Ⓟ
 |                                  |                               |
 HCOH                               HCOH                            HCOH
 |                                  |                               |
 C=O     ———————————→               C=O     ————————————→           C=O
 |              HS-enz              |                               |
 S-enz                             O Ⓟ                             OH
```

The 3-phosphoglycerate that results from these reactions is converted into 2-phosphoglycerate through the doubly phosphorylated intermediate. The enzyme phosphoglyceromutase uses this doubly phosphorylated compound as a kind of bound coenzyme:

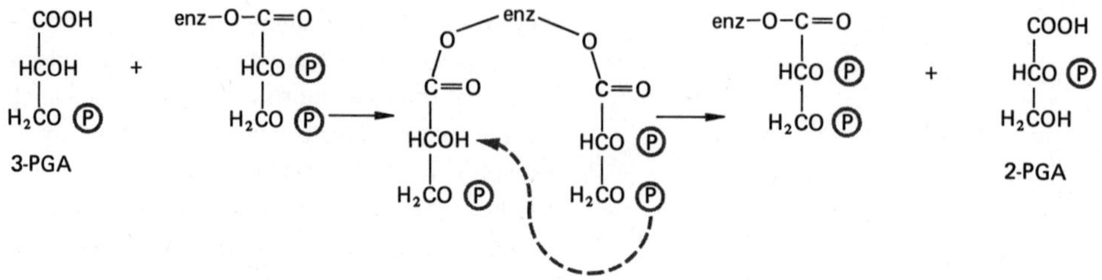

$$
\begin{array}{ccccccc}
\text{COOH} & & \text{enz--O--C}{=}\text{O} & & \text{enz--O--C}{=}\text{O} & & \text{COOH} \\
| & & | & & | & & | \\
\text{HCOH} & + & \text{HCO} \ \textcircled{P} & \longrightarrow \quad \longrightarrow & \text{HCO} \ \textcircled{P} & + & \text{HCO} \ \textcircled{P} \\
| & & | & & | & & | \\
\text{H}_2\text{CO} \ \textcircled{P} & & \text{H}_2\text{CO} \ \textcircled{P} & & \text{H}_2\text{CO} \ \textcircled{P} & & \text{H}_2\text{COH} \\
\text{3-PGA} & & & & & & \text{2-PGA}
\end{array}
$$

Finally, the 2-phosphoglycerate is converted into phosphoenolpyruvate and thence to pyruvate, through reactions that we have already seen. Pyruvate is the key compound in most of the important fermentations. For example, all animal tissues and many microorganisms can reduce pyruvate directly to lactate:

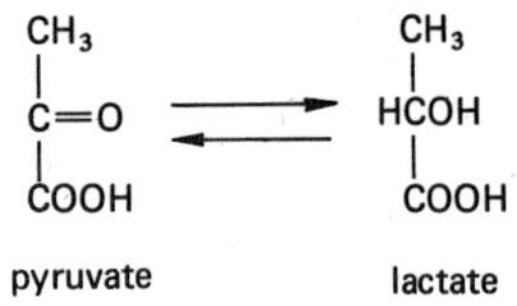

$$
\begin{array}{ccc}
\text{CH}_3 & & \text{CH}_3 \\
| & & | \\
\text{C}{=}\text{O} & \rightleftarrows & \text{HCOH} \\
| & & | \\
\text{COOH} & & \text{COOH} \\
\text{pyruvate} & & \text{lactate}
\end{array}
$$

This is what happens in your muscles when they catabolize glucose so fast that the mitochondrial electron-transport system can't keep up; rather than going into the TCA cycle, the excess pyruvate is converted into lactate. In this case, you will see that the electrons that were removed during glycolysis as NADH are transferred to pyruvate; remember that fermentation is defined as an oxidation process in which an organic compound is the electron acceptor.

The bacteria that produce lactic acid by fermentation in growing on various sugars are very important economically, for they are primarily responsible for the formation of cheeses, yogurts, and similar foods by fermentation of the lactose of milk. There are two major types of lactic fermentations; homolactic bacteria produce only lactate, while heterolactic bacteria produce ethanol as well.

The propionic bacteria conduct a fermentation of pyruvate to propionic acid that is the major reaction in the production of Swiss cheese. It begins with one of the Wood–Werkman reactions from pyruvate to oxalacetate and leads to propionate by partial reversal of the TCA cycle:

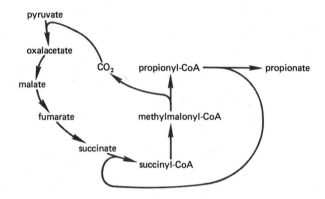

Most of the other fermentations utilize all or part of the pyruvic dehydrogenase system, so we may begin with the intermediate hydroxyethyl-TPP. The ethanol fermentation, which is so interesting to much of mankind, requires only the removal of acetaldehyde and its oxidation to ethanol:

$$CH_3HCOH\text{-}TPP \longrightarrow CH_3\overset{O}{\overset{\|}{C}}H + TPP \longrightarrow CH_3COOH + TPP$$

Many bacteria spoil foods by carrying out a butylene glycol fermentation, in which the hydroxyethyl-TPP is condensed with another pyruvate:

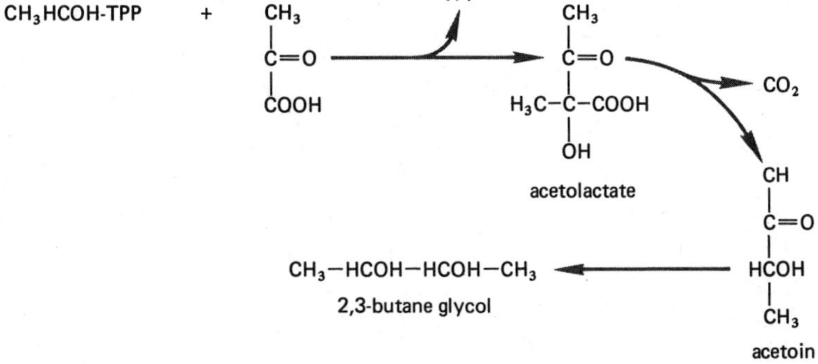

$$CH_3HCOH\text{-}TPP \quad + \quad \begin{matrix} CH_3 \\ | \\ C=O \\ | \\ COOH \end{matrix} \quad \xrightarrow{\ TPP\ } \quad \begin{matrix} CH_3 \\ | \\ C=O \\ | \\ H_3C-C-COOH \\ | \\ OH \end{matrix}$$

acetolactate

$$\longrightarrow CO_2$$

$$\begin{matrix} CH \\ | \\ C=O \\ | \\ HCOH \\ | \\ CH_3 \end{matrix}$$

acetoin

$$CH_3-HCOH-HCOH-CH_3 \longleftarrow$$

2,3-butane glycol

In other organisms, pyruvate is carried all the way to acetyl-CoA, as in aerobic respiration, and this is either converted into acetyl phosphate, which is used to form ATP, or oxidized to ethanol:

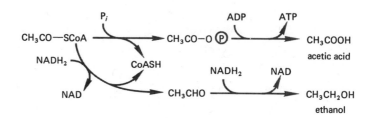

Two molecules of acetyl-CoA may also condense to make acetoacetyl-CoA. This reaction is the beginning of fatty acid biosynthesis, but it can also lead to butyric acid, the principal noxious substance in rancid butter, or to acetone and isopropanol:

$$2CH_3CO-SCoA \longrightarrow CH_3CO-CH_3CO-SCoA$$

CO₂ branch:
CH_3CO-CH_3 acetone $\xrightarrow{NADH_2 \quad NAD}$ $CH_3HCOHCH_3$ isopropanol

HSCoA branch:
$CH_3CH_2CH_2COOH$ butyrate $\longrightarrow$ $CH_3CH_2CH_2HCOH$ butanol

Through such simple reactions the bacteria make our lives both easier and harder. Many industries are based on carefully controlled fermentations by bacteria and yeasts. The results from some of the food industries are major delights for the palate (or major abominations, depending on your own cultural background). Many fascinating aspects of fermentation are discussed in Kenneth V. Thimann's book, *The Life of Bacteria* (New York: Crowell-Collier and Macmillan, 1963, 2nd ed.)

**bacterial
genetics**

9

In this chapter we will see how the principles of genetic analysis developed in Chapter 5 can be applied to bacteria and how sets of mutants can be obtained that will permit us to perform a penetrating analysis of any bacterial system.

A. General principles

9-1 MUTANTS AND PATHWAY ANALYSIS

It is hard to overemphasize the importance of finding appropriate mutants for analyzing cellular systems. You may be entirely unaware that some function exists until you find a mutation that abolishes or alters it. Every new mutation leads, potentially, to the discovery of a new structure or function, particularly to a new RNA or protein molecule whose existence was not even suspected. This has been particularly true for mutations that affect regulatory mechanisms. Moreover, even though it may be clear that a particular function must be fulfilled, an appropriate mutation can help to establish that it is associated with some particular RNA or protein molecule.

Mutants for metabolic pathways have been particularly useful in analyzing the pathways themselves. Since this is probably the most important application of mutational analysis in bacteria, we will discuss it here in some detail. The possibilities of analyzing metabolic pathways by means of mutants were first explored systematically by George Beadle and Edward L. Tatum with the common bread mold *Neurospora*. Normal *Neurospora* will grow on a simple, minimal medium consisting only of salts and glucose (plus the vitamin biotin in this case, which we will ignore just to simplify matters); since it makes all of its own components from these compounds, it is said to be **prototrophic.** A mutant that cannot make some compound is **auxotrophic** for that compound. Beadle, Tatum, and their associates collected many auxotrophs for different amino acids and analyzed several pathways with them. The same general methods they established can be applied to bacteria and many other organisms.

Suppose we want to find out something about the *E. coli* system for synthesizing histidine. We must first find some histidine auxotrophs, so we treat a growing bacterial culture with a mutagen or irradiate it with ultraviolet light. We now have a somewhat better chance of finding a mutant in this culture than in an untreated culture, but since mutation rates range from about 10^{-6} mutations per cell per generation downward, there is still only a tiny fraction of interesting cells in the whole culture. To avoid a task that would be tedious at best and probably impossible in most cases, we employ the Darwinian principle of selection.

A very clever and useful selective technique was devised by Bernard Davis; knowing that the antibiotic penicillin kills only growing bacteria (by disrupting cell-wall synthesis, so any growing cell makes a defective wall and lyses by osmotic shock), Davis realized that it could be used to select cells that are not growing because they are auxotrophs. To select histidine auxotrophs, we grow the mutagenized culture to a fairly high density (say, 10^8 cells/ml) with histidine and then wash the cells into fresh medium that contains a variety of amino acids, vitamins, and nucleotide bases, but does *not* contain histidine. We incubate this culture with penicillin. Prototrophs or cells that are auxotrophic only for one of the added nutrients will try to grow, but they will commit suicide because of the penicillin. After about 90 minutes, most of the cells will have killed themselves and the remainder will be highly enriched for histidine auxotrophs. Other antibiotics can be used in a similar way to select yeast mutants, and one could probably devise an analogous procedure for almost any organism. Even the human mutants we will discuss are generally selected out of the population because they come into clinics and hospitals with specific medical complaints.

A further refinement of the selective technique, called **replica plating,** was devised by Joshua and Esther Lederberg. After we wash the culture out of penicillin, we plate small samples on plates containing histidine so the auxotrophs can grow. We then fix a piece of sterile velvet on a form that just fits inside the Petri plate, press it lightly onto the plate, and then stamp it lightly onto a clean plate that does not contain histidine (Fig. 9-1). This makes the second plate a *replica* of the first, except that histidine auxotrophs will not grow there. A quick comparison of the two plates will reveal all of the potential auxotrophs, which can then be picked and tested further. We designate all of the mutants *his⁻* and number them arbitrarily *his*-1, *his*-2, and so on.

On the basis of the operations we performed with phage mutants in Chapter 5, you will recognize that there are two general things to be done with these bacterial mutants. First, we must try to establish a map and locate the sites of mutation relative to one another. Second, we must try to establish some kind of functional test to determine the function that is defective in each mutant and to assign the mutations to different complementation groups or genes. We could perform these operations with phage because it is easy to infect bacteria simultaneously with different types of phage so their genomes can interact. The minimum requirement for genetic analysis in bacteria is therefore a system that will permit different cells to exchange genetic material with one another.

9-2 BACTERIAL MATING

While it was clear from Luria and Delbrück's work that bacteria have an ordinary hereditary mechanism just like other organisms, no one had managed to find genetic exchange between them, although many people had tried. Tatum and his student Joshua Lederberg were finally successful in 1946 because they realized that genetic exchange must be rare, if it occurs at all, and so they employed a selective technique to find the few cells in a culture that might have received new genetic material from other cells. They grew two different strains of *E. coli;* one was auxotrophic for substances A and B, and the other was an auxotroph for substances C and D. Neither of them would grow on a minimal glucose-salts agar. Lederberg and Tatum mixed the

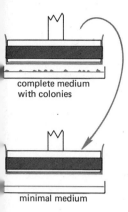

complete medium
with colonies

minimal medium

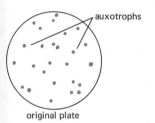

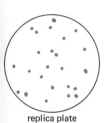

auxotrophs

original plate

replica plate

FIGURE 9-1

Selection of auxotrophs by replica plating. Comparison of the original plate and the replica shows two colonies that do not replicate; these may be auxotrophs.

FIGURE 9-2

Davis's U-tube experiment, which illustrated that contact between cells was necessary for genetic exchange.

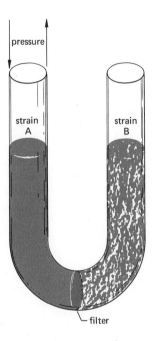

pressure

strain A

strain B

filter

two strains, let them incubate together for a while, and then plated samples on minimal agar. Prototrophic colonies appeared that could only have been recombinants due to an exchange of genetic material between cells. In other words, the recombinant type $A^+B^+C^+D^+$ had been selected from exchanges between the parents $A^+B^+C^-D^-$ and $A^-B^-C^+D^+$. In other experiments, different recombinants could be selected. For example, using the same two parents, only A^+C^+ recombinants could be obtained and among these the markers B^+/B^- and D^+/D^- occurred in all possible combinations. That is, these markers **segregate** among the recombinants, and the frequencies with which various markers segregate from one another is a measure of the (presumed) map distance between them. By performing crosses with a variety of different markers, Lederberg and his associates started to draw some simple linkage maps for *E. coli*.

Davis was properly skeptical; he wanted to rule out the possibility that the genetic exchange was a transformation of live cells by DNA from dead cells in the medium, so he performed the experiment shown in Fig. 9-2. The two strains of auxotrophic bacteria were placed on opposite sides of a filter through which DNA but not cells could pass. Liquid was forced back and forth from one side to the other and then cells were plated to select for prototrophs; none appeared, indicating that the process of exchange was specific and required contact between cells. Although no one had studied the process microscopically, it was assumed that mating between bacteria is similar to the *conjugation* that was already well known in some protozoa, where two different cells come in contact with one another and exchange nuclei, after which some genetic processes can occur. However, with this view in mind it became difficult to reconcile all of the data on frequencies of recombination between bacteria; it became difficult to draw consistent maps. The process of genetic exchange between bacteria appeared to be different from anything that had been seen in other organisms.

The resolution of the problem came primarily from experiments by William Hayes which demonstrated that the two parents in a bacterial cross do not play identical roles. In some of the crosses he was performing, Hayes used the markers for sensitivity and resistance to the antibiotic streptomycin, designated sm^s and sm^r. Call his parent strains X and Y. He found that if he performed the cross X sm^s × Y sm^r and plated on agar containing streptomycin, he obtained a normal yield of various recombinants. However, if he performed the reciprocal cross X sm^r × Y sm^s and plated on the same agar, he got no recombinants. This result indicates that strain X can be dispensed with after mating, but strain Y cannot be. Therefore, instead of a two-way exchange of material between cells there must be a one-way transfer of material from strain X, which acts as a *donor,* to strain Y, which is a *recipient.*

The Lederbergs and L. L. Cavalli found that only certain strains of *E. coli* are fertile with one another. Fertile strains, F+, will give recombinants with one another and with nonfertile F− strains, but F− strains will not recombine with one another. In a cross of F+ by F−, the F+ always acts as the donor and the F− as the recipient. Furthermore, mutations can occur from the F+ to F− condition; but when an F− culture is mixed with an F+ culture, the F− cells very quickly become F+. Thus, fertility in bacteria is like a disease. It is caused by a transmissible factor called F, the **fertility factor,** that can be transferred from cell to cell; F clearly has some of the characteristics of a virus.

The fact that mating occurs through a nonreciprocal transfer explained some of the difficulties, but others soon appeared. In all crosses, a few markers were used to select those cells that had participated in a mating; the rest were allowed to segregate in order to obtain data on frequencies of recombination between them. It was soon shown that the frequency with which one of these markers appears among recombinants depends upon whether it was in the F+ or F− parent; many markers in the F+ appear rarely or at low frequency. Hayes was therefore led to postulate that the transfer of genetic material is not only one way, but also incomplete, so any F− receives only a part of the F+ genome.

The frequencies of recombination in all of these experiments were very low. Cavalli finally opened the door to more quantitative results when he found a mutant of one of the F+ strains which produced about 1000 times as many recombinants as its parent; he called it Hfr (now Hfr C) for **h**igh **f**requency of **r**ecombination. Hayes found another strain that we now call Hfr H, and many other Hfr's have now been isolated. Each one acts like a very efficient donor, but it transfers only a limited set of markers at high frequency. Each of them transfers the F factor itself with very low frequency and most of the recombinants from an Hfr × F− cross are F−.

Francois Jacob and Elie Wollman found that the frequency with which markers from Hfr H appear in recombinants depends upon the markers used in the selection. Table 9-1 shows that if they selected only those cells that have received the markers

TABLE 9-1

constitution of recombinants from an Hfr × F⁻ mating under different regimes of selection

MARKERS SELECTED	FREQUENCY OF RECOMBINANTS	PERCENT RECOMBINANTS WITH Hfr MARKER*							
		$T^+ L^+$	azi^s	$T1^r$	lac^+	gal^+	mal^+	xyl^+	man^+
$T^+ L^+ sm^r$	9.3	100	91	72	48	27	0	0	0
$gal^+ sm^r$	2.4	83	78	79	81	100	0	0	0

*The markers *lac, gal, mal, xyl,* and *man* refer to the abilities to ferment lactose, galactose, maltose, xylose, and mannose, respectively. *azi* and *T1* are resistance to azide and to phage T1.

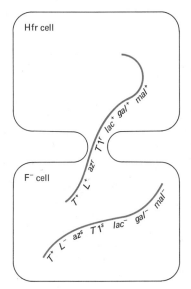

FIGURE 9-3
Markers are transferred from the Hfr to the F⁻ in a linear sequence, with T⁺ and L⁺ entering first and gal⁺ much later. If selection is made only for the first markers, the frequency with which the later ones are transferred decreases with their distance from the origin.

T^+ and L^+ from the Hfr, they obtained a gradient of recombination for the other markers. But if they selected those that received *gal⁺* they obtained comparable numbers of recombinants for these markers. This suggested the picture of mating shown in Fig. 9-3. It is assumed that the T^+ and L^+ markers are the first to be transferred by Hfr H and that the others enter in a linear sequence. If matings are selected only by the first markers to enter, the frequencies with which the other markers appear decrease as a function of their distance from the origin simply because there is a greater chance that something will accidentally interrupt the mating before they can be transferred. However, if matings are selected by using one of the later markers, such as *gal⁺*, then all of the markers that enter before *gal⁺* will already have been transferred and their frequencies among recombinants will depend upon events that occur inside the F⁻ cell after transfer is complete.

To substantiate this picture, Jacob and Wollman repeated the cross, but at intervals after mixing the parent strains they removed samples to a Waring blender and spun

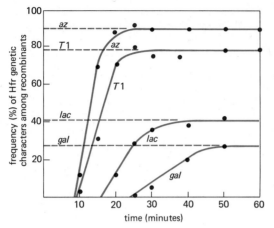

FIGURE 9-4

Frequency of Hfr markers appearing among recombinants in an interrupted mating experiment. Markers farther from the origin enter later than those near the origin and do not appear in as great a fraction of the recombinants.

TABLE 9-2

sequences of marker transfer by various Hfr strains

Hfr H	Hfr 1	Hfr 2	Hfr 3	Hfr 4
T	L	pro	ade	B₁
L	T	T1	lac	ilu
azi	B₁	azi	pro	mal
T1	ilu	L	T1	trp
pro	mal	T	azi	gal
lac	trp	B₁	L	ade
ade	gal	ilu	T	lac
gal	ade	mal	B₁	pro
trp	lac	trp	ilu	T1
mal	pro	gal	mal	azi
ilu	T1	ade	trp	L
B₁	azi	lac	gal	T

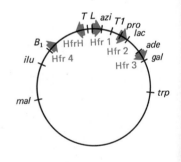

the cells vigorously to interrupt the matings artificially. Figure 9-4 shows the frequencies with which several Hfr markers appeared among the recombinants. It is clear that the Hfr genophore is being transferred linearly to the F⁻ cell, for each Hfr marker makes its first appearance at a specific time. In fact, one of the best ways to determine the main features of the bacterial map is to measure the number of minutes between the times at which bacterial markers are transferred.

Jacob and Wollman then took one F⁺ strain and isolated several different Hfr's from it. As Table 9-2 indicates, they found that each strain transfers a particular part of the genome in a specific sequence and direction. But all of these sequences can be derived from a *circular map* by cutting it at some arbitrary point. Jacob and Wollman therefore proposed that the *E. coli* map and genophore are both circular. An F⁺ strain carries the F factor somewhere in its cytosome, but it can undergo random transformations that are very much like mutations, in which the F factor attaches itself randomly to some point on the genophore. The cell is then an Hfr, and the point of attachment of the F factor determines the point where the circle will be broken during mating and transfer will begin. The origin and the F factor itself are on opposite sides of the break, so the F factor is attached on the far end of the genophore and is transferred with the lowest frequency of all markers.

9-3 TEMPERATE PHAGE AND LYSOGENY

So far, we have used T4 exclusively as an example of a phage because a great deal is known about it and because it behaves in a rather simple manner, as do all phages of the T1–T7 series. They can only multiply in a **lytic cycle:** They adsorb to their hosts, inject their DNA, and then there is rapid intracellular DNA replication and synthesis of new phage proteins, which ends when mature phage virions are made and the cells lyse. Phage that can do only this are called **virulent.**

The virulent phages are probably in the minority; most bacterial viruses are **temperate,** and when they infect cells they have an alternative to the lytic cycle. Temperate phage infect in the same way that virulent phage do, and sometimes when their DNA is injected it begins to replicate rapidly and goes into a lytic cycle. But somewhat more often, it enters the **lysogenic** cycle instead. Rather than replicating, it attaches itself to the bacterial genophore and actually becomes a part of the genophore; in this condition, it is called a **prophage.** The prophage is replicated along with the rest of the genophore and so is passed along to all of the progeny of the original cell. Bacteria that carry prophages have two principal characteristics. First, they are **immune** to further infection by other phages of the same type and of many related types. Second, they have the capacity to produce a burst of phage occasionally through spontaneous lysis; for this reason they are called **lysogenic** (and individual cells may be called **lysogens**). What actually happens is that a prophage occasionally leaves its stable association with the genophore and enters the lytic cycle; this happens in roughly 10^{-3}–10^{-5} of the cells in the culture per generation, so there are always mature phage virions in the growth medium with the cells.

Because the well-known T phage with which Delbrück and his school were so successful are all virulent, for a long time there was some reluctance to believe in lysogeny. However, lysogeny was one of the phenomena that many of the older phage workers, before Delbrück, were most concerned with. Back in the 1920s, Elie Wollman's parents had shown clearly that lysogenic bacteria do not contain mature phage virions that can be released by artificial lysis. They proposed that the phage must exist in some noninfectious form that can be transformed to an infectious condition. However, no one was able to prove that lysogenic cells could really exist in the absence of free phage, nor was there any information about the mechanism whereby free phage were liberated. The definitive demonstration of lysogeny came from Andre Lwoff and Antoinette Gutmann about 1950. They isolated a single cell and followed its growth under a microscope. Each time it divided, they removed one of the daughter cells and plated it on agar. The cell went through 19 divisions without lysing and without liberating any phage; but every daughter cell produced a colony of lysogenic cells, proving that lysogeny is a real condition that persists in the absence of free phage. They went on to follow small clones of cells under the microscope and observed that at intervals a cell would occasionally lyse and liberate phage.

Lwoff therefore proposed that temperate phage exist in a prophage state in lysogens. They can be **induced** by various agents, including ultraviolet light, to enter the vegetative state and go through a lytic cycle. However, the nature of the association between prophage and the host genophore did not become clear until some years later. Esther Lederberg found that a strain of *E. coli* she had been working with, K12, is lysogenic for a phage that she called λ. To indicate that a strain carries a certain prophage, we write the symbol for the phage in parentheses, so this strain is designated K12(λ). She could also isolate nonlysogenic derivatives, and so she used the character of lysogeny like any other marker in matings between F$^+$ and F$^-$ cells, designating it *ly$^+$* or *ly$^-$*. She found that the *ly* marker is closely linked to *gal* but that the crosses were strangely nonreciprocal. A cross of *ly$^-$* F$^+$ × *ly$^+$* F$^-$ gave normal recombination, but it was virtually impossible to transfer the lysogenic character from an F$^+$ to an F$^-$ cell. Jacob and Wollman then studied this phenomenon with Hfr × F$^-$ crosses. They found that λ lysogeny maps close to *gal* and that crosses from lysogenic Hfr strains proceed perfectly well up to the time when the λ-*gal* segment of the genophore is first transferred to the F$^-$. Once the λ-*gal* region is transferred, the cell that receives it *lyses* and liberates phage λ. This can only mean that the prophage itself resides at a site very close to *gal*. When it enters the F$^-$ cell it is induced; this phenomenon is called **zygotic induction** since the F$^-$ cell in which recombination with Hfr material occurs can be called a zygote (p. 273). Furthermore, since induction occurs only on transfer of a lysogenic genophore to a nonlysogenic cell, there must be something in a lysogenic cell that prevents induction and lytic growth; for now, we will call it the "immunity substance." This hypothetical substance explains both immunity and zygotic induction, for it is assumed that when a λ genophore enters an immune cell—either by injection from a free phage virion or transfer during mating— the "immunity substance" keeps it from replicating. A λ genophore can replicate only when it enters a cell that has no "immunity substance," or when this substance is destroyed by ultraviolet light or some other agent. These relationships are summarized in Fig. 9-5.

9-4 TRANSDUCTION I

In 1952, Lederberg's student Norton Zinder tried to find genetic exchange in *Salmonella,* which is a close relative of *E. coli.* He picked two auxotrophic strains, LT-2 and LT-22, mixed them, and then plated to find prototrophic recombinants. Lo and behold, recombinants did appear, showing that *Salmonella* also have a mechanism of genetic exchange. Zinder then tried Davis' U-tube experiment, expecting to get the same results that had been obtained with *E. coli,* but in this case, strangely enough, recombinants appeared without cell contact, indicating that a filterable material was passing from cell to cell and carrying genetic information. But the agent was not DNA; further analysis showed that strain LT-22 carries a temperate phage, P22, that is responsible for carrying bits of bacterial DNA from one cell to another. This mode of genetic transmission is called **transduction.**

We now know that many temperate phage in a variety of bacteria can transduce

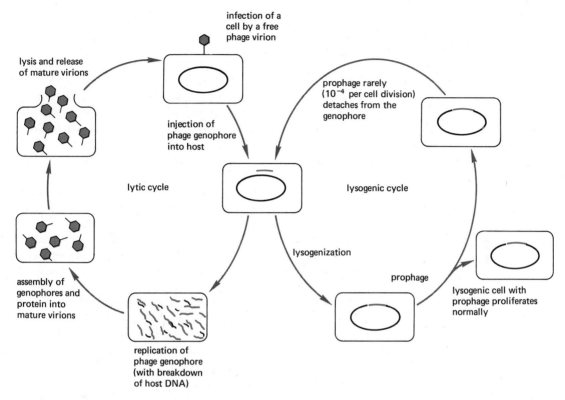

FIGURE 9-5

The lytic and lysogenic cycles as alternative modes of multiplication for a temperate phage.

genetic markers from cell to cell. These transductions are divided into two groups. Phage such as P22 in *Salmonella* and phage P1 and P2 in *E. coli* perform *generalized* transductions; apparently they can pick up any marker in their host cells. However, phage like λ carry out *specialized* transductions; each of these phage can be mapped at a specific point on the standard map and can only pick up markers that lie very close to these points. Phage λ, for example, can only transduce markers of the *gal* genes on one side of its prophage position and *bio* (biotin synthesis) genes on the other side. We shall see, shortly, that transduction permits us to perform detailed genetic analyses of bacteria.

B. Episomes and genetic exchange

9-5   THE EPISOME CONCEPT

A bacterium can be either fertile or infertile, F^+ or F^-. It can carry the F factor, but it clearly does not need it. However, once inside a cell the F factor can exist in two states. There may be one or a few F factors somewhere in the cytosome that can replicate independently and transfer themselves from cell to cell very efficiently; the cell is then F^+. Or the cell may be Hfr, with the F factor attached stably to its genophore; in this state, the F factor apparently can effect a transfer of bacterial markers into an F⁻ cell.

A bacterium can also be either lysogenic or nonlysogenic for a temperate phage. That is, it may carry a prophage, but it clearly does not need it. However, once a phage genophore has gotten inside a cell, it can apparently exist in two states. It may become a vegetative phage that replicates independently, makes its own protein, and eventually produces a burst of mature virions that can infect other cells. Or the genophore may attach itself stably to the bacterial genophore and become a prophage that replicates along with the bacterial DNA.

Jacob and Wollman noticed the close parallelism between the F factor and temperate phage. It was obvious to them that these might be considered two examples of the same general phenomenon and that other examples might be discovered. It is now clear that bacteria—and probably organisms in general—can harbor many extrachromosomal genetic elements, or genetic elements that can behave independently of the bacterial genophore. These elements are now called **plasmids,** and there are probably many more plasmids than we know, even though new ones are always being discovered. Plasmids can probably be defined best as *nonessential* or *dispensible* genetic elements that can exist separately from the host genophore. However, Jacob and Wollman defined a class of plasmids, the **episomes,** that have one other property; episomes are dispensible, but if an episome does exist in a cell it can be in one of two states: *autonomous,* replicating at its own pace as an independent element, or *integrated* into the bacterial genophore and replicating along with the rest of the genophore.

These are the definitive characteristics of plasmids and episomes, but they have other common characteristics. First is **conversion;** a plasmid is a genetic element that carries specific information and some of this information is invariably expressed in the host cell, giving it new properties. The most important changes seem to be in cell-surface properties. For example, many bacteria produce long, thin protein tubes called **fimbriae** or **pili** on their surfaces; they look somewhat like flagella, but they do not move. Cells that carry F factors have special **F-pili** that are required for the mating process and for DNA transfer from Hfr to F⁻ cells. Lysogenized cells exhibit other surface changes which we will discuss in Chapter 20.

Cells that harbor plasmids typically exhibit **immunity** and **exclusion** toward that plasmid and related ones. A cell lysogenized by a certain phage is immune to further infection ("superinfection") by that phage and related ones; these phages can inject their DNA, but it will not replicate. A cell that carries an F factor will not accept other F factors. Remember also that *r*II mutants of phage T4 do not multiply in strain K; actually, the strain is K(λ), and these mutants cannot multiply in any bacteria that carry λ prophage.

For our purposes, the most significant feature of episomes is their ability to mediate genetic exchanges between bacteria. With the exception of transformation, which requires only naked DNA, every genetic exchange between bacteria is mediated by some kind of episome. We will develop a general picture of episome-mediated transfer that can then be applied to particular cases.

There is now a great deal of physical evidence, from electron microscopy and elsewhere, that all episomes are circular DNA molecules or can become circular at some point in their multiplication. On the basis of genetic data that indicated a circular map for phage λ, Allan Campbell proposed an important model for interactions between episomes and bacterial genophores, which is illustrated in Fig. 9-6. Each episome is supposed to have an attachment region att^e that is similar in structure (base sequence) to a corresponding region on the genophore, att^B. The two circles join at their common attachment sites and a crossover occurs that fuses them into one circle. Thus, the episome becomes physically integrated into the bacterial genophore.

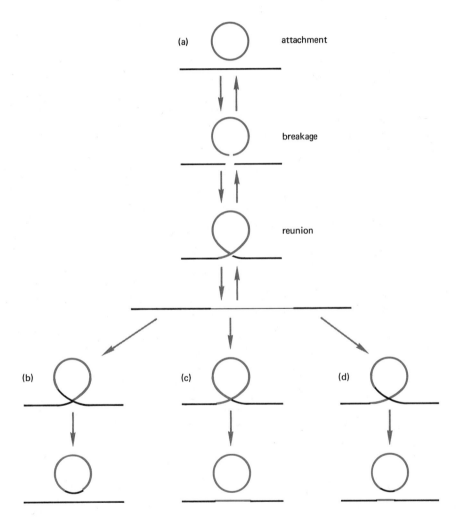

FIGURE 9-6
Interaction of an episome with a bacterial genophore. (a) Attachment of the episome to a complementary site on the genophore and insertion through a break-reunion crossover. The process may be reversed precisely or it may be reversed imprecisely to make (b) a larger episome carrying some bacterial DNA, (c) a smaller, possibly defective episome that has left some of its DNA behind, or (d) an episome that has left some of its DNA and picked up some of the bacterial DNA.

As Fig. 9-6 shows, the process is completely reversible, but it will sometimes occur in reverse imperfectly, so that either a piece of the episome is left in the genophore, or a piece of the genophore is carried away by the episome, or both. We shall now see how this model can explain most of the episome-mediated exchanges between cells.

9-6 FERTILITY FACTORS

Francois Jacob, Sydney Brenner, and Francois Cuzin proposed a model for the transfer of F factors from cell to cell. Just as the bacterial genophore is attached to the cell membrane, it is proposed that the F factor is attached at a similar membrane site. Transfer of the F factor occurs through a replication stimulated by contact with the F⁻ cell (Fig. 9-7).

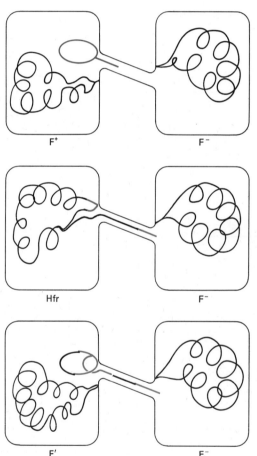

FIGURE 9-7

Genetic transfer by F factors. The F factor transfers itself by replication initiated by some stimulus from the F⁻. In an F⁺ cell, the F factor just transfers itself. In an Hfr cell it transfers part of itself and as much of the bacterial genophore as it can. In an F' cell it transfers itself along with whatever small piece of bacterial genophore is attached.

This is the condition in an F⁺ cell. If Campbell's model is right, the F factor should be integrated into the genophore as shown in the figure. If this Hfr cell comes in contact with an F⁻ cell the F factor will again be stimulated to replicate and begin transferring itself, but now it is attached to the entire bacterial genophore and so will transfer bacterial DNA until something breaks off the mating process. The piece of F DNA on the far end of the genophore will rarely be transferred in this process.

If the reversal of F integration into a bacterial genophore is imperfect, the F factor may carry away bacterial markers. Such a factor is designated F′, with a notation of the markers that are carried. One of the most important is an F-*lac* factor that carries markers for the lactose system that has been a model for the study of genetic regulation. If F factors can attach to the whole genophore (Hfr) and to a small piece (F′), they should be able to attach to intermediate pieces. To test this possibility, Alvin Clark made a strain that carries two F factors, each of which is attached to about half of the genome. This strain has two linkage groups that are transferred independently in mating.

The Jacob–Brenner–Cuzin model makes at least two important predictions: (1) Genetic transfer should be inhibited by inhibitors of DNA replication and (2) one strand of the DNA transferred to the F⁻ should be old and the other should be synthesized at the time of transfer. The first prediction has been borne out by several experiments in which DNA synthesis is inhibited in various ways. Mark Ptashne performed a nice experiment that apparently confirms the second prediction. He infected bacteria carrying phage λ with the related phage 434 and found some mature λ virions among the 434 phage; many had exclusively prophage DNA in both strands, showing that they had been made from prophages released intact from the cell DNA without replication. Ptashne then found an F′ that carries the λ prophage region and grew this strain in a medium with the heavy isotopes ^{13}C and ^{15}N, to label all of their DNA. He mated these cells in normal, light medium to F⁻ cells that were infected with 434 and then examined the λ progeny. Many of these phage contained DNA that was labeled in one strand, but none were labeled in both strands; therefore, the F′ DNA must have been replicated at the time of its transfer.

The transfer of markers mediated by an F′ factor is generally called **F duction** or **sexduction.** It is not obviously different in mechanism from the transfer by an Hfr, but it has different genetic consequences; to describe these, we need to devise a bit of terminology. Several of the terms we use for bacteria come from the classical genetics of eucaryotes. In a sexual, eucaryotic system, the cell in which chromosomes from the two parents are combined is called a **zygote.** One complete set of chromosomes is a **haploid set,** and through sexual processes cells can be made that have two, three, four, or more such sets; these are then called **diploid, triploid, tetraploid,** and so on. We take over many of these terms to describe bacterial genetics, but we often use the prefix *mero-* (from the Greek *meros,* a part) because only a fragment of the donor genome is usually transferred. The recipient cell in which recombination takes place is therefore a **merozygote.**

In an Hfr × F⁻ cross, the fragment of donor DNA that is transferred is a **merogenote.** It cannot replicate by itself and its only possible genetic effect is to recombine some of its markers into the F⁻ genophore so they appear among the progeny of the cross. However, an F′ factor not only carries bacterial markers but can also replicate itself autonomously. Therefore, an F′ cell is a **merodiploid**—it contains two copies of certain genes in a stable association.

Cells that carry one set of genes in their genophores and another set duplicated in an episome are generally called **syngenotes.** The set of genes in the genophore is then the **endogenote** and the set in the episome is the **exogenote.** If both exo- and endogenote carry the same markers, the cell is a **homogenote;** otherwise, it is a **heterogenote.**

The great importance of syngenotes, in general, is that they present opportunities for complementation studies. Two different mutations affecting the same general function can be combined in this way and tested for their effect upon one another. These tests are generally performed with transduction systems, which we will consider next.

9-7 TRANSDUCTION II

Let's first consider a well-known system of specialized transduction to establish some principles and then see how a generalized transduction differs. We take a culture of KI2 (λ) that is prototrophic for the genes of galactose fermentation (*gal$^+$*) and irradiate it with ultraviolet light to induce the prophages into the lytic cycle. We then take the lysate of λ particles and infect nonlysogenic KI2 cells that are *gal$^-$*. About 10^{-5} of these cells will be able to ferment galactose because they have received a *gal$^+$* marker by transduction. The phage lysate that was used is therefore an LFT (low frequency of transduction) lysate.

Some of these *gal$^+$* cells (called **transductants,** since they have been transduced) appear to be very stable, as if their *gal$^-$* mutation had been replaced by a *gal$^+$* marker. But others appear to be unstable; they occasionally segregate *gal$^-$* clones during growth, so they appear to be syngenotes that carry *gal$^+$* genes in a prophage that is lost occasionally and *gal$^-$* genes in their genophores. When these syngenotes are irradiated with ultraviolet light, a lysate is produced in which about half of the phage particles carry the *gal$^+$* marker and can transduce other cells. This is an HFT (high frequency of transduction) lysate.

When the phage in an HFT lysate are centrifuged in CsCl, they form two bands. One band has the same density in every preparation, but the other differs from one HFT lysate to another and may be either more or less dense than the standard. The HFT lysate consists of two different kinds of phage particles. One is a normal λ. The other is the transducing phage; it has been formed by the kind of error shown in Fig. 9-6d, where some of the phage DNA has been replaced by bacterial DNA carrying *gal$^+$* genes. Since it has had to leave some of its own genes behind, it does not have enough information to multiply by itself, and is therefore said to be **defective.** Hence it is called λ *dg* (for *d*efective *g*alactose-transducing). Therefore, a transducing lysate of any kind can be obtained only from cells that are *double* lysogens and carry both a λ *dg* and a normal λ; the latter supplies the functions missing in the former.

A specialized transducer like λ is obviously useful for exploring limited regions of the genome, particularly if an HFT lysate can be obtained. However, a generalized transducing phage like P1 can be used to explore many regions of the genome. The difference between the two seems to be in the type of DNA that can be wrapped up in the capsid protein of each phage. Thus, the Campbell model appears to be valid for all temperate phages and every one can pick up bacterial markers nearby through an error in excision from the genophore. In this way, phage ϕ80 can pick up nearby tryptophan markers (*trp*) and become a defective, transducing ϕ80*dt*, and phage P1 can pick up nearby lactose markers (*lac*) and become a P1 *dl*. However, for these generalized transducers another mechanism, generally called "wrapping choice," must be superimposed. Whenever any temperate phage grows lytically in a cell, the bacterial genophore is broken up into little pieces. Specialized transducers, like λ, ignore these pieces of DNA, but generalized transducers can pick them up and wrap them in capsid proteins as if they were pieces of phage DNA. These transducing phage particles obviously contain no phage DNA at all, but the phage capsids can still carry the bacterial DNA into other cells. These bacterial pieces still retain all the linkage relationships and functions that they had originally, and so they are enormously valuable in genetic analysis.

9-8 BACTERIOCINS

Many strains of bacteria engage in chemical warfare with their neighbors by producing proteins called **bacteriocins** that kill susceptible cells. The best known are the **colicins** produced by *E. coli* and some of its relatives. We consider them here briefly because they are produced by plasmids.

The colicins are proteins or protein complexes, including some lipoproteins; some of them look like very defective phages. Just as a few cells in a lysogenic culture lyse randomly and produce phage, some cells in a colicinogenic culture burst randomly and liberate colicins. Similarly, colicinogenic cells are immune to their own colicins. A sensitive cell has specific adsorption sites on its surface to which colicins attach; these are sometimes identical to phage adsorption sites, and they can be lost by mutation.

The plasmids that carry the genetic information for colicins are **colicinogenic factors (Col factors)**. They are as varied as any other plasmids. Some, such as Col E1, Col E2, and Col K, cannot transfer themselves. However, Col I can transfer itself efficiently and it can transfer Col E1 and Col E2; it also acts like a sex factor in *Salmonella* and has been used to map this host. Then there is a class of Col V factors that are incompatible with F factors and act rather like F factors with colicin genes. Some of these Col factors may behave episomally, but others are clearly not integrated into the host genome.

9-9 RESISTANCE FACTORS AND OTHERS

In 1957, Japanese physicians and health officials began to notice an alarming increase in the number of dysentery cases they encountered that were resistant to one or more of the antibiotics streptomycin, chloramphenicol, sulfonamide, and tetracycline. This type of dysentery is caused by *Shigella,* a close relative of *Escherichia;* investigation soon revealed that these bacteria were carrying plasmids which were called resistance transfer factors, but are now known as **R factors.** These plasmids carry genes that confer drug resistance upon their hosts; they spread through a population as rapidly as F factors, and any cell that picks up such a plasmid is obviously at an enormous selective advantage. R factors now enjoy a virtually world-wide distribution, and they are a serious problem in modern medicine.

The catalog of R factors is now very long; it includes factors with genes that confer resistance to many antibiotics, including some that are not yet widely used. Interestingly, many R factors carry genes that confer resistance to heavy metals, such as bismuth, cadmium, lead, and mercury; these may have a great selective advantage because so many of our natural waters are polluted with heavy metals from industrial processes. Still other R factors carry a type of colicin resistance.

Two very interesting plasmids of *E. coli* have been discovered in relation to disease production by this organism. *E. coli* is not generally pathogenic, but those strains that are apparently carry one of two plasmids—*Hly,* which produces a hemolysin that breaks down red blood cells, and *Ent,* which produces an enterotoxin.

The F factors, R factors, Col factors, temperate phages, and other plasmids are part of an enormous class of genetic elements. There is a whole spectrum, from nontransmissible, extrachromosomal elements at one end of the scale to virulent viruses at the other end, and all the shades and variations in between seem to be represented. We must expect to find such genetic elements associated with all cellular systems. Those associated with bacteria have been explored more than any others, but a variety of genetic elements probably exist even in our own cells. Such elements may explain many phenomena that are hardly understood today, including diseases such as cancers. It is valuable to keep this in mind when trying to understand the genetic structure of organisms.

C. Genetic analysis

9-10 COMPLEMENTATION AND GENE FUNCTION

Now that we have a basic system for carrying out genetic exchanges between different bacteria, we can return to the original problem outlined at the beginning of the chapter. We have collected a set of mutants—histidine auxotrophs, to continue the example—and now we wish to investigate the functional relationships between them and to establish a genetic map showing, if possible, the positions of the mutations and the genes they define.

As we indicated earlier, we assume that the system for histidine biosynthesis consists of a set of enzymes that catalyze the steps of a biosynthetic pathway from somewhere in the vicinity of the TCA cycle intermediates to histidine. If we indicate metabolites by italic lowercase letters, enzymes by Greek letters, and genes by capital letters, the system looks like this:

$$
\begin{array}{cccc}
\text{genes:} & A & B & C & D \\
& \downarrow & \downarrow & \downarrow & \downarrow \\
\text{enzymes:} & \alpha & \beta & \gamma & \delta \\
\text{metabolites:} & a \rightarrow b \rightarrow c \rightarrow d \rightarrow \text{histidine}
\end{array}
$$

We have made the assumption—perhaps an oversimplification—that each enzyme is informed only by a single gene, or, in other words, that each enzyme consists of only one polypeptide. If this is an oversimplification, the detailed genetic and biochemical analysis will soon reveal this fact.

The first problem, then, is to separate the various *his* mutants into functional classes or complementation groups, each corresponding to one gene. We suggested earlier that this might be done by making F′ merodiploids, and indeed this is possible in some cases; but the F′ merodiploids can only be made if a suitable F′ factor can be found and this is not simple. The general method is to use a kind of transduction, since a generalized transducing phage can be found that will carry any part of the genome. It happens that quite frequently the merogenote carried in by a transducing phage does not establish contact with the recipient genophore and recombine with it. If recombination occurs, then transductants can be detected because they grow into large colonies that have markers from the donor strain; for example, if we are transducing *his*⁺ markers into *his*⁻ recipients, we plate the recipients on agar without histidine and count the number of transductants that have arisen from recombination, since these are the only bacteria that will grow. But when this experiment is performed, many tiny **microcolonies** frequently appear on the plates. These colonies are the result of an **abortive transduction.** It can be shown that the process shown in Fig. 9-8 has occurred. The merogenote carried in by the transducing phage does not interact with the genophore, it is not destroyed, but it does not replicate, either. We know it is present because it is transcribed and directs the synthesis of whatever proteins it codes. When the original cell divides, the merogenote goes to only one daughter cell, and there it continues to direct the synthesis of some protein. The other daughter cell contains some of the enzymes, but only enough to sustain it for another division or two; then it stops growing. After several hours, a clone is formed in which only one cell is producing new enzyme; the other cells grow only a little and then stop. This clone forms a microcolony on minimal agar.

Abortive transduction allows us to perform a general complementation test between any two *his* auxotrophs. We grow transducing phage on one auxotroph and transduce the other with them; if microcolonies appear, the two mutations must lie in different complementation groups, but if no microcolonies appear, the mutations cannot complement each other and they are in the same complementation group. With this technique, we can perform the same kind of functional analysis for bacteria that we did for phage and assign each *his* mutant to a group, arbitrarily designated *A, B, C,* and so on. Our mutants are now designated *hisA*-1, *hisC*-2, *hisA*-3, *hisD*-4, and so on.

A fairly simple test can now be used to establish the sequence of enzymatic steps in the pathway, without knowing anything about the intermediates *a, b, c,* and so on. If a cell lacks enzyme γ, for example, it will be unable to convert *c* to *d; c* will accumulate in the cell, often to very high levels, and it will generally spill out into the surrounding medium. Similarly, if a cell lacks enzyme β, it will be unable to convert *b* to *c,* and *b* will accumulate in the cell and the medium. This phenomenon allows different mutants to **cross feed** one another, as shown in Fig. 9-9, for each mutant will feed those blocked at an earlier step and will in turn be fed by those blocked at later steps. This analysis will show that the sequence of biosynthesis is $a \rightarrow b \rightarrow c \rightarrow d \rightarrow$ histidine.

A further refinement of this analysis is to identify the intermediate compounds themselves. Sometimes intermediates can be guessed and tested for their ability to support the growth of these mutants. In general, the structures of the compounds

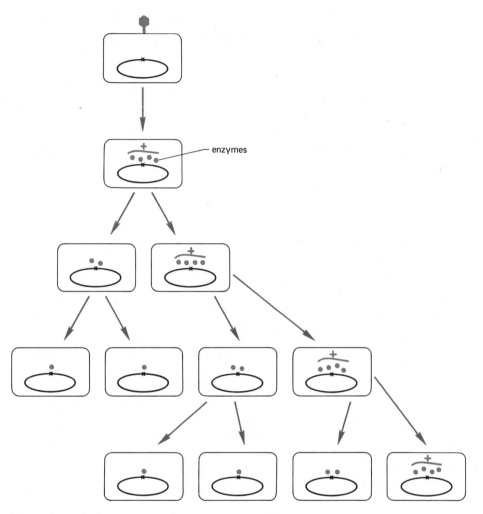

FIGURE 9-8

*Abortive transduction of a his⁻ cell by a his⁺ transducing phage. Histidine enzymes
are made only in the one cell that contains the merogenote; they are diluted out in the
other cells of the clone, which divide only a very few times more.*

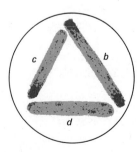

FIGURE 9-9

*Pattern of cross feeding with three mutants. The bacteria are plated on agar with just
enough histidine to support a light growth. Heavier growth occurs at certain points
because of cross feeding; mutant d can feed c and b, and mutant c can feed b,
which proves that b is blocked before c and c is blocked before d in the biosynthetic
pathway.*

are unknown, and they exist in such low concentrations in wild-type cells that it is virtually impossible to isolate and identify them. Appropriate mutants can therefore be used to accumulate enough of the material for chemical analysis.

exercise 9-1 Italine, francine, and germanic acid have been identified as intermediates in the biosynthesis of swedic acid, which is one of the chief components of Venusian proteins. A stock of *Venerius kennedyii,* a common Venusian bacterium, is plated on minimal agar supplemented with swedic acid and swedate auxotrophs are selected by replica plating. When plated on minimal agar supplemented with each of the four compounds in turn, the mutants fall into four categories with these characteristics (+ = growth, 0 = no growth):

PLATES SUPPLEMENTED WITH

GROUP	ITALINE	FRANCINE	GERMANATE	SWEDATE
A	+	+	0	+
B	0	0	0	+
C	0	+	0	+
D	+	+	+	+

(a) Write down the biosynthetic pathway of swedic acid.
(b) Which enzyme is defective in each group?
(c) If you were to plate a mutant of group A and a mutant of group B next to each other on a plate with a small amount of swedic acid, which mutant would feed the other, and what compound would be cross fed?

exercise 9-2 Mutants are collected that are unable to synthesize γ-saraboside. Complementation tests place them into three groups, A, B, and C. Mutants of classes A and C are able to grow on plates containing sarabasol, a compound that is similar to γ-saraboside. When grown on limiting amounts of γ-saraboside, class A mutants will feed class C mutants.

(a) What do you predict about the ability of class A and C mutants to feed class B mutants?
(b) What is the most likely sequence of mutational blocks in this system, and where does sarabasol fit into the pathway?

9-11 MAPPING BY TWO-FACTOR CROSSES

A rough map of the bacterial genophore can be obtained very easily by noting the time at which each marker enters in an interrupted mating experiment. However, the most interesting genetic data appear in analyses of fine structure, where individual loci and small blocks of very close genes are analyzed. We will now examine some of the general principles of mapping at this level.

The principles of mapping established in Chapter 5 for phage work carry over directly to bacterial mapping. For example, several fine maps have been established by using deletion mutants of the type Benzer used for the *r*II region. However, it is more common to perform crosses using two or more point mutations and then analyze the segregants from these crosses. Two classes of factors must always be considered. The *prezygotic* factors affect the transfer of genetic material from donor to recipient. Many factors can reduce the frequency with which a piece of DNA gets across from one cell to another, including the difficulty of picking up an appropriate piece of donor DNA, the factors that can destroy this DNA, the factors that can depreciate the recipient bacteria, and for all we know, the mood of the experimenter and the phase of the moon. The only way to take account of these factors is to run an appropriate control with every cross. The other factors are *postzygotic*, and affect the frequency of recombination between the merogenote and the recipient genophore; these, hopefully, are the factors that are really interesting and tell us something about the genetic structure of the genophore.

Remember that the merogenote does not replicate and that the only way markers from it can appear among progeny is for a recombination to occur in which donor markers are incorporated into a complete recipient genophore. This always requires an *even number of crossover events*, for if there were an odd number the recipient genophores would be broken and might not be able to replicate. Now suppose we bring a merogenote from a *his*-1 donor into a *his*-2 recipient and select for prototrophic recombinants by plating on minimal medium. We are selecting for merozygotes in which the following crossover has occurred:

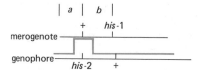

One crossover has occurred in region *a* and the other in region *b*, but the frequency with which prototrophs appear does not measure anything in particular by itself because it reflects both prezygotic and postzygotic events.

We now eliminate all events relating to transfer by using a wild-type donor with the same recipient and again selecting for prototrophs. In this case, we are selecting for an exchange of the type

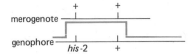

In this cross, there must still be two crossover events, but they are relatively unrestricted; one must occur on the left of the *his*-2 marker and the other must occur on the right, but neither of them has to occur in a limited region between two markers, as in the first cross. The second cross should yield considerably more recombinants than the first and the ratio

$$\frac{\text{prototrophs obtained in cross 1}}{\text{prototrophs obtained in cross 2}}$$

should be a measure of the distance between *his*-1 and *his*-2. Obviously, if *his*-1 and *his*-2 were at the same site, it would be impossible to obtain recombinants in cross 1 and this ratio would be zero; the farther apart the two markers are, the greater the ratio will be and it can therefore be used as a relative measure of the distances in this region.

At the risk of being tedious, let's reemphasize one basic point. When we perform one of these crosses with a few million organisms, many, many different patterns of recombination actually occur. The crossover points are placed entirely at random on the genomes. Most of the recombinants are not interesting to us—or, at least, we have no simple way of finding out what they are. Out of all these millions, we *select* the very few recombinants that have interesting properties; all the rest die on the Petri plate because we have created conditions where they cannot grow. We then say that the number of recombinants we obtain with a given genotype is a function of the probability that certain events occur, so by counting recombinants we measure probabilities. We have no way of *making* certain recombinants or of ensuring that any specific class of recombinants will appear.

Let us emphasize also that the merogenote in these crosses can be either a piece of Hfr DNA carried in through mating or a piece of DNA picked up by a transducing phage. The methods of performing the experiment may be different in these cases, but the principle of mapping is the same.

To establish a map with two-factor crosses, we have to depend upon the recombination frequencies obtained from many separate experiments, using two markers at a time. We can't really depend on these data; the numbers obtained in two separate crosses using the same two markers may even be quite different from one another, simply because so many variables can creep into the experiments. We can eliminate many of these problems by using three markers at a time in each cross; these experiments permit us to establish an unambiguous sequence for the markers, even if we cannot establish distances very precisely.

9-12 MAPPING BY THREE-FACTOR CROSSES

The sort of data we have just illustrated can be obtained very easily by using a third marker outside the region of the two being mapped, but still close enough to them so that all of the markers can be cotransduced by a single merozygote. Suppose, for example, that there is a methionine (*met*) marker close to *his*; we will place it to the left of the *his* region arbitrarily. As a recipient, we use the double mutant *met*⁻ *his*-2 and as a donor we use *met*⁺ *his*-1:

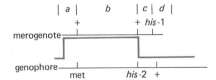

We plate recombinants on minimal medium and also on minimal medium supplemented with histidine. In both cases, bacteria that form colonies must be *met+*, having picked up the *met+* marker from the merogenote through a crossover in region *a*. In addition, all those cells that grow on minimal medium must have experienced a crossover in region *c*, to pick up the two wild-type *his* markers. But cells that grow on the supplemented plates may have had their second crossover anywhere in regions *b*, *c*, or *d*. Therefore, the ratio

$$\frac{\text{colonies on minimal medium}}{\text{colonies on supplemented medium}}$$

in a measure of the length of region *c*—the distance between the two *his* markers. A series of crosses performed with the same outside marker gives more reliable map distances than a series of two-factor crosses.

The fundamental theorem of three-factor mapping can also be applied in bacterial crosses to establish the *order* of a series of markers when the relative distance between them is not so important. This is shown in Fig. 9-10. We make two reciprocal crosses, first using one type of cell as the donor of the merogenote and then the

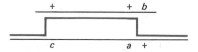

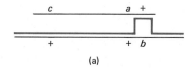

(a)

FIGURE 9-10

Establishing the sequence of three markers with reciprocal crosses. In (a) only two crossovers are required regardless of which strain is the donor, but in (b) four crossovers are required when the single mutant is the donor.

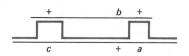

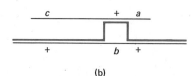

(b)

other. If the markers are relatively close to one another, the probabilities of crossovers in all regions will be about the same; the important factor becomes the *number* of crossovers required to establish a certain type of prototroph. In the example shown, if the order of markers is *c-a-b*, then the number of recombinants in the two crosses will be about equal. However, if the order is *c-b-a*, four crossovers will be required to produce a prototroph when *b* is in the merogenote and there will be many fewer prototrophs in this cross than in the reciprocal. This gives an unambiguous method of determining the order of a set of markers; the problem is to create some of the double mutants; since one is generally interested in mapping mutations that produce the same phenotype, it is difficult to tell that a strain is carrying two mutations rather than one.

exercise 9-3 Demerec collected several cysteine auxotrophs and numbered them *cys*-10, *cys*-12, and so on. He grew phage P22 on each one and on the wild-type cells and obtained stocks of transducing phage. He then attempted transductions of the mutants with these stocks in all possible combinations and observed the following numbers of wild-type transductants per 10^8 phage:

TRANSDUCED INTO	PHAGE GROWN ON					
	cys-10	*cys*-12	*cys*-14	*cys*-16	*cys*-18	+
cys-10	0	45	25	15	10	1200
cys-12	40	0	40	35	35	2000
cys-14	30	40	0	2	4	2000
cys-16	15	20	2	0	2	1000
cys-18	10	20	1	3	0	1000

(a) Draw the best map you can on the basis of these data.

(b) To obtain a better map, Demerec made the double mutant *cys*-14 *cys*-18 and grew transducing phage on it. When this stock was used to transduce *cys*-16, seven prototrophs were found, but when phage carrying *cys*-16 were used to transduce the double mutant, 35 prototrophs were found. Now use this result to improve your map.

exercise 9-4 Jacob selected eight *lac⁻* mutants and then attempted to order them with respect to the outside markers *pro* (proline) and *ade* (adenine) by performing a pair of reciprocal crosses for each pair of *lac* markers:

$$\text{Hfr:} \quad pro^- lac\text{-}x\ ade^+ \times \text{F}^-: pro^+ lac\text{-}y\ ade^-$$
$$\text{Hfr:} \quad pro^- lac\text{-}y\ ade^+ \times \text{F}^-: pro^+ lac\text{-}x\ ade^-$$

In all cases, prototrophs are selected by plating on minimal medium with lactose as the only carbon source. The following data give the number of colonies in the two crosses for each pair of markers, for example: x/y 100: y/x 50. Determine the relative order of the markers. (Hint: Take the crosses one by one in the order given, and establish a rule for yourself, such as "there will be more recombinants when the Hfr marker is on the right of the F⁻ marker," or some similar statement.)

1/2	173	1/4	46	1/6	168	5/7	199
2/1	27	4/1	218	6/1	32	7/5	34
1/3	156	1/5	30	3/6	20	1/8	226
3/1	34	5/1	197	6/3	175	8/1	40
2/3	24	4/5	205	1/7	37	2/8	153
3/2	187	5/4	17	7/1	215	8/2	17

exercise 9-5 The Martian bacterium *Martibacillus novellus* is green and can make phlizic acid. A mutation *wht* makes the bacteria white. A number of phlizic acid auxotrophs (*phl⁻*) are selected: *wht* and *phl* are cotransducible by the phage pm-22, and we arbitrarily place *wht* on the right. A series of transductions are performed using donors that are *wht⁺ phl-x* and recipients that are *wht phl-y*; the transductants are plated on minimal agar to select *phl* prototrophs, but no selection is made for color. The following table gives two kinds of data; first, the ratio of green/white colonies among the transductants, and second, the presence (+) or absence (0) of microcolonies from abortive transductions. From these data:

(a) Place the six markers on a map in their proper order.

(b) Divide them into complementation groups and mark the lines between cistrons on the map.

			DONORS			
RECIPIENTS	*phl-1*	*phl-2*	*phl-3*	*phl-4*	*phl-5*	*phl-6*
phl-1 wht	– –	0.1 +	0.2 +	0.1 +	0.2 +	0.1 +
phl-2 wht	3.0 +	– –	2.7 +	0.1 0	2.6 +	2.4 0
phl-3 wht	2.4 +	0.1 +	– –	0.1 +	0.2 0	0.1 +
phl-4 wht	3.1 +	3.4 0	3.0 +	– –	2.8 +	2.6 0
phl-5 wht	2.7 +	0.2 +	2.4 0	0.1 +	– –	0.1 +
phl-6 wht	2.7 +	0.2 0	2.5 +	0.1 0	2.4 +	– –

If you have trouble with this problem after you have studied it, read on.

Exercise 9-5 was inserted ahead of a full explanation as a kind of challenge. The principle is really very simple; we set up a cross in which we select for a crossover event between two markers and then look at the way outside markers segregate:

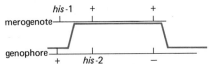

If the markers are in the sequence shown, there is a very good chance that the outside marker in the merogenote will be incorporated through a second crossover to the right of this marker. However, if the markers are in the opposite order,

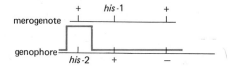

another pair of crossovers will be required to pick up the outside marker from the merogenote. Now look at the problem again from this point of view and see how the white/green marker segregates.

9-13 THE RESULTS OF MAPPING

Maps are now available that show the arrangement of gene loci in *E. coli* and *Salmonella* in impressive detail, and maps are being developed for other bacteria. While these may be impressive, they are not particularly enlightening at this stage; if you are interested in seeing some, you may turn to the references. It is somewhat more enlightening to calculate how much we *don't* know about the map.

exercise 9-6 Assume that a single *E. coli* genophore weighs 2×10^9 daltons.

(a) How many nucleotide pairs does it contain? _____

(b) If three nucleotide pairs code for one amino acid, how many amino acids does the whole genophore code? _____

(c) If the average protein subunit contains 500 amino acids (probably a bit high), how many different genes are there in the genophore?

_____

(d) A recent summary (late 1968) listed 221 known gene loci in *E. coli;* what percent of the genome was known at that time? _____

If this calculation is correct—and it is at least a good approximation—we still have about 90% of our sheer cataloguing work left. And after that comes the job of investigating all of the interesting gene systems to learn something about their function, their regulation, and so on. This field is not quite closed yet!

The truly interesting results are from some of the systems that have already been investigated in detail. When Miloslav Demerec and his group began to do detailed mappings, they discovered that very closely related genes tend to map close together. One of the most spectacular examples was the histidine region of *Salmonella*

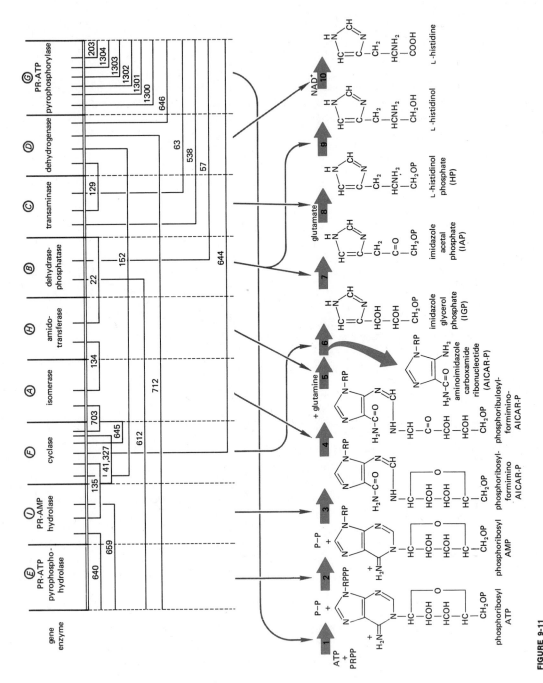

FIGURE 9-11

The histidine system of Salmonella. Above, Hartman's map of the region, established primarily with the deletion mutations below the line. Below, the corresponding biosynthetic pathway, established primarily by Bruce Ames' work.

(Fig. 9-11), where Philip Hartman was able to show that a block of eight genes all map together, and their order is practically the same as the biosynthetic sequence. The same pattern has appeared in many other gene systems in bacteria, and it seems to be the general rule.

The principal reasons for this clustering of related genes lie in the architecture of enzyme systems and the mechanisms of evolution, which we will discuss in Chapter 15. One of the principal lines of thought elicited by these mapping results had to do with the mechanism of genetic regulation of protein synthesis, and we will now turn to this vast subject to see how control mechanisms are superimposed on these gene-enzyme systems.

10

protein synthesis and its control

With the background we have acquired, we can finally come to grips with the central biological problem of heterocatalysis. In this chapter we will discuss the biochemical mechanism of protein synthesis and examine the genetic elements that regulate the process.

A. Genetic regulation

10-1 THE GENERAL MECHANISM OF PROTEIN SYNTHESIS

In Chapter 4 we outlined a general mechanism for the synthesis of a specific polypeptide, using the information in a part of the genome. It is clear that the DNA message is transcribed into an RNA message and that this serves as a template on which a series of amino acids are aligned by the protein factories (the ribosomes). The fit between the RNA codons and the amino acids requires a series of adaptor molecules; these were postulated by Crick, who assumed that they would be small RNA molecules. It was therefore very exciting when work by Paul Zamecnik, Mahlon Hoagland, and their associates revealed the "soluble RNA" molecules that have molecular weights of about 30,000 daltons and seem to have the properties of adaptors. From now on, we will use the more descriptive term **transfer RNA** (tRNA) for these molecules.

Since protein synthesis is an endergonic process, it must be driven by coupling to exergonic reactions. Hoagland and Zamecnik also showed that there are **activating enzymes** that combine amino acids with ATP to make an **amino acyl adenylate.** Notice that here we again find the general mechanism of pyrophosphate release in driving a synthetic reaction.

The 3' end of each tRNA has the base sequence -C-C-A. The amino acyl adenylate exchanges with the terminal adenylate of the tRNA, releasing AMP, so the amino acid is properly incorporated into its adaptor in covalent linkage:

$$\text{aa-AMP} + \text{tRNA-C-C-A} \rightarrow \text{tRNA-C-C-A-aa} + \text{AMP}$$

This amino acyl-RNA complex then goes to the ribosomes, where the RNA finds its proper position in some sequence as directed by an RNA template, and the amino acid is polymerized into a polypeptide. The whole process is outlined in Fig. 10-1.

Further work in several laboratories confirmed the main features of this scheme. The fact that there is no competition between different amino acids for binding to tRNA indicates that each amino acid binds to its own specific adaptor. It is now clear that there is at least one tRNA for each amino acid, and sometimes more. There is also at least one activating

amino acyl adenylate

(P)OCH₂ ... adenine ... O OH ... O=C—CH—R ... NH₂

289

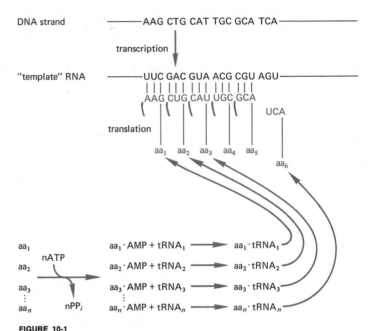

DNA strand ————AAG CTG CAT TGC GCA TCA————

transcription

"template" RNA ————UUC GAC GUA ACG CGU AGU————

AAG CUG CAU UGC GCA

UCA

translation

aa_1 aa_2 aa_3 aa_4 aa_5

aa_6

aa_1
aa_2
aa_3
$\vdots$
aa_n

nATP

nPP_i

$aa_1 \cdot AMP + tRNA_1 \longrightarrow aa_1 \cdot tRNA_1$
$aa_2 \cdot AMP + tRNA_2 \longrightarrow aa_2 \cdot tRNA_2$
$aa_3 \cdot AMP + tRNA_3 \longrightarrow aa_3 \cdot tRNA_3$
$\vdots$
$aa_n \cdot AMP + tRNA_n \longrightarrow aa_n \cdot tRNA_n$

FIGURE 10-1
The general mechanism of protein synthesis.

enzyme for each amino acid that catalyzes both the formation of amino acyl adenylates and their binding to the tRNAs.

Therefore, by 1960 we had a comfortable model for protein synthesis. Proteins were known to be made on ribosomes; it was known that RNA is an intermediate in protein synthesis and that it can function as a template for aligning amino acids in the proper sequence through their tRNA molecules; since the ribosomes contain about 80% of the total cellular RNA and the tRNA molecules account for essentially all of the rest, it was assumed that the template RNA molecules are built into ribosomes. However, it soon became clear that this model could not account for all of the observations; in particular, it did not allow for the rapid changes in the enzymes of bacterial cells that can be induced by growing them under different conditions. We will now see what features a proper model must incorporate.

10-2 THE LACTOSE SYSTEM

The β-galactosidase system of *E. coli* has been studied more thoroughly than any other as a model for the regulation of protein synthesis. Two types of mutants can be found that have the *lac⁻* phenotype. The *lacZ* mutants are all deficient in the enzyme β-galactosidase itself, and the various *lacZ* mutations all map in one locus. The *lacY* mutants have the enzyme but cannot use it for growth because they lack a *permease* that carries lactose across the cell membrane; permeaseless mutants are often called *cryptic*. The *lacY* mutations map in another locus very close to *lacZ*.

A third and more interesting class of mutants are **constitutive** rather than inducible; that is, their differential rate of β-galactosidase synthesis is high and relatively constant even when there is no inducer present. These interesting mutations affect the regulation system; they are called *lacR* and they map in another locus close to *lacZ*.

The Z and Y genes could be regulated in two ways. They might be normally inactive except when they are *activated* by some agent; this is a **positive control system,** and the inducer would serve as part of the activating agent. Or they might be normally active, except when they are *repressed* by some agent; this is **negative control,** and the inducer's role would be to inactivate the repressing agent. Arthur Pardee, Francois Jacob, and Jacques Monod performed a famous experiment to distinguish between these models. They mated an Hfr with the genotype R^+Z^+ with an F$^-$ with the genotype R^-Z^-. In the absence of inducer, the Hfr cannot make enzyme, and the F$^-$ cannot make enzyme in any case because it has no functional Z gene. However, when the two are mated the merozygotes make enzyme with the kinetics shown in Fig. 10-2. This result is compatible only with a negative control model. The R gene

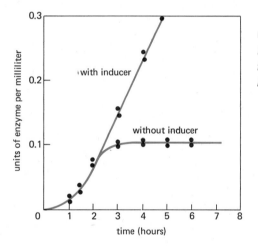

FIGURE 10-2
Kinetics of enzyme synthesis and repression following bacterial mating, with and without inducer.

must make a product we call a **repressor,** which diffuses through the cell and prevents the synthesis of β-galactosidase. A suitable inducer, such as lactose or IPTG, inactivates the repressor. When the R^+Z^+ genome is injected into the R$^-$ cell, it suddenly finds itself in the absence of repressor, so the good Z$^+$ gene can start to make the enzyme. Synthesis continues in the absence of inducer until the R$^+$ gene has made enough repressor to turn it off.

The defect of the R$^-$ mutant is that it does not make a functional repressor. Therefore, when an R$^-$ and an R$^+$ gene are combined in one cell, the R$^+$ is **dominant** because the repressors it makes mask the defective R$^-$ products. On the other hand, suppose that the R gene were a positive control element that made an activator. A constitutive R^c mutant would be one whose activators were functional even in the absence of an inducer, and this mutant gene would be dominant to an R^+ in a merozygote. (To put it the other way, the R^+ gene would be **recessive** to the R^c.) Therefore, an experiment of the Pardee–Jacob–Monod type is a potentially powerful tool in determining the action of regulatory elements.

10-3 MESSENGER RNA

With this general model in mind, Francois Jacob and Sydney Brenner reexamined the classical view of protein synthesis. The classical picture assumed that the ribosomal RNA contains templates for protein synthesis, but there were several good reasons for questioning it.

1. The DNA base ratio $(A + T)/(G + C)$ varies considerably from one species to another, over a range of about 0.3–3.0. Yet the corresponding base ratio of RNA is remarkably constant. If the bulk of the ribosomal RNA were carrying information from DNA, it should have approximately the same base ratio as the DNA.

2. The base analogue 5-fluorouracil (5-FU) is incorporated into the RNA (but not the DNA) of growing bacteria, as shown by Francois Gros and Shiro Naono. While growing with 5-FU, the bacteria manufacture defective, inactive enzymes. But the cells can be washed free of 5-FU and after a short time they again manufacture normal protein at a normal rate. These experiments suggest that the information for protein synthesis is carried in RNA, but that the template is of short duration and is quickly turned over, so that faulty information, due to the incorporation of 5-FU, is soon discarded. However, Cedric Davern and Matthew Meselson have shown that the ribosomes are stable and their RNA does not turn over. Ribosomal RNA itself is therefore not the template.

3. When bacteria are infected by a virulent phage, the synthesis of bacterial proteins ceases and only phage-specific proteins are made. In these cells, Lazarus Astrachan and Elliott Volkin detected a fraction of RNA that turned over very rapidly and had a base composition similar to that of the phage DNA.

4. The Pardee–Jacob–Monod experiment shows that synthesis of β-galactosidase begins at a maximum rate very soon after the Z^+ gene is injected. To account for this rapid change on the classical model, one would have to assume that each gene makes only a small number of ribosomes, each of which is capable of an extremely high rate of protein synthesis; in fact, one would have to assume that 0.01% of the ribosomes make about 5% of the cell protein. If this were so, then most of the ribosomes in the cell would be inactive; but this cannot be reconciled with the observation that when cells are required to make more protein they must also make more ribosomes.

Jacob and Brenner therefore postulated that the templates for protein synthesis are unstable RNA molecules that are made on the DNA, last long enough to direct the synthesis of a few molecules of protein, and are then broken down. They called this hypothetical molecule **messenger RNA** (mRNA). The ribosomes themselves are assumed to be stable factories that align all of the components and synthesize the peptide bonds.

Brenner, Jacob, and Meselson then performed a critical experiment to examine the unstable RNA that Astrachan and Volkin had found, since this seemed to have the properties of the hypothetical messenger. Their method depended upon making some ribosomes more dense than normal by growing cells in ^{13}C and ^{15}N, so they first had to find out how normal and dense ribosomes would band in a CsCl gradient. Figure 10-3 shows that normal ribosomes form a dense A band that consists of 30S and 50S units and a lighter B band of whole 70S ribosomes. Dense ribosomes also form two bands, and by coincidence the B band of dense ribosomes is just as dense as the A band of light ones.

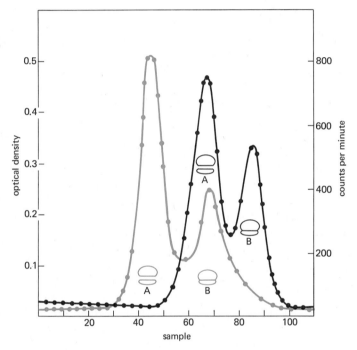

FIGURE 10-3

Distribution of ribosomes in a CsCl gradient. The black line shows the distribution of light ribosomes as determined by optical density and the colored line shows the distribution of dense ribosomes as determined by radioactivity.

To investigate a substance that is presumably made rapidly, we generally label a cell culture with a **pulse** of radioactivity by adding the tracer for only a very short time, comparable to the time it should take to make, say, half the total amount of the substance in the cell. When Brenner, Jacob, and Meselson gave infected cells a pulse of radioactive uridine to label their RNA, they found that all of the radioactivity went to the B band of whole ribosomes, as it should on either the classical or the messenger model. Then they performed the critical experiment. They grew a small sample of bacteria in heavy medium, infected it with phage and immediately transferred it to light medium, and then gave it a pulse of radioactivity. Now, if the cells are making new ribosomes after infection with information for specific phage proteins, these ribosomes should be light and all of the radioactivity should be in the B band of light ribosomes. However, if they are making new messengers after infection and these messengers are being attached to old ribosomes made before infection, then the radioactivity should all be in the B band of dense ribosomes. Figure 10-4 shows that the latter is actually found. Hence there are no ribosomes made after infection and the information for making specific phage proteins is carried in messengers.

Francois Gros and his colleagues examined uninfected cells at the same time by giving pulse labels of radioactive uridine. They showed that this radioactivity sedimented in RNA fractions that are very different from the 16S and 23S RNA from ribosomes. They are generally smaller than the stable ribosomal RNA, although some large messengers (19S, 30S, etc.) can be detected. These messengers associate with ribosomes in high Mg^{++} concentrations and dissociate in very low concentrations. Furthermore, a single messenger associates with several ribosomes to make a **polyribosome** complex; after careful extraction from cells, polyribosome fractions can be separated on a sucrose gradient with from two to seven or more ribosomes per messenger, and the size of the complex seems to be limited simply by the length of the messenger.

The most definitive characteristic of messenger RNA should be its base sequence, which must be complementary to a region of DNA. Benjamin Hall and Sol Spiegelman developed a **hybridization** procedure for showing that a given RNA is complementary to a given DNA, and the original method has been developed to a fine art with several variations. It depends upon the discovery by Julius Marmur and Paul Doty that DNA molecules can be melted into separate strands by heating them to temperatures from about 60–90°C. If the DNA solution is then cooled slowly, the double-stranded molecules reform and have normal biological activity, but if the solution is cooled quickly there is not enough time for them to find appropriate complementary partners with which to hydrogen bond and they remain single stranded. Hall and Spiegelman heated T2 DNA and RNA from T2-infected cells labeled with ^{14}C-uridine. They showed that hybrid DNA-RNA molecules did form and could be separated as a characteristic peak in a CsCl gradient (Fig. 10-5). Therefore, the RNA must have been made on phage DNA templates.

We now know that all RNA molecules are made on DNA; tRNA and rRNA (ribosomal RNA) molecules can also be hybridized specifically with DNA. RNA molecules are made by the enzyme **DNA-dependent RNA polymerase** in the 5' to 3' direction by adding nucleoside 5'-triphosphates and splitting out PP_i. They are made on only one strand of the DNA in any locus; theoretically it would be very difficult to make functional RNAs, particularly messengers, on both strands at any locus, since the two molecules would be complements of one another and would carry very different messages. The best experimental evidence that only one strand is transcribed comes from studies with certain phage whose strands have such different densities that they can be separated in a CsCl gradient. Labeled RNA made during infection with

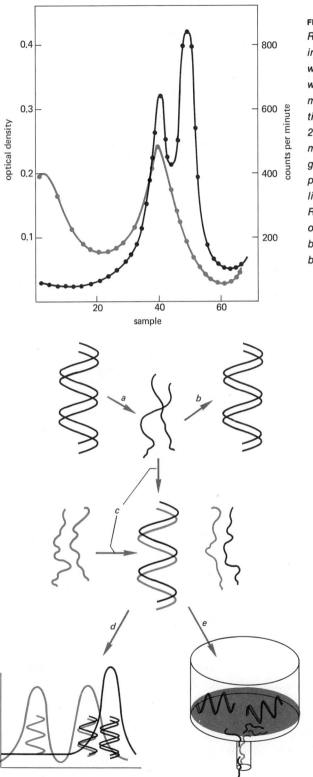

FIGURE 10-4

RNA synthesis after isotope transfer in phage-infected cells. Bacteria were grown with heavy isotopes, washed, and transferred to light medium at the time of phage infection. A pulse of ^{32}P was added from 2 to 7 minutes. The culture was mixed with phage-infected cells grown in light medium to mark the positions of light ribosomes (black line). The colored line shows the RNA made after infection as counts of ^{32}P; it is all associated with a B band of heavy ribosomes made before infection.

FIGURE 10-5

DNA-RNA hybridization. If double-stranded DNA is heated (a), it melts into single strands. If these are cooled slowly (b), they will reanneal into double-stranded molecules. If the melted DNA is mixed with radioactive RNA (c), some of the molecules will anneal as DNA-RNA hybrids, which can be separated from double-stranded DNA and from RNA either by its characteristic density in CsCl (d) or because it sticks to certain nitrocellulose filters while the unannealed RNA does not (e).

these phage hybridizes with only one of the strands. With phage such as λ it can be shown that some messengers hybridize with one strand and some with the other, but a given section of DNA is still transcribed from only one of its strands.

DNA-dependent RNA synthesis can be blocked with the antibiotic **actinomycin D,** which binds to the DNA so it inhibits the polymerase. Using this drug, Levinthal and

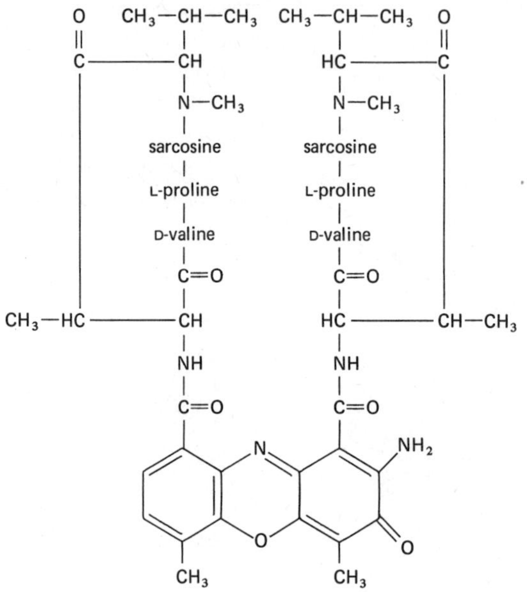

actinomycin D

his associates followed the decay of mRNA in *Bacillus subtilis.* They found that its half-life is about 2 minutes in cells with a 1-hour generation time. A messenger is therefore made quickly, is used to direct the synthesis of a few molecules of protein, and is then degraded. Moreover, this RNA is critical for the control of protein synthesis.

exercise 10-1   Since you should now know the average molecular weight of a nucleotide and of an amino acid, calculate how much a messenger must weigh in relation to the protein it codes, assuming that each codon is three nucleotides.

10-4   THE OPERON MODEL

On the basis of data from many different sources, Jacob and Monod proposed a general model for the regulation of protein synthesis. They noted that the *lacZ* and *lacY* loci map very close together and that the products of these two genes are always induced coordinately. They pointed out that there are many similar cases; for example, all of the enzymes for histidine biosynthesis are specified by a block of contiguous genes and these enzymes are also induced and repressed coordinately, so the differential rates of synthesis of all of the enzymes are approximately proportional to one another.

Jacob and Monod therefore postulated that the genome is divided into minimal units of regulation, which they called **operons.** An operon consists of one or more structural genes that specify proteins and a control element called an **operator** that acts like a switch to turn these genes on or off. The operator is under the control of a repressor that is specified by a **regulator** (R) gene. The operator switch is "on" when the repressor is not bound, and under this condition the structural genes can be transcribed as messenger RNA molecules.

Induction and repression are seen as two sides of the same coin. In an inducible system (Fig. 10-6), the repressor binds to the operator *unless* it is combined with an inducer. In a repressible system (Fig. 10-7), the **aporepressor** binds to the operator *only* if it is combined with a **corepressor.** But in both cases the regulation is negative; it is assumed that messenger RNA will be synthesized from the structural genes if no repressor is bound to the operator.

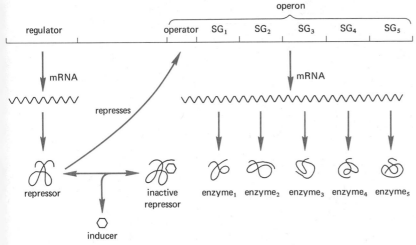

FIGURE 10-6
The Jacob–Monod model for regulation of an operon consisting of five structural genes (SG) that is induced by a small inducer molecule.

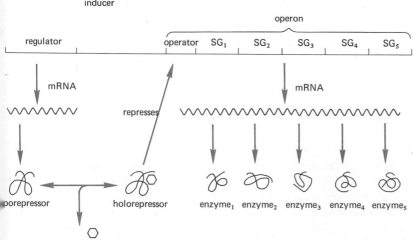

FIGURE 10-7
The Jacob–Monod model for regulation of an operon consisting of five structural genes (SG) that is repressed by a small repressor molecule.

The critical feature of the operator is that it affects only genes *cis* to it, not those *trans* to it in the same cell. In contrast, the product of the regulator gene, the aporepressor, binds both cis and trans. Based on this property, Jacob and Monod predicted that there should be two types of operator mutants. An operator-constitutive mutation O^c is one in which the switch is permanently open. An operator-negative mutation O^0 is one in which the switch is permanently closed. These mutants were looked for and found. Table 10-1 gives the properties of some of these mutants and of F genotes in which they are combined. You should study this table long enough to convince yourself that operator mutations act only cis and that they have the predicted properties.

TABLE 10-1

characteristics of lac mutants and merodiploids*

| | ENZYME ACTIVITY | | | |
| | UNINDUCED | | INDUCED | |
GENOTYPE	GALACTOSIDASE	PERMEASE	GALACTOSIDASE	PERMEASE
$R^+O^+Z^+Y^+$	< 0.1	< 1	100	100
$R^-O^+Z^+Y^+$	120	120	120	120
$R^+O^cZ^+Y^+$	25	25	100	100
$R^+O^0Z^+Y^+$	< 0.1	< 1	< 0.1	< 1
$\dfrac{R^+Z^-Y^+}{F\ R^-Z^+Y^+}$	2	2	200	250
$\dfrac{R^-Z^-Y^+}{F\ R^-Z^+Y^+}$	250	250	200	250
$\dfrac{O^+Z^-Y^+}{F\ O^cZ^+Y^+}$	75	75	250	300
$\dfrac{O^+Z^+Y^-}{F\ O^cZ^-Y^+}$	1	75	100	250
$\dfrac{O^0Z^+Y^+}{F\ O^+Z^-Y^+}$	< 0.1	< 1	< 0.1	250

*All values are in percent of wild-type fully induced culture.

The operon model predicts that induction and repression of specific enzymes should be accompanied by induction and repression of the corresponding messengers. This prediction was tested and confirmed in several laboratories. Giuseppe Attardi, working with Jacob, Brenner, and Gros, showed that a messenger RNA can be isolated from bacteria that are induced for the enzymes of galactose metabolism (the *gal* operon) that will hybridize with DNA from λ *dg* but not from normal λ. Aaron Novick and I took a different approach; we gave cells that were induced for the *lac* operon a pulse of ^{14}C-labeled uridine and at the same time gave uninduced cells a pulse of

³H-labeled uridine; we then separated the labeled RNA fractions. If there were no difference between the messenger fractions of the two cells, the ratio of $^{14}C/^3H$ should have been constant, but as Fig. 10-8 shows, this ratio rises sharply in the fraction that sediments at 30S in a sucrose gradient, indicating the presence of a messenger that is made in induced but not in uninduced cells. Masaki Hayashi and Naomi Franklin, working with Spiegelman and Luria, obtained similar results and showed that part of the RNA in this fraction would hybridize with DNA from phage P1 *dl* (carrying *lac* DNA) but not with normal P1 DNA. It was therefore clear that induction is accompanied by the changes in messenger RNA demanded by the operon model.

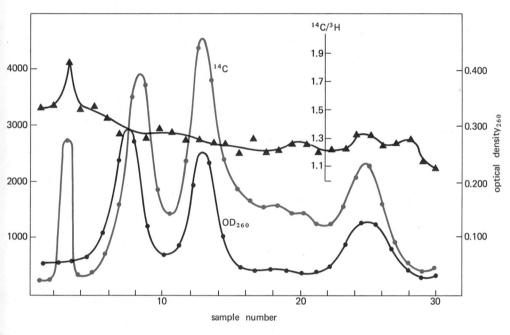

FIGURE 10-8

Localization of the lac *messenger. Induced cells were labeled with ^{14}C-uridine and uninduced cells with ^3H-uridine. Extracted RNA was separated on a sucrose gradient; radioactivity is mostly in messenger peaks that are distinct from stable RNA peaks. The ratio $^{14}C/^3H$ rises over the 30S peak, indicating that* lac *mRNA is there.*

To our surprise, the specific messenger RNA associated with induction of the *lac* operon was very large—too large to code for merely the β-galactosidase, but certainly large enough to code for the entire operon. Similarly, Robert G. Martin showed that induction and repression of the enzymes of the histidine operon in *Salmonella* is accompanied by changes in a 38S messenger RNA that is large enough to code for the entire operon. We must therefore say that every operon is transcribed as a *single, large messenger RNA molecule.*

Through a rather involved series of discoveries, we now know of one other element, the **promotor**. *lacP* mutations, which define the promotor, map between the operator and regulator regions and are characterized by a very low rate of enzyme synthesis

(1–5%) even when fully induced. The promotor is presumed to be the place where the RNA polymerase begins messenger synthesis. In fact, we now know that this polymerase consists of a core of three proteins, $\alpha\beta\beta'$, that can start to transcribe anywhere on the DNA, plus a fourth protein, σ, the sigma factor, that confers specificity on the core and restricts it to initiation at promotor regions. The action of repressor then becomes more understandable, for it binds to the operator just to the right of the promotor and blocks the polymerase. Promotor regions have been found in other operons as well.

exercise 10-2 The following system is inducible by substance XX:

$$\text{XX} \xrightarrow{\text{ENZYME 1}} \text{XY} \xrightarrow{\text{ENZYME 2}} \text{XZ}$$

Enzymes 1 and 2 are specified by genes A and B, respectively; they make an operon with a regulator gene (R) and operator gene (O). For the following F genotes, with and without inducer, specify which enzymes will be made; estimate 100%, 50%, < 0.1% of wild type.

| | WITH INDUCER | | WITHOUT INDUCER | |
F GENOTE	ENZYME 1	ENZYME 2	ENZYME 1	ENZYME 2
$R^+O^+N^+B^+$				
$R^-O^+A^+B^+$				
$R^-O^+A^-B^+$				
$R^-O^+A^+B^-$				
$R^+O^cA^-B^+$				
$R^+O^0A^+B^+$				
$R^-O^0A^+B^+$				
$R^+O^+A^-B^-$				

Just because we have concentrated on the lactose system and similar repressible systems as examples of control, you must not get the impression that all systems operate in the same way. We also find positive control systems and more complex situations. Your job right now should be to understand the systems illustrated here as well as you can.

B. Coding and translation

10-5 DIRECTION OF PROTEIN SYNTHESIS

One can imagine two general ways in which a polypeptide might be synthesized. On the one hand, all of the amino acyl tRNA molecules might line up along a messenger simultaneously and the ribosomal enzymes would zip them into a polymer

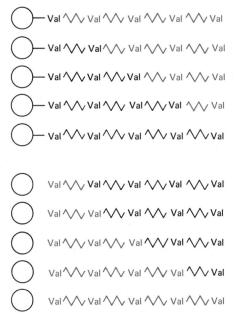

Consequences of synthesis from the N terminus of the polypeptide. Ribosomes as isolated from ordinary reticulocytes have unlabeled N-terminal amino acids and are completed by more C-terminal labeled amino acids. The situation is reversed by incubating first with the label and then finishing each chain with ordinary amino acids.

all together. On the other hand, each amino acyl tRNA might come to the messenger alone and transfer its amino acid to a polypeptide growing residue by residue from one end. In the latter case, one should be able to detect incomplete oligopeptides. John Bishop, John Leahy, and Richard Schweet showed that the second model is correct and that polypeptide synthesis proceeds from the N-terminal end.

Bishop, Leahy, and Schweet used the mammalian cells (reticulocytes) that make hemoglobin; in fact, they hardly make any other protein, so essentially all of the polypeptides made in the reticulocyte system have the same sequence. The N-terminal residue of hemoglobin is valine and there are valines elsewhere in the molecule. Surprisingly, when reticulocyte ribosomes were isolated and protein synthesis was followed by using ^{14}C-valine as a tracer, almost no radioactivity was found in the N-terminal amino acids. This suggested the mechanism shown in Fig. 10-9; most of the isolated ribosomes were making hemoglobin and have oligopeptides of various lengths attached, so their N-terminal valines are unlabeled. ^{14}C-Valine will be incorporated only into the more C-terminal valine positions, except for the very few that will be incorporated into the N-terminal positions of new polypeptides. To test this hypothesis, they reversed the situation by incubating reticulocyte ribosomes with ^{14}C-valine for a long time and then adding large amounts of ^{12}C-valine. As they expected, the N-terminal valines were now almost all labeled and the more C-terminal positions were unlabeled.

Howard Dintzis came to the same conclusion almost simultaneously by doing similar experiments. It is characteristic of science that several people get the same good idea at the same time.

10-6 GENERAL NATURE OF THE CODE

All of the clever suggestions that were once made about the coding relationship between RNA and protein are now no more than intellectual exercises. The actual coding mechanism was elucidated experimentally by Crick and Brenner with Leslie Barnett and R. J. Watts-Tobin. They had previously shown that the acridine dyes, particularly *proflavin*, produce mutations by causing the addition or deletion of a

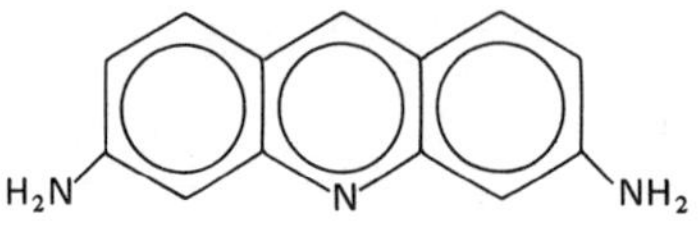

proflavin

nucleotide. Leonard Lerman had shown that these dye molecules insinuate themselves between the bases of DNA; Fig. 10-10 shows how this might cause a mutation during DNA replication. Crick and his colleagues then collected a number of *rII* mutations induced by proflavin in phage T4 and used them for some very clever experiments, which can really be understood only if their postulated mechanism for coding is correct.

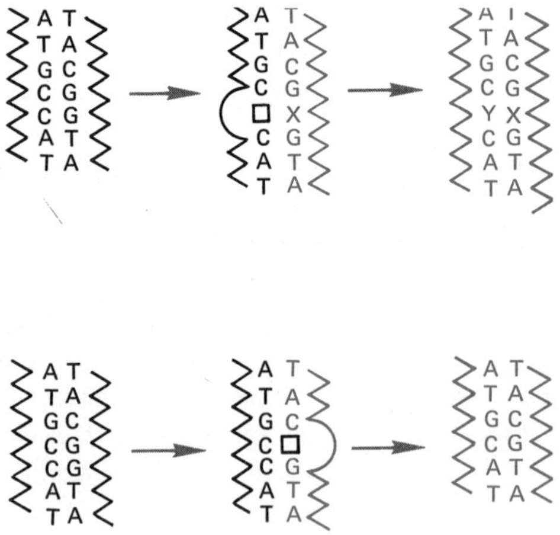

FIGURE 10-10

Mechanism of proflavin mutagenesis. If a proflavin molecule (box) is inserted in the nonreplicating strand, it can cause an insertion (XY = any base pair). If it is inserted into the strand being synthesized, it can cause a deletion.

They proposed that the RNA message is translated by a **reading-frame** mechanism that starts at one end and marks off three bases at a time, associating the appropriate amino acid with each triplet. However, there are no "commas" in the message between triplets; the only way to read the message correctly is to start at the right point and keep moving by groups of three. If anything gets the reading frame out of proper phase, it will stay out of phase and produce the wrong polypeptide. Suppose that the messenger is supposed to be read like this:

1 2 3 1 2 3 1 2 3 1 2 3 1 2 3 1 2 3 1 2 3 1 2 3

What if proflavin induced a mutation here by deleting one base pair, so the corresponding RNA message reads

Then the triplet within which the mutation occurred and all the triplets to the right of it would be read incorrectly; the proflavin therefore is said to cause a **frame-shift mutation**.

Now suppose that a second frame-shift mutation occurred to the right of the first, but that this was an *addition* of a base pair. The messenger made from this double mutant would be

The second mutation reverses the frame shift caused by the first. Now only the short segment between the two mutations is incorrectly translated, and if this region of the protein is not too critical the wild-type phenotype is expressed. In general, a second mutation that counters the effect of an initial mutation is called a **suppressor**.

The set of proflavin-induced mutations that Crick et al. collected is in a small segment of the *r*IIB cistron that does not seem to be too critical for *r*II function. They started with a single mutation which was arbitrarily assumed to be an addition and numbered 1_+. They then found 18 independent suppressors of 1_+ which were assumed to be deletions and numbered 2_- through 19_-. All of them map very close to 1_+ and can be separated from it by recombination with a wild-type phage. Since each of the 18 by itself behaves just like any other proflavin-induced *r*II mutation, Crick and his friends looked for proflavin-induced suppressors of them. Each suppressor of a suppressor is presumably another addition, so these can be numbered beginning with 20_+. Each suppressor of a suppressor can again be separated by recombination and a series of suppressors of suppressors of suppressors collected. In this way, a large set of proflavin-induced *r*II mutations were collected, each of which could be designated (still arbitrarily) plus or minus. On the whole, any plus mutation in the set can suppress any minus mutation, but two pluses or two minuses together make a double mutant that still has the *r*II phenotype.

These data are compatible with the hypothesis, but they do not indicate the number of bases read at each time by the reading frame. Crick et al. then made several triple mutants by combining three plus mutations or three minus mutations; in theory, a messenger made from DNA carrying three plus mutations close together would look like this:

The first mutation shifts the reading frame by one position, the second shifts it by two positions, and the third shifts it by three positions, which puts it back in phase. As predicted, the triple mutants have the wild-type phenotype, which shows that the code is really triplet. Of course, it is always possible that every proflavin-induced mutation is an insertion or deletion of two nucleotides, so the code is really sextuplet or even more complicated, but this is a more complex hypothesis and is harder to understand. Now we will see that the code is really triplet.

10-7 THE RNA CODE

With the general nature of the code determined, the problem of actually breaking the code still remained. The general method for solving the problem suddenly became apparent through the work of Marshall Nirenberg and J. Heinrich Matthei. For some time, artificial polyribonucleotides with known sequences had been available for structural studies. Nirenberg and Matthei decided to see how some of these would behave as artificial messengers in a protein-synthesizing system. They added a polyuridylic acid (poly U) and to their great delight found that it would direct only the synthesis of polyphenylalanine. Thus they determined the first word of the code: UUU means phenylalanine.

TABLE 10-2
the RNA dictionary

Second Letter

First Letter	U	C	A	G
U	UU U/C Phe; UU A/G Leu	UC U/C/A/G Ser	UA U/C Tyr; UA A/G ochre amber	UG U/C Cys; UGA umber; UGG Trp
C	CU U/C/A/G Leu	CC U/C/A/G Pro	CA U/C His; CA A/G Gln (basic)	CG U/C/A/G Arg
A	AU U/C/A Ilu; AUG Met (large aliphatic)	AC U/C/A/G Thr	AA U/C Asn; AA A/G Lys (basic)	AG U/C Ser; AG A/G Arg
G	GU U/C/A/G Val	GC U/C/A/G Ala	GA U/C Asp; GA A/G Glu (dicarboxylic)	GG U/C/A/G Gly

This discovery was vigorously pursued in Nirenberg's laboratory and elsewhere, so the composition of several codons was soon known. However, this method cannot determine the actual sequence of bases in a codon because a polymer made of, say, C, G, and U has a random sequence and one can only associate amino acids with codons on a statistical basis. Nirenberg and Phillip Leder then took advantage of the observation that amino acyl tRNA molecules will bind to ribosomes quite specifically in the presence of messengers. They therefore made trinucleotides of RNA with known sequences and tested them for binding of specific amino acyl tRNAs. In this way, they could associate each amino acid with certain sequences of RNA bases, and extension of this work by Nirenberg and others has led to the codon assignments shown in Table 10-2. These codon sequences have almost all been confirmed with several kinds of evidence, and we now feel that we know the code.

The code has some remarkable properties whose full implications are still to be realized. Notice that the codons for the groups of amino acids outlined in Chapter 4 tend to fall into regular patterns. Furthermore, the first two bases, reading from 5' to 3', carry almost all of the information; in several cases the third base is irrelevant and in almost all cases U = C and A = G in this position.

The most elegant demonstration that the Crick–Brenner coding scheme is correct and that several of these codons are also correct comes from work by Eric Terzaghi, George Streisinger, Yoshimi Okada, Akira Tsugita, M. Inouye, and their associates at the University of Oregon and the University of Osaka. The Oregon group collected proflavin-induced mutants for the lysozyme of phage T4, which they could purify and analyze by using the skills of the Japanese protein sequencers. They analyzed the protein made by the wild-type phage and by a phage carrying a presumed minus mutation and a presumed plus suppressor. The lysozyme made by a mutant phage is defective, but that made by the double mutant is only slightly defective; its primary structure differs from the wild type in only five amino acids. The altered sequence can be derived from the wild type by using the Nirenberg code. The wild-type sequence and its presumed messenger are

- Thr - Lys - Ser - Pro - Ser - Leu - Asn - Ala - Ala - Lys -

ⒶGU CCA UCA CUU AAU GC?

By deleting A
we obtain   GUC CAU CAC UUA AUⒼGC?   and inserting G,

- Thr - Lys - Val - His - His - Leu - Met - Ala - Ala - Lys -

There is only one sequence of bases and only one way of altering the sequence that will give these results. There can be no clearer demonstration of the correctness of the coding scheme developed here. Moreover, this experiment shows that messenger RNA is read in the direction from 5' to 3' as the protein is made from the N-terminal end to the C-terminal end.

It is also obvious from the work of this group that the DNA (or RNA) base sequence is collinear with the amino acid sequence of the protein. Charles Yanofsky and his group had shown this to be true by analyzing mutant forms of the enzyme tryptophan systhetase and showing that the sites of mutations are collinear with the amino acid changes they produce.

exercise 10-3 The Oregon–Osaka group later analyzed some other frame-shift mutants and found the following changes:

Trp-Tyr-Asn → Met-Val-Tyr

Ala-Val-Arg → (Gly, Cys, Cys)   (sequence not determined)

Write the sequence of a messenger for each of the wild-type proteins that can be mutated by frame shifts to give the mutant sequence. Although the sequence is not specified in the second mutant, it should not be hard to deduce this.

exercise 10-4 Yanofsky found two independent mutations affecting the glycine residue in position 47 of tryptophan synthetase. In mutant A23, Gly → Arg and in mutant A46, Gly → Glu. These mutants recombine with a frequency of 0.002% to restore the wild-type glycine residue. From Table 10-2, determine what mutations have occurred and how the recombination occurs.

exercise 10-5 The following amino acid replacements have been found in mutants for the capsid protein of tobacco mosaic virus. In each case, specify the mutational event that probably caused the change. Are all of these transitions?

Asn → Ser   Ser → Leu

Ilu

Asp → Ala   ↗

↘   Thr → Met

Gly   ↘

Ser

10-8 TRANSFER RNA AND THE RIBOSOME

Since Crick had postulated the existence of short adaptors, perhaps three to six nucleotides long, it came as a surprise to find that tRNA molecules may be up to 85 nucleotides long. From its size, it is apparent that the tRNA must have several different specificities, and we can begin to identify these. Robert Holley and his group first determined the primary structure of the yeast alanine tRNA (Fig. 10-11) which is shown here in its most probable configuration as a basis for discussion; about 10 different tRNA molecules have been sequenced now, and they all conform to this same general "cloverleaf" pattern.

Transfer RNAs all have an unusual base composition. While they are probably all transcribed from DNA as strings of the usual four nucleotides (A, G, C, U), many bases are soon modified, chiefly by methylation. About a dozen different methylated nucleotides have been found, including thymidine (5-methyluridine). Inosine and 4,5-dihydrouridine are also common. The most interesting of these nucleotides is pseudouridine (Ψ), in which uracil is bonded to ribose through C-5 rather than C-1.

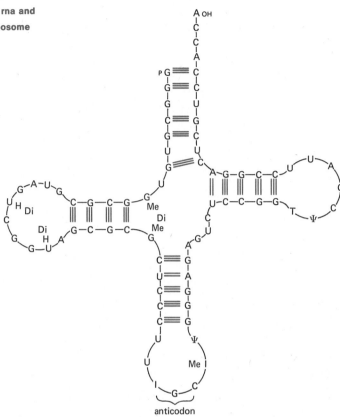

FIGURE 10-11
*Structure of alanine tRNA; Me =
methyl-, DiMe = dimethyl-, and
DiH = dihydro-.*

Pseudouridine is a strong acetylating agent that could bind an amino acid thus:

pseudouridine

amino acyl pseudouridine

All tRNAs examined to date contain the sequence -G-T-Ψ-C- on the right-hand loop,
and this may be a common binding site for the amino acid at some stage of protein
synthesis.

By definition, an adaptor must carry an **anticodon** region that can hydrogen bond
to the codon of the messenger. On all known tRNAs, the bottom loop contains an
appropriate anticodon sequence; in the alanine tRNA, this is -I-G-C-. The codon
for alanine is GCx, where the third base carries no information; this can clearly
hydrogen bond (antiparallel) to the anticodon for the tRNA if it is possible for inosine
to bond to several bases. Crick has proposed that whenever the third base carries
less information than the other two, the corresponding anticodon base will be able
to "wobble" somewhat to accommodate different pairings. According to the "wobble"

theory, inosine can pair with U, C, and A; U can pair with A and G; G can pair with C and U; and C can pair only with G. If all four bases must be recognized in the third position, the theory predicts that there will be two tRNA species; either one will pair with U and C and the other with A and G, or one will pair with U, C, and A and the other will pair with G. It is significant that in the only two cases where the third base is critical (Trp and Met), that base is G.

So far, the rules of the "wobble" theory seem to be holding very well. For example, phenylalanyl tRNA, which must recognize UU$_C^U$, has the anticodon 2-*O*-methylguanosine-A-A. Tyrosyl tRNA, which must recognize UA$_C^U$, has the anticodon A-Ψ-G; Ψ can form hydrogen bonds just like U.

The tRNA must be able to bind specifically to parts of the ribosome complex. Its

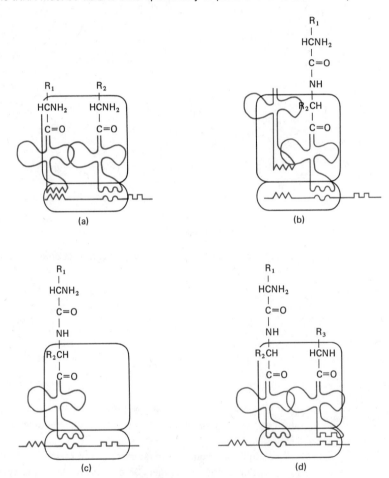

FIGURE 10-12

Mechanism of peptide synthesis. (a) The first two amino acyl tRNA molecules occupy two ribosomal sites. (b) The carboxyl group of the first bonds to the amino group of the second and the first tRNA is released. (c) The dipeptidyl tRNA is transferred to the first site. (d) A third amino acyl tRNA moves into the second site, ready to accept the dipeptide.

left-hand loop always carries the sequence -A-G-dihydrouridine-x-G-G, which is probably designed for some common binding site. At any time there are apparently two sites on the ribosome to which tRNA molecules can bind. An incoming amino acyl tRNA first binds nonenzymatically to site 1, providing that the proper codon of a messenger is in position there; this is the binding reaction that Nirenberg and Leder took advantage of in deciphering the code. There is then an enzymatic reaction requiring GTP for energy in which the tRNA is moved to site 2. In fact, we don't know whether this is a real movement of the tRNA, a rearrangement of the ribosome, or something else.

A peptide synthetase actually forms the peptide bond between two amino acids as in Fig. 10-12. The first two amino acyl tRNAs reside on neighboring ribosomal sites and a bond is formed from the amino group of the first to the carboxyl group of the second, with the release of the first tRNA. A third amino acyl tRNA is brought in and the cycle is repeated. Thus the growing peptide chain is always bonded at its C terminus to a tRNA. There is an interesting antibiotic, **puromycin**, that looks very much like an amino acyl tRNA:

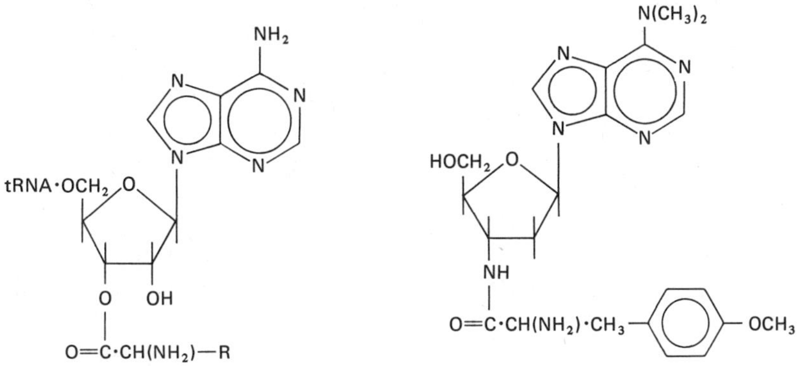

amino acyl RNA   puromycin

When puromycin is added to an active ribosomal complex, it enters the peptide synthetase system and the polypeptide is transferred to it, but it cannot bind to the ribosome so a polypeptidyl-puromycin is released and protein synthesis is inhibited. There are natural factors that release finished polypeptides from the ribosome in the same way when the proper signal appears on the messenger.

10-9 SUPPRESSION, TERMINATION, AND INITIATION

In principle, suppression could occur in many special ways; the system of proflavin-induced suppressors is just one example. The more interesting types for our present purposes are generalized translational suppressors that can translate a mutant mRNA into a protein that is at least partially functional. Such a suppressor might act on a *mis-sense* mutation to restore the wild-type amino acid; however, mis-sense suppressors must be very rare and very weak. Remember that a codon that is mis-sense at some mutated position is sense everywhere else; a mechanism that corrects it at the point of mutation will insert a wrong amino acid elsewhere and this would be highly detrimental to the cell. Only a few mis-sense suppressors have been found; they are very inefficient and the cells that carry them grow more slowly than normal.

However, a *nonsense* suppressor, which inserts an acceptable amino acid at a codon that is not in the amino acid dictionary, might be very efficient. Several of these suppressors have been discovered by a rather devious route which began with the discovery of the amber (*am*) mutants of phage T4, the suppressor-sensitive (*sus*) mutants of phage λ, and some other cases. Seymour Benzer and Sewall Champe found some "ambivalent" *r*II mutants of T4 that would multiply in one strain of *E. coli* B but not in another. They showed that a host in which they could multiply differed by a single mutation from a restrictive host; the mutation is therefore a kind of suppressor.

Benzer and Champe had found an interesting *r*II mutant, *r*1589, that carries a long deletion covering parts of the A and B cistrons, but still has B activity and will complement mutants in A. However, a double mutant carrying both *r*1589 and an ambivalent *r*IIA mutation that maps to the left of it (Fig. 10-13) has neither A nor B activity. Champe and Benzer proposed a simple explanation. *r*1589 joins the A and B cistrons by abolishing the codons that normally mark the place where synthesis of the A protein is to stop and synthesis of the B protein is to begin, so the protein made from *r*1589 contains the N-terminal sequence of the A protein and the C-terminal sequence of the B, and has B activity. The ambivalent mutants are *nonsense* types; the mutation creates a triplet that is not in the dictionary, so peptide synthesis stops at that point. When such a nonsense mutation is combined with *r*1589, synthesis of the hybrid A-B protein stops somewhere in the A part.

When Epstein and Edgar found the amber mutants of T4, defined by their ability to grow in strain CR63 but not in B, it became clear that the ambivalent mutants are amber mutants. Strain CR63 carries a suppressor mutation (*supD*) that corrects the nonsense mutation, and several other amber suppressors are now known. Mutants for various proteins were studied in suppressor strains and the amino acid compositions of these proteins were compared with the wild type. Only glutamine and tryptophan residues were ever affected by amber mutations, and these were replaced by serine (by *supD*), glutamine (by *supE*), or tyrosine (by *supF*). From this pattern of mutation, Gln → amber and Trp → amber, it was inferred that the amber triplet is UAG. (Which bases are changed in the two mutations?)

Brenner and his associates then found another class of nonsense mutations, which they called *ochres,* that are suppressible by a distinct class of ochre suppressors. (Ochre suppressors will correct amber mutations, but not vice versa.) From the patterns of mutation, the ochre triplet appears to be UAA. A third triplet, UGA (*umber*) is also a chain terminator and suppressors have been found for it.

Proof that the amber mutations stop peptide synthesis was provided by Anand S. Sarabhai and his associates in Brenner's laboratory. Eighty percent or more of the protein made during a T4 infection is the principal head subunit, so this polypeptide can be analyzed with little purification. Sarabhai et al. analyzed the polypeptides made by a series of amber mutants in the gene that specified the head protein. When the wild-type protein is digested with trypsin and the peptides are separated chromatographically, they form a certain pattern on the chromatography paper—a so-called **fingerprint** of the protein. Each amber mutant makes some of these peptides and not others. As Fig. 10-14 shows, the number of peptides is a function of the map

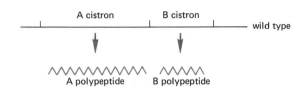

FIGURE 10-13

Effects of r1589 and ambivalent mutations. The wild-type rII cistrons make two separate polypeptides (top); r1589 makes one polypeptide (middle). In the double mutant with an ambivalent mutation to the left of r1589, peptide synthesis stops at the site of the mutation.

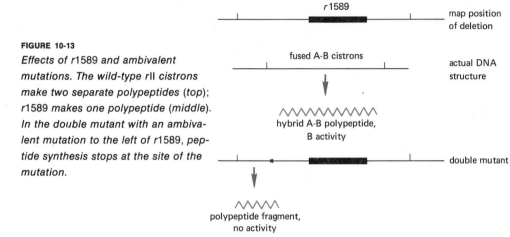

FIGURE 10-14

Correspondence between map position of an amber mutation and the length of polypeptide made. The mutations are listed from top to bottom in map order. Each one has the peptides listed for it; the closer the mutation is to the right end of the locus, the more identifiable peptides it has.

position of the mutation, as it must be if protein synthesis proceeds from the N terminus up to the point of the mutation.

What, then, is a suppressor mutation? In most cases, it is a change in a tRNA molecule which gives it the ability to pair with a nonsense triplet and insert an amino acid into the polypeptide. The tyrosyl tRNA of a *supF* strain has been compared with that of the *sup*[+] strain; the wild-type tyrosyl anticodon is G-U-A, which can bond to the codons U-A-^U_C. In *supF,* this has been changed to C-U-A, which can recognize the amber triplet U-A-G. Other nonsense suppressors are probably similar mutations of the protein-synthesizing apparatus.

Now two related questions arise: (1) What is the natural chain terminator that indicates where protein synthesis is supposed to stop; and (2) Is there an analogous chain-initiator triplet that shows where protein synthesis is supposed to begin? The natural chain terminator is probably not amber; amber suppressors are very efficient and can restore half to two-thirds of wild-type activity. If they were translating this fraction of natural terminators into amino acids, they would probably be lethal. Ochre, however, could be the natural terminator; ochre suppressors are only about 1–5% efficient and any that are more efficient may be lethal.

The three chain terminators may be used in different ways in natural messengers. Mario Capecchi found a release factor R that hydrolyzes the bond between the last amino acid of the polypeptide and the last tRNA molecule when one of these triplets occurs on the messenger. E. Scolnick, R. Tompkins, and T. Caskey, in Nirenberg's laboratory, have now shown that there are two factors; R_1 corresponds to UAA and UAG, and R_2 corresponds to UAA and UGA. The R factors are evidently enzymes that bind to the mRNA and release the polypeptide from the ribosome.

A first clue to the initiation mechanism came when Kjeld A. Marcker and Frederick Sanger separated two very different tRNA molecules from *E. coli.* One of them, tRNA_F, carries only N-formylmethionine (fMet), while the other, tRNA_M, carries ordinary Met. In an amino acid incorporating system with artificial messengers, the fMet appears only in the N-terminal position. There is also a convenient class of natural messengers that can be used in such systems; some phage, such as R17, have a single-stranded RNA genome that weighs about 10^6 daltons and codes only three proteins, and recognizable parts of these proteins can now be made artificially with the extracted genomes as messengers. Jerry Adams and Capecchi found that the N-terminal residues of these proteins are always fMet. In fact, in both infected and uninfected cells a significant share of the proteins have N-terminal sequences such as Met-Ala or Met-Ser while others have only Ala or Ser. It appears that all phage and bacterial proteins are made with an initial fMet residue and that some enzyme then removes either the formyl group or the fMet residue. (The preponderance of Ala and Ser residues in the second position has no explanation.)

It is now clear that tRNA_F will bind to the codons AUG and GUG only when they appear at the *beginning* of a message. In this position, one of these codons combines with tRNA_F plus several initiation factors, and this complex apparatus ensures that the reading frame starts translating the messenger in the proper phase. At any internal positions, after the message is initiated, AUG binds only tRNA_M and GUG binds only valyl-tRNA, just like any ordinary codons.

Presumably natural messengers that carry the information from several cistrons have alternate initiator and terminator codons. As we go to press, the sequences of

```
          CH3
          |
          S
          |
          CH2
          |
  O       CH2
  ||      |
HC—N—C—COOH
    H   H
```

N-formylmethionine

two such messengers have appeared. Matthew M. Rechler and Robert Martin have
determined the sequence at the end of the *hisD* cistron by comparing the normal
protein informed by this gene with that from a frame-shift mutant. They infer the
sequence:

```
  Glu    Gln    Ala    term     possible init. sites   wild type
 ┌──┐   ┌──┐   ┌──┐   ┌──┐           ┌──┐ ┌──┐
               U    A              A   A
GAG GAA GG   C U  G  A C  X U    G  G  X X
 └──┘   └──┘  └─┘  └──┘    └──┘  └──┘
  Ala    Ser    Leu    Thr    term            mutant
```

Note that the terminator is either UAA or UGA, but not UAG. Also, one cannot tell
just where the next initiator is, but there is a space of at least one base, C, between
the two genes.

The nucleotide sequence of the R17 RNA is now known around the end of the
B cistron and the beginning of the C cistron. At the end of B there is a *double*
terminator UAA UAG; perhaps this is a safety factor in case an amber or ochre
suppressor strain is infected. The next sequence is AUG CCG GCC AUU CAA ACA
UGA, which could code for fMet-Pro-Ala-Ilu-Gln-Thr, but apparently this peptide is
never made. Then there is another space, GGA UUA CCC, and finally the C cistron
begins with AUG. It looks as if spaces between cistrons are essential for proper
translation.

With this information in mind, we have to take another look at the O^0 mutants of
the *lac* operon. There have always been certain so-called **polarity mutants** that
carry a mutation in the Z gene that also have little or no Y (permease) activity; this
phenomenon reflects the polarity of transcription and translation. Jonathan
Beckwith found that these mutants are all suppressible by amber and ochre
suppressors, which means that they are simply nonsense mutations. He then found
that the O^0 mutations are also suppressible; in other words, they are just extreme
polarity mutations. In fact, nonsense mutations at the operator end of the Z gene
have virtually no permease activity, but the amount of permease increases to near
100% as the nonsense mutations get closer to the Y gene. The general interpretation
is rather simple; it is assumed that when a ribosome finds a nonsense mutation and the
polypeptide it was working on is terminated, the ribosome becomes rather likely to
fall off the messenger. Normally, the terminator falls shortly before another initiator
so the ribosome hardly has to move at all before it begins to make another polypeptide.
But when a terminator comes in the middle of a gene, the ribosome has to traverse all
of the rest of the gene—without translating it—until it comes to the next initiator. The
farther it has to travel, the more likely it is to come off, so it is less likely to make any
more polypeptides on the same messenger.

So operators cannot really be defined by apparent O^0 mutations. The only
legitimate operator mutations known are O^c, and all of them appear to be short
deletions, so there is simply no DNA for the repressor to bind to.

It has now become clear that transcription and translation are intimately coupled.
The translation of messengers begins as soon as they are synthesized, while they are
still attached to the DNA. This can be seen sometimes as a sequential induction of
the enzymes in an operon; the enzyme informed by the first gene in the operon
appears slightly before the next, and so on, so the enzyme informed by the gene
farthest from the operator appears last. A more direct demonstration comes from

recent electron micrographs by Oscar L. Miller, Jr., Barbara Hamkalo, and Charles A. Thomas, Jr. They have caught some operons in gently disrupted bacteria with polyribosomes still attached (Fig. 10-15). The longest polyribosomes are presumably farthest from the operator; the messengers in them must contain almost a complete transcription of the operon. The shortest polyribosomes must be translating only the first genes of the operon.

10-10 RIBOSOMAL SUPPRESSORS

There is now good evidence that the ribosome plays an active part in mRNA-tRNA pairing and that mutations affecting the ribosome can affect the translation of a messenger. The relative concentrations of various ions (Mg^{++}, NH_4^+, etc.) are critical for translation; each cell probably has a normal concentration of these ions in its cytosol and the ribosomes are adapted to this condition. Artificial changes in the concentrations of these ions can change the ribosomes enough to cause different translations of artificial messengers.

Antibiotics such as streptomycin can also affect the ribosome. Most bacteria are streptomycin sensitive (sm^s); Julian Davies has shown that in these bacteria streptomycin binds to the 30S subunit and interferes with protein synthesis. Mutations are known that make the bacteria streptomycin resistant (sm^r) and even streptomycin dependent (sm^d). These mutations presumably affect the structure of ribosomal proteins.

Luigi Gorini discovered a class of auxotrophs that he called **conditional streptomycin-dependent** (*CSD*). These bacteria carry a mutation that can be corrected by a sm^r mutation in the presence of streptomycin. The sm^r mutation, in the presence of the drug, acts like a suppressor of a mis-sense mutation; since streptomycin interacts with the 30S subunit, this suggests that the ribosomes themselves play a part in translation and can correct errors under some conditions. Davies, Gorini, and Walter Gilbert have now shown that amino acid incorporation with sm^s ribosomes is strongly influenced by streptomycin, but that ribosomes from sm^r cells are not affected. It is therefore assumed that ribosomes have some ability to recognize tRNAs and that they play some specific role in translation; this is presumably a mechanism that has evolved for ensuring greater fidelity in translation than could be achieved by mere codon-anticodon pairing. However, the ribosomes can mutate so that their part in the translation is altered, and they can be affected by antibiotics of the streptomycin class. They can therefore be forced to make errors in translation, but these errors can also show up as corrections of mis-sense mutations.

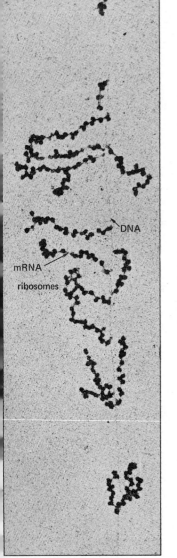

DNA

mRNA

ribosomes

FIGURE 10-15

An operon of E. coli *that is about 3 μm long and bears a series of polyribosomes in which messengers are being translated as they are transcribed. (Courtesy of Oscar L. Miller, Jr.)*

exercise 10-6 Suppose that an *E. coli* cell contains 7.5 million nucleotide pairs of DNA. The 16S rRNA molecule weighs 5.5×10^5 daltons. Spiegelman and Hayashi find that 0.1% of the total DNA is complementary to the 16S rRNA.

(a) How many nucleotides are in the 16S rRNA molecule?

(b) How many genes are there in the cell for the 16S molecule?

(c) Let's round the last answer off to four genes. Suppose the cell has to synthesize 10^4 new ribosomal RNA molecules per hour in order to double itself; if all four genes work together, how many seconds should it take each one on the average to make its RNA? How long should it take to add each nucleotide to the molecule?

SUPPLEMENTARY EXERCISES

exercise 10-7 Several cultures of *E. coli* K are grown under different conditions and tested for their specific activity of enzymes A, B, C, and D which are involved in methionine metabolism. Here are the results:

CULTURE	ENZYME A	ENZYME B	ENZYME C	ENZYME D
1	0.78	1.54	16.3	10.1
2	1.62	3.26	15.4	20.2
3	2.05	4.15	14.8	26.5
4	2.65	5.08	13.7	33.4
5	3.65	7.28	13.0	47.0
6	4.10	8.33	12.9	52.0

Which enzymes are probably informed by genes in a single operon?

exercise 10-8 Indolyl-β-galactoside is a substrate for β-galactosidase but not an inducer. When the enzyme acts on this compound, blue indole is released. If you plate bacteria on agar containing indolyl-galactoside, what kinds of bacteria will form blue colonies?

exercise 10-9 Phenyl-β-galactoside is a substrate for β-galactosidase but not an inducer. One product of the reaction with this substrate is phenol, which kills cells. For what kinds of mutants will you select by plating bacteria on agar containing phenyl-galactoside?

exercise 10-10 A cell has the following biosynthetic pathway for amino acids D and E:

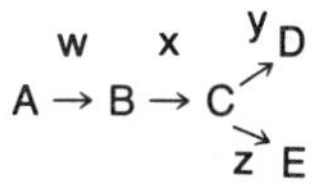

$$A \xrightarrow{w} B \xrightarrow{x} C \underset{z}{\overset{y}{\rightrightarrows}} \begin{matrix} D \\ E \end{matrix}$$

The enzymes w, x, y, and z are informed by genes W, X, Y, and Z, which form a single operon that is repressed only by E and not affected by D.

(a) Find at least one condition under which the cells could not grow.

(b) Do you consider these cells very well adapted? In other words, would you expect to find this system in any real cells?

exercise 10-11 Francois Chapeville and his friends did a beautiful and informative experiment. They knew that Raney nickel attacks the sulfhydryl group of cysteine and converts it into alanine. They attached cysteine to its tRNA (cys-tRNA), which normally binds to messengers containing UCU triplets, and then attacked this amino acyl RNA with Raney nickel, effecting the reaction

$$\text{cys-tRNA}^{cys} \rightarrow \text{ala-tRNA}^{cys}$$

They found that the altered compound still bound to UCU messengers and transferred alanine as if it were cysteine. What did this experiment prove?

exercise 10-12 To prove yourself that there are other kinds of control mechanisms, consider some of Ellis Englesberg's studies on the *E. coli* system for metabolizing arabinose. The pathway, enzymes, and genes are shown here:

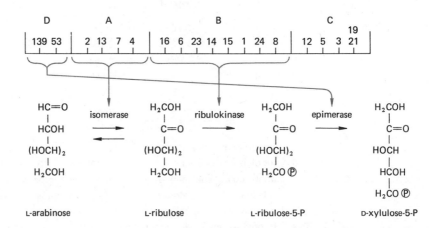

Mutations in *araC* are of two types: C^- mutants are negative for all three enzymes and C^c mutants are constitutive for all three.

(a) The *C* alleles show trans-dominant effects in F′ merodiploids; this shows that, in spite of its position, the *C* gene cannot be the same as a regulatory element in the *lac* system; which element?

(b) The possibility remains open then that *C* is the same as the regulator (*R*) gene; then C^c is R^- constitutive and C^- is like an R^s ("superrepressed") mutant in which the repressor binds to the operator so tightly that it does not come off. However, Englesberg made appropriate merodiploids and obtained the following relationships (where > means "is dominant to"):

ARABINOSE	LACTOSE
$C^+ > C^-$	R^s ? R^+
$C^c > C^-$	R^- ? R^s
$C^+ > C^c$	R^+ ? R^-

Compare these relationships with those in the *lac* operon and show that they are not the same.

(c) The last result above shows that the *ara* system may be complicated, but one simple model compatible with the data is a positive control system where *C* makes an *activator* which acts on an *initiator* region *I* and is normally active only when bound to arabinose. Deletions of *C* which extend across the hypothetical *I* region are all *ara⁻*; what would you expect *ara⁺* revertants of them to be?

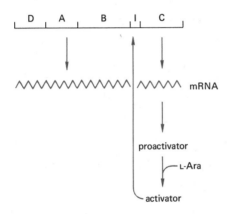

PART

11

the eucaryotic cell

In this chapter we will examine the structure and general function of a "typical" eucaryotic cell. This typical cell is somewhat like the average American; by reading the census figures you can learn all about him, but you never expect to meet him on the street. These are useful fictions; each one is a composite of many features from different individuals. We will picture a eucaryotic cell with features that you can expect to find in one form or another in most of the cells you will study; but you never expect to meet this cell under a microscope.

Just to keep our feet on the ground, we will begin by outlining briefly a few of the major plant and animal cell types, to see what our typical cell is typical *of*. We will ignore most of the protista here, because we will examine their features in more detail in Chapter 13.

A. A broad view

11-1 PLANT AND ANIMAL CELLS

Some of the major cell types found in animals are outlined in Fig. 11-1. **Epithelial cells** form layers that cover surfaces, line cavities, and protect more delicate tissues. Depending on their length, they may be columnar, cuboidal, or squamous, and the exposed surfaces of the first two are often ciliated; for example, your nasal passages are lined with a ciliated epithelium that is constantly moving a stream of mucus down into your throat. Epithelium may be stratified, creating an outer layer of thin, tough cells, as on the surface of your skin. Some epithelial cells are more specialized. **Gland cells** make products called secretions which they export to other parts of the body. **Sensory cells** are specialized for receiving stimuli; these include the cells in the retina of your eye that absorb light and those in your tongue that detect tastes.

Connective tissue consists of comparatively small numbers of cells embedded in a *matrix,* often one that the cells themselves have made. The tissues themselves include tendons, ligaments, bone and cartilage; their matrices are primarily protein and polysaccharide fibers, some very stiff and some very elastic. The cells that secrete these fibers are **fibrocytes,** which remain embedded in the matrix. **Osteocytes** are specialized cells that lay down the calcium phosphate crystals of bone. Connective tissue also includes **histiocytes** (macrophages), which can engulf and destroy infecting bacteria and other foreign bodies, and **chromatophores,** which are colored cells.

Blood is a special type of connective tissue whose matrix consists of water, dissolved proteins, and other materials.

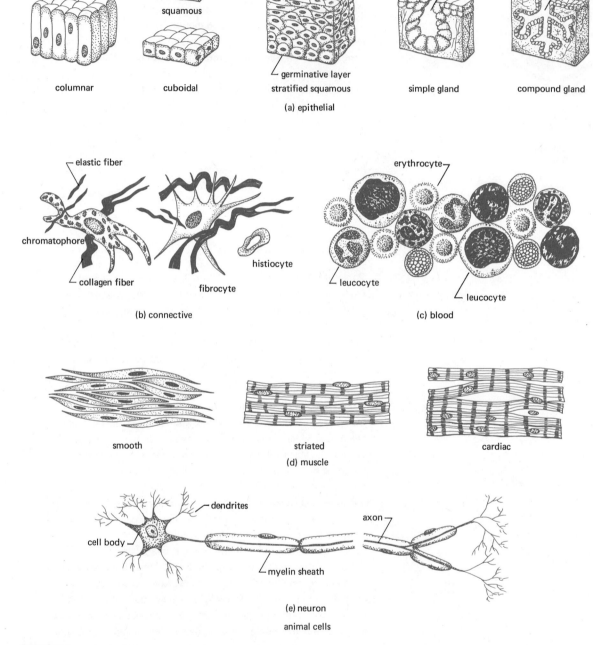

FIGURE 11-1

A set of representative animal cells.

There are many specialized blood cells, but we will distinguish only two major groups: **erythrocytes,** or red cells, which carry gases to and from body cells, and **leucocytes** or white cells, including many different specialized types.

There are three major types of **muscle cells,** all of which contain fibrous protein structures that can contract quickly. **Smooth muscle** cells are spindle shaped and form body walls, blood vessels, etc. **Striated muscle,** which is the major tissue animals use for movement (the muscles of our arms and legs, for example), consists of long, multinucleate cells filled with contractile proteins and divisible into short segments or sarcomeres. Cardiac muscle is only slightly different and is the substance of heart tissue.

There are many types of **nerve cells** or **neurons.** The most typical neuron has a long, thin axon that can carry signals quickly over great distances.

Most of the rectangular, relatively unspecialized cells of organs such as the liver are called **parenchymal cells.** This term is also used for most of the cells in green plant tissue (Fig. 11-2), which are hollow and polygonal, with thin walls and large central vacuoles. Somewhat more specialized plant tissue is **collenchyma,** which has slightly thickened walls and is used where more support is needed. **Epidermal cells** are similarly shaped, but produce a protective layer of wax on one side. **Sclerenchyma** cells have very thick walls and are used for strengthening tissues.

The **vascular cells** of plants are used to transport water and dissolved nutrients. Tracheids are spindle shaped and may be up to 5 mm long; they have very heavy walls, but no nuclei or cytosomes, and neighboring tracheids connect to each other through little holes to make long conducting tubes. Vessels are also formed from cells that have thick walls and lose their contents, but here the cells lose their entire cross walls to form long, continuous tubes. Sieve tubes are formed by cells that retain their contents and communicate with each other through sieve plates.

11-2 TEARING DOWN THE CELL

Differential centrifugation of eucaryotic cells is somewhat more interesting than the same process carried out with procaryotes because there are several membranous components of different sizes that can be separated easily by varying the gravitational force. Figure 11-3 shows a regime that yields four interesting fractions: nuclei, mitochondria, lysosomes, and microsomes.

One of the first important lessons is that the objects we can isolate are primarily little bags—little membranous vesicles with quite characteristic sizes and contents. Since we can perform many different chemical analyses and enzymatic assays, we can learn about their contents rather easily. Let's see what each of these fractions contains.

The nuclei contain almost all of the cellular DNA; we already knew this from staining experiments and from the role of the nucleus in determining the form of cells (Section 4-7). They contain some RNA, and the polymerases for making DNA and RNA, but relatively little else in the way of enzymes.

The mitochondria have a very characteristic form. They have two layers of membrane and a number of internal cristae that are invaginations of the inner membrane. There

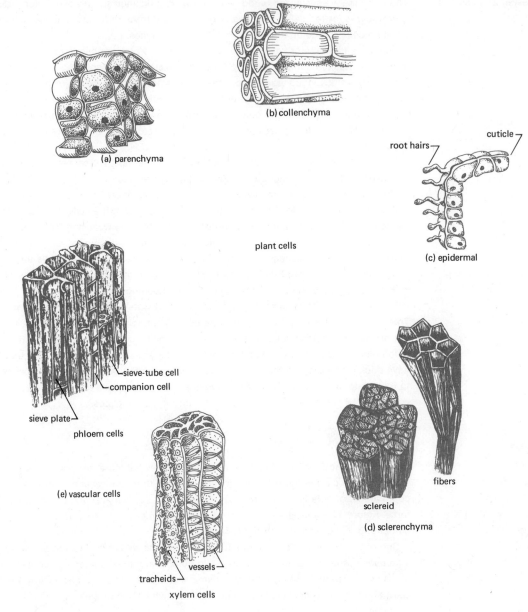

(a) parenchyma

(b) collenchyma

plant cells

cuticle

root hairs

(c) epidermal

sieve-tube cell

companion cell

sieve plate

phloem cells

(e) vascular cells

vessels

tracheids

xylem cells

fibers

sclereid

(d) sclerenchyma

FIGURE 11-2
A set of representative plant cells.

is an extensive matrix space between the cristae. Mitochondria have about the same mass as bacteria, but about 40% of their dry weight is phospholipid. They are very rich in enzymes and contain, at the very least, (1) all the enzymes of the TCA cycle and

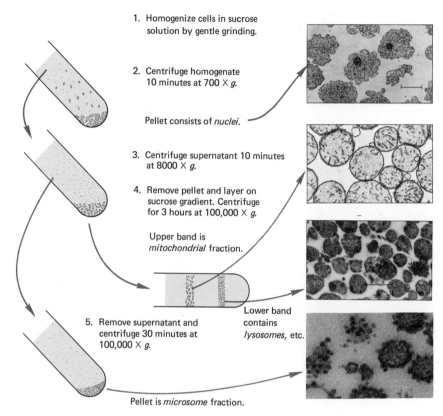

1. Homogenize cells in sucrose
   solution by gentle grinding.

2. Centrifuge homogenate
   10 minutes at 700 × g.

   Pellet consists of *nuclei.*

3. Centrifuge supernatant 10 minutes
   at 8000 × g.

4. Remove pellet and layer on
   sucrose gradient. Centrifuge
   for 3 hours at 100,000 × g.

   Upper band is
   mitochondrial fraction.

Lower band
contains
lysosomes, etc.

5. Remove supernatant and
   centrifuge 30 minutes at
   100,000 × g.

Pellet is *microsome* fraction.

FIGURE 11-3

*A regime of differential centrifugation with some of the pellets it produces. (Nuclei
courtesy of Donald E. Slagel; mitochondria from Charles Hackenbrock; lysosomes
from Henri Beaufay; and microsomes from George Palade.)*

the Wood–Werkman reactions, (2) all of the cytochromes and associated lipids of the
electron-transport system, (3) most of the enzymes for tetrapyrrole biosynthesis, and
(4) several other oxidative enzymes, including the enzymes for oxidation of fatty acids.

The **lysosomes** are very interesting. In electron micrographs they look quite
heterogeneous, and they contain an important set of degradative enzymes that can
hydrolyse many polymers into their monomers (Fig. 11-4). Their function, clearly, is to
digest macromolecules.

The **microsomes** contain most of the cell RNA, primarily as ribosomes that are
membrane bound. The ribosomes can be removed and the membranes that remain
behind have a number of enzymes, including (1) cytochromes that can oxidize NADH
and NADPH and (2) enzymes for the biosynthesis of triglycerides, phospholipids, and
sterols.

We can already begin to see some organization of cell units, but this comes out
most clearly when we start to look at whole cells.

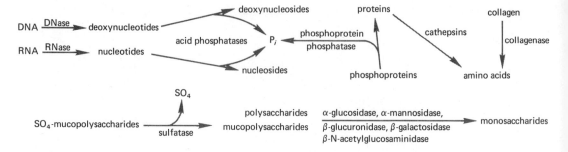

FIGURE 11-4

Some of the enzymes found in lysosomes and the reactions they catalyze. The list of known enzymes is growing daily and there are many others known in different tissues.

11-3 PUTTING THE CELL TOGETHER

When a cross section is taken through our typical cell, we see a pattern like that shown in Fig. 11-5. There is a prominent nucleus, there are mitochondria, and there are many membranes that fill most of the space between these. Keith Porter first found these when he began to do early high-resolution electron microscopy; since they appeared to form a vast network and were concentrated in the inner ("endoplasmic") regions

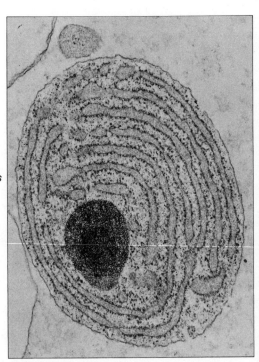

FIGURE 11-5

A human plasma cell that shows a fairly typical arrangement of membranes. There is a dark nucleus and a series of concentric endoplasmic reticulum membranes with ribosomes on one surface and a gray lumen on the other side. This section does not include any mitochondria. (Courtesy of C. Phillip Holland.)

around the nucleus, he called them **endoplasmic reticulum** (ER). Most of the ER membranes we see are "rough" because of a layer of ribosomes all along one face. It was not hard to show that the "microsome" fraction of broken cells consists of ER membranes with their attached ribosomes that have simply rounded up into little vesicles at the time the cells were broken. (In fact, formation of vesicles is one of the most characteristic features of most membranes.) The "microsomes" are therefore a complete artifact, and it is the ER that is native and interesting.

The reticulum takes many forms. The ER membranes are usually roughly parallel and double, with ribosomes only on one face (Fig. 11-6). Rather extensive continuity can be seen between them in some pictures, and it is reasonable to say that most of the ER membranes in a cell are connected to one another in a vast network. Furthermore, Michael Watson discovered that connections can be found between the innermost

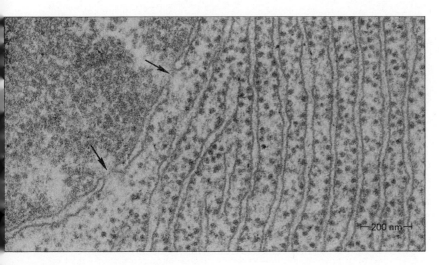

FIGURE 11-6
The nuclear and ER membranes of a mouse parotid gland cell. Notice the trilaminar structure of the membranes and the very distinct ribosomes attached outside the nucleus. There are two pores in the nuclear membrane (arrows) with a light septum across each one and a characteristic lack of chromatin nearby. Lead hydroxide stain, 75,000X. (Courtesy of Harold F. Parks.)

ER membrane and the nuclear membrane (Figs. 11-6 and 11-7). Thus, the nuclear membrane is, in a sense, merely the innermost ER membrane, and it is difficult to distinguish one type from another.

In some cells, the ER has very few ribosomes attached and is said to be "smooth" (Fig. 11-8). It has been seen as vesicles, tubules, and various odd shapes. The quantity of ER varies considerably from one cell type to another. Embryonic cells generally have the least and adult cells that are making a lot of protein have the most. The amount of membrane is clearly related to the function of the cell; and now we must remember the general principle we discussed in Chapter 8 in relation to the amount of internal electron-transport membrane in bacteria. At that time, we pointed out that

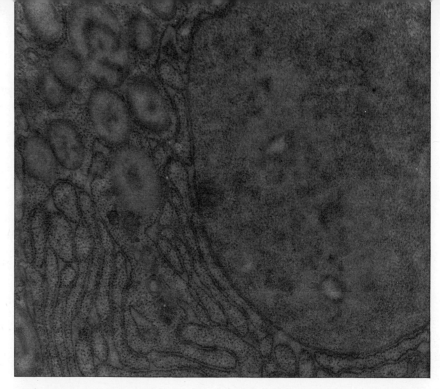

FIGURE 11-7

*Nuclear and ER membranes of a mouse parotid gland cell showing the complete
continuity between the two types of membranes. Trace some membranes to identify
the lumen of the ER and see that it is continuous with the space between the layers
of the nuclear envelope. Notice where the ribosomes are. (Courtesy of Harold F.
Parks.)*

electron-transport systems are built into membranes and their ability to make ATP is
therefore a function of how much membrane they have. Since the ribosomes are the
sites of protein synthesis and are attached to the ER membranes in eucaryotic cells,
the amount of protein made is a function of the amount of ER membrane; cells that
are making vast amounts of protein are literally filled with rough ER membranes. In
other cells, the ER membranes contain enzymes for making products other than
protein, and they become filled with membranes in order to make large amounts of
these products. If we can establish the general concept that enzyme systems of almost
all kinds are largely built into or bound onto membranes, there will be no mystery at
all in the extent to which membranes appear in cells.

   One of the most definitive features of a eucaryotic cell is its nucleus, which is always
bounded by two typical membranes; these have naturally been called the inner and
outer nuclear membranes, but a careful search reveals that the two membranes join
at many points. Fig. 11-6 shows that the points of continuity come in pairs separated
by about 800 Å, creating apparent pores in the nuclear membrane. While "pore" is the
standard term for these regions, they are generally not open holes. The best cross-
sectional pictures show a very thin line across the pore; this is visible in Figs. 11-6 and
11-8. Figure 11-9 is an elegant surface view of pores taken by Joseph G. Gall, showing
that these are very real, regular structures, with an octagonal shape, but there is clearly

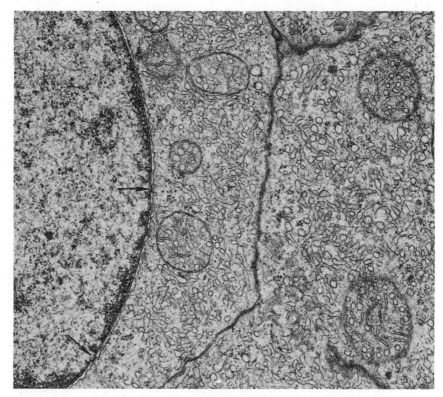

FIGURE 11-8

Smooth ER membranes in the adrenal gland cells of the Siamese tree shrew. Several mitochondria are visible and there are pores in the nuclear membrane (arrows). This vesicular appearance is quite typical of smooth ER. Glutaraldehyde and OsO_4 fixation, stained with uranyl acetate and lead hydroxide, 46,500X. (Courtesy of Jeptha R. Hostetler.)

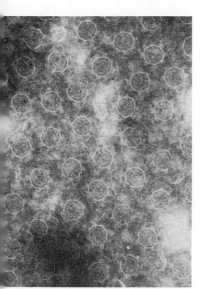

FIGURE 11-9

Surface view of pores in the nuclear membrane of an oocyte of the salamander Triturus. The pores appear to be closed by a septum and have a generally octagonal shape. Negative stain with phosphotungstate, 75,000X. (Courtesy of Joseph G. Gall.)

something fairly solid across them. Although they are not wide open, the pores probably allow the free passage of something; as the pictures show, there is typically an open region, devoid of chromatin, on their inner surface, as if something were streaming through them. C. M. Feldherr has shown that colloidal gold particles up to about 85 Å in diameter will pass the pores very quickly, through a space in their center, but that larger particles are held back. However, we should note that isolated nuclei exhibit osmotic properties, as if they are rather tight bags; and it can be shown that nuclei transport amino acids, and so on, in and out more or less like intact cells. On the whole, nuclei should be considered cellular compartments that communicate with the rest of the cell through specific transport systems, some of which are located in the pore regions where transportation may be associated with holes that are large in molecular terms. What passes through these pore regions is unknown. Since the genome is inside the nucleus and the ribosomes are outside, we would like to imagine a flow of RNA from the nucleus to the cytosome through these pores; while this has been suggested by some experiments, others suggest a different mechanism that involves changes in the nuclear and ER membranes themselves.

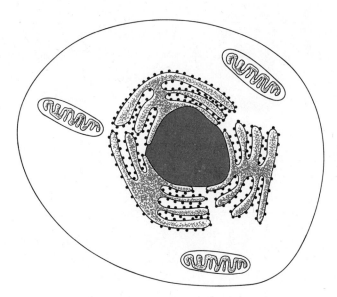

FIGURE 11-10
General topology of a eucaryotic cell. The lumen of the ER forms one space (color); the true inside of the cell consists of the nucleus, the spaces that contain ribosomes, and the mitochondria. The outside of the cell is beyond the cell membrane.

From this brief discussion, we can conceive of most eucaryotic cells as having the topology shown in Fig. 11-10. There are three regions: inside, outside, and middle. The *outside* is everything beyond the cell membrane. The *inside* is a region that includes the nucleus and the space between the ER membranes that is in contact with the ribosomes; the mitochondria must be seen as islands in this space. The *middle* is generally called the *lumen* of the ER. It is a space on the opposite side from the ribosomes and is continuous with the space between the two layers of the nuclear membrane. With this three-phase cell in mind, we can begin to understand how a eucaryotic cell functions.

B. Internal membrane systems

11-4 PROTEIN SYNTHESIS AND THE ENDOPLASMIC RETICULUM

We can understand the function of the ER in relation to biosynthesis—particularly protein biosynthesis—when we realize that most of the cells in which we find an extensive reticulum are making materials to be exported. For example, connective tissue is layed down by fibrocytes that make protein and polysaccharide molecules internally and then export these subunits to the outside, where they crystallize into solid masses. Most of the small protistans are covered by hard, protective layers of protein and polysaccharide, which they make internally and then carry to their surfaces to make walls. Animals contain many organs and cells whose function is to make secretions to be passed into the blood for use in other parts of the body (endocrine glands) or to be used in the digestive system (exocrine glands). The principal endocrine secretions are hormones that regulate the metabolism of the whole animal, and several of these are proteins. The exocrine secretions are principally digestive enzymes made in the pancreas, the walls of the stomach, and nearby tissues, that are secreted into the stomach and intestines for food digestion. So, in general, the synthesis of proteins and other molecules in these cells must be accompanied by a *transport system* to the outside, and this seems to be the main function of the ER.

Protein synthesis occurs on the ribosomes attached to the inner face of the reticular membranes (in relation to the picture of Fig. 11-10). Calvin Redman and Phillip Siekevitz have now shown that in gland cells the proteins are transported across the membrane into the middle phase (the lumen of the ER) *as they are synthesized.* Thus these proteins, which will eventually appear outside the cell, pass through a major barrier to the outside during biosynthesis. The precise mechanism is still unknown, but we do not have to invoke a complicated mechanism for gathering the protein molecules from the cytosol, sorting them out from all other proteins there, and then pushing them across a membrane.

Once in the lumen of the ER, the proteins can move rather freely through open channels toward the outside of the cell; perhaps they are driven outward by contortions of the ER membranes. The proteins then appear in special membranes of the **Golgi apparatus,** which may be considered a part of the ER system. About 1900, Camillio Golgi first described some structures in barn owl nerve cells that take up a silver stain very readily; many people refused to believe in their existence for the next half century, until electron microscopy showed that they are real and have a characteristic form. In cross section they look like parallel, flattened vesicles that broaden out at their tips (Fig. 11-11). They are usually parallel to the nuclear membrane and closely associated with ER membranes. Hilton H. Mollenhauer and D. James Morre have isolated some of the Golgi vesicles and shown that they consist of central cisternae connected to a mass of tubules (Fig. 11-12).

All of the observations on secretion identify the Golgi membranes as packaging and exporting centers of cells. The path seems to run from the lumen of the ER through small vesicles into the cisternae of the Golgi membranes closest to the ER and nucleus; this is sometimes identified as a "forming face" of the stack of Golgi membranes. The secretion then proceeds through the tubules around the edges of the membranes and into vesicles that form there and move toward the outside of the cell;

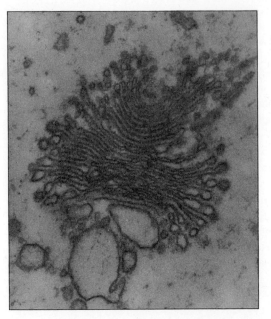

FIGURE 11-11

Golgi membranes of the flagellate Trichomonas, *showing their characteristic appearance in cross section. (Courtesy of A. V. Grimstone.)*

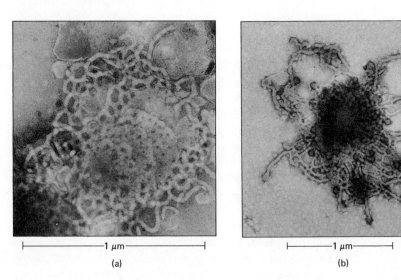

|—————1 μm—————| |—————1 μm—————|

(a)   (b)

FIGURE 11-12

Isolated dictyosomes (Golgi vesicles) from (a) radish and (b) cauliflower, showing the central disc and its associated tubules. Notice, in (b), the little thorny balls; their function is unknown. Glutaraldehyde fixation, stained negatively with phosphotungstate. (Courtesy of H. H. Mollenhauer and D. J. Morre, reprinted from Journal of Cell Biology *by permission of Rockefeller University Press, New York.)*

the points where the vesicles form has been called a "maturation face" of the Golgi membranes. The vesicles formed consist of dense secretions surrounded by a single membrane; these packets of digestive enzymes are easily seen even with a light microscope and have been called zymogen granules.

The vesicles migrate to the cell membrane and there they empty their contents to the outside by a process of **exocytosis** (Fig. 11-13) through coalescence of their membranes with the cell membrane.

A vesicle containing
material to be secreted
approaches the cell envelope

FIGURE 11-13
The ejection of material in a vesicle by exocytosis.

and fuses with it so the
inside of the vesicle becomes
continuous with the outside
of the cell

and the vesicle's contents
can be deposited outside

Lucien Caro and George Palade have traced enzyme secretion from pancreatic cells by autoradiography. They allowed cells to incorporate a pulse of labeled amino acids and at various times they located the label in tissue slices. First, it appeared in the region of rough ER, then over visible protein in the lumen of the ER, and finally in the Golgi membranes and in vesicles derived from them. Palade has shown that collagen subunits are secreted by fibrocytes in the same way. This general pathway seems to be clearly established. One of the principal uncertainties is in the matter of "connections" between the ER and Golgi membranes. Permanent connections probably do not exist; if they did, they would have been found more regularly. However, transient connections almost certainly exist; ER membranes must pinch off vesicles that are full of whatever material is in the ER lumen, and these vesicles must then move out and fuse into Golgi membranes. We should therefore picture the membranes themselves, as well as their products, being synthesized in the inner regions of the cell and flowing outward. It is significant that the enzymes for manufacturing lipids have been localized in the ER membranes, where significant membrane synthesis must occur.

11-5 THE LYSOSOME SYSTEM

Most eucaryotic cells perform another activity that is essentially the opposite of secretion; they pick up particles and pieces of debris from their environment and digest them. Figure 11-14 shows how they are picked up by **endocytosis,** the reverse of exocytosis; the cell surface invaginates to make a little pocket and this pinches off as an **endocytic vesicle** or **phagosome** because of the rhythmic contraction of the invagination toward the inside. In histiocytes and some leucocytes that clean out bacteria and other foreign matter, the process is called phagocytosis. It is also given many other names (pinocytosis, arthrocytosis, etc.), but these are distinctions without a difference and we will not use these terms.

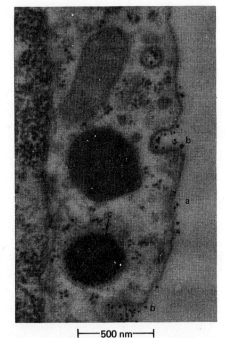

FIGURE 11-14

Endocytosis in a mouse liver cell. Heavily stained granules adsorb to the cell surface (a), induce invaginations (b), which pinch off to make endocytic vesicles (c). (Courtesy of Harold F. Parks.)

├──500 nm──┤

Endocytic vesicles move inward from the cell surface; their common fate is fusion with lysosomes. Lysosomes abound in cells that are active in endocytosis; they contain the hydrolytic enzymes necessary for digestion of most biological macromolecules, and they merely have to fuse with the endocytic vesicles to bring these enzymes in contact with their substrates. The growth of vesicles during this phase can be seen easily.

Lysosomes are apparently made in the Golgi apparatus. Alex Novikoff has traced typical lysosomal enzymes from their synthesis in the ER, through the Golgi membranes, and into lysosomes. Lysosomes are therefore little different from any other protein-filled vesicles in their formation, but rather than being secreted they are held in readiness for intracellular digestion.

This explains why lysosomal pellets look so heterogeneous. Pure lysosomes, which contain merely enzymes, are just dense-staining bodies. Other vesicles in the lysosomal fraction are phagosomes or digestive vacuoles, which have been combined with the lysosomal enzymes. After digestion has gone as far as it can, indigestible residues can be seen in some vesicles. These residual bodies empty themselves by simple exocytosis at the cell membrane. This whole process can be seen very prominently in some of the large protozoa that form food vacuoles by endocytosis; these large vacuoles circulate around the cell and their residues are finally ejected in a process that can be seen by light microscopy.

11-6 THE CELL SURFACE

Two features of cell surfaces become very important in any discussion of the synthesis of products for export and the intake and digestion of extracellular materials. First, if it is important for cells to communicate with their environment, then elaborations of the surface to create a greater area for interactions become very significant. Many features of multicellular organisms can be understood as mechanisms for increasing surface/volume ratios to facilitate transport. The epidermal cells of plant roots usually send out long, thin root hairs that can take up water and nutrients from the soil. Leaves have little openings to the air (stomata) so that air can get in and out and gas exchange with cells inside the leaf can occur. Lungs are basically masses of tiny air pockets where exchange can occur between the atmosphere and the blood. The intestine is a long tube with a folded surface and most of it is covered with tiny finger-like **villi** which increase the surface where transport of digested food into the blood can occur; however, each of these villi is made of many cells, and the surface of each cell is contorted into **microvilli** (Fig. 11-15) which increase the cell surface enormously.

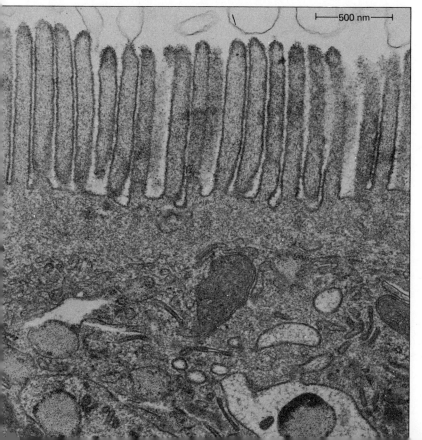

├─────500 nm─────┤

FIGURE 11-15
A border of microvilli in epithelial cells of mouse intestine. Fixed in glutaraldehyde and OsO$_4$, stained with lead hydroxide. (Courtesy of Roger O. Lambson.)

The form and regularity of microvilli suggests that they are very stable structures, in contrast to the transient extensions that are seen on many other cells. Many materials are specifically transported across the great surface they create. Other materials are pulled into the cells by endocytosis at the base of the clefts between microvilli.

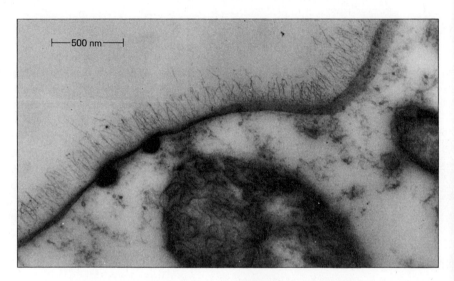

FIGURE 11-16
The fine, polysaccharide filaments on the surface of an ameba, Amoeba proteus. (*Courtesy of Philip W. Brandt.*)

When surfaces of the epithelial cells of intestines, other cells that have one free face bordering on a cavity, or the surfaces of free protista are examined with high-resolution techniques, they appear to be very rough and fuzzy. Staining reveals that the fuzz is a polysaccharide, and in some cases it can be seen as a mass of long, thin strands (Fig. 11-16). This polysaccharide fuzz is probably important in specific interactions between cells, for it may allow cells to recognize one another and bind to one another. It also seems to be very important in endocytosis, for particles attach to the fuzz first and are trapped there before they induce invagination of the cell membrane. Not all materials will induce endocytosis, and those that do must interact specifically with the outer layers of the membrane.

To summarize the major features of eucell structure, Fig. 11-17 shows a very general cell with a set of internal membranes. The cell is shown making both packets of enzymes for export and lysosomes for intracellular digestion. Try to picture the general topography and topology in relationship to the simplified drawing of Fig. 11-10.

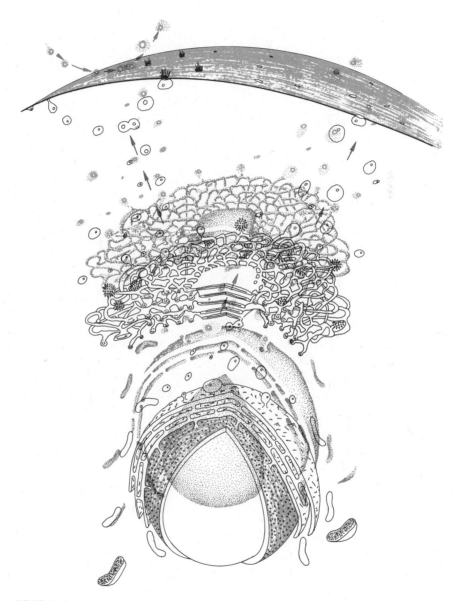

FIGURE 11-17

General structure of a eucaryotic cell. A flow of materials is indicated from the lumen of the ER membranes through the Golgi membranes, and out toward the cell surface. At left, lysosomes are being synthesized which fuse with endocytic vesicles. At right, vesicles full of enzymes are being made for export.

exercise 11-1 What do you see designated by capital letters in each of these electron micrographs?

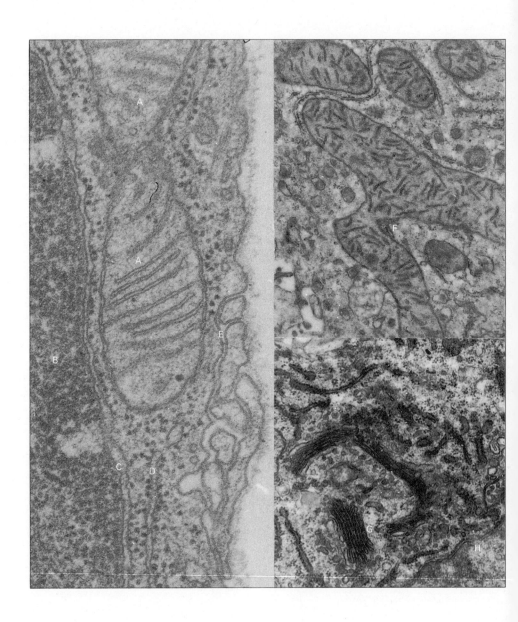

exercise 11-2 What is happening near F here?

exercise 11-3 What is the possible relevance of D to E here?

C. Motions and contractions

Among the characteristics of eucaryotic cells are a rather amazing variety of movements. These include the movements of cell division that we will discuss in Chapter 12; we will try to summarize these movements here and give a general picture of the mechanisms responsible for them.

11-7 SOME GENERAL PRINCIPLES

This is a vast and complicated subject that is just beginning to sort itself out in molecular and mechanistic terms, and so it is best to begin with a summary of some of the principles we will establish.

First, we will find that practically all of the movements we can analyze at this time are associated with elongated or fibrous protein structures and that these take one of two forms: **filaments** or **microtubules.** The proteins of these filaments or microtubules have a great deal in common with one another; that is, these structures are all composed of polypeptide structural units that fall into two classes:

1. Long, fibrous **myosins.**
2. Short, globular **tektins.**

Second, we will find that most movements that can be understood at all generally conform to a model in which filaments or the subunits of filaments and microtubules slide past one another into new positions, but not in which the polypeptide chains themselves undergo extensive conformational changes within each subunit.

Third, wherever there is enough information to point to an energy source for the movement, that energy source is ATP, which is not at all surprising. There has been a long history of discussion about this matter, particularly in the vast literature of muscle physiology, but the point seems to be established now. In systems that have been well analyzed, there is some protein that contains an ATPase activity. This is not as enlightening as you might think. An ATPase is simply an enzyme that hydrolyzes ATP into $ADP + P_i$ (or perhaps $AMP + PP_i$); the fact that it exists does not tell us how the free energy of the ATP is coupled into the movement.

Fourth, we have a great deal of information now about mechanisms that regulate movements in particular systems, especially in muscle. The mechanism always seems to be an uptake and release of cations, particularly Ca^{++} ions, by specialized cell membranes in intimate contact with the active protein filaments. However, we will not elaborate on this until after we have considered membrane structure and function in more detail.

11-8 STRIATED MUSCLE

As a point of departure, we will first examine the structure of vertebrate striated muscle, which is now known better than any other strongly motile system. Figure 11-18 shows that a muscle consists of muscle fibers, each of which is a long, multinucleate

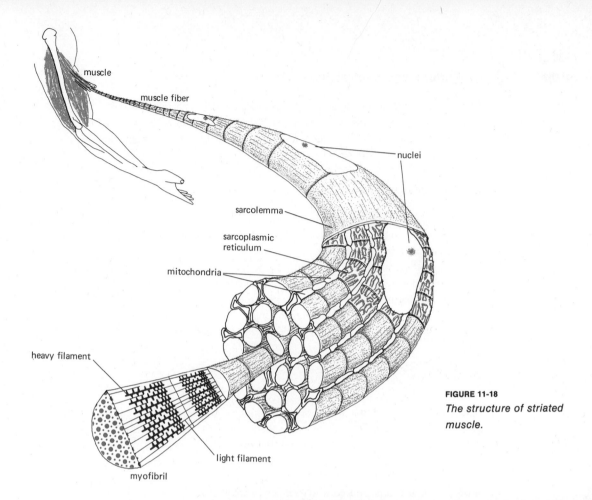

muscle

muscle fiber

nuclei

sarcolemma

sarcoplasmic
reticulum

mitochondria

heavy filament

light filament

myofibril

FIGURE 11-18
*The structure of striated
muscle.*

FIGURE 11-19
*The banding pattern of
myofibrils from wasp flight
muscle. Letters were assigned
on the basis of polarizing
microscopy—I means isotropic
and A means anisotropic.
Notice the mitochondria
packed between myofibrils
where they provide ready free
energy for contraction.
(Courtesy of David S. Smith.)*

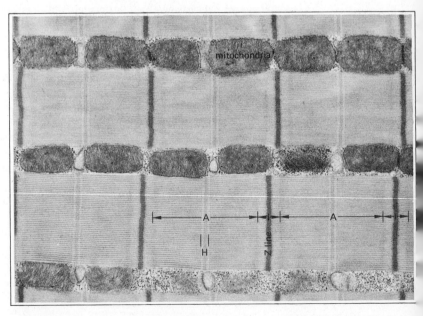

mitochondria

A

A

H

Z line

cell with a diameter between 10 and 100 μm, depending on the animal and the muscle. The fibers contain mitochondria, a complex endoplasmic reticulum (often called the sarcoplasmic reticulum), and a cell membrane (generally called the sarcolemma). The contractile units or **myofibrils,** about 2–3 μm in diameter, run through the fiber; each of them is a bundle of contractile proteins, the **myofilaments.**

Each myofibril has a characteristic pattern of repeating bands which were given letter designations on the basis of early microscopy (Fig. 11-19). Somewhat arbitrarily, the repeating unit between two Z lines is called a **sarcomere.** We now know a great deal about the distribution of proteins in this structure.

Myosin is the major structural protein of muscle, accounting for about 55–60% of its dry weight. It is one of the classical α-helical k-m-e-f proteins; each myosin molecule weighs 530,000 daltons and is 1400 Å long. When exposed briefly to proteolytic enzymes, myosin breaks quite specifically into two fragments, the light and heavy meromyosins (LMM and HMM). Light meromyosin is a long, thin fiber that weighs 140,000 daltons, while heavy meromyosin is a more compact, globular piece that weighs about 300,000 daltons. Native myosin has an ATPase activity that is confined to the HMM fragment. The whole molecule can now be seen as a fibrous unit with a globular end.

When myosin is treated with alkali to pH greater than 10, three small (20,000 dalton) globular (g) units are released from each molecule. When the remainder is subjected to a 8-M guanidine, it dissociates into two fibrous (f) units that weigh about 215,000 daltons each. All of these data together suggest that the whole myosin molecule is a two-stranded coiled coil with a globular end to which the three g units are bound.

Actin is one of the tektins. The basic structural unit (G-actin) weighs about 60,000 daltons and is quite spherical. It polymerizes into fibers (F-actin); Jean Hansen and J. Lowy have found that these fibers are helices of two filaments, each of which is a string of G-actin units, which wind around each other with a very gentle pitch, 13–15 subunits per turn, where each subunit is about 55 Å long.

Tropomyosins constitute only about 15% of the muscle protein. Each molecule weighs 53,000 daltons; it is 400 Å long and is almost entirely α-helix.

Work by Hugh Huxley and Jean Hanson, by Andrew F. Huxley, and by Andrew Szent-Gyorgi and H. Holtzer have led to a consistent picture of the arrangement of these proteins in the myofibril. There are always two types of myofilaments:

1. Heavy filaments of myosin, that make the A band.
2. Light filaments of actin that make the I bands.

In cross section, the heavy and light myofilaments usually interdigitate with one another in a hexagonal pattern (Fig. 11-20). The light actin molecules are anchored on the Z lines. Tropomyosin is associated with the actin molecules in the I bands, perhaps wound around them in additional helices.

Bridges between the heavy and light myofilaments can be seen in both longitudinal and cross sections. Hugh Huxley has shown that the highly asymmetrical myosin molecules from dissociated filaments will reassociate as shown in Fig. 11-21 to make heavy filaments with the globular regions projecting. Since other experiments have shown that the HMM molecules have an affinity for actin, we can say with some

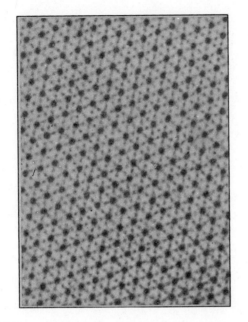

FIGURE 11-20
Cross section of aphid flight muscle showing the hexagonal pattern of heavy and light myofilaments. (Courtesy of David S. Smith.)

FIGURE 11-21
Structure of myosin. Two fibrous (f) units and three globular (g) units form one myosin fibril. The fibrils then associate to make a symmetrical filament with their globular regions projecting. The approximate position where the fibril is split into meromyosins is also shown.

FIGURE 11-22
While the details are unknown, it is clear that the light filaments are pulled together by the globular parts of the heavy filaments, which probably attach, pull, reattach, and pull again.

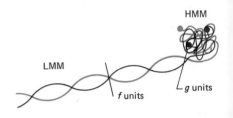

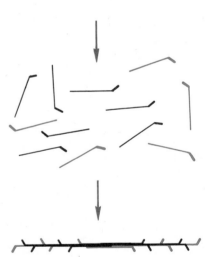

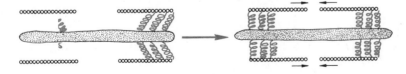

certainty that the bridges seen by electron microscopy are the globular or HMM regions of myosin, which reach over and attach on the actin filaments.

Careful observations of muscles in various stages of contraction by Huxley and Hanson and by Andrew Huxley and R. Niedergerke have shown that the individual filaments do not change their length during contraction. Instead, the two sets of filaments slide past one another. The general mode of action between the filaments is rather obvious, since the bridges between the heavy and light filaments are connected only by the globular regions of the myosin and the latter contains all of the ATPase activity. We do not know exactly what is happening, but it is evident that one set of filaments can pull the other past it in the same way that a centipede pulls itself along by taking many tiny steps (Fig. 11-22).

11-9 MICROTUBULES

The introduction of glutaraldehyde as a fixative in 1963 suddenly opened a new world of electron microscopy, for it revealed that much of the apparently structureless space of many cells was actually highly structured by long, thin microtubules. Keith R. Porter and Myron Ledbetter found them in a variety of cells; they generally seem to be present wherever cell shapes must be maintained that the cell membrane by itself might not be able to hold, and in this sense they are somewhat equivalent to a "cytoskeleton" which has often been postulated for maintaining cell shapes. They run parallel to and slightly below the cell membranes of many cells like a set of girders and are particularly prominent in plant cells. Many microtubules run the length of the axon in nerve cells. They have been seen in many cells running parallel to the direction of cyclosis or streaming, and they are strongly associated with cell motility.

The typical microtubule is 230–250 Å in diameter. Porter has analyzed some by Markham's rotational method (Section 6-2) and found them to have thirteen-fold rotational symmetry. When seen from the side there are clear indications of a subunit structure and a slight helical pitch.

Some of the best examples of microtubules come from protistan cells. L. E. Roth has found several examples of microtubules running below the surfaces of protozoa. These bundles are especially dense near regions that change their shape rapidly, such as the orifice of the *contractile vacuole* that constantly pumps water out of the cell (Fig. 11-23). Some of the protozoa generally classified as amebas have very unusual extensions; all of these extensions are called pseudopodia, but in the case of the heliozoa or "sun animalcules" the pseudopodia are more like flagella. J. A. Kitching has found that each one contains a bundle of microtubules arranged in two spirals (Fig. 11-24). The bundle of microtubules runs all the way to an anchor on the nuclear membrane. These pseudopodia can move and contract very much as cilia and flagella do, and we will see shortly that organized microtubular structures are also found in these organs.

One more example will suffice for now. Several ciliated protozoa have long, ribbon-like *axostyles* running the length of their bodies, and they change shape as the axostyle bends. A. V. Grimstone has shown that axostyles are made of a spectacular bundle of microtubules (Fig. 11-25).

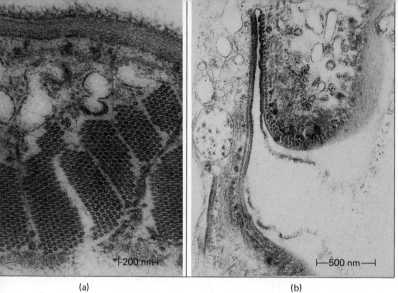

(a)   (b)

FIGURE 11-23

Microtubules in the flagellate Diplodinium.
*(a) The extensive bundles under the cell
surface that apparently hold the cell in
shape. (b) Microtubules around the mouth
of the contractile vacuole, which is
constantly moving to pump water out of the
cell. [Courtesy of L. E. Roth, reprinted from*
Primitive Motile Systems in Cell Biology
*(R. D. Allen and N. Kamiya, eds.) by
permission of Academic Press, New York.]*

FIGURE 11-24

A cross section of a pseudopod (actinopod) of the heliozoan Actinosphaerium
*(drawing). The bundles of microtubules, which terminate on the nuclear membrane,
are in two concentric whorls. (Courtesy of Keith Porter and Lewis Tilney.)*

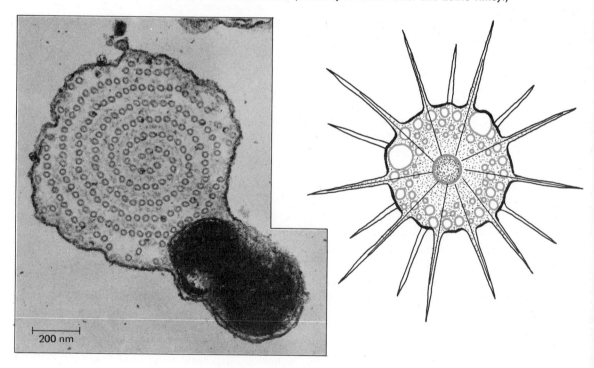

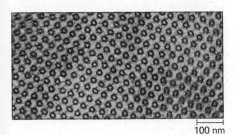

100 nm

FIGURE 11-25

Cross section and longitudinal section through the axostyle of Saccinobaculus. *Notice the regular hexagonal packing, the fine extensions between microtubules, and the periodic spacings, about 15 nm, in the longitudinal section. (Courtesy of A. V. Grimstone.)*

11-10 NONAD OBJECTS: KINETIDS

Cilia and flagella are among the most widely distributed and important locomotory structures in eucells. Cilia is the more general term; flagella are long cilia but there is no sharp distinction between them. They may also be called **kinetids,** and their **basal bodies** (Fig. 11-26) are **kinetosomes.** All of the kinetids in a cell constitute the kinetome and if they are arranged in orderly rows, each row is a kinety.

The most striking feature of a kinetid is its nonad symmetry. Structures like this are often called "9 + 2" objects because of their nine pairs of microtubules in a ring with two more in the middle; however, the "9" are usually 18 or 27 and the "2" are often missing. In any case, these microtubules have the typical form seen elsewhere, although they may have only 11 units in a ring instead of 13. The stalk of a kinetid contains nine pairs of microtubules with two more in the middle; in the kinetosome the outer doublets generally become triplets and the inner two disappear. There are other fine elements in a kinetid that do not always show up clearly. The two central microtubules are joined by very thin lines and there is a clear "cartwheel" structure near and within the kinetosome formed by radiating fibrils. These fibrils probably constitute a system for coordinating the activities of the microtubules. Several different patterns have been found in different organisms.

The immediate source of free energy for ciliary movement is ATP. The ATPase in kinetids is clearly associated with the outer microtubules, and I. R. Gibbons has shown that when ATPase activity is removed by dialyzing cilia against a buffer, the tiny "arms" on the outer microtubules disappear. When ATPase is restored to the cilia, the "arms" reappear. The membrane of the kinetid appears relatively unimportant for motion; Peter Satir and Frank Child have shown that cilia that have been extracted with glycerol and consist of little more than the microtubules will still contract in response to ATP.

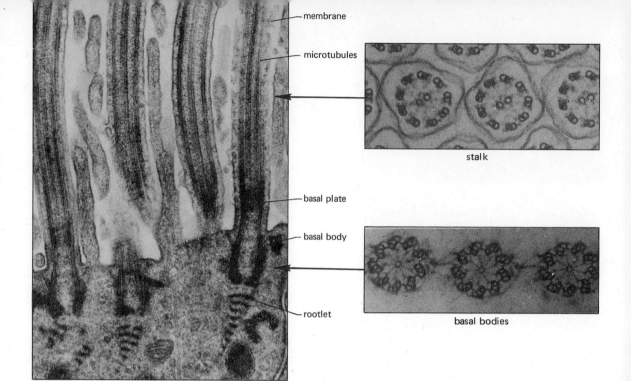

membrane

microtubules

stalk

basal plate

basal body

rootlet

basal bodies

FIGURE 11-26

Ciliary structure. The longitudinal section (courtesy of Roger Lambson) shows the microtubules running into basal bodies, with banded rootlets below. The cross sections (courtesy of Peter Satir) show the typical nonad patterns—nine doublets with tiny arms plus two central microtubules in the stalks, and nine triplets with a cartwheel pattern of fibrils in the basal bodies.

FIGURE 11-27

Some typical beating patterns of kinetids. (a) Movement of a cilium, showing a stiff power stroke from right to left and a relaxed return stroke, so the water moves from right to left. (b)—(e) Several patterns of flagellar movement. Arrows indicate the direction the organism moves.

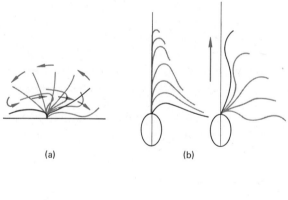

(a)

(b)

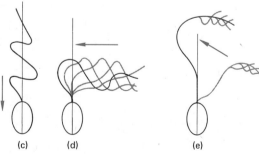

(c)   (d)   (e)

Kinetids can execute some remarkable beating patterns (Fig. 11-27). There is little point in trying to catalog the variations here, but it should be pointed out that contraction can start from either the kinetosome or the tip, so one cannot postulate that the beat is initiated exclusively at the kinetosome and travels up the shaft, nor that the energy input is always at the kinetosome. It is generally assumed that the pattern of the beat depends upon the ways the microtubules are connected to one another and that a stimulus passes from one to another in a regular pattern. For example, if the microtubules on opposite sides were to be alternately stimulated to contract (or otherwise change their shape), the kinetid would beat in a plane, while if the stimulus passes around in a circle from one to the next the kinetid would beat in a circle.

The other nonad structure that is widely distributed in eucells is the **centriole.** Figure 11-28 shows that this has the typical nonad symmetry of a kinetosome, with triplet microtubules outside. The centriole is an essential feature of eucell division wherever it is found and we will discuss its precise role in this process later. The resemblance between centrioles and kinetosomes is not trivial. These objects have a complicated evolutionary history that we are just beginning to decipher, but it is already clear that one of them has evolved from the other and there are many protista in which a single object spends part of its time acting as the kinetosome of a kinetid and part as a centriole.

├───500 nm───┤

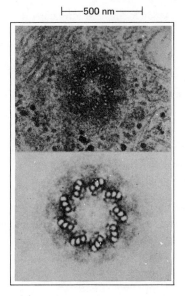

FIGURE 11-28
Centriole of a Chinese hamster cell in cross section, showing the nonad structure. Below, slightly enlarged, is the result of a nine-fold Markham rotation analysis of the original picture. (Courtesy of Elton Stubblefield.)

11-11 TEKTINS

As data accumulated on a variety of structural proteins from many sources it became obvious that there is a class with very similar amino acid compositions, molecular weights, and other structural features; Daniel Mazia and Albert Ruby call them tektins. The list of proteins that fall in this family includes muscle actin, the protein of several different microtubules, the flagellin of bacterial flagella, actin-like proteins from the cytosols of various cells, and a growing list of structural proteins from membranes. They have molecular weights of about 50,000–60,000 daltons and are globular molecules which can polymerize into long fibers. Each globular unit may itself be a dimer of two polypeptides held together by sulfhydryl bonds.

The tektins are excellent examples of sub- and self-assembly processes. We emphasized the two concepts in Chapter 6 and we can now expand on them with some real examples. Fumio Oosawa and his colleagues have studied the assembly of actin and flagellin filaments, with really remarkable results. In both cases, the native filaments can be dissociated to subunits by heating. The subunits can be reconstituted into filaments, but a solution of subunits polymerizes very slowly. If broken fragments of the native filaments are added, polymerization occurs explosively; the added fragments act as **nucleating agents** for the monomers, and one of the first lessons that emerges from this work is the importance of nucleating agents in *initiating* the assembly process.

The second lesson is that the nucleating pieces determine the structure of the whole filament. Oosawa and his associates used flagella from strains of *Salmonella* that have normal flagella and from a mutant strain that has curly flagella. When the flagella are spread out for electron microscopy, the normal ones form a regular sine curve and the curly ones form a sine curve with twice the frequency. However, if normal fragments are added as nucleating agents to curly subunits, the subunits assemble into normal flagella; and similarly, normal flagella subunits can be induced to form curly flagella by simply adding fragments of curly flagella as nucleating agents.

The rate of assembly can be influenced by many factors, including concentrations of salts and nucleotides. G-actin subunits will polymerize very rapidly when ATP or ITP are added, more slowly when ADP is added, and very slowly when no nucleotides are bound. G-actin containing magnesium can also polymerize faster than G-actin containing calcium.

The fact that subunits can associate in at least two different forms, depending upon the structure of the nucleating agent, indicates that the flagellum has at least two forms with very similar free energies and that it could switch from one to the other very easily. This is important in thinking about the action of a flagellum or any tektin structure that has to move.

As we pointed out in Chapter 6, the only way that identical subunits can associate into a linear structure is by making a helix. The bonds between the subunits of the helix are weak (hydrogen bonds, van der Waals attractions, etc.) and the structure as a whole holds together through the *cooperative action* of many of these weak bonds at once. We call the set of bonds between two subunits a **domain of bonding.** A set of identical subunits that have at least one domain of bonding between them can assemble themselves into a stable structure with relatively low energy. Now suppose that there are two alternative domains of bonding between the subunits; for example, each subunit may be capable of undergoing a slight conformational change and they might be able to bond to one another in a slightly different way when they are in their two states. In general, there will be a small energy barrier to the transition from one state to the other. The structure can pass reversibly from one state to the other whenever anything nudges it over the barrier. Both structures must be helices, but the two helices can have slightly different pitches, widths, and lengths. A dislocation induced at one end of the helix—perhaps requiring only a single molecule of ATP—will then pass rapidly along the helix as each molecule in turn nudges its neighbor into a slightly different bonding pattern (Fig. 11-29). The change does not have to be very drastic; a change in pitch of only a few degrees may be enough to cause a considerable bending of a cilium.

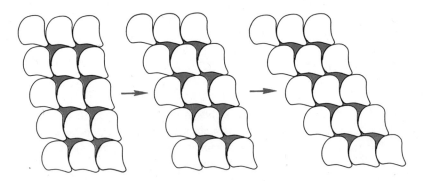

FIGURE 11-29
A change in the shape or orientation of a helix made of protein subunits can be produced by a slight change in the domains of bonding between the subunits.

In most cases we would expect the change to be readily reversible so the structure can move repeatedly. One of the cases we know best is not reversible; the contraction of the phage T4 sheath conforms very well to the pattern we have just described, but this just has to happen once for each virion and the two different forms are preserved stably for examination. The sheath consists of 144 subunits of 55,000 daltons each and from their amino acid composition they appear to be tektins. Figure 11-30 shows the change in configuration that occurs when the sheath contracts. Except for its irreversibility, this is a perfect model for the behavior of tektin helices.

11-12 AMEBOID MOVEMENT

Some of the most spectacular changes to be seen in any organism appear in the large amebas that can be found in pond water. These protista range from about 10–100 μm in diameter; they flow around on the substratum by pushing out great *pseudopods* and flowing into them so that in 5 minutes or less a large ameba can move its own length. Figure 11-31 shows the pattern of flow in a monopodial ameba, which extends in only one direction. However, some amebas can extend in several directions at once, and they may push out pseudopods to engulf food particles at the same time that others are moving the cell. There appears to be a *cortex* or *cortical layer* just inside the membrane which forms a tube through which the rest of the cell fluid flows. The cortex is sometimes called "ectoplasm," while the more mobile fluid inside is the "endoplasm." However, a very careful analysis of cell structure, particularly with

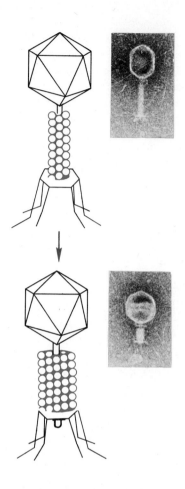

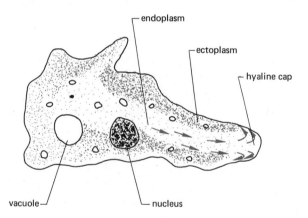

FIGURE 11-30
The tail sheath of phage T4 contracts and gets thicker in response to a stimulus that changes the orientation of its subunits. The change is obvious in electron micrographs (Courtesy of Alex Lielausis.)

FIGURE 11-31
The general pattern of flow in an ameba.

endoplasm

ectoplasm

hyaline cap

vacuole

nucleus

phase contrast microscopy, shows that one cannot make a very firm distinction between them. Both regions are highly structured, and both have a somewhat sponge-like appearance, with many fibrous strands running through them.

The dominant theory of ameboid movement, first stated by S. O. Mast and developed more recently by R. J. Goldacre and Douglas Marsland, is the *contraction-hydraulic* theory. In outline, it assumes that the tube of gel-like cortex contracts at the rear of the cell, forcing the cytosol through the tube to the front, where it condenses into more gel. The cell membrane is supposed to be synthesized at the rear and pushed forward, while old membrane is broken down and resorbed at the front. Movement thus depends very largely on pressures generated within a closed membranous cell.

Robert D. Allen made the accidental observation that an ameba inside a capillary tube will exhibit perfectly normal ameboid movements even when its membrane is broken. On the basis of this and other observations, he suggested an alternative explanation, since it is hard to see how any hydraulic pressure can be built up without a completely closed membrane. Allen suggested that the force of movement be placed at the front of the pseudopod, so that contraction of cell fluid into a gel state there will *pull* additional fluid from the rear; this is the *front-contraction* theory.

It is hard to collect critical evidence that would exclude either of these theories; most observations on amebas are compatible with either one. However, some data are now available on the rate of membrane turnover in amebas. Lewis Wolpert, C. M. Thompson, and C. H. O'Neill have measured the rate of membrane synthesis and C. Chapman-Andresen has measured the time it takes amebas to rebuild their cell surface after exhausting it through endocytosis. Both sets of data give an upper limit of about 0.2% per minute. Considering that an ameba that moved its own length in 5 minutes would have to synthesize new membrane at a rate of 20% per minute, these results seem to rule out the contraction-hydraulic theory, which was already dealt a severe blow by Allen's observations. However, Allen's model has also raised many objections. In order to improve the model, we will consider some phenomena closely related to ameboid movement.

11-13 CYTOPLASMIC STREAMING

Many cells that do not move like amebas still show very rapid internal motions of their cytosol and of particles in the cytosol. This is generally called cyclosis or cytoplasmic streaming. Enormous rates have been measured; for example, the slime mold *Physarum* is a large ameboid protistan whose cytosol has been found to flow at a rate of 1350 μm/second. The cytosol in the hyphae of many molds flows at 250 μm/second. In most cells, streaming rates of 3–6 μm/second are common and 30–60 μm/second is an upper limit.

There are many similar phenomena. Chloroplasts, nuclei, and similar membranous components are often seen *spinning* rapidly, sometimes as fast as 5 μm/second at their surfaces. There are *saltatory* movements in cells; Lionel Rebhun has described this strange phenomenon in which small particles suddenly move 20–30 μm in a few seconds and then stop just as suddenly. The particles do not change shape and their movement is not affected by nearby particles.

Some of the most rapid movements in cells may be due to hydraulic pressure caused by major changes in the cell or ER membranes. Some people have tried to implicate microtubules in other movements because they are often oriented in the direction of streaming; however, there is no good evidence that microtubules can move in any way that could cause streaming and it seems more likely that they are simply holding the cell in shape and perhaps keeping the streaming oriented.

Noburo Kamiya and his students have studied the rapid streaming in *Physarum* with a clever device that holds the cell in a tiny hole between two chambers; by exerting pressure on either side, they can force the cell to move one way or the other and measure the force involved. The cytosol shuttles back and forth every 2–3 minutes and it may exert a force of up to 5000 dynes/cm^2. The flow is very much like that of water moving through a pipe; that is, the velocity in the center of the stream is greater than that at the outside. This suggests that it is simply being pushed by hydraulic pressure. When Kamiya adds ATP to the medium in one chamber, the half of the cell in that side develops enormous pressures, up to about 35,000 dynes/cm^2.

Kamiya and K. Kuroda have also studied cyclosis in the alga *Nitella*. Here the situation is different; rather than flowing like water, the inner cytosol moves as a mass relative to the outer cortex, much as in ameboid movement. The mechanism here seems to be a shearing force at the interface between these regions. Similarly, the cytosol near a rotating chloroplast is moving in the opposite direction, as if in reaction to a shear force being generated at the interface between them. It is hard to escape the conclusion that these processes depend upon shearing, contractile mechanisms like those in striated muscle.

The necessary elements exist in the cytosol. K. E. Wohlfarth-Botterman has shown that the cytosol of amebas and other active cells is filled with visible 80-Å fibers that are most dense in regions where there is the most streaming activity. Fibrous proteins can be extracted from the active cells. Paul T'so and Jerome Vinograd extracted a protein from *Physarum* that they called myxomyosin; it is similar to the myosin of muscle and has ATPase activity. Oosawa has extracted a soluble actin that is very similar to muscle actin. The complex of these proteins should be able to contract and move regions of the cell just like muscles. The muscles of animals may be seen as highly organized, efficient developments of the primitive contractile elements of protista.

Wolpert, Thompson, and O'Neill have obtained preparations of cytosol without the surrounding membrane; this material contains many 80–100-Å fibrils, it is intrinsically motile, and it responds to ATP. It exhibits saltatory movements, streaming, and will contract into a more gel-like state, during which part of it separates as a rather clear fluid (syneresis). Syneresis is apparently the process that forms the hyaline cap at the tip of a pseudopod. Lowy has calculated that muscle filaments exert a force of about 3×10^{-5} dynes on one another, and Wolpert has shown that the same force between free filaments in a cell is more than enough to account for the observed tensions in streaming and related movements. A single pair of filaments is enough to move a 1-μm particle through the cytosol at a speed of 5 μm/second, assuming that the viscosity is no more than about 10 poises. A mere seven filaments per square micron can account for streaming and the density of filaments is even greater than this.

In summary, a relatively complete picture of cell movement can be obtained by considering three mechanisms. Some movement probably depends upon large-scale membrane changes. Many organized movements are associated with helical objects made of tektins that can probably change their configurations in small, reversible ways. And a large number of movements are caused by myosin and actin filaments that pull against one another.

D. A look ahead: tissue culture

We have concentrated on bacterial cells through most of this book because so much is known about them; we have acquired our knowledge of these cells primarily because they can be grown in large quantities under many conditions, so it is easy to experiment with them. We obviously hope to have just as much information about the cells of plants and animals in the near future, and one of the greatest hopes for this work lies in the ability to grow these cells by themselves in relatively pure culture and handle them just like bacteria. The methods of tissue culture are now being developed to a fine art; it is appropriate to mention this subject here, since so much of the information we will present about eucaryotic cell growth and division and about the viruses of eucaryotic cells comes from tissue-culture work.

Most of the cells in a healthy, adult animal are not proliferating, although they have the potential to divide and heal wounds or damage caused by infections. A bit of tissue extracted from an animal will generally start to proliferate if it is planted in a sterile nutrient medium that is basically like a bacterial medium except that it must be supplemented with a variety of amino acids and vitamins, and a complex base such as blood serum must sometimes be added. The cells in the tissue may be dispersed by treating them with trypsin, which breaks the cementing materials holding them together, and they will proliferate with a division time of about 12–24 hours. Every few days, some cells from an older culture bottle must be transferred to a new one. The cells in the first few bottles constitute a **primary culture;** if they survive and continue to proliferate they become a **cell strain.** A strain may continue to proliferate for many months, but after about a year of continuous transfer it starts to die out.

However, during this long series of serial transfers, many cell strains produce mutants that can become **cell lines.** These are not simple point mutants; they are cells in which the normally diploid set of chromosomes has become grossly abnormal. If a haploid set contains n chromosomes, then animal cells are normally diploid, $2n$. In some genetic process, cells that are triploid, $3n$, or tetraploid, $4n$, can be produced; all of these cells are **euploid,** since they have an integral number of haploid sets. The cells of lines are generally **aneuploid;** that is, there are abnormal numbers of individual chromosomes. They might be less than $2n -$ **hypodiploid** $-$ or more than $2n -$ **hyperdiploid.** They might have three copies of one chromosome **(trisomy)** and four copies of another chromosome **(tetrasomy).** Individual cells in the culture may have very different chromosome sets from one another. Chromosomes may also appear with gross chromosomal abberrations, such as deletions and inversions (in which a piece of chromosome is turned around the wrong way). These gross changes in the genome have resulted in changes in normal growth properties, and cell lines resemble tumor cells in many ways. Many of them have been carried in laboratories for years and they show no sign of dying out. Their names are as individualistic as their chromosome sets. You will hear about a common line called **HeLa** that came from the cervical cancer of a woman named Helga La_____. Strain BHK came from hamster kidney cells, Detroit-6 from human bone marrow, and KB from a human oral cancer; there are many others. In addition to their use in experiments on growth and division, many of them serve as hosts for animal viruses.

One further line of experimentation has been started by Boris Ephrussi and his colleagues. They have found that some cells in tissue culture can be induced to hybridize with one another; in some cases, mating is helped by infecting them with certain strains of virus. This valuable line of research holds the possibility of doing regular genetic work with ordinary somatic (nonreproductive) cells; and if this can be done, there is no limit to the amount that can be learned about the tissue cells of plants and animals by using all the techniques applied to bacteria.

the genome and the cell cycle

With the background of the last two chapters, we can now reexamine the subject of replication and its control in procaryotes. We will then go on to the eucaryotic genome and consider the structure of the nucleus and the events of the regular cell cycle.

12-1 THE REPLICON MODEL

Jacob and Brenner outlined a fundamental model for DNA regulation that must underlie all other considerations; it is based on two kinds of evidence. First, consider the behavior of episomes contrasted with merogenotes. Both of them are pieces of DNA, but a merogenote, such as a fragment of Hfr genophore, never replicates in the recipient cell, even though it may recombine with the recipient genophore. Yet an episome can replicate in its host cell autonomously, sometimes so rapidly and independently that it kills the host, but it can also be integrated into the genophore where its replication is controlled by the host. This means that not every piece of DNA can replicate; but if a piece can replicate, its replication can be controlled quite precisely.

The second line of evidence comes from studies on the inhibition of bacterial syntheses, particularly with a convenient *E. coli* strain designated 15TAU that requires thymine, arginine, and uracil (which are essential components of DNA, protein, and RNA respectively, you notice). Some important facts about this strain come from work by Phillip C. Hanawalt and Ole Maaløe. First, if a culture of 15TAU is deprived of arginine and uracil, DNA synthesis continues for another hour or so until the DNA per cell has increased by about 50%. This means that DNA synthesis can go on to some extent in the absence of protein and RNA synthesis.

Second, Seymour Cohen and Hazel Barner discovered that when 15TAU is grown with arginine and uracil but without thymine, the cells go into **thymineless death (t.l.d.)** after about 30 minutes and die rapidly and exponentially. The mechanism of t.l.d. is not known, nor is it important right now; the important point is that only cells that are making DNA are susceptible to t.l.d., so susceptibility can be used as an indication of DNA synthesis. Maaløe and Hanawalt found that when a glucose-grown culture of 15TAU is deprived of all three nutrients at once, 3% of the cells are immune to t.l.d. (Fig. 12-1), so DNA synthesis must take up 97% of a generation time and the only cells that are immune to t.l.d. are those that were caught in the short interval when they are not making DNA. Maaløe and Hanawalt then removed arginine and uracil from a culture—

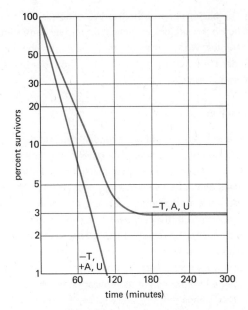

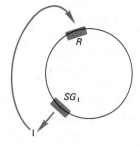

FIGURE 12-1

Kinetics of thymineless death. Cells grown in the absence of thymine only die exponentially. However, if thymine, arginine and uracil are removed simultaneously, about 3% of the cells are immune to thymineless death.

FIGURE 12-2

The replicon model. SG_1 makes an initiator molecule I, which acts on the replicator R.

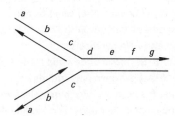

FIGURE 12-3

In a population of genophores being replicated regularly from one end to the other, there will be twice as many copies of a as of g, on the average, and the number of copies of b, c, d, e, and f will decrease steadily in that order.

which allows DNA synthesis to continue—and at various times tested samples for immunity to t.l.d. They found that a larger and larger proportion of the population becomes immune to t.l.d. as time goes on, until all of the cells are immune when DNA synthesis has stopped. The interpretation of these experiments is that DNA synthesis that is in progress can proceed in the absence of protein and RNA synthesis, but that it takes new protein and/or RNA synthesis to initiate another round of replication.

Jacob and Brenner interpreted these results in terms of a regulatory mechanism. First, they said, it is obvious that DNA synthesis is regulated positively, not negatively as is mRNA synthesis. Maaløe and Hanawalt's results show that a new material must be made to initiate DNA synthesis; they called this hypothetical material an **initiator.** Second, all pieces of DNA cannot be replicated and those that are replicable must be able to control their own synthesis independently of other such units. The minimal unit of DNA replication is a **replicon,** which must contain two genetic elements (Fig. 12-2): a structural gene SG_I that specifies the initiator, and a **replicator** where the initiator acts. Each replicon responds only to its own initiator; thus, if phage λ and the F factor, which are both replicons, inhabit the same cell, the replication of one of them does not trigger replication of the other.

In outline this model must be correct, although many of its details are still being disputed. Jacob, Brenner, and Cuzin obtained evidence for an initiator by experimenting with merodiploids that have a lac^- genophore and carry an F' lac^+ factor. These cells will be lactose fermenters only if their lac^+ episomes continue to replicate, so when Jacob et al. selected cells that are lac^- at 42°C, they found several thermolabile mutants of the F initiator. These cells grow at 42°C, but their F factors cannot replicate, so the population becomes lac^- exponentially, in proportion to the growth rate. Hfr cells in which the F factor has been integrated into the genophore continue to be lac^+ at 42°, presumably because the F factor is now under the control of the genophore's replicon. Finally, some of these mutants do not transfer the lac^+ marker to F^- cells efficiently at 42°, which implicates the F replication system in mating.

There also appears to be a single point in each strain where replication begins, although mutations in a replicator element at this point have not been found. This initiation point has been demonstrated by several people in different ways, but the basic idea is always the same. If DNA synthesis really progresses linearly along the map from an origin to a terminus (Fig. 12-3), there should be twice as many copies of genes at the origin as at the terminus, and the frequency of markers between the two should decrease smoothly from 2 to 1. Hiroshi Yoshikawa and Noburo Sueoka extracted DNA from a *Bacillus subtilis* prototroph and used it to transform a multiple auxotroph. When DNA is extracted from cells in stationary phase, the frequencies of transformation by all markers are about the same. However, when DNA is extracted from a culture in exponential phase, cells are transformed by the *met* marker with the lowest frequency and by the *ade* marker with the highest. The ratio *ade/met* is close to 2, and the ratios of other markers relative to *met* fall between 1 and 2. From these data, *ade* can be placed close to the origin, *met* close to the terminus, and a map can be drawn for other markers between them. Experiments using the same general principle have been used to establish maps for several bacteria; all of them confirm the concept of a unique sequence of replication for each strain.

12-2 THE MAALØE–KJELDGAARD PRINCIPLE

Maaløe and Kjeldgaard have argued convincingly for a general principle for the regulation of macromolecular biosynthesis: The overall synthesis of polymers is regulated by mechanisms that control the frequencies with which individual polymer chains are initiated. Let us see how this applies to specific cases.

*Recently proteins
known as Ψ factors
have been found which
bind to the RNA
polymerase and
stimulate it to
transcribe rRNA and
tRNA cistrons.*

RNA synthesis apparently is regulated by operators and promotors which determine where and with what frequency the synthesis of messengers may be initiated. Once the synthesis of a messenger begins, there is no evidence that additional controls are superimposed, and the DNA-dependent RNA polymerase apparently moves along at a characteristic rate at any given temperature. Presumably there are similar controls on transfer and ribosomal RNA synthesis, although regulatory elements for their genes have not been identified yet. In some cases, the rate of initiation may be determined entirely by the structure of a promotor region for each gene; for example, the genes for various coenzymes needed in very small amounts may simply be regulated by promotors for which the RNA polymerase has a rather low affinity. To demonstrate that this is possible, Jonathan Beckwith and Arthur Pardee developed strains of bacteria that make some enzymes constitutively at different rates over a thousand-fold range.

The synthesis of polypeptide chains may very well be regulated by mechanisms that control the formylmethionyl tRNA. Additional initiation factors that are part of this system are being discovered, and by manipulating these elements the rate of peptide chain initiation could be regulated.

The replicon model also places emphasis on initiation of DNA synthesis. This is a very clear case of the initiation principle, for it is obvious that DNA synthesis continues at a characteristic pace once initiated, but that new synthesis of control elements is necessary to start another round of replication.

With this principle in mind, we shall now try to develop the concept of a regular cycle of events in the life of each cell, each event being the synthesis of some new material. If regulation is achieved primarily by initiation of chain synthesis, then the cycle must be a sequence of triggering events, each of which sets off the next in turn. We begin by considering the timing of DNA replication.

12-3 THE DNA REPLICATION CYCLE

The bacterial cell cycle as we now see it has the following features. First, an initiator material is synthesized in every cell at a rate proportional to the general growth rate; when a critical amount of this initiator has accumulated, a round of DNA replication begins. Second, each round of replication requires a fixed time at a given temperature; in *E. coli* at 37°C, this is about 40 minutes. Third, the completion of a round of replication initiates a series of other events that result in cell division. These events also take a fixed time at a given temperature, and in *E. coli* at 37°C this is about 20 minutes. Now it is easy to see what the cell cycle is like in bacteria with a division time of 60 minutes, but bacteria can grow much faster than that. Cycles in bacteria with division times of 60, 40, and 20 minutes are shown in Fig. 12-4. At 40 minutes, initiator is made fast enough so the gap between one round of replication and the next is only about 3% of a cycle or about a minute, so one hardly ever sees a cell that is not making DNA. However, it takes 20 minutes after the end of a round of replication for cell division to occur, so the cell divides halfway through its replication cycle. At 20 minutes, initiator accumulates so fast that by the time one round of replication is half finished, a new one is initiated. There are three growing points moving along any genophore at a given time. As soon as the first round has finished, the next one is halfway through and cell division occurs at this time so the initial condition is restored and another round of replication can begin. Now we must examine the evidence for this scheme.

This model comes mostly from the work of Charles Helmstetter, who developed a very elegant way of obtaining a population of cells that have just divided. Helmstetter found that if he absorbed *E. coli* cells onto a membrane filter and eluted continuously

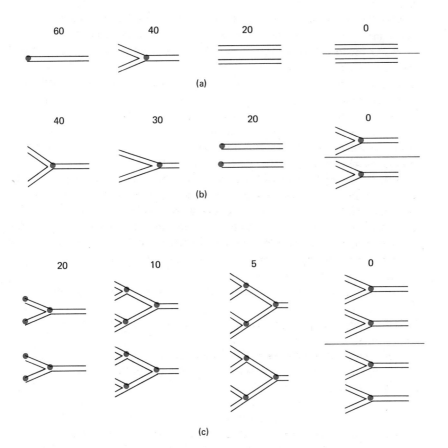

FIGURE 12-4

DNA replication cycles in bacteria growing with division times of (a) 60 minutes, (b) 40 minutes, and (c) 20 minutes. Heavy dots are replication points and times are given in minutes before division.

with fresh medium, only the cells that have just divided become detached from the filter, while their sister cells are still bound. Helmstetter's basic experiment is to give a pulse of tritiated thymidine to a culture, bind the cells to the filter, and collect eluted cells at various times. The radioactivity in each sample of cells is a measure of the rate of DNA synthesis in those cells at the time of labeling. In these experiments,

Helmstetter finds a regular rise and fall in radioactivity, due to initiation and cessation of rounds of replication. These experiments indicate that DNA is being replicated according to the pattern outlined above. In slowly growing cells, there is always one period of about 40 minutes during each cell cycle when DNA is made at a standard rate. When the cell cycle becomes shorter than 40 minutes, there are clearly periods when new rounds of replication are initiated and the rate of thymidine incorporation rises.

When DNA synthesis is inhibited, Helmstetter and Olga Pierucci find that cell division continues for 20 minutes. The only cells that divide are presumably those that had finished a round of DNA replication less than 20 minutes earlier and had thereby set the division apparatus in motion. Together with other data, this indicates that completion of one round of replication is a necessary and generally a sufficient condition for division.

The 40-minute replication time and the 20-minute time between replication and cell division are numbers that come primarily from Helmstetter's work, but they agree generally with other data from Maaløe's laboratory. The principal point of contention now is the mechanism of initiation. Any specific statement about this subject would probably be wrong by the time you read this, but the data generally say that replication is initiated only when a cell has grown to an integral multiple of some minimum size and that initiation occurs when there is exactly one minimal cell mass per initiation point. We understand this generally by saying that the initiator is a complex protein (such as a complex of replication enzymes) that is made at a rate proportional to the general growth rate, so the time it takes to complete the initiator is proportional to the rate of protein synthesis.

We have identified a cycle of events that must follow one another in this order:

1. A period of initiator synthesis. This period is variable and is a function of growth rate.
2. A period of DNA replication that is quite constant.
3. A period of events (nuclear separation, cell wall formation, etc.) that lead to cell division.
4. Cell division.

The fact that some of these events may overlap during rapid growth is secondary. We will see shortly that this cycle is almost exactly like the cycle of events in eucell growth and division.

12-4 THE CYCLE OF PROTEIN SYNTHESIS

We may now ask whether the other known proteins (enzymes) are also synthesized cyclically. The question was first asked and answered affirmatively by Millicent Masters, Peter L. Kuempel, William D. Donachie, and Arthur Pardee. The major features of this cycle may be summarized as follows.

In a synchronized cell culture, in which all cells divide at essentially the same time, the specific activity of each enzyme doubles at a characteristic time (Fig. 12-5a). The sequence of doubling times is the same as the sequence of the genes for these

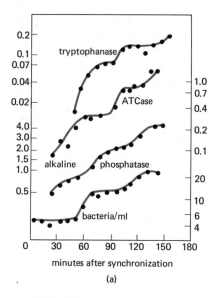

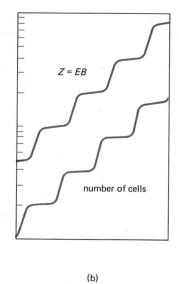

(a)

(b)

FIGURE 12-5

Bursts of enzyme activity during the bacterial cell cycle. (a) Experimental results for tryptophanase, aspartic transcarbamylase, and alkaline phosphatase in E. coli, *with numbers of cells per milliliter (lower right scale × 10⁷). (b) Computer graph of numbers of cells and activity Z of one enzyme, derived by solving the differential equations in the text.*

enzymes on the map. However, any enzyme can be induced at any time during the cell cycle. Furthermore, the doubling of enzyme activities will continue for some time even in the absence of DNA synthesis.

A naive interpretation of these results might suggest that the mRNA for a gene can be made only at the time the gene is replicated, but this is obviously wrong, especially in view of the continuing cycle without DNA synthesis. Pardee and his colleagues provided a more realistic interpretation. They recognized that periodic protein synthesis is an intrinsic property of the complex network of DNA, RNA, and protein that lies at the heart of cell behavior. Several versions of the model can be devised, but they are all basically very similar; we will consider a simple model devised by Brian C. Goodwin.

A single repressible enzyme system has the general form

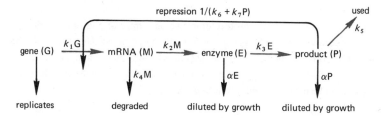

The values on the arrows represent the rate of each process; some relatively simple equations can be written for this network using these values. We assume that mRNA is made at a rate proportional to the number of gene copies and inversely proportional to the concentration of the repressing end product; messengers are also degraded rapidly, so a proper equation for mRNA synthesis is

$$\frac{dM}{dt} = \frac{k_1 G}{k_6 + k_7 P} - k_4 M$$

Equation 8-1 may be rewritten with $\sigma = k_2 M$, so

$$\frac{dE}{dt} = k_2 M - \alpha E$$

The end product P will be made at a rate proportional to the amount of enzyme, and it will be diluted by growth and used for additional biosynthesis:

$$\frac{dP}{dt} = k_3 E - \alpha P - k_5$$

You are not expected to solve these equations; in fact, Goodwin let a computer do the job with the provision that G doubles periodically, and he obtained the graph shown in Fig. 12-5b. Compare this with the experimental result of Fig. 12-5a.

Every enzyme in the cell that is regulated by this type of network will exhibit periodicities of biosynthesis; but since each network has different rate constants, the enzymes will double with different periods. The orders in which they double will therefore be different from one generation to the next and they will not double in a sequence collinear with the map sequence. Goodwin pointed out that the synthesis of a very small amount of mRNA from each gene at the time it is replicated will keep the major cycles of doubling collinear with the map order. This is the result of a common physical phenomenon called *entrainment*. If two mechanisms are oscillating with very similar periods and if the two can interact, then one of them will "capture" or entrain the other and force it to oscillate with an imposed phase and frequency rather than its own. Entrainment is often seen in the more complex cycles in plants and animals, and the capacity for entrainment appears to be an important feature of biological rhythms. The regular solar cycle of day and night seems to be one of the most important entraining mechanisms; many organisms have internal cycles that are approximately 24 hours long, and these are brought into phase with the local environment by daily light signals from the sun. You are familiar with the problems of being out of phase with the local environment if you have ever taken a quick airplane trip across several time zones; it may take a few days to get back to a normal pattern of living.

The normal cell cycle therefore emerges as a natural and necessary consequence of the negative feedback regulation of protein synthesis. All proteins in the cell must be made in the same sort of cycle, including the proteins necessary for DNA replication and cell division. All phases of the cell cycle can be understood in terms of the same general pattern of regulatory mechanisms.

With the outline of events in the bacterial cycle in mind, we can now understand the eucaryotic cycle. The similarities are really much greater than anyone had thought a few years ago, and it is easy to envision the eucaryotic cycle as a development from the procaryotic cycle.

Since we can interrupt a cycle arbitrarily at any convenient point, we will begin with a cell that has just finished dividing in two and is about to begin a new round of growth to the point where it can divide again. The chromosomes are just barely visible as long threads in the nucleus, but they are becoming a diffuse mass of chromatin again. The cell has the typical form that we described in the last chapter. This period during which very little appears to be happening is called G_1 (for Gap$_1$); it is a gap between more obvious events, but the cell is actively metabolizing and growing.

G_1 ends when the **S period** of DNA synthesis begins. All of the nuclear DNA, at least, is replicated during a relatively short period and the DNA content of the cells essentially doubles; the replication of mitochondrial and chloroplast DNA is not necessarily synchronous with that of the chromosomes. Other chromosomal materials, however, such as histones, are synthesized during S.

The end of S marks another gap, G_2, during which again little appears to be happening, although we know that preparations are being made for cell division. Division entails two processes. First, the replicated chromosomes must be precisely divided in half so each daughter cell receives the same information; the processes that accomplish this are called **mitosis.** Second, the daughter nuclei formed by mitosis must be separated by formation of new cell membrane (and wall, if this exists); this process is **cytokinesis.** There are some cells in which mitosis occurs repeatedly in a cytosome without cytokinesis, leading to a multinuclear condition.

The essential features of mitosis are shown schematically in Fig. 12-6 for a cell that has only two chromosomes, in parallel with a series of photographs of a cell that has 14. During **prophase,** the chromosomes shorten, thicken, and become visible. E. J. DuPraw and others have shown that the chromatin fibers are attached to the inside of the nuclear membrane, just as the bacterial genophore is attached to its membrane. The chromosomes are held in position by this stable attachment and they condense by coiling up against it. As they become thicker, we can see that each chromosome consists of two equal **chromatids** which are joined at a point called the **centromere** or **kinetochore.** As the chromosomes reach their most condensed state, the nuclear envelope disappears, marking the end of prophase and beginning of metaphase.

At the same time, a **spindle** (not visible in the photographs) begins to form around the chromosomes; it consists of many fibers that terminate in two **division centers** or **poles** on opposite sides of the cell. In animal cells and many protista, there is a centrosome region near the nucleus that contains a **centriole,** the little nonad structure we discussed in Section 11-10. The centriole is double at prophase, and it separates into two units that move toward opposite poles as they begin to organize a spindle. Higher plant cells have no centrioles and the spindle merely forms between the poles; there are many other variations on spindle formation, and for our purposes now they are not very significant.

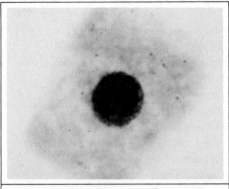

interphase
(a)

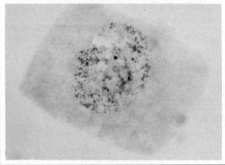

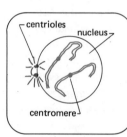

centrioles
nucleus
centromere

early prophase
(b)

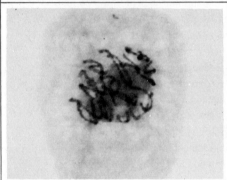

late prophase
(c)

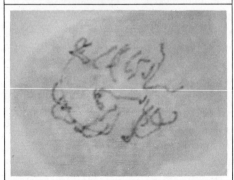

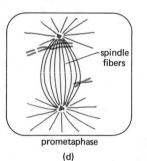

spindle
fibers

prometaphase
(d)

FIGURE 12-6

Mitosis in Lolium multiflorum (*annual ryegrass*). *These photographs of typical stages in mitosis, taken by Gilbert T. Webster, are accompanied by stylized drawings of a cell with only two chromosomes to show events that cannot be captured by this photographic process. (a) Inter-phase. The nucleus is compact and dark, with undifferentiated chromatin. (b) Early prophase. The nucleus en-larges and patches of chromatin begin to emerge. The centrioles, if they exist, begin to separate. (c) Late prophase. The nuclear membrane is disappearing; separate chromosomes are clearly visible as long, thin strands. The centrioles are approach-ing opposite poles of the cell. (d) Prometaphase. The chromosomes are shorter and more compact, and here and there they can be seen as double structures (chromatids). The centrioles are at opposite poles and there is a well-formed spindle be-tween them. (e) Metaphase. The chromosomes are at their maximum* density; they are clearly separated into chromatids which are joined at a constriction, the centromere or kinetochore. They are lined up on a plane, the metaphase plate, midway between the two poles. (f) Anaphase. The two chromatids of each chromosome separate suddenly and move to opposite poles. Some of the spindle fibers are shortening and dragging the chromosomes along, which makes their arms trail out behind and form V or J shapes. Other spindle fibers remain between the two poles. (g) Telophase. The chromatids, which are now chromosomes themselves, are at the poles. They are beginning to uncoil and assume their interphase condition. New nuclear membranes are forming around them and the remaining spindle fibers are starting to disappear. (h) Interphase. The nuclei of the daughter cells have become uniformly dense and compact and their membranes are intact. A new cell envelope has formed between them, creating two new cells.

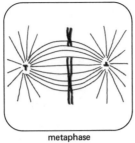

metaphase
(e)

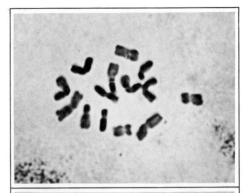

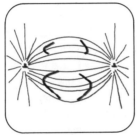

anaphase
(f)

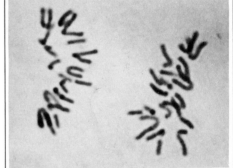

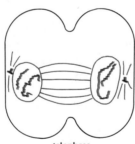

telophase
(g)

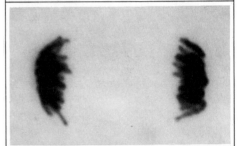

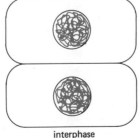

interphase
(h)

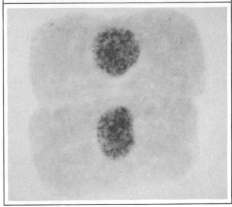

During **metaphase** the chromosomes may sweep rather dramatically back and forth across the cell, but they eventually settle down with their centromeres more or less in a plane, the **metaphase plate,** halfway between the poles and perpendicular to the spindle axis. **Anaphase** begins when the chromatids of each chromosome separate at the centromere and start moving toward opposite poles. Some spindle fibers are attached to the centromeres and the anaphase movement is clearly related to them. While the chromatids are moving toward the poles, the poles are moving away from each other, apparently by means of continuous fibers of the spindle that run from one pole to the other. Finally, the chromatids—which should now be considered independent chromosomes—form nuclei again during **telophase,** as simultaneously the spindle disappears and the nuclear envelope reforms. At the end of telophase the cells are back at G_1 again. The period $G_1 + S + G_2$ is sometimes called **interphase.**

Now notice just how much this cycle of events resembles the division cycle of bacteria with a division time of an hour or more. There is a G_1 period with no DNA synthesis, followed by a period of synthesis. After measuring the durations of the G and S periods of cultured mammalian cells, Jesse Sisken has concluded that most of the variation in the cell cycle is due to differences in the length of G_1; this is exactly the situation in bacteria, where more slowly growing cells take longer to initiate DNA replication. The S period is quite constant for a given cell type, and this is followed by a G_2 period that is also quite constant. Synthetic events can be identified in the G_2 period that are essential for mitosis. D. F. Peterson, R. A. Tobey, and E. C. Anderson have shown that in mammalian cells with a division time of about 8 hours, there is a period of essential RNA synthesis up to 1.9 hours before mitosis and a period of essential protein synthesis up to 1 hour before mitosis. Proteins are made during this time that are necessary for both the initiation and completion of mitosis. Sisken has also shown that during almost all of G_2, up to a few minutes before mitosis begins, an essential division protein is being made. When cells in this period are given *p*-fluorophenylalanine they incorporate it instead of phenylalanine and make a faulty protein so they cannot go past metaphase. Whatever the protein is, it is conserved for several generations, at least, because cells that enter metaphase two and three cycles later are still affected slightly.

A variety of other experiments indicate that, as in bacteria, the completion of DNA synthesis is a necessary and sufficient condition for mitosis, assuming that synthesis during G_2 is not inhibited artificially. In general, the cycle of events in the eucaryotic cell cycle resembles the events in bacteria, although there are some interesting differences. John Gorman, Patric Tauro, and Harlyn Halvorson challenged synchronized yeast cultures with enzyme inducers at various times and showed that each enzyme is made at a specific time in the cell cycle and cannot be induced at other times. The time of synthesis seems to be correlated with the position of the gene for the enzyme on its chromosome; it appears that each chromosome is read from one end to the other, so there is a burst of enzyme synthesis correlated with transcription. In general, then, the eucaryotic genome is not as accessible to inducers and repressors as the bacterial genome. This may be partially a result of the greater compartmentation in eucaryotic cells; we may find that regulation at the level of transcription is less important here than are translational controls and controls on the activity of existing enzymes.

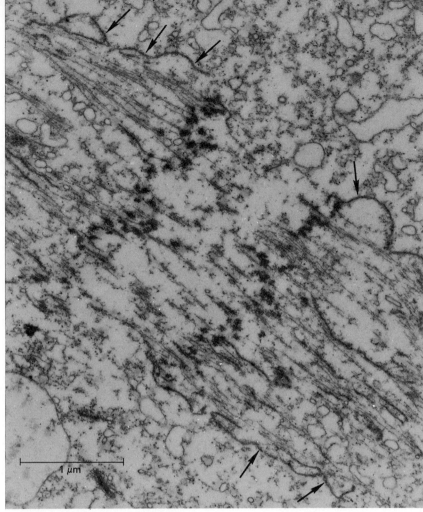

FIGURE 12-7

Mitotic spindle fibers of the ameba Pelomyxa carolinensis *after about 5 minutes of anaphase separation. The microtubules are obvious; they are attached to dark pieces of chromosomes and fragments of the nuclear envelope can be seen around them* (arrows). (*Courtesy of L. E. Roth.*)

12-6   THE MITOTIC APPARATUS

A number of electron micrographs have now shown quite clearly that the fibers of the mitotic spindle are really microtubules (Fig. 12-7). This revelation casts a new light on the whole matter of cell organization and the mechanism of mitosis. For a time, the mitotic spindle seemed to appear mysteriously during prophase out of nowhere; this is no longer so, and a number of studies now indicate that the changes we see in mitosis are a simple reorganization of existing microtubular structures. The interphase cell has a lot of microtubules as supporting and shaping elements, particularly underneath its cell membrane. In this position, they hold the cell firmly in shape.

During mitosis, it is simply necessary to move some of these microtubules into the center of the cell and make them into a spindle. This puts them into a position where they can actively separate chromatids and at the same time loosens the supporting framework of the cell so it can divide in half and perform all of the motions that mitotic cells go through.

If the spindle fibers are microtubules that have the structure we outlined in Chapter 11, we can also see a simple, obvious mechanism of mitosis. You must realize, first, that the anaphase movement of chromatids is really not very fast, no matter how it looks in time-lapse films taken of dividing cells or how it seems in relation to the relatively calm periods of the cell cycle. The chromatids typically move at about $1 \mu m$/minute; this is 1 cm/week! A snail's pace is much faster. Now we know that tektin structures such as microtubules can polymerize and depolymerize very easily; we assume that there is a cellular mechanism, such as changing intracellular ionic conditions, that will promote either addition of protomers to the microtubules or their removal into the cytosol. The movement of chromatids in anaphase could then be caused simply by depolymerizing spindle microtubules at the division center, while leaving them connected to the chromatids. This would not have to happen very fast, as you can easily prove to yourself.

exercise 12-1   Assume that a spindle microtubule consists of protomers that are 30 Å long, with 13 units per turn of the helix. How fast must protomers be removed from the microtubule to shorten it by 1 μm each minute?

In terms of the rates that other reactions occur in cells, it is not hard to conceive of 70 protomers/second being added to or removed from a microtubule. Notice, also, that depolymerization of the microtubules that run from each division center to a chromatid can be accompanied by polymerization of the strands that run between division centers, and this will explain the movement of the poles away from each other during mitosis. So far, there are no good experiments that prove that the movements of anaphase can be accounted for in this way, but this is quite a reasonable picture in view of everything else we know about tektin structures and self-assembly processes.

The drugs colchicine and colcemid have been used in many experiments to inhibit the mitotic apparatus; treated cells proceed through metaphase and remain there because the mitotic spindle has not formed normally. These drugs are therefore very useful for trapping chromosomes in a condition where they can be examined carefully. More recently, some investigators have been able to inhibit the formation of cilia and flagella with the same drugs; this indicates that both kinds of structures are built of the same type of proteins (tektins).

The centrioles, in cells where these objects exist at division centers, are very interesting little objects. Inside of their 27 outer microtubules there are several fine structures, including a cartwheel at one end, a helix, and some kind of central body; these are intriguing, but there is no information yet about their function. The microtubules themselves can serve as nucleating agents for the synthesis of mitotic microtubules, although they are obviously not always necessary considering the number of cells that divide without centrioles. Part of the centriole's structure must be designed as a nucleating agent for new centrioles, for new ones are assembled on the old ones in a regular cycle of events shown in Fig. 12-8.

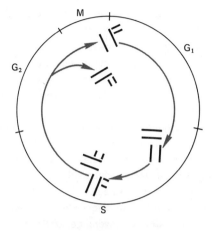

FIGURE 12-8

The centriole cycle (after Elton Stubblefield). New centrioles are assembled orthogonal (at right angles) to the old ones, beginning near the middle of S. The division centers of mitosis are orthogonal pairs and the new centrioles finally mature early in the following S period. (Photograph courtesy of Roger Lambson and Jesse Sisken.)

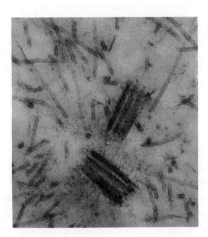

12-7 CHROMOSOME STRUCTURE

The whole subject of chromosome structure in eucaryotes has one of the longest and most confusing histories of any subject in biology; we will not try to delve into it very far here. After considering all the evidence, which has sometimes been interpreted in terms of extremely complicated chromosome structures, it is probably fair to say that chromosomes are actually about as simple as one could imagine them and that each of the chromatids seen at metaphase in the typical eucell is a single, long double helix of DNA with associated proteins, etc.

The chromosome of an older era in biology looked like a long string of beads—chromomeres—and in this era one could imagine that each bead was a single gene. As pictures improved, one could see that the chromomeres were really tightly coiled regions and a fairly continuous thread appeared to run the length of the chromosome with various degrees of coiling in different places. In some pictures, the coil appeared to have minor coils. With electron microscopy, Hans Ris identified a fiber about 200 Å wide that he considered the fundamental unit of the chromosome; he found some evidence that this was separable into two 100-Å fibers and that each of these consisted of two 40-Å fibers that might be DNA molecules with protein. Other investigators have found similarly complicated patterns. However, it has now become possible to sort out some of the tangled masses seen in the electron microscope, to digest their protein away carefully with proteolytic enzymes, and to spread the DNA out at interfaces. The best evidence from all of these sources says that a chromatid is made out of a strand of about 35–40 Å that is probably a single DNA helix. In the condensed, highly coiled chromosomes visible at metaphase, this long fiber is wound back and forth in a very highly coiled arrangement.

J. Herbert Taylor and his colleagues performed a set of autoradiographic experiments that are the chromosomal counterpart of the Meselson–Stahl experiments on DNA replication. They grew cells in ³H-thymidine for several generations to label their DNA uniformly, and then washed them out into unlabeled medium. As Fig. 12-9 shows, both chromatids are uniformly labeled at the first metaphase after an S period without label. But at the next metaphase, radioactivity is confined to one of the two chromatids in every case. This pattern of isotope transfer is most easily reconciled with the simplest model of chromosome structure, where the conserved units that run the length of the chromosome are single strands of DNA.

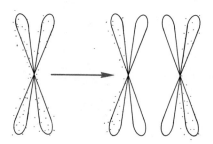

FIGURE 12-9
The appearance of chromosomes in Taylor's experiments. After labeling for several generations, both chromatids are uniformly radioactive at metaphase. After an additional generation of growth in cold medium, only one chromatid is typically labeled at the next metaphase.

exercise 12-2 Suppose that a chromatid really consists of two double helices of DNA bound together:

Let the chromatid drawn here consist of labeled strands only. Now let it go through a replication without label and separate as follows:

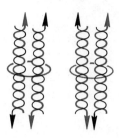

Are the two chromatids labeled equally? Now let this chromatid go through a second round of replication without label:

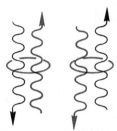

Draw the new strands in black.

Are the two chromatids still labeled equally? Is this compatible with Taylor's result? Let this chromatid go through a third round of replication without label.

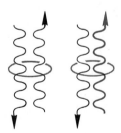

Draw the new strands in black.

Now how are the chromatids labeled?

A third line of evidence about chromosome structure comes from studies on the huge **lampbrush chromosomes,** up to 0.8 mm long, found in the cells (oocytes) that are destined to become eggs in certain salamanders. They have dense chromomere

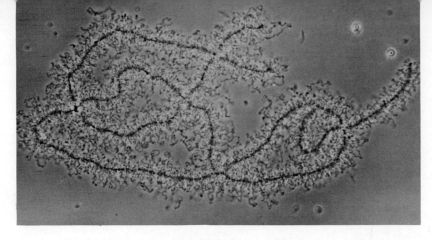

FIGURE 12-10

Phase-contrast micrograph of a Triturus *lampbrush chromosome freshly isolated in saline.* (*Courtesy of Joseph G. Gall.*)

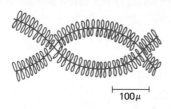

The lampbrush chromosomes look like this with low magnification

100μ

FIGURE 12-11

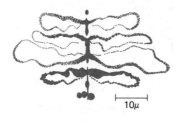

and like this with higher magnification. When the chromosome is pulled with fine needles,

10μ

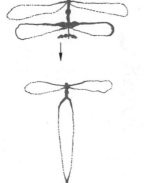

it breaks apart like this,

suggesting that each chromatid is really wound up like this at the base of each loop and that the chromosome consists of two very long, continuous DNA fibers with attached material.

regions from which loops protrude, with short regions between chromomeres (Fig. 12-10). H. G. Callan observed that when these chromosomes are stretched with needles, the chromomeres with their loops separate as shown in Fig. 12-11. The obvious interpretation is that each chromatid is one continuous thread that is highly coiled in the chromomere region. Joseph G. Gall has followed the breakage of lampbrush chromosomes during treatment with DNase. He found that the kinetics of breakage indicate two strands of DNA in the loops and four strands in the interchromomere regions. These results are consistent with a model in which each chromatid is one double helix.

It is obvious that the DNA in a chromosome must be highly coiled; if it were spread out completely it would be several centimeters long in some cases. It can be made to coil and contract by binding histone molecules across it, and this presumably happens during prophase. The typical metaphase chromosome must then consist of a DNA molecule with associated protein and RNA that is wound back and forth on itself in a very dense, irregular bundle, as shown in Fig. 12-12.

Most chromosomes (the exceptions are in some protista, a few odd insects, etc.) have a centromere somewhere; it could be located anywhere, but it is convenient to divide the continuum into *metacentric* chromosomes with a centromere near the middle, *telocentrics* with one at the end, and *acrocentrics* with a centromere somewhere in between. When chromatids of these three types are pulled toward the poles by their centromeres, they form Vs, Is, and Js, respectively. Because of their centromeres and other characteristic constrictions, the chromosomes in any set generally can be identified individually, and this is very useful in genetic studies. For example, human abnormalities can be detected by culturing the white cells from a small blood sample for several days and then stopping them in metaphase with colchicine. The chromosomes are then spread out, photographed (Fig. 12-13) and arranged in an **idiogram** or **caryotype,** so each one can be identified and abnormalities can be found.

The kinetochore region of a chromosome is particularly interesting, but it has been very hard to get any good pictures of it. What information there is suggests that this region is not greatly different from the rest of the chromosome in the sense that it must be able to break and replicate much like the rest. B. R. Brinkley and Elton Stubblefield have managed to get some pictures (Fig. 12-14) that show a dense core 200–300 Å in diameter containing a pair of fibrils 50–80 Å wide. Microtubules of the spindle can be seen attaching to these fibrils, but this part of the chromosome still remains basically a mystery.

FIGURE 12-12
DuPraw's model for the structure of a chromosome.

12-8 CHROMOSOME REPLICATION

The problems of DNA replication in eucaryotic chromosomes can be seen by doing some simple arithmetic. It takes 40 minutes to replicate a single bacterial genophore. If a eucell has 100 times as much DNA and this is distributed in 10 chromosomes, then it would take 400 minutes to replicate all the chromosomes from one end to the other simultaneously. That is almost 7 hours; in fact, S periods may last for about 2 hours. This is not entirely incompatible with a mechanism like that in bacteria, given a few assumptions to bring these values more into line with one another. But consider also that the bacterial genophore is over a millimeter long when spread out, and that a

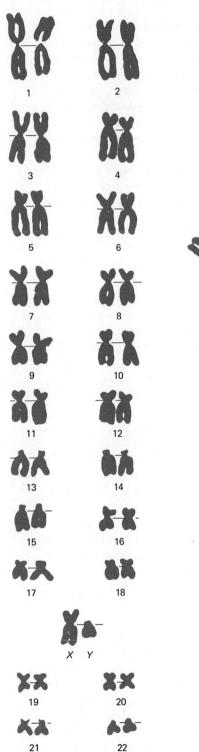

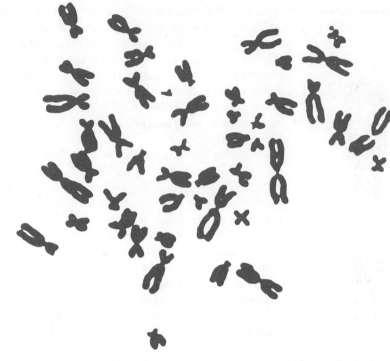

FIGURE 12-13

Metaphase chromosome set of a normal human male. The 23 pairs of chromosomes, including the sex chromosomes X and Y, have been assembled from this photograph into an idiogram so they can be checked for abnormalities. (Courtesy of C. Phillip Holland.)

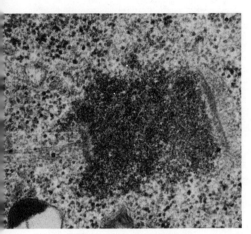

FIGURE 12-14

Kinetochore of a Chinese hamster chromosome. The dense core is bounded by two heavy fibers parallel to the chromosome axis. Some spindle microtubules can be seen, apparently attaching to them. (Courtesy of B. R. Brinkley.)

eucell chromosome would then be at least a centimeter long. If it is hard to imagine the bacterial DNA being replicated as it is and spinning on its axis, it is much harder to imagine the eucell chromosome doing the same thing.

Among Taylor's early experiments were some which indicated that each chromosome is being replicated simultaneously at many points. Joel Huberman and Arthur Riggs have followed up these experiments with some very fine autoradiography. Their pictures show (Fig. 12-15) that the chromosome does indeed contain a number of origins spaced on the order of 100 μm apart and that replication points travel in both directions from each initiation point, terminating when they meet neighboring replication points traveling in the opposite direction. The chromosome can thus be divided into a number of replicating units; we will not call them replicons because this term implies a good deal more and because all the units are probably under the control of the same initiation mechanism.

The control of replication in eucells is a much more complex matter than in procells. In the latter, the only thing the cells do is grow and proliferate, so replication must occur repeatedly in all cells. Most eucells are part of multicellular organisms, and there must be many controls on replication and proliferation; a cell that replicates when it is not supposed to is going to become a tumor. We can, however, identify a few relatively simple mechanisms.

Yasuo Hotta and Herbert Stern have shown that the kinases that make the triphosphates of deoxyadenosine, deoxyguanosine, and deoxycytosine are always present in plant cells. However, the kinases that convert deoxythymidine into its triphosphate do not appear until about the time that DNA synthesis begins. The appearance of this enzyme may be a simple enzyme induction, perhaps induced by the accumulation of deoxythymidine. Moreover, the kinase activity disappears several hours after it is induced, probably because a specific inhibitor is made. Other investigators have found

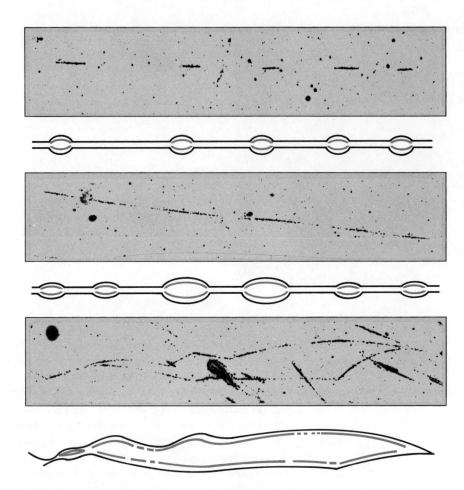

FIGURE 12-15

Autoradiographs of Joel Huberman's DNA labeling experiments. Each dark streak is a region of DNA synthesis, presumably where the two old strands (continuous black lines) separate and two new strands are being made (colored lines). From top to bottom, three stages are shown with increasing pulse times; after each pulse of ³H-thymidine, cold thymidine was added as a chaser and the decreased density of the tracks at their ends are presumably the result of diluting the intracellular thymidine pool. (Courtesy of J. Huberman.)

a similar pattern of enzymes in animal cells that are not growing; the thymidine kinases appear only after the tissues are injured and forced to regenerate.

G. E. Stone and David M. Prescott tried to inhibit DNA synthesis in the ciliate *Tetrahymena* by depriving it of required amino acids. If amino acids are withheld just before the S period, synthesis is initiated and continues until about 20% of the DNA has doubled; the next round of replication is completely inhibited. Noting that the pool of thymidine nucleotides in these cells is just sufficient to support about 20% of DNA replication, Stone and Prescott also suggested that new thymidine kinases must be made to support continuing DNA synthesis.

12-9 THE NUCLEOLUS AND RIBOSOMAL SYNTHESIS

Although the nucleolus was one of the first structures to be identified in cells and it can be seen easily with a light microscope, its function remained a mystery until only a few years ago; we now know that the nucleolus is the site of ribosomal RNA synthesis.

Nucleoli appear very highly structured in electron micrographs; they have a central region that is rather dense and a more diffuse, reticular outside, and often a large thread appears to run through them. However, none of this has really told anything significant about their function. Evidence that the nucleolus was the site of rRNA synthesis began to come only about 1960 when Robert P. Perry and J. E. Edström began to trace RNA in animal cells autoradiographically.

The details of RNA synthesis in the nucleolus and nucleus have come primarily from a long series of studies by James E. Darnell, Sheldon Penman, Jonathan R. Warner, and their associates. Most of this work has been done by tracing the flow of labeled RNA through fractions that can be separated by sucrose gradients; the basic ideas of this analysis were set out by Brian McCarthy and Roy Britten in their studies on biosynthesis of bacterial ribosomes. They showed, for example, that a small RNA (about 14S) is first made and that radioactivity in the peak of this material is transformed gradually by addition of protein units into a 30S unit and into a 43S unit, the latter then becoming a 50S unit. The flow from one to another is seen either by labeling steadily for various times or by labeling with a pulse and then chasing this pulse with unlabeled precursor for various times.

Eucaryotic ribosomes are a little larger than those of procaryotes. They consist of a 40S unit (1.4 million daltons) and a 60S unit (2.7 million daltons); the small unit contains an 18S RNA (0.7 million daltons) and a 28S RNA (1.9 million daltons). (There is also a small, 5S RNA in the ribosomes that weighs about 4×10^4 daltons). The major RNA species are derived from a huge 45S precursor (4.5 million daltons) which is degraded gradually into the 18S and 28S pieces, as shown in Fig. 12-16. RNA methylation occurs on the 45S unit, and all of the RNAs gradually acquire their protein subunits so that mature 40S and 60S particles can be found in the nucleus. These subunits are also transferred out of the nucleus, but they first appear joined together as parts of polyribosomes on mRNA molecules; the only free 80S ribosomes in the cell seem to be those that have just finished protein synthesis as part of a polyribosome.

Barbara McClintock first observed in 1934 that nucleoli are always associated with a constriction on a particular chromosome, the **nucleolar organizer** region. The nucleolar organizer must contain at least the genes for the rRNA; it is interesting to calculate how many there are.

exercise 12-3 F. M. Ritossa and Sol Spiegelman hybridized rRNA from the fruit fly *Drosophila* with DNA from *Drosophila* strains containing different numbers of nucleolar organizers. They found that 0.135% of the DNA hybridized when there was one organizer, 0.270% when there were two

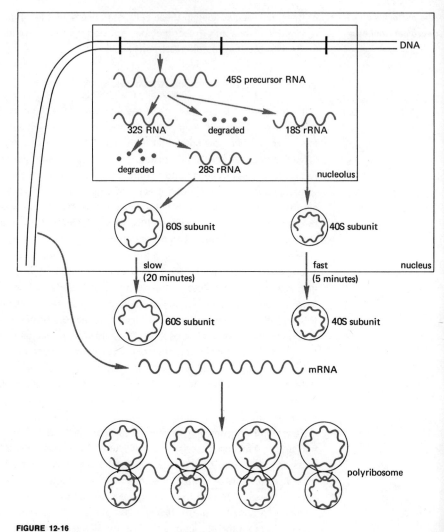

FIGURE 12-16

A summary of the course of ribosome synthesis in eucaryotes.

organizers, and proportional amounts for three and four organizers. If a normal *Drosophila* cell with two organizers contains 0.24 picograms of DNA, how much DNA is there in one organizer? How many nucleotide pairs is this?

exercise 12-4 How many nucleotides are there in the combined 18S and 28S rRNAs? How many gene copies must there be for each of these RNAs in a nucleolar organizer region?

The numbers derived from these exercises agree well with those from other sources; a nucleolar organizer contains about 100–200 copies of the rRNA genes. This presumably takes the form of many regions that code for the 45S precursor, so the regions for 18S and 28S RNA alternate along the DNA with some extra material in between. (Genes for the 5S RNA are not in this alternating sequence). Oscar L. Miller, Jr., and Barbara R. Beatty have spread out the nucleolar region and taken some beautiful pictures of RNA synthesis (Fig. 12-17). The nucleolar organizer DNA consists of many alternating segments, presumably genes for the 45S RNA. About 100 molecules of RNA polymerase are moving along each one and making a series of RNA molecules that are in different stages of completion from one end of a segment to the other. The fibrils

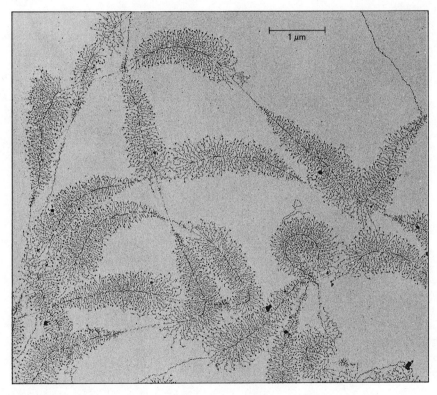

1 μm

FIGURE 12-17
Synthesis of rRNA on nucleolar genes of Triturus. *The nucleolus was dispersed in deionized water, centrifuged through a sucrose solution onto the microscope grid, and stained with 1% phosphotungstate. (Courtesy of Oscar L. Miller, Jr., and Barbara R. Beatty.)*

can be removed with either ribonuclease, which splits the RNA itself, or with proteolytic enzymes, which must destroy the polymerase. The lengths of the DNA segments are compatible with everything we have said before about rRNA synthesis; the RNA molecules don't look as long as they should because they are not spread as completely as the DNA.

13

some
organisms
and their
reproduction

The purpose of this chapter is to present a few of the organisms that we will use later to illustrate genetic principles in eucaryotes and to outline some of the evolutionary relationships between them. A catalog of different creatures is not one of the major purposes of this book, but we can illustrate some interesting and important features of cell structure and reproduction in this way.

A. Evolutionary guidelines

Even if it is too early to produce the kind of comprehensive classification and summary of evolution that systematic biologists have been trying for years to achieve, we can indicate some of the major lines of evolution. This depends upon a few simple guidelines that indicate very broad relationships between different groups.

13-1 PIGMENTS AND CHLOROPLASTS

Most of the organisms whose relationships are in doubt are photosynthetic, and the pigments they use to trap light (and for unknown purposes) are useful indicators of their relationships.

The chlorophylls are very useful. Their general structure is shown in Fig. 13-1. The bacteriochlorophylls are found only in the purple and green bacteria; the others are found in the various "algae." It would be useful to know the structure of chlorophyll c, since it is one of the most widely distributed.

Type	R_1	R_2	R_3	R_4	R_5	3, 4
Chl a	$-CH=CH_2$	CH_3	$\overset{O}{\overset{\|}{C}}-OCH_3$	phytyl	H	==
b	$-CH=CH_2$	$CH=O$	$\overset{O}{\overset{\|}{C}}-OCH_3$	phytyl	H	==
c						
d	$-CH=O$	CH_3	$\overset{O}{\overset{\|}{C}}-OCH_3$	phytyl	H	==
BChl a	$-\overset{O}{\overset{\|}{C}}-CH_3$	CH_3	$\overset{O}{\overset{\|}{C}}-OCH_3$	phytyl	H	dihydro
b						
c	$-HCOH-CH_3$	CH_3	H	farnesyl	CH_3	==
d	$-HCOH-CH_3$	CH_3	H	farnesyl	H	==

FIGURE 13-1

Structures of the chlorophylls (Chl) and bacteriochlorophylls (BChl).

381

The biliproteins are found in only a few groups. Their prosthetic groups are open tetrapyrroles of the hemoproteins. Phycoerythrin contains the red pigment phycoerythrobilin and phycocyanin contains the blue pigment phycocyanobilin. There are several different types of each protein.

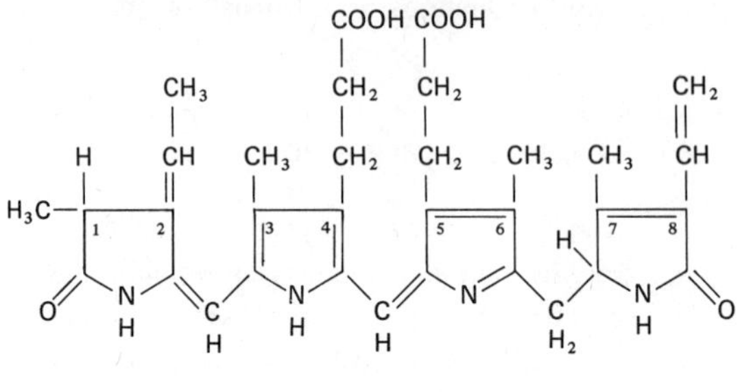

phycoerythrobilin

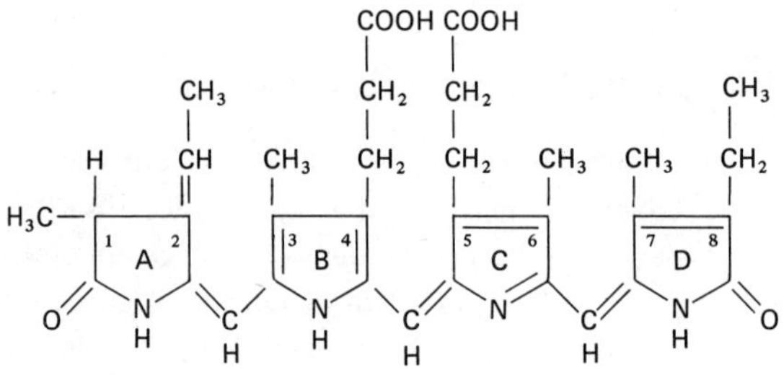

phycocyanobilin

The membranes into which these pigments are built for photosynthetic electron transport are also very distinctive. The most primitive type, in the blue-green algae (Fig. 13-2), are not very different from these in photosynthetic bacteria. We assume that all eucaryotic algae came from a procaryote rather like the blue-green ones, since these are the first cells in which chlorophyll a appears and this pigment is found in all eucaryotic phototrophs.

The red algae seem to be one of the more primitive groups of eucaryotes. Figure 13-3 shows that, while they have a eucaryotic structure, their photosynthetic lamellae are very much like those in the blue-green algae. In particular, they have tiny particles attached all along these lamellae, which Elizabeth Gantt and S. F. Conti have shown to be the phycobiliproteins that give the cells their distinctive color. The rather similar granules in the blue-green algae of Fig. 13-2 may also be biliproteins; the fact that the two groups have these pigments has been taken to indicate a close relationship between them.

Figure 13-4 shows a different red alga with smaller, more localized chloroplasts; here you can see more clearly that there is a boundary membrane surrounding a number of more isolated membranes called **thylakoids.** This is the general pattern of organization that developed in the other eucaryotes, along two major lines—a **brown line** and a **green line**—that are happily associated also with two different chlorophyll compositions.

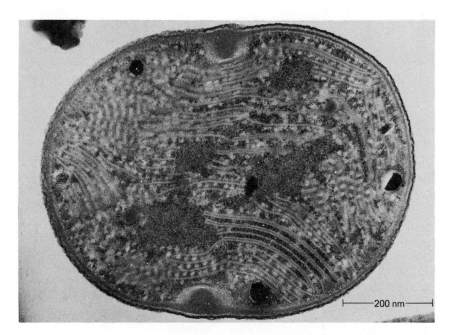

200 nm

FIGURE 13-2

The blue-green alga Gleocapsa
alpicola. *Most of the cell is filled
with photosynthetic lamellae
whose surfaces are covered with
large, diffuse granules. Large,
dark granules are polyphosphates
and tiny dark granules are ribosomes, mostly superimposed on a
dark, diffuse nuclear body.
Glutaraldehyde-osmium fixation.
(Courtesy of Mary M. Allen.)*

FIGURE 13-3

The red alga Porphyridium
cruentum. *Most of the cell is
occupied by a large chloroplast;
the thylakoid surfaces are covered
with granules. Endoplasmic reticulum and Golgi membranes and
tiny mitochondria can be seen.
Glutaraldehyde-osmium fixation,
stained with uranyl acetate and
lead hydroxide. (Courtesy of
S. F. Conti.)*

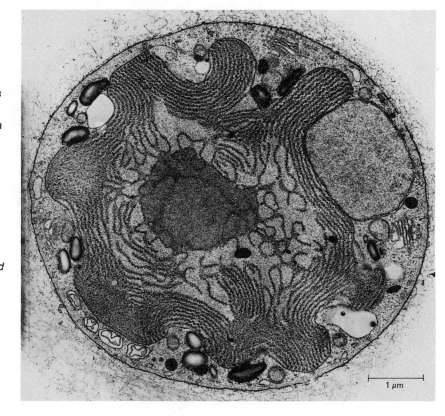

1 μm

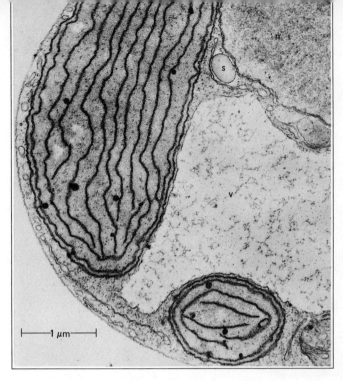

1 μm

FIGURE 13-4

Chloroplast of the red alga Rhodymenia pal-
mata *showing the widely spaced thylakoids.
Note the starch granule (s), nucleus (n), and
vacuole (v). (Courtesy of Leonard V. Evans.)*

FIGURE 13-5

Chloroplast of a zoospore of the brown alga
Pylaiella littoralis, *showing the triple bundles
of thylakoids that run into the large pyrenoid
(py). Its cap of starch (c) is nearby.
(Courtesy of Leonard V. Evans.)*

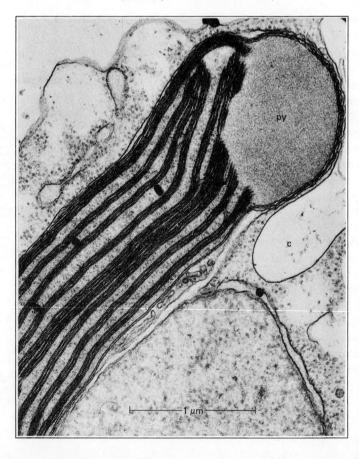

1 μm

The brown-line algae are generally colored yellow-brown, golden, or brown and have chlorophylls *a* and *c*. Their chloroplasts are very distinctive and contain triple stacks of thylakoids (Fig. 13-5).

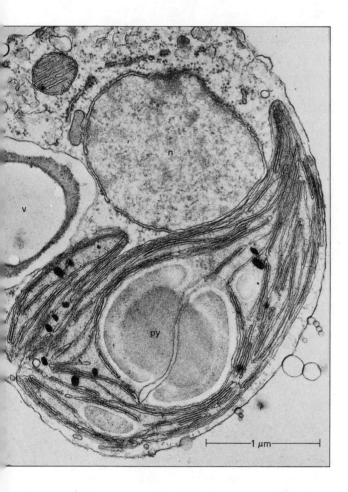

FIGURE 13-6

Chloroplast of a zoospore of the green alga Enteromorpha intestinalis, *showing the less ordered arrangement of thylakoids surrounding a pyrenoid and some starch deposits. (Courtesy of Leonard V. Evans.)*

The green-line algae are typically grass green and contain chlorophylls *a* and *b*. Their chloroplasts (Fig. 13-6) contain many thylakoids arranged more randomly. Notice that in both of these groups, the thylakoids are organized around—and sometimes run directly into—a large, structureless **pyrenoid.** This is a mass of protein that probably contains the enzymes for making starch, a storage form of the carbohydrates made in photosynthesis. Large starch granules can often be seen near the pyrenoid.

The green algae evolved quite directly into the true plants. Here the chloroplasts become organized in another characteristic way (Fig. 13-7) with long thylakoids interdigitated with short ones. Figure 13-8 shows the dense **grana** region of these chloroplasts in more detail.

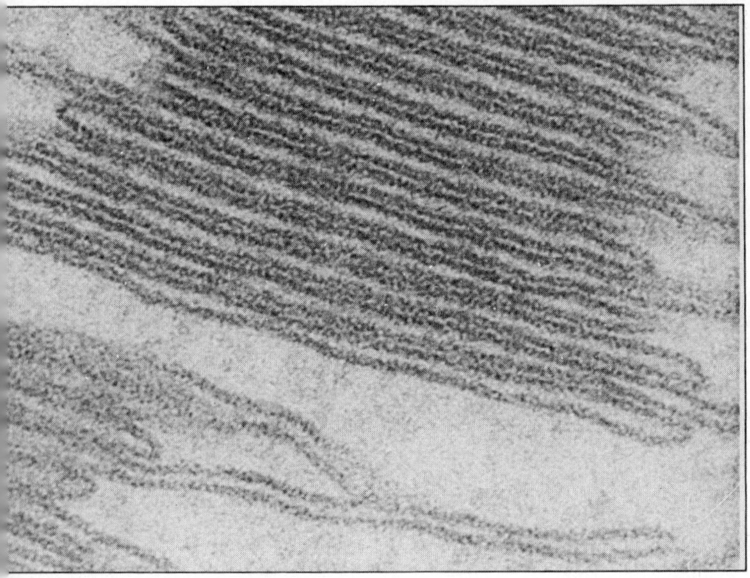

FIGURE 13-7

Chloroplast of corn (Zea mays), *showing the general arrangement of thylakoids. The grana regions are the dense stacks of closely opposed thylakoids; less dense thylakoid regions are intergrana, and the structureless background is stroma. (Courtesy of J. Rosada-Alberio and T. Elliot Weier.)*

FIGURE 13-8

High-magnification view of the thylakoids in a grana region of the chloroplast from a bean plant (Vicia faba). *Trace some of the membranes to see how the thylakoids fit together. (Courtesy of L. K. Shumway and T. Elliot Weier.)*

13-2 FLAGELLA

While all eucaryotic cilia and flagella have the same basic nonad structure, there are some structural embellishments that are (potentially, at least) useful clues to evolutionary lines. In the past, some technical terms have been used to distinguish different types of flagella; more careful electron microscope studies have shown that there are really more distinct types than had been thought, that some of the distinctions are quite subtle, and that it is too early to devise rigorous categories. The adjectives we use are meant to be suggestive and descriptive, not technical.

Broadly, there are three distinct types of flagella: whiplash, tinsel, and scaly (Fig. 13-9), with several variations among them. Not all whiplash flagella have a very thin end. Tinsel flagella may have fairly stiff hairs or rather delicate ones; the hairs generally run along opposite sides of the flagellum, but in some cases there is only a single row.

We can distinguish several types of basic flagellate cells. First, there is the type characterized by two whiplash, anterior flagella; this is typical of all of the motile green algae. Second, there is the type characterized by one anterior tinsel flagellum and one posterior whiplash flagellum. This is typical of the brown-line algae, except that in some cases the posterior flagellum has been lost. Third, there is the type with two or four anterior flagella that have scales and, generally, fine hairs; this is typical of a group of small green algae, the Prasinophyceae.

The green line remains quite easy to define on the basis of flagellation. The brown line is also fairly easy to define on this basis if we take into account that in some cases (brown algae proper) only the reproductive cells are flagellated and that in other cases the whiplash flagellum has been lost. When flagellation and some chemical criteria are used, some groups that used to be considered true fungi because they look generally like molds can be seen to be brown-line algae that have lost their pigments.

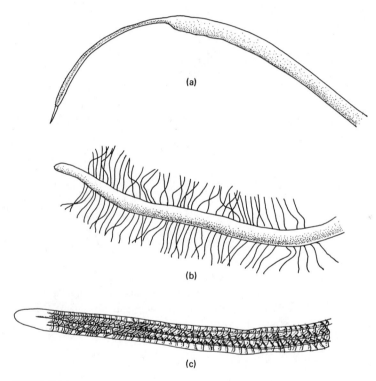

FIGURE 13-9
Three major types of flagella. (a) whiplash, (b) tinsel, and (c) scaly.
(Courtesy of Irene Manton.)

We ought to point out, in passing, that one of the more advanced groups of the brown line, the choanoflagellates, possess cells with a single tinsel flagellum surrounded by a collar (Fig. 13-10). These choanocytes are identical to those found in the sponges, so here, at least, a direct line of evolution is quite clear; in defining categories, we would probably not want to say that sponges are either plants or animals.

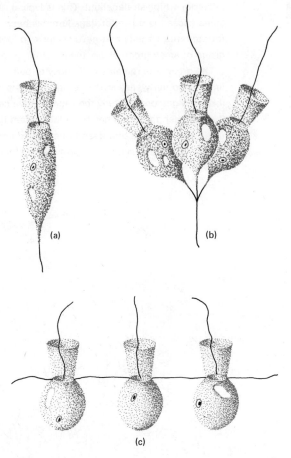

FIGURE 13-10
Two choanoflagellates, Monosiga (a) *and* Codosiga
(b), *compared with choanocytes of a sponge* (c).

13-3 WALLS AND STARCHES

Two kinds of polysaccharides are typically used for walls in the protista: chitin and
cellulose. *Chitin* is a poly-N-acetyl-glucosamine while cellulose is a polyglucose.
Cellulose is found in the walls of both green-line and brown-line algae, but chitin is
almost confined to the true fungi. This characteristic correlates very well with the type
of flagellation and constitutes another criterion for distinguishing true fungi from
colorless brown algae.

Many of the brown-line algae have developed the pathways for metabolism of silicon
and they build silicate cell walls. These are most highly developed in the beautifully
sculptured, lacy walls of the diatoms, which consist of two overlapping halves, rather

like a Petri plate. This construction forces diatoms to undergo a unique cell division in which successive generations of cells become smaller and smaller (Fig. 13-11); after several divisions, the reduced cells must go through a sexual process resulting in larger cells again.

The type of starch formed as a storage product also correlates very well with other features of the green-line and brown-line algae. The green algae generally have amylose and amylopectins that are α:1 $\rightarrow$ 4 glucans, sometimes with α:1 $\rightarrow$ 6 branches. The brown algae generally have β:1 $\rightarrow$ 3 glucans, also with occasional 1 $\rightarrow$ 6 branches, that are called paramylum, leucosin, and laminarin.

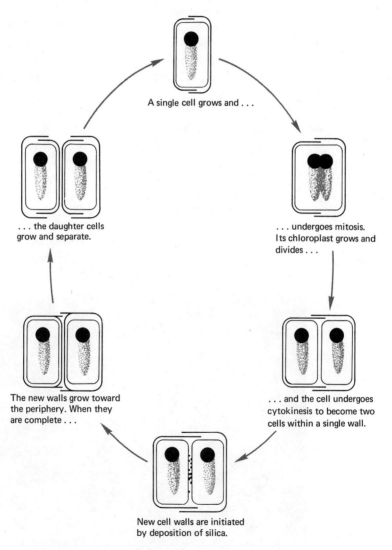

A single cell grows and . . .

. . . undergoes mitosis. Its chloroplast grows and divides . . .

. . . the daughter cells grow and separate.

The new walls grow toward the periphery. When they are complete . . .

. . . and the cell undergoes cytokinesis to become two cells within a single wall.

New cell walls are initiated by deposition of silica.

FIGURE 13-11
Cell division and wall formation in a diatom.

13-4 MITOSIS

The mechanism of mitosis would be one of the most powerful indicators of lines of evolution, if we only knew more about it. Our conception of the "ordinary" pattern of mitosis comes from studies of plant and animal cells. But more recent studies of protista indicate that there are many intermediate types of organization between the typical procaryotic and the typical eucaryotic cell and that many of these protista have intermediate types of mitosis. We can hardly begin to count and classify the different patterns; much of this work was done years ago with fairly primitive light microscopy, and it needs to be repeated with modern techniques.

The dinoflagellates have recently become interesting because of their unusually primitive features. The "dinos" are mostly unicellular, either colorless or golden-brown to greenish-brown, all motile forms with a unique shape and flagellar pattern (Fig. 13-12). They may be naked or covered with a cellulose armor divided into plates, but the cell surface is always split into two grooves containing flagella—one transverse and the other trailing.

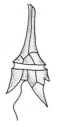

Ceratium   Gymnodinium   Peridinium

FIGURE 13-12
Some representative dinoflagellates.

FIGURE 13-13
Chromosomes of the dinoflagellate Prorocentrum, *showing the fine, fibrillar structure reminiscent of the bacterial nuclear bodies. (Courtesy of Peter Giesbrecht.)*

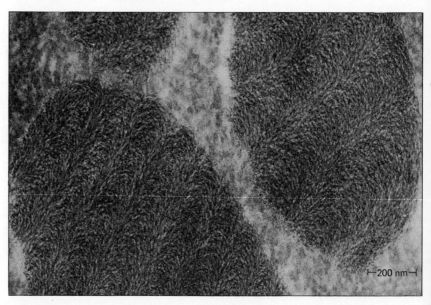

⊢200 nm⊣

John D. Dodge and Peter Giesbrecht have shown independently that dinoflagellate chromosomes (1) do not contain histones, (2) do not undergo a coiling and uncoiling cycle during mitosis, and (3) do not look like eucaryotic chromosomes in electron micrographs. Figure 13-13 shows that their chromosomes have the same fine texture as the bacterial nuclear body; this is consistent with their lack of histones.

As the chromosomes begin to replicate, invaginations of the nuclear envelope appear which become . . .

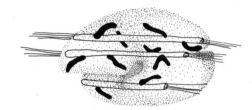

. . . channels through the nucleus that are filled with microtubules. The daughter chromosomes attach to these channels . . .

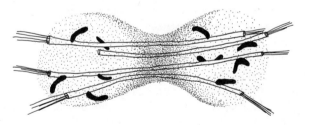

. . . and move to opposite poles as the channels elongate. The nuclear envelope then constricts to form two nuclei.

FIGURE 13-14

The mechanism of nuclear division in a dinoflagellate.

The dinoflagellates do not undergo the typical eucaryotic mitosis. Their nuclear division, which we will call **promitosis,** occurs without a breakdown of the nuclear envelope, and it cannot be divided into stages comparable to those in mitosis. It begins as the nucleus enlarges somewhat and the chromosomes start to split lengthwise into "chromatids" starting at one end, so each chromosome becomes progressively a Y, then a V, and then two separate bodies. Donna Kubai and Hans Ris have shown that the nuclear envelope is simultaneously undergoing the changes shown in Fig. 13-14. The channels that appear in the nucleus are entirely extranuclear and are filled with microtubules. The chromosomes attach to the nuclear envelope on these channels and they are separated to opposite poles as the nucleus elongates and pinches into two separate nuclei. There are no centrioles, though some of Giesbrecht's pictures show a structure that might have some of the functions of a centriole. Moreover, the process is not inhibited by colchicine; the microtubules presumably help to shape the nuclear envelope, but they are not involved in chromosome separation.

Notice that this pattern is in many ways intermediate between the typical procaryotic and typical eucaryotic division patterns. In bacteria, the genophore is attached to the cell membrane and nuclear division occurs through elongation of the cell and a directed movement of mesosomes. In eucells, the chromosomes are attached to the inside of the nuclear membrane; when the nucleus disappears, they become attached to mitotic microtubules and are separated by changes in microtubular structure. In the dinoflagellates, the nucleus behaves rather like a bacterial cell dividing inside a eucaryotic cell; the chromosomes are attached to the nuclear envelope and are separated by elongation of the envelope under the influence of microtubules.

Because their nuclear apparatus is so similar to that of bacteria, Giesbrecht wondered if the dinoflagellates might have a bacterium-like cell wall. Bacteria are sensitive to penicillin, D-cycloserine, and other drugs that interfere with cell wall biosynthesis and are inactive against typical eucaryotes. He found that these drugs induced changes in cell structure in dinoflagellates similar to those induced in bacteria.

The dinoflagellates thus appear to be the most primitive known eucaryotes. By definition, they are certainly eucaryotic, but they retain many procaryotic features. Their somewhat specialized cell architecture and flagellar structure makes them rather enigmatic; they have probably retained their procaryotic features while undergoing a specialized evolution. At least they illustrate some transitional features between the procaryotes and the earliest eucaryotes. Because of their nuclear structure, we will call them *mesoprotista,* in contrast to the more advanced *metaprotista.*

There are probably many other mesoprotista that have not been identified yet. Gordon F. Leedale has found that the euglenoids, a group of green algae, are very similar to the dinoflagellates in their mitosis. Their chromosomes divide in half and move toward opposite poles as the nucleus constricts. However, the chromosomes do undergo the typical coiling-uncoiling cycle, do contain histones, and look like typical eucaryotic chromosomes.

Many of the simple amebas may also be mesoprotistan. The nuclei of radiolarians and heliozoans, like the ones we discussed in Section 11-9, contain on the order of 2000 tiny chromosomes which are separated in cell division by a process like promitosis.

Mogens Westergaard and Dieter von Wettstein have recently investigated the nuclear structure of a true fungus, *Neottiella,* and found that the chromosomes remain in a highly contracted state all through interphase, relaxing slightly during early prophase when they appear to be replicating. Thus the fungal nucleus resembles the dino- flagellate and euglenoid nucleus in many respects. This is interesting because there is a very good metabolic reason to suspect that fungi are related to euglenoids. There are basically two pathways of lysine biosynthesis; one of them is found in all bacteria and in most of the plants, algae, and so on that have been tested. However, the other one has so far been found only in the euglenoids and fungi, which indicates a very close relationship. (The test has *not* been applied to a number of important groups, such as the red algae, that may be related to the true fungi. This would be a rather simple way of deciding some critical points; which of you will be the first to do the proper experiments?)

Other organisms have a type of mitosis intermediate between the mesoprotistan and the more typical plant and animal type. Ursula G. Johnson and Keith R. Porter have shown that in the green alga *Chlamydomonas* the nuclear envelope remains more or less intact, but a spindle of microtubules forms inside (Fig. 13-15). The flagellar basal bodies detach and move into the cytosome, but although they look like typical centrioles, they do not serve as mitotic division centers; they probably help to orient the subsequent cytokinesis. Other bands of microtubules orient division of the nucleus into daughter nuclei.

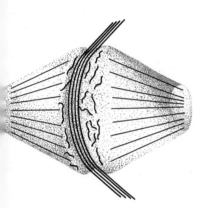

FIGURE 13-15
Mitosis in Chlamydomonas: *a section through a late metaphase-early anaphase nucleus with its intranuclear spindle (courtesy of Ursula G. Johnson and Keith R. Porter) and an interpretive drawing showing the whole nucleus with its orienting microtubules.*

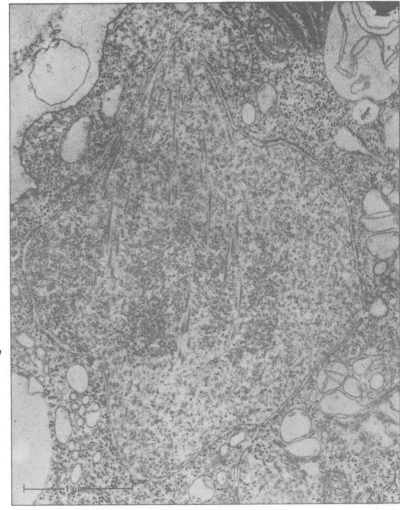

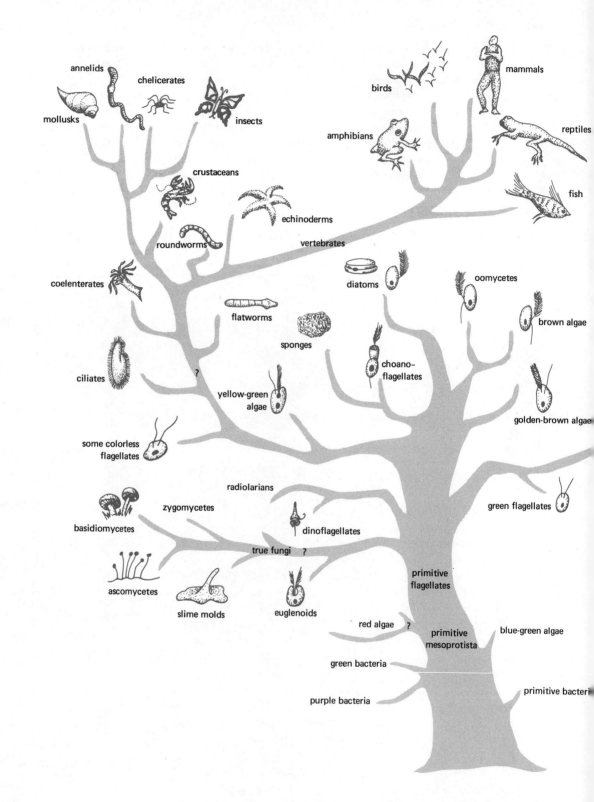

annelids

chelicerates

mollusks

insects

crustaceans

echinoderms

roundworms

vertebrates

coelenterates

flatworms

sponges

diatoms

oomycetes

brown algae

choano-flagellates

ciliates

yellow-green algae

golden-brown algae

some colorless flagellates

green flagellates

radiolarians

zygomycetes

basidiomycetes

dinoflagellates

true fungi ?

primitive flagellates

ascomycetes

slime molds

euglenoids

red algae ?

primitive mesoprotista

blue-green algae

green bacteria

purple bacteria

primitive bacteria

birds

mammals

amphibians

reptiles

fish

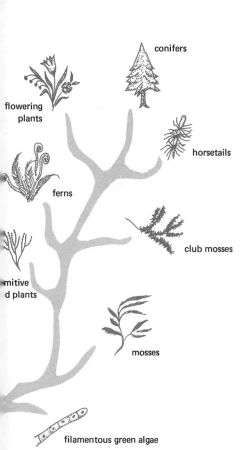

conifers

flowering
plants

horsetails

ferns

club mosses

mitive
d plants

mosses

filamentous green algae

FIGURE 13-16
A family tree of terrestrial organisms.

Another mitotic feature in some organisms is an aster around the centriole, apparently formed of microtubules that radiate in all directions. The aster is characteristic of animal cells—at least those that have been studied carefully—but is also found in some of the brown-line algae; this is one reason why we tend to look for ancestors of the animals among this branch of the protista, but it may not be a very critical feature. In general, it is still too early to sort these mitotic patterns and use them as criteria for lines of evolution. However, comparative studies of mitosis, taken along with other data, will probably be very useful in the near future, and may also reveal something about the mechanism of mitosis itself.

Figure 13-16 is a very broad outline of the family tree of organisms, with the protista emphasized. There ought to be question marks almost everywhere on this tree, but only the most uncertain points are so marked. The origin of the true animals is one of the most debatable points. A very reasonable hypothesis (but certainly not the only one and probably not the most popular one today) comes from Jovan Hadzi, who pointed out that the most primitive flatworms are very similar to some of the ciliates. We will consider this again in Section 13-10.

B. Cycles of reproduction

13-5 GROWTH CYCLES

Cells such as *E. coli* that multiply only by binary fission are said to be **vegetative.** Most bacteria have no other condition, but those in the genera *Bacillus* and *Clostridium,* which are common soil organisms, are different. Quite regularly, particularly toward the end of their exponential growth phase, cells begin to wall off one of their nuclear bodies with a strong, impermeable coat to make a **spore;** the process is called **sporulation.** Spores are extremely resistant to heating and drying; they may be scattered by the wind and may survive for long periods, until they come to rest in some moist place where they **germinate** into single vegetative cells that proliferate as usual. The processes of vegetative growth, sporulation, germination, and vegetative growth again form a **growth cycle** for these bacteria.

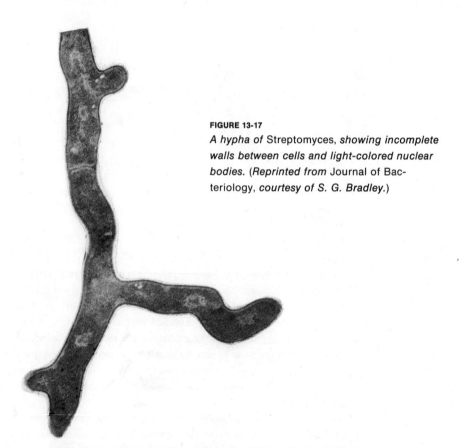

FIGURE 13-17
A hypha of Streptomyces, *showing incomplete walls between cells and light-colored nuclear bodies. (Reprinted from* Journal of Bacteriology, *courtesy of S. G. Bradley.)*

Other procaryotes grow in similar ways. The actinomycetes, for example, have a vegetative form called a *mycelium;* it consists of a mass of long cells called *hyphae* (singular, hypha) whose organization is **coenocytic** or **syncytial,** which means they have many nuclear bodies not separated by cross walls (Fig. 13-17). When grown on solid medium some of the hyphae grow upward to make an aerial mycelium. In one group, the *Nocardias,* an old mycelium forms cross walls between its nuclear bodies and breaks up into a mass of short rods, each of which can swim away to form a new mycelium. In another type, the *Streptomycetes,* the ends of many hyphae curl into spirals within which cross walls form to make spores called **conidia** (singular, conidium). Each conidium can form a new mycelium.

The term spore is used to describe many different structures. We will use it for any cell that is itself incapable of vegetative growth (generally meaning binary fission) but can develop into a vegetative cell. The term is used particularly for cells that have a tough, protective outer layer. Most of the growth cycles of eucaryotes we will discuss involve stages called spores; these have been given an incredible variety of special names that you will meet in more specialized books, but we will use only the most general terms for them.

13-6 THE SEXUAL CYCLE

Most eucaryotes reproduce sexually through a general pattern of events that forms a **sexual cycle** of four phases; there are many variations on the cycle, but they are all quite easy to understand once you know the basic principles. In essence, the cycle is an alternation between a **haplophase** during which each cell nucleus is *haploid* and contains a single set of chromosomes ($1n$, where n is the basic chromosome number), and a **diplophase,** where every cell nucleus is *diploid* and contains two sets of chromosomes ($2n$).

The transition from haplophase to diplophase is **syngamy** or **fertilization,** in which two haploid cells fuse with each other, followed by fusion of their nuclei, **karyogamy,** to produce a diploid nucleus. The specialized haploid cells that fuse are called **gametes,** and their immediate product after syngamy is a **zygote.** The two gametes always differ in at least one characteristic, possibly a chemical difference in their walls, and are said to be different **mating types,** designated arbitrarily $+$ and $-$. (In some cases, they are designated A and a, or a and α.) The gametes may also differ greatly in size and shape, and are frequently designated different *sexes* — male ($\male$) and female ($\female$). We frequently distinguish three situations:

1. **Isogamy:** The gametes are identical in appearance; usually they are small, flagellated cells.

2. **Anisogamy:** One gamete is visibly larger than the other.

3. **Oogamy:** The female gamete, or **ovum,** is large and nonflagellated, while the male gamete, or **sperm,** is small and flagellated.

The gametes are not always motile and not always visibly different from the vegetative haploid cells. In many fungi, for example, they are simply specialized regions of a mycelium that grow toward each other and fuse. Vegetative cells are frequently induced to become gametes by putting them into a special medium, such as plain water instead of a rich broth.

The opposite transition from diplophase to haplophase occurs in a process called **meiosis,** which may be considered a variation of mitosis designed to separate the two mixed sets of chromosomes in a diploid nucleus precisely into two haploid sets. Thus, the sexual cycle is always a repetition of the four phases: haplophase, syngamy, diplophase, meiosis.

The vegetative cells of any organism may be haploid, diploid, or both; consequently, we distinguish three types of sexual cycle—**haplontic,** in which only haploid cells undergo mitosis; **diplontic,** in which only diploid cells undergo mitosis, and **haplo-diplontic,** in which both types undergo mitosis.

Before going further, let us consider the process of meiosis in detail.

13-7 MEIOSIS

Everyone who has studied biology at some time is sorry that the terms "mitosis" and "meiosis" are so much alike but refer to such different processes; at least you can take some comfort from the fact that you are not alone. Remember that in mitosis, which can occur in both haploid and diploid cells, each set of doubled chromosomes is split into two sets of chromatids. However, in meiosis, which can occur only in diploid cells, the pairs of doubled chromosomes are separated precisely into four sets of chromatids; each set of four cells derived by meiosis from one original cell is a **tetrad.**

Meiosis is also divisible into arbitrary stages, whose names, ending in *-nema,* describe the conditions of the thread-like chromosomes. The corresponding adjective for each stage ends in *-tene.* The stages are illustrated in Fig. 13-18 with photographs of real cells and drawings of a diploid cell containing four chromosomes—that is, two haploid sets of two chromosomes each. Now, in order to produce this cell there must have been a syngamy at some time in the past between two gametes, each carrying two chromosomes. In the figure, one of these haploid sets is shown in black and the other in color. Each set contains one rather long metacentric chromosome and one shorter acrocentric chromosome. The two metacentric chromosomes are **homologues** or **homologous chromosomes,** as are the two acrocentrics. The homologues are superficially identical, although they could carry different alleles which the microscope could not detect.

The first two stages of meiosis have been called leptonema and zygonema, since the classical studies of H. von Winiwarter and others around 1900. In the first stage, the chromosomes become visible threads and in the second, homologous chromosomes begin to pair with one another. However, this distinction is probably unrealistic, for there is now good evidence that the homologues have been paired with one another since some time during the last premeiotic mitosis. (In those cycles in which meiosis follows syngamy without any intervening cell divisions, the homologues probably pair with one another as soon as they are mixed in one nucleus.) The chromosomes certainly behave rather differently in this last mitosis than they do in previous cell divisions, and the short, condensed chromosomes of mitotic metaphase could pair with one another much more easily than the extended thread-like chromosomes of interphase.

By pachynema, the homologues are held together tightly in the form of bivalents. Synapsis of homologues is initiated from the ends and from the centromeres, and it is very precise. The distinctive knobs and constrictions that can be picked out on each chromosome come to lie right next to one another. The forces involved in synapsis are unknown, although its specificity suggests that at least one factor must be the melting of hydrogen bonds within chromatids and their reformation between chromosomes at some points.

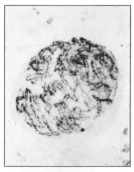

leptonema-
zygonema

(a)

pachynema

(b)

diplonema

(c)

FIGURE 13-18

*The process of meiosis. The photographs (courtesy of Grace M. Donnelly and Arnold
H. Sparrow) show the large chromosomes of the salamander* Amphiuma, *which has
a diploid set of 28. The drawings show meiosis in a cell with a diploid set of four
chromosomes; the black chromosomes were originally derived from one parent and
the colored set from the other parent, to illustrate the genetic consequences of
meiosis. (a) Leptonema-zygonema. The chromosomes are just becoming visible with
the light microscope; they appear to be single, and homologous chromosomes are
already beginning to pair or* **synapse** *with each other. (b) Pachynema. The homologous
chromosomes have now paired with one another to make* **bivalents,** *and they are
visibly shorter and thicker. Homologues appear to be held together tightly and
specifically. (c) Diplonema. The chromosomes are now visibly double, consisting
of two chromatids. They are much more compact and appear to be pulling away from
each other, but are held together by* **chiasmata** *(singular,* **chiasma***) where there is
strong interaction between homologous chromatids.*

diakinesis
(d)

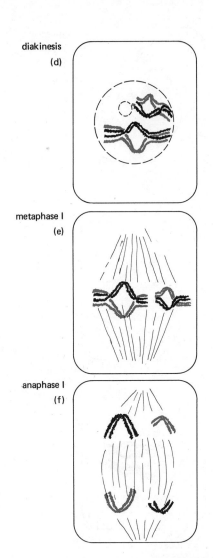

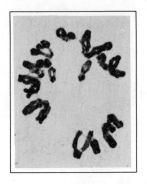

metaphase I
(e)

anaphase I
(f)

Figure 13-18 (cont.)

(*d*) *Diakinesis. This stage cannot be sharply differentiated from diplonema. The chromosomes are shorter and thicker; they are still pulling away from each other, and the chiasmata are terminalizing or moving toward the ends of the chromosomes.* (*e*) *Metaphase I. The nuclear membrane has broken down and the spindle is complete. The bivalents orient themselves on the metaphase plate with their centromeres actively pulling away from each other; there appears to be considerable tension on the chromosomes.* (*f*) *Anaphase I. The centromeres remain intact (in contrast to mitosis) and the two homologues in a bivalent move toward opposite poles.*

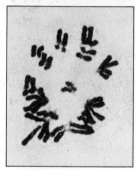

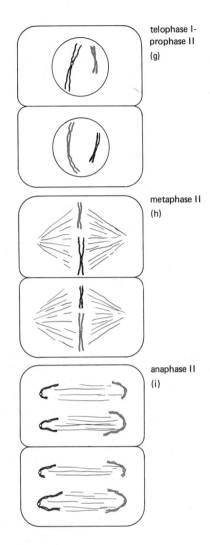

telophase I-
prophase II
(g)

metaphase II
(h)

anaphase II
(i)

Figure 13-18 (cont.)

(g) *Telophase I-prophase II. There is an* **interkinesis** *period between the first and second meiotic divisions, during which the nuclear membrane reforms and the chromosomes become longer and thinner. At prophase II, the chromatids start to condense again and the nuclear membrane disappears.* (h) *Metaphase II. The chromosomes are again very short and thick; their chromatids are visibly separated and appear to be pulling away from each other. The chromosomes line up as in mitosis with their centromeres along the metaphase plate.* (i) *Anaphase II. The sister chromatids separate to opposite poles.*

telophase II
(j)

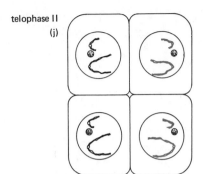

Figure 13-18 (cont.)

(j) Telophase II. Nuclear membranes reform and the chromatids—which are now independent chromosomes— return to their interphase condition.

As the chromosomes separate from diplonema on, the chiasmata can be seen where homologous **chromatids** cross each other; these are apparently the points where crossing over occurs. We will examine the genetic consequences of these events later. The chiasmata move toward the ends of the chromosomes. You can see what is happening with a simple experiment: Twist two pieces of string around each other several times, and then pull them apart from the middle and notice that the points of contact between them terminalize as they unwind.

The critical difference between mitosis and meiosis becomes apparent during metaphase I and anaphase I. In mitosis, the chromosomes lined up in preparation for a *division of their centromeres* and separation of the chromatids. In meiosis, *the centromeres remain intact* and the two homologues move toward opposite poles during anaphase. This constitutes the **first meiotic division.** The **second meiotic division** is essentially a mitosis in a haploid cell, for the chromosomes line up again in metaphase II and sister chromatids separate from one another in anaphase II.

There is now a tetrad of four cells, each of which generally develops into a gamete or a spore of some kind. Each cell of the tetrad contains one chromosome of each type—a single haploid set. But there may be a subtle difference between them. Look at the metaphase I of Fig. 13-18e. This was drawn arbitrarily with the metacentric bivalent oriented black:up and colored:down and the acrocentric bivalent oriented the other way. The key word is "arbitrarily." This metaphase configuration could just as well have been drawn with both black chromosomes oriented one way and both colored chromosomes the other. In fact, there are four different ways to pick a haploid set out of these chromosomes and they are all equally probable.

If we consider a large number of meiotic cells, we are faced with exactly the same situation shown in Fig. 5-1a for a set of phage particles whose genomes consist of two pieces. Remember that the cells we are looking at were made from haploid gametes, one with a black set of chromosomes and one with a colored set. In meiosis, we could again produce these **parental sets** by selecting chromosomes of the same color; as in the phage case, the probability of picking an all-black or all-colored set is 0.25. However, it is just as easy to form a **recombinant set** by selecting one black and one colored chromosome.

It should now be obvious that meiosis presents a mechanism for recombination in sexual organisms. The colored markers we have been using can be replaced easily by real genetic markers and their patterns of transmission from one generation to the next can be traced through meiosis. However, the only way to understand the pattern of inheritance in an organism is to understand that organism's sexual cycle. Therefore, we will now outline the cycles of a few representative organisms, to give us a basis for genetic analysis.

A very general sexual cycle is shown in Fig. 13-19. The complete cycle would be haplodiplontic, with the addition of spore formation in both phases. Any real cycle will be some selected part of this. For example, a haplontic cycle can be obtained by following the arrow in the diplophase directly from the zygote through meiosis. The two side cycles, with formation of mitospores, may occur in some organisms as additional means of proliferation, but they are not part of the sexual cycle itself.

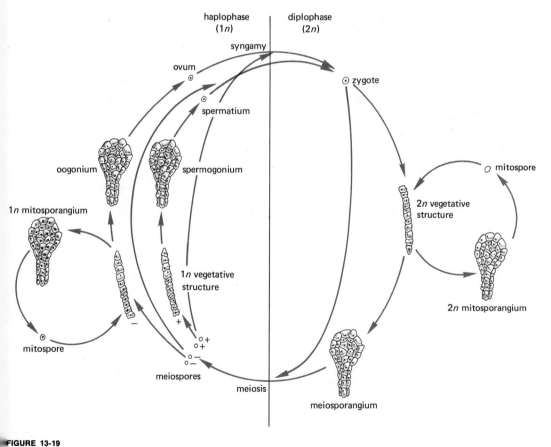

FIGURE 13-19
The general sexual cycle; every real cycle can be derived from this by selecting the right path. On each side is an asexual cycle which begins with the formation of **mitospores** *by a mitotic division.*

13-8   THE HAPLONTIC CYCLE

In this cycle, the zygote is the only diploid cell and we may conveniently begin with it. It has been formed from two haploid gametes; it may persist for a time, but it does not divide mitotically and eventually it undergoes meiosis. However, it may change its shape considerably and become a **meiosporangium** that can be recognized visually; within the meiosporangium, meiosis occurs, yielding 4, 8, or 16 **meiospores.** Each meiospore then undergoes mitosis and forms vegetative cells.

If the vegetative form of the organism is unicellular, then each meiospore becomes a clone of unicells. Since there are two mating types, + and −, which combine to produce the zygote, two types of meiospores result from meiosis and there are therefore + and − vegetative cells. Under appropriate conditions, some of these become gametes (again + and −) and the cycle repeats itself.

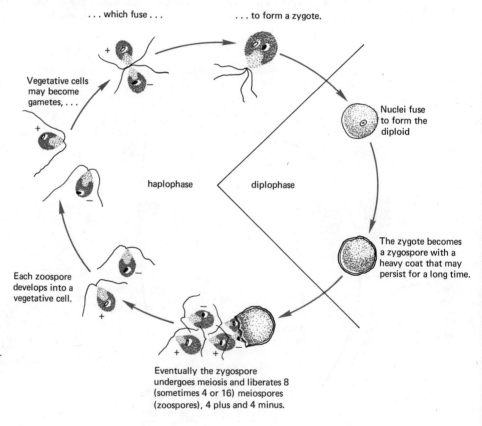

FIGURE 13-20

The sexual cycle of Chlamydomonas.

In many plants, fungi, etc., the vegetative form is a mycelium, a small plant, or some similar multicell, which is generally called a **thallus.** There are now two variations on the haplontic cycle.

In *heterothallism*, + spores and − spores produced in meiosis develop into distinct + and − thalli. The + thallus produces only + gametes and the − thallus produces only − gametes.

In *homothallism,* only one kind of spore and one kind of thallus are produced. These thalli are **hermaphroditic,** which means they produce both + and − gametes.

If there is a specialized structure on the thallus that produces gametes, it is called a **gametangium.** If ova and sperm can be distinguished in the species, the gametangia that produce these gametes are called **oogonia** and **spermogonia,** respectively.

The little green alga *Chlamydomonas* has the simple haplontic cycle shown in Fig. 13-20. The two mating types are designated mt^+ and mt^-; cells of opposite type fuse to form a zygote which undergoes meiosis, and in the strains usually used the four cells in the tetrad derived from this meiosis undergo one mitotic division to make eight meiospores (called zoospores because they are motile).

Another haplontic cycle is seen in the bread mold *Neurospora* (Fig. 13-21). There are again eight meiotic products, and one of the beauties of this organism is that they are lined up in the ascus in the order in which they come through meiosis, so it is very easy to analyze genetic events.

The fungus *Mucor* shows a heterothallic cycle (Fig. 13-22). The gametes are merely specialized arms of the mycelium.

An example of an oogamous cycle is shown in Fig. 13-23 for the filamentous green alga *Oedogonium.* Some cells of the filament produce sperm while others become ova.

13-9 THE HAPLODIPLONTIC CYCLE

The yeast *Saccharomyces cerevisiae* has a haplodiplontic cycle (Fig. 13-24) and is very useful because either haploid or diploid cells can be induced to undergo a simple vegetative growth. Notice, however, that proliferation is not an equal division of one cell into two; yeasts proliferate by **budding** and produce small cells which grow larger and then produce more buds. In spite of this variation, however, growth is still exponential and can still be described by the equations we have already written for bacteria.

The plants exhibit a variety of haplodiplontic cycles. One of the most interesting features of plant evolution is a tendency toward enlargement of the diplophase and a corresponding decrease in the size of the haplophase. For example, in mosses, which are the most primitive living plants, the haplophase is the obvious, leafy ground cover that is commonly called the moss itself; the diplophase consists of the delicate, wheat-like stalks that most people miss unless they look closely. In the little capsule at the top of each stalk, a **spore mother cell** undergoes meiosis, and the spores liberated from the capsule then grow into leafy plants that produce tiny gametangia (Fig. 13-25). However, the millions of years of later evolution that led to the seed plants have seen a reversal of this condition. The large, obvious plants we see around us are all diplophases; the haplophase is confined to a short series of mitoses in the development of the spores. As Fig. 13-26 shows, there are two types of spores, produced in separate meioses. The **microspores** grow into **pollen** grains and the **megaspores** grow into **ovules,** within which the fertilized ovum develops. In **pollination,** a pollen grain is transferred (by wind or gravity, or by birds and insects) to the

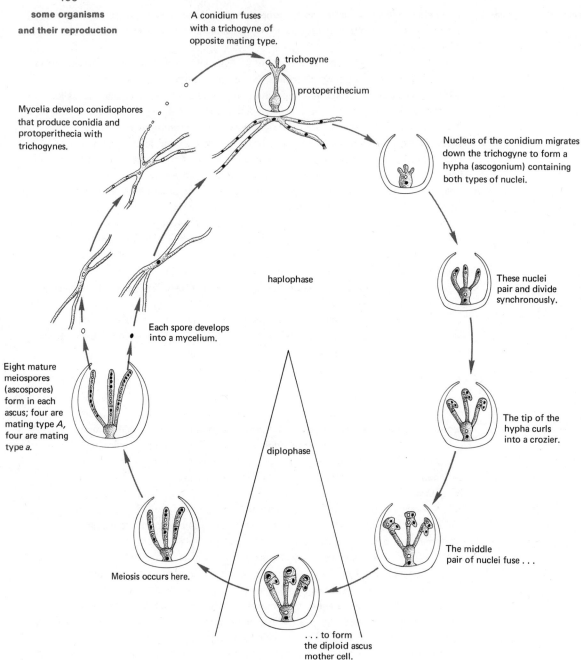

A conidium fuses
with a trichogyne of
opposite mating type.

trichogyne

protoperithecium

Mycelia develop conidiophores
that produce conidia and
protoperithecia with
trichogynes.

Nucleus of the conidium migrates
down the trichogyne to form a
hypha (ascogonium) containing
both types of nuclei.

haplophase

These nuclei
pair and divide
synchronously.

Each spore develops
into a mycelium.

Eight mature
meiospores
(ascospores)
form in each
ascus; four are
mating type *A,*
four are mating
type *a.*

diplophase

The tip of the
hypha curls
into a crozier.

The middle
pair of nuclei fuse . . .

Meiosis occurs here.

. . . to form
the diploid ascus
mother cell.

FIGURE 13-21

The sexual cycle of Neurospora.

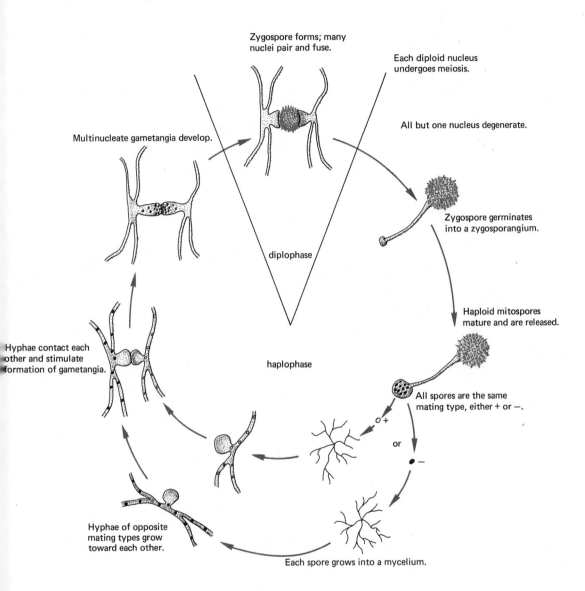

Zygospore forms; many nuclei pair and fuse.

Each diploid nucleus undergoes meiosis.

Multinucleate gametangia develop.

All but one nucleus degenerate.

Zygospore germinates into a zygosporangium.

diplophase

Haploid mitospores mature and are released.

Hyphae contact each other and stimulate formation of gametangia.

haplophase

All spores are the same mating type, either + or −.

○ +

or

● −

Hyphae of opposite mating types grow toward each other.

Each spore grows into a mycelium.

FIGURE 13-22
The sexual cycle of Mucor.

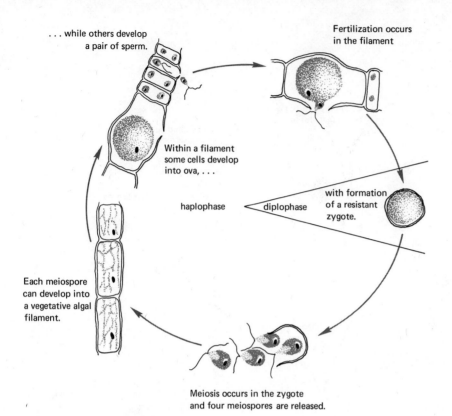

... while others develop a pair of sperm.

Fertilization occurs in the filament

Within a filament some cells develop into ova, ...

haplophase   diplophase

with formation of a resistant zygote.

Each meiospore can develop into a vegetative algal filament.

Meiosis occurs in the zygote and four meiospores are released.

FIGURE 13-23
The sexual cycle of Oedogonium.

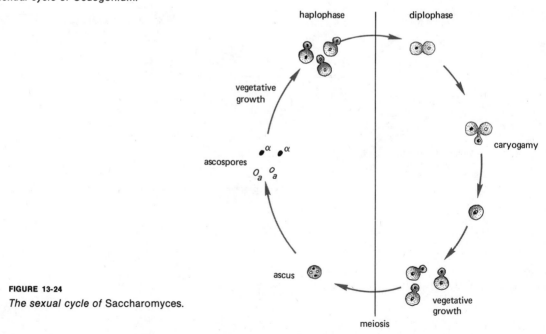

haplophase   diplophase

vegetative growth

caryogamy

ascospores

ascus

vegetative growth

meiosis

FIGURE 13-24
The sexual cycle of Saccharomyces.

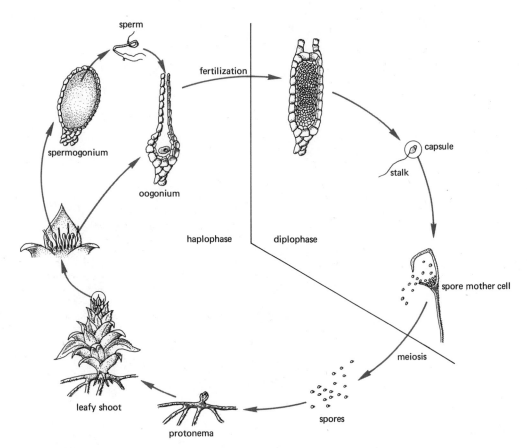

FIGURE 13-25
The sexual cycle of a moss.

stigma of the flower, from where its sperm nuclei migrate down to the ovule. A double fertilization occurs in the ovule to make a diploid zygote that grows into a new plant (a) and a triploid endosperm tissue that makes food for the young plant (b).

13-10 THE DIPLONTIC CYCLE

Relatively little need be said about the diplontic cycle since it is the common animal type that is so familiar to all of us. There are still two major variations; while most

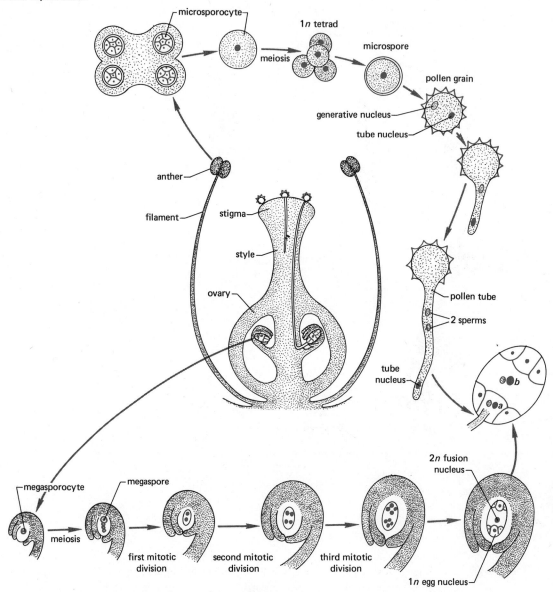

FIGURE 13-26

*The sexual cycle of a flowering plant. The zygote develops into a new diploid plant
that bears the flowers in which these events are repeated.*

species have distinct male and female sexes that produce only sperm or only eggs, some animals are hermaphroditic and contain both types of gonads in a single body.

Figure 13-27 shows the processes of spermatogenesis in the **testes** (male gametangia or male **gonads,** as they are generally called in animals) and oogenesis in the **ovaries** (female gonads). Notice that the terms used to describe cells in meiosis are parallel in the two processes: spermatogonium and oogonium, primary spermatocyte and oocyte, secondary spermatocyte and oocyte.

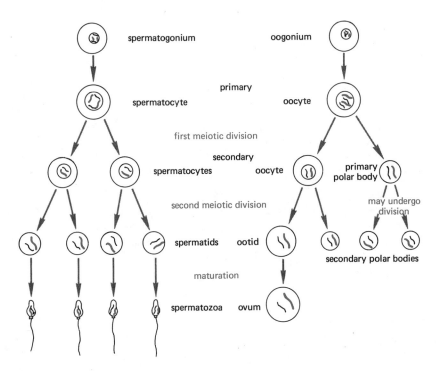

FIGURE 13-27

Spermatogenesis and oogenesis in animals. The last meiotic events in oogenesis sometimes occur only after fertilization.

The reproductive cycles of most protozoa have not been studied carefully, but the little ciliate *Paramecium,* which has been used for genetic experiments, has a very interesting, basically diplontic cycle (Fig. 13-28). These cells have a **micronucleus** (sometimes several micronuclei) that functions in reproduction and a polyploid **macronucleus** that controls most cellular synthesis. Vegetative cells are diploid, and there is only one haploid mitosis preceding nuclear exchange. Autogamy is an unusual variation in which the haploid nuclei of a single cell fuse with one another after meiosis.

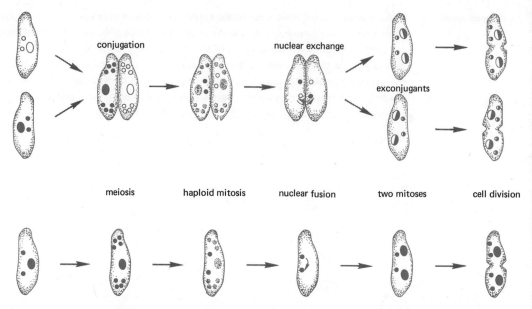

FIGURE 13-28

Reproduction in Paramecium. *All micronuclei undergo meiosis; seven of the eight haploid products disintegrate and the eighth divides mitotically. Conjugating cells exchange micronuclei which then fuse into a diploid nucleus. In autogamy, the two haploid micronuclei of one cell fertilize each other. Meanwhile, the macronuclei have disintegrated. The diploid micronucleus divides twice mitotically and two of its four products become polyploid macronuclei. The other two divide again as cell division begins.*

According to the hypothesis of Hadzi and others, organisms of this type evolved into the primitive animals. Figure 13-29 compares a primitive ciliate type with a very primitive flatworm. The similarities are quite remarkable. The flatworm has not even achieved a strictly cellular organization yet; it is really a syncytium, even though different parts of the body have become rather specialized and differentiated into three general layers. In the innermost layer, there is cyclosis and circulation of food vacuoles in which solid food is digested by lysosomal enzymes, and the undigestible residue is then expelled — just as in the ciliates. These flatworms are hermaphroditic; a pair of them copulate by lying with their gonopores together and fertilizing each other; this is very reminiscent of ciliate conjugation. One of the few difficulties with Hadzi's hypothesis is that mitosis in ciliates is intranuclear and very unlike that in animals. However, mitosis has not been studied carefully in these primitive flatworms; for all we know, it is similar to the ciliate type.

From these simple beginnings have come the entire world of animals, including ourselves. In most cases, there are two distinct sexes, and along with this development has gone the evolution of the great variety of sexual behavior that embellishes the simple act of copulation. However, the essential features of the diplontic cycle remain the same in almost all cases, and there is little point in discussing all the variations in form, development, and behavior in the animals at this time. To do justice to the marvelous world of animals you should really treat yourself to a good zoology course— ideally, near the ocean where their incredible beauty and variety can be appreciated.

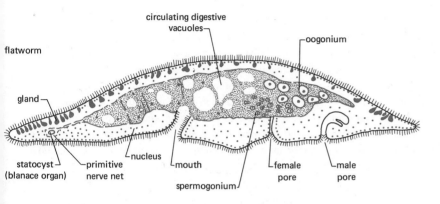

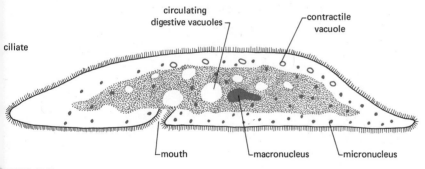

FIGURE 13-29
A comparison of an idealized acoel flatworm with an idealized ciliate.

13-11 THE COURSE OF EVOLUTION

We have just seen that in both plants and animals there has been a rather consistent tendency toward dominance of the diplophase; in fact, the haplophase is virtually absent in animals except for gametes and an occasional odd group of insects in which haploid individuals develop. Why should this be? The answer is really quite simple.

In a haploid cell, every gene is exposed to scrutiny by the environment and the forces of natural selection. If a lethal mutation occurs, every individual bearing the mutation dies. If a mutation occurs that changes the color of the organism and it is not particularly advantageous to have that color, the organism will probably be selected against and the mutation will not be propagated. However, in a diploid cell every gene may be protected by another just like itself; if a mutation occurs in one of the two genes, the other is still properly functional. Because of the phenomenon of dominance, which we will explore in Chapter 14, the effects of a mutation may never be seen when the wild-type gene is present. This allows the population to build up an extensive supply of mutations and carry them along in hidden form from one generation to the next. These mutations may be useless or detrimental to the population, but as conditions change and the population must adapt itself, they may become highly beneficial. Furthermore, through the sexual process new combinations of genes are being produced continually, and some of them may have great selective advantages in the proper environment. Therefore, the combination of diploidy and sex provides a major bank of variability in a population and a mechanism of continuing survival and further evolution. It is not hard to understand why the major groups of organisms have become diplontic and evolved complex sexual mechanisms. In Chapter 14 we will see how we can understand these systems genetically and we will see how sexual mechanisms operate in more detail.

14

genetic
analysis of
eucaryotes

Once you understand the mechanism of meiosis, all of eucaryotic genetics follows quite directly. In this chapter we will examine the patterns in which various characteristics are transmitted from generation to generation in eucaryotic organisms, starting with the simplest cases. Although the problems we encounter later are more complex and must often be solved with less information than is available for the earlier ones, you will see that we never have to make a big conceptual jump from one case to another.

A. Haploid systems

14-1 ANALYSIS OF ORDERED TETRADS

As we promised before, the perfect elementary lesson in genetics is provided by fungi like *Neurospora* and *Sordaria,* which belong to the class of **ascomycetes.** The zygote in ascomycetes develops immediately into a closed sac or **ascus** where meiosis takes place, and in these genera the ascus is long and narrow so the products of each meiotic division remain together and form a linear, ordered tetrad (Fig. 14-1). After the second meiotic division, each cell in the tetrad undergoes one mitotic division, so the ascus contains a set of eight ascospores; but since each pair of ascospores is identical, from now on we will simply write down the sequence of the tetrad from which they come, treating a pair of spores as one spore.

The ascospores can be dissected out of an ascus without changing their sequence. When placed on nutrient agar they will grow into fresh mycelia whose characteristics can be determined. Moreover, we can often work with genetic markers that affect the color of the spores themselves, so we can tell immediately what markers they carry. For the sake of simplicity, let us begin with a marker w, which makes white spores, in contrast to w^+, which makes black spores. We will cross a strain of *Neurospora* carrying w with one carrying w^+ and examine the asci that emerge in the light of the meiotic events that produce their ascospores.

The marker w must reside somewhere on a chromosome and w^+ must reside at the same locus on a homologous chromosome. These two chromosomes are put into the same nucleus at the time of syngamy; they pair up during prophase I of meiosis and at metaphase I they separate from one another. At anaphase I, each chromosome goes to one pole of the cell without dividing its centromere, and at this point the chromatids bearing w have separated from those bearing w^+ (Fig. 14-2). After a second meiotic division there are four nuclei, the two in one half of the

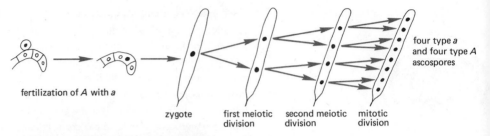

fertilization of *A* with *a*

zygote   first meiotic   second meiotic   mitotic
division   division   division

four type *a*
and four type *A*
ascospores

FIGURE 14-1

Segregation of meiospores in Neurospora *to
make an ordered tetrad. The photograph shows
mature asci that David R. Stadler obtained
in working with a mutant that produces light-
colored spores rather than normal black spores.
Notice the different arrangements of spore pairs.*

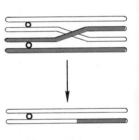

FIGURE 14-3

*Reciprocal exchange
between chromatids of
homologous
chromosomes during
chiasma formation
in meiosis.*

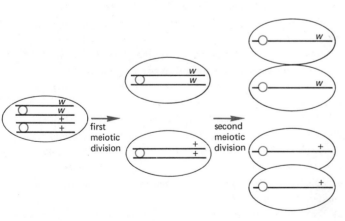

first
meiotic
division

second
meiotic
division

FIGURE 14-2

Segregation of chromatids bearing markers for white and black spore color.

ascus carrying *w* markers and the two in the other half carrying *w⁺*; and after the mitotic division has occurred and the spores have had time to develop their color, there are four black ones in one half of the ascus and four white ones in the other half. Notice that the black spores may be in either the proximal or the distal half of the ascus, since the chromosome marked *w⁺* is as likely to go one way as the other at anaphase I. In either case, just as long as there are four black spores in one half and four white ones in the other half we say that the color markers segregated from one another at the first meiotic division. We will abbreviate this as **M$_I$ segregation.**

However, when we examine many asci from a cross like this we find some with a different pattern, such as black-white-black-white (referring to *pairs* of spores, remember) or black-white-white-black. How can we explain these asci? Remember that during diplonema and diakinesis we observed that the chromatids of each bivalent are involved in tight, intimate interactions that we called chiasmata. All of the available evidence indicates that chiasmata are crossover points, in which there is a physical break and a *reciprocal* recombination so that each one acquires one or more pieces of the other (Fig. 14-3). Suppose such a crossover occurs between one chromatid of each homologue in the cross of the markers *w* and *w⁺*:

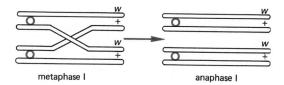

metaphase I   anaphase I

Now each chromosome bears one chromatid marked *w* and one marked *w⁺*; when the chromosomes separate at anaphase I, the two markers do not segregate from one another. Segregation does not occur until anaphase II; and when the spores mature so their color can be seen, each *half* of the ascus will contain a pair of black spores and a pair of white spores. Within each half, the spore pairs may be oriented in either direction, of course, since the chromatids are just as likely to go one way as the other at anaphase II. In any case, when we see two black spores and two white spores in each half of the ascus we say that the alleles segregated from one another at the second meiotic division **(M$_{II}$ segregation).**

We therefore find six types of asci in a cross of *w* by *w⁺*:

1	2	3	4	5	6
w	*w⁺*	*w*	*w⁺*	*w*	*w⁺*
w	*w⁺*	*w⁺*	*w*	*w⁺*	*w*
w⁺	*w*	*w*	*w⁺*	*w⁺*	*w*
w⁺	*w*	*w⁺*	*w*	*w*	*w⁺*

Types 1 and 2, which should be about equal in number, result from M$_I$ segregation. The other four result from a crossover and subsequent M$_{II}$ segregation. Therefore the fraction of asci showing M_{II} segregation is a measure of the distance between the locus and its centromere.

Theorem: If a locus is sufficiently close to its centromere to ignore multiple crossover events, the number of map units between the locus and the centromere is half the fraction of tetrads showing M_{II} segregation.

Proof: The frequency of recombination between two markers, *R*, is defined as the fraction of recombinants emerging from a cross. In this case, individuals that may or may not be recombinant are the haploid cells emerging from a meiosis. In order to have a second marker, we place an imaginary marker *x* or x^+ on the centromere of each chromosome:

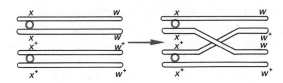

Let *r* be the fraction of meiotic cells in which a crossover occurs between the *x* and *w* loci. Every crossover between *x* and *w* produces a tetrad in which *w* shows M_{II} segregation, as

$$x \quad w$$
$$x \quad w^+$$
$$x^+ \quad w$$
$$x^+ \quad w^+$$

but only half of the cells in this tetrad are recombinant. Therefore $R = 0.5r$.

In tetrad analysis with ordered tetrads, we treat the centromere like another marker; the centromere may be defined genetically as the unique marker that always shows M_I segregation. In principle, if we are willing to take the trouble we can establish very complete maps for organisms with ordered tetrads and determine precisely where each marker is on its chromosome. However, it is not always worth the extra work involved; while these ordered tetrads make perfect didactic tools and are useful in certain serious genetic studies, most genetic work is done with other systems. But before going on to the other systems, let us see what else can be learned from ordered tetrads.

We will consider a cross between strains carrying two alleles at each of two loci, first on the assumption that the loci are not linked and second on the assumption that they are. In the first case, when there are no crossovers between a marker and its centromere, the two markers behave entirely independently of one another. At the first meiotic division, each cell receives either a pair of chromatids marked *A* or a pair marked *a*, because of the behavior of one pair of homologues, and receives either a pair of chromatids marked *B* or a pair marked *b*, because of the behavior of the other homologue:

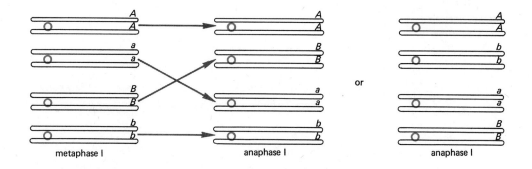

metaphase I   anaphase I   or   anaphase I

Therefore the tetrad formed after second meiotic division can have only two types of cells in it; it is called a **ditype.** However, the two markers may be in either the parental or the recombinant combination, and these are designated

A B	A b
A B	A b
a b	a B
a b	a B
parental ditype (PD) or	**nonparental ditype (NPD)**

If the two markers are really unlinked, the frequency of PD should equal the frequency of NPD. Notice that this situation is exactly the same as the ones we discussed in Chapter 5 using a phage model.

Now suppose we consider the same situation, with the markers unlinked, except that we permit a crossover to occur between the *B/b* locus and its centromere:

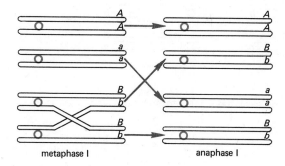

metaphase I   anaphase I

At the first meiotic division, each cell still gets either a pair of chromatids marked *A* or a pair marked *a*, but each cell also gets one chromatid marked *B* and one marked *b*. The tetrad that is formed after the second meiotic division therefore contains four types of spores and is called a **tetratype (T):**

A B
A b
a B
a b

In principle, we can also obtain a tetratype from a crossover between the *A/a* locus and its centromere. In general, from a cross of this sort we can obtain three pieces of information:

1. The fact that PD = NPD shows that the markers are unlinked.
2. The frequency of tetrads in which *B* shows M_{II} segregation gives the distance between *B* and its centromere.
3. The frequency of tetrads in which *A* shows M_{II} segregation gives the distance between *A* and its centromere.

Now let us consider the case in which the loci are linked. They might be on either arm of the chromosome, and it is interesting to consider first the case in which they are on the same arm:

A crossover can occur either between the centromere and *A* or between *A* and *B*. In the first case, the single crossover will lead to M_{II} segregation for *A* and *B* together:

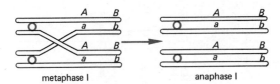

In the second case, the single crossover will lead to M_{II} segregation for *B* alone:

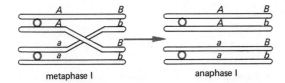

Therefore we should notice the following points that will emerge from the data:

1. The fact that PD > NPD shows that the markers are linked.

2. The frequencies of M_{II} segregation for A and B are still good measures of the distances between these loci and the centromere.

3. A rather large frequency of tetrads in which A and B show simultaneous M_{II} segregation indicates that they are on the same arm of the chromosome.

If the loci are on opposite arms of the chromosome, there will be very few tetrads in which they show simultaneous M_{II} segregation. Notice, incidentally, that we could describe recombination in these crosses entirely in terms of PD, NPD, and T frequencies, but since we have the more informative data on M_I versus M_{II} segregation, it is more reasonable to use these data.

example 14-1 A white spore marker in *Sordaria* shows M_{II} segregation in 24 out of 100 asci; how far is the marker from its centromere? From the theorem we proved above, the marker is 12 map units from the centromere, since it shows M_{II} segregation in 24% of the asci.

exercise 14-1 Mating type in *Neurospora* is designated A or a. In a cross between two strains of *Neurospora*, 16.2% of the asci show M_{II} segregation. How far is mating type from its centromere?

exercise 14-2 In another *Neurospora* cross, the marker *pale color*, p, shows M_{II} segregation in 26.0% of its asci. How far is p from the centromere?

exercise 14-3 In a cross involving the marker *fluffy*, f, and its allele f^+, the following patterns were observed upon dissection of 110 asci:

PATTERN	ASCI	PATTERN	ASCI
$+ + f\ f$	20	$f\ f + +$	23
$+ f + f$	25	$f + f +$	7
$+ f\ f +$	16	$f + + f$	19

What is the distance between f and the centromere?

exercise 14-4 A marker q is used in a cross with q^+. Only one out of 82 asci examined shows M_{II} segregation. What do you conclude about the map position of q?

exercise 14-5 In a cross of a pale mutant of mating type A ($p\ A$) with a wild type of mating type a ($+ a$) in *Neurospora*, the following ascus segregations were observed:

$+ a$	$+ a$	$+ a$	$+ a$	$+ A$	$+ A$		
$+ a$	$p\ A$	$+ A$	$p\ A$	$p\ a$	$p\ a$		
$p\ A$	$+ a$	$p\ a$	$+ A$	$+ a$	$+ A$		
$p\ A$	$p\ A$	$p\ A$	$p\ a$	$p\ A$	$p\ a$		
410	120	60	2	5	3	Total:	600

Draw a map for these markers. Consider the distance between each marker and the centromere and then consider whether the markers are on the same arm or different arms.

example 14-2 Consider a cross of *Neurospora* strains *A ade* (adenine auxotrophy) × *a ade⁺*. The following numbers of asci are obtained:

A ade	A +	A +	A ade	A ade	A +	A +
A ade	A +	A ade	a ade	a +	a ade	a ade
a +	a ade	a +	A +	A ade	A +	A ade
a +	a ade	a ade	a +	a +	a ade	a +
1014	0	1	97	49	0	0

In tabulating these results, the order of ascospores within each *half* of the ascus has been ignored, since the only information we need for this analysis is the number of asci in which each marker segregates at the first or second meiotic division. We can separate the asci into four classes on this basis and calculate a frequency for each:

$$M_I \text{ for } A, M_I \text{ for } ade: \quad w = 1014/1161 = 0.8740$$
$$M_I \text{ for } A, M_{II} \text{ for } ade: \quad x = 1/1161 = 0.0001$$
$$M_{II} \text{ for } A, M_I \text{ for } ade: \quad y = 97/1161 = 0.0836$$
$$M_{II} \text{ for } A, M_{II} \text{ for } ade: \quad z = 49/1161 = 0.0422$$

We must now find a map order consistent with these data. From the large value of *w*, the markers are clearly linked. Since *z* is also quite large, it appears that a single crossover can make both markers segregate at M_{II} simultaneously, so they are probably on the same arm. Mating type shows M_{II} segregation with a frequency $y + z = 0.1258$, while *ade* shows M_{II} segregation with a frequency $x + z = 0.0423$, so the former is 6.29 map units from the centromere and the latter is 2.13 map units from the centromere. As an additional check, the two markers separate from each other in $x + y = 0.0837$ of the asci and are therefore 4.18 map units from each other. These numbers give a consistent map.

14-2 A BIT OF BACKGROUND

The first serious genetic experiments ever performed on organisms that undergo a regular meiosis were those of Gregor Mendel, around 1865. Mendel used peas, which are diplontic and therefore show the complication of dominance. Nevertheless, by a penetrating analysis of his data he showed that the characteristics he examined (green versus yellow color, smooth versus wrinkled peas, etc.) were inherited as if they were determined by genetic elements that segregate from one another at random. Mendel's results went unnoticed for a long time; they were rediscovered and confirmed in about 1900 by Karl Correns, Hugo de Vries, and A. von Tschermak. Shortly thereafter, Theodor Boveri and William S. Sutton pointed out that the behavior of the chromosomes in meiosis is exactly parallel to the presumed behavior of the genetic elements.

Therefore they postulated that these elements (genes) are carried on the chromosomes. Their hypothesis gained support from many experiments and is now an established principle. It must be emphasized that the behavior of the eucaryotic genome in meiosis is the physical basis for the genetic phenomena we observe. Whenever possible, we will relate the genetic results to chromosome behavior. The study of chromosome behavior in itself might well have led to Mendel's results if these had not been obtained by breeding experiments.

Mendel formulated two laws to summarize his data. They are only partially useful in a modern genetic analysis, but they are worth looking at briefly. The *law of segregation* states that the alternative versions of any character (alleles) *segregate* from one another in reproduction and behave as independent units. We now understand that this is a consequence of the *separation* of homologous chromosomes in meiosis. This law is clearly true except in some very unusual cases where bivalents may not separate properly.

The *law of independent assortment* states that the units which carry two different characters always behave independently of one another and *assort* at random among the progeny. This law is only true if the genes for the characteristics in question occur on nonhomologous chromosomes; linked genes, by definition, do not follow the rule. Mendel arrived at this law on the basis of his crosses, but he was blessed with the incredibly good luck that the eight characteristics of pea plants that he used are all carried on different chromosomes. This allowed him to perform simple crosses and arrive at simple results that he could analyze.

The first exception to the law of independent assortment appeared in some of the earliest experiments performed at the turn of the century. Thomas Hunt Morgan, around 1910, observed deviations from a random assortment of markers in *Drosophila* and suggested that the markers were carried on homologous chromosomes. The analysis of a number of markers revealed four linkage groups, presumably corresponding to the four chromosome pairs in *Drosophila*. Morgan's student Sturtevant then showed that one can unambiguously map the markers on a chromosome by assuming that they occupy unique positions and that the frequency of recombination between them will be proportional to distance on the chromosome. This work layed the foundations for all future mapping experiments.

14-3 ANALYSIS OF UNORDERED TETRADS

For many experiments, it is sufficient to have tetrads in which the products of meiosis do not stay lined up in a row. In these, we can only count the number of PD, NPD, and T tetrads, but this gives a great deal of information.

First, it should be clear that all of the spores in NPD tetrads are recombinant, by definition, as are half of those in T tetrads. Therefore the frequency of recombination between two markers is

$$R = \frac{NPD + 0.5T}{PD + NPD + T} \tag{14-1}$$

For example, mutants unable to make *para*-aminobenzoic acid (*pab*) and other mutants unable to make thiamine (*thi*) can be isolated in *Chlamydomonas*. A cross of *pab thi*$^+$

× *pab*⁺ *thi* yields 151 tetrads, of which 83 are PD, 2 are NPD, and 66 are T. Satisfy your-self that there is 23% recombination between these two markers.

Given two markers in a cross, the first question that must be answered is, simply, are the markers linked? Remember that two markers are said to be unlinked if they recombine with a frequency of 50% and they cannot be shown to be linked to a com-mon marker. Given 50% recombination, it is still possible that the two markers are very far apart on the same chromosome, so we first require some evidence that this is not the case. We will obtain PD, NPD, and T tetrads from a cross. Since T tetrads con-tain equal numbers of parental and recombinant spores, it is clear that any evidence of linkage must come from a comparison of PD and NPD tetrads, for if the two markers are linked PD > NPD.

In some cases, the excess of PD over NPD tetrads will be obvious and one can be quite sure that there is some linkage, but if PD ≈ NPD a more definitive test is neces-sary. The most certain way to establish linkage is to fill in the gap between the markers with a number of other, more closely linked markers. However, there are at least two ways to reduce our uncertainty. First, there is the χ^2 test which we have used before. This is obviously an appropriate test and it is often used in genetic work. Second, the frequency of T tetrads can provide some evidence *against* linkage, even though it cannot provide evidence *for* linkage. If the PD/NPD ratio is close to 1, the markers may really be unlinked and the T frequency may have any value, depending on the distance between each marker and its centromere. However, if the markers are on the same chromosome a tetratype can only result from a crossover between them, and we shall prove shortly that if they are far apart the T frequency must approach $\frac{2}{3}$. Therefore, a low frequency of tetratypes argues against linkage.

When more than two markers are used in a cross, the recombination frequency between any two of them can be determined by scoring each tetrad according to its type for those two alone, ignoring all others. Assuming that the distances between linked markers are approximately additive, a reasonably consistent map can be drawn. It must always be assumed that the rarest classes of tetrads are due to the most un-likely combinations of crossovers. We will now consider some of the factors that can contribute to difficulties in interpreting recombination data.

exercise 14-6 Suppose a cross is made between individuals carrying two markers and the number of PD and NPD tetrads is *x* and *y*, respectively. On the hypothesis of no linkage, prove that $\chi^2 = (x - y)^2/(x + y)$.

exercise 14-7 In a two-factor cross, suppose the number of PD and NPD tetrads are 417 and 397, respectively. Calculate χ^2 and determine the approximate probability that the markers are unlinked.

14-4 CROSSING OVER AND MAPPING FUNCTIONS

There are at least three phenomena that could lead to difficulties in understanding recombination. First, some crossovers might occur between sister strands of a bivalent; these lead to no genetic exchange that we can observe, but they might interfere with other crossovers. We will assume that sister-strand crossing over does not occur and try to justify the assumption later. Second, one crossover might make a second nearby crossover less likely (or more likely, for that matter); this is called **chromosome inter-**

ference. We have already seen examples of this in phage recombination and should be prepared for it in eucaryotes. Third, the occurrence of a crossover between two of the strands in a bivalent might alter the probability that the same two will cross over again or that they will exchange with the other two; this is called **chromatid interference.** The evidence suggests that this does not occur and we will begin by making the assumption that it does not.

Let us consider two markers, *a* and *b,* located on the same arm of a chromosome. Since we assume there is no sister-strand crossing over, a single exchange between them will lead to a tetratype tetrad and hence to 50% recombination.

Suppose there is now a second crossover in the same region. This can happen in four ways, as shown in Fig. 14-4. If there is no chromatid interference, 25% of the second crossovers will occur between the same two strands, thus cancelling each other and making a PD tetrad. Another 25% will be four-strand doubles; these will yield NPD tetrads. The remaining 50% will consist of the two kinds of three-strand doubles, which will yield T tetrads.

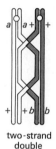
two-strand
double

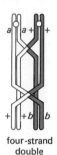
four-strand
double

FIGURE 14-4
The three types of double crossovers that can occur between two markers on the same arm of a chromosome. Trace the chromatids to verify the descriptions of tetrads given in the text.

three-strand doubles

We can continue this analysis for any number of crossovers; the frequencies of the three tetrad types will change as follows:

NUMBER OF CROSSOVERS	PD	T	NPD
0	1	0	0
1	0	1	0
2	1/4	1/2	1/4
3	1/8	3/4	1/8
4	3/16	5/8	3/16
5	5/32	11/16	5/32
many	1/6	2/3	1/6

Thus, the tetratype frequency fluctuates considerably for small numbers of crossovers, but it approaches $\frac{2}{3}$ for larger numbers. Let us say that the distance between a and b is x map units ($x\%$ recombination). This means that the frequency of recombination between the two sites is x and therefore the frequency of crossovers is $2x$. If crossovers are Poisson distributed on the chromosome, the probability of n crossovers in the region between a and b is the nth term of the Poisson distribution and the frequency of tetratypes is

$$P_n = \frac{2}{3} \sum_n \frac{(2x)^n}{n!} e^{-2x} \left[1 - \left(-\frac{1}{2} \right)^n \right] \tag{14-2}$$

$$= \frac{2}{3}(1 - e^{-3x})$$

exercise 14-8 Derive the simplified form of equation 14-2.

exercise 14-9 Write analogous expressions for the frequencies of PD and NPD tetrads and simplify them.

exercise 14-10 By combining these expressions, prove that the frequency of recombination for two markers that are far apart on a chromosome (i.e., when n is large) is $\frac{1}{2}$ (remember equation 14-1).

Equation 14-2 can be used as a kind of mapping function, but if you draw a graph of P_n against x you will see that it is hardly useful if x is greater than 1; it is approximately linear only for short distances and, of course, can be used only when the frequencies of tetrad types can be determined.

We can derive a similar expression for a more general cross in which only the frequency of recombination can be determined. This is the case in all diplontics and most haplontics, where, for example, one simply counts the number of spores of each kind in a random population, where all the tetrads from many meioses are mixed. If we again say that the average number of crossovers between two sites is $2x$, then the probability that no crossover occurs is e^{-2x} and the probability that at least one occurs is $1 - e^{-2x}$. Since half of the gametes from a meiosis in which a crossover has occurred are recombinant, the probability of a recombination is

$$P = \frac{1}{2}(1 - e^{-2x}) \tag{14-3}$$

This is the Haldane equation (5-1) we derived before. As before, the frequency of recombination is proportional to map distances for short distances only, approaching 50% for long distances. Again, in dealing with eucaryotes we must use markers that are moderately close together to construct a map.

Incidentally, we can use some of the above results as evidence that there is no sister-strand crossing over. There are four possible exchanges that do not involve sister strands and two that do; therefore, if there were sister-strand crossing over, a single crossover would result in recombination only 66.6% of the time, rather than 100%. Furthermore, the frequency of tetratypes would never rise above $\frac{2}{3}$; however, since this frequency has been found to be more than $\frac{2}{3}$, sister strands apparently do not exchange.

Other experiments have been designed to measure the ratios of two-strand:three-strand:four-strand doubles, which should be 1:2:1 if there is no chromatid interference. With some doubtful exceptions, most measured ratios are approximately what one would expect. If there is any deviation, it is toward an excess of two-strand doubles, which is negative chromatid interference. The assumptions we made above are therefore reasonably well justified by the data.

We have already indicated how a map can be established by certain simple crosses. It is difficult to obtain consistent data from two-factor crosses alone, since a variety of extraneous influences cause differences in the frequencies of recombination determined under supposedly identical conditions. A cross using three or more factors allows some comparison between different experiments. A three-factor cross also permits one to order the three markers unambiguously by applying the fundamental theorem of three-factor mapping—the class of recombinants for the middle marker must always be the least frequent.

Even though we have concentrated on ordered and unordered tetrad analysis for pedagogical reasons, you should realize that a great deal of genetic analysis can be done without examining individual tetrads. Most work is probably done now with random spores because genetic analysis with them provides twice as much information per spore as does analysis with tetrads. This is because recombination is generally reciprocal, so the determination of the genotypes of half the spores in a tetrad automatically determines the other half, whereas each spore in a random mixture can be considered an independent observation. The extra work necessary for tetrad analysis is generally performed only when special problems demand it.

B. Diplontic systems

Having seen how a single genome behaves in a cell and how its structure can be determined, we can now discuss some of the complications that appear when two different genomes are put together in a stable association in a diploid organism.

14-5 DOMINANCE AND RECESSIVENESS

If a man and a woman have many children (definitely *not* recommended in these crowded times), about half of them will be boys and half girls. Thus, the characteristic of sex, which is the equivalent of mating type in the protista, behaves according to the simplest Mendelian principles, and the factor (or factors) that determines it seems to segregate just like the mating-type character in the organisms we have already studied.

Now, suppose the man comes from a line of brown-eyed people and the woman from a line of blue-eyed people. (If we were talking about domestic animals, we would call these people **purebred,** at least for this one characteristic.) If their genomes behaved like those we have already studied, about half of their children would be brown-eyed and half would be blue-eyed. However, all of their children are brown-eyed; this is an unexpected result, which we must now try to explain.

To avoid some social complications that might result from carrying this experiment through another generation, let's switch to a laboratory animal—the common fruit fly,

Drosophila. Normal flies have red eyes, but purple-eyed mutants (*pr*) are occasionally found. We select a red-eyed male and a purple-eyed female and mate them. We say that they belong to the **parental (P) generation** and that the first generation produced by breeding them is the **first filial generation (F₁).** Later generations will be called **F₂, F₃,** and so on; all of these generations will be produced by **inbreeding** — crossing males and females in each generation to one another, without introducing any new individuals from outside, and since this involves many random matings between cousins or between brothers and sisters, we obviously cannot do these experiments with humans. In this experiment, we find that all of the F_1 flies have red eyes. However, when we mate the F_1 flies to one another we get an F_2 population of both red-eyed and purple-eyed flies, in the approximate ration of 3 red:1 purple.

This experiment shows that the marker for purple eyes was not lost somewhere during these matings, but was merely hidden. The allele for red eyes is *dominant* to the allele for purple eyes (the *recessive* allele); even though an individual may carry both alleles, only the dominant one is evident. We have encountered this before; for example, we made artificial merodiploids for certain operons in bacteria in order to see whether a given allele is dominant or recessive. In diploid eucaryotes, diploidy is a regular part of the sexual cycle and we must generally be concerned with the dominance or recess-iveness of particular alleles. In this case, we can generally understand the dominance of pr^+ over pr by assuming that there is a biosynthetic pathway for the red eye pigment and that the *pr* gene informs one enzyme of the pathway; a mutation in this gene blocks the synthesis after a purple pigment has been made, and the pigment accumu-lates in cells that contain only the *pr* allele to make purple eyes. However, if an individual has both a pr^+ allele and a *pr* allele, the former can make a good enzyme and the latter will have no visible effect.

Some special terms and symbols must be used for diploids. We would like to denote the two homologous chromosomes that carry every pair of alleles by a little picture like this:

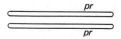

However, it is clumsy to introduce this picture everywhere, so we simplify the two lines denoting the chromosomes to a single line, and then we often write this whole thing with a diagonal slash instead of a horizontal line, in just the same way that we write fractions. We can denote the purple-eyed female by *pr/pr,* for example. When we consider linked genes, we will write one linkage group before the slash and the other linkage group after, or we will write this as a group of symbols above and below a horizontal line if the other notation will cause confusion. Now notice that the purple-eyed female carries identical alleles at the *purple* locus; she is said to be **homozygous.** An individual with the genotype pr/pr^+ is said to be **heterozygous.** Naturally, an individual may be homo- or heterozygous for every locus independently.

If an individual has one copy of a dominant allele, we often cannot tell whether he is a homozygote or a heterozygote, and sometimes we simply don't care. We then use a dot notation for the other allele; for example, the notation $pr^+/\cdot$ means that the second allele may be either pr^+ or *pr.*

To denote dominance and recessiveness, we generally capitalize the first letter in the symbol for a dominant allele. For example, humans have a set of blood types denoted by A, B, AB, and O. The gene for this characteristic is designated I (for isoantigen); it has three alleles, I^A, I^B, and i^O. I^A and I^B are **codominant** with one another (they are expressed equally) and both of them are dominant to i^O. Therefore, a person with the genotype I^A/I^B has the phenotype AB and a person with the genotype I^B/i^O has the phenotype B, just like one who has the genotype I^B/I^B.

Now let us go back to the original example, eye color in *Drosophila*, and consider the matings among F_1 flies, which are all heterozygous, $+/pr$, because their fathers could produce only sperm carrying the pr^+ allele and their mothers could produce only eggs carrying the pr allele. The $+$ and pr alleles segregate from one another during meiosis in the F_1, so every F_1 male produces some sperm carrying $+$ and some carrying pr and every F_1 female similarly produces two kinds of eggs. In a large enough population, all the matings will be random and there will be equal numbers of the four kinds of gametes, so we may consider the F_2 generation to be formed by randomly selecting gametes out of a common pool. There is a probability of $\frac{1}{4}$ that two gametes carrying the $+$ allele will fertilize each other, so $\frac{1}{4}$ of the F_2 progeny will be homozygous $+/+$. There is a probability of $\frac{1}{4}$ that a $+$ sperm will fertilize a pr egg and a probability of $\frac{1}{4}$ that a pr sperm will fertilize a $+$ egg, so $\frac{1}{2}$ of the F_2 progeny will be heterozygous $+/pr$, but will still have red eyes. Finally, there is a probability of $\frac{1}{4}$ that a pr sperm will fertilize a pr egg, so $\frac{1}{4}$ of the progeny will be homozygous pr/pr and will have purple eyes. The ratio of 3 red:1 purple in the F_2 generation is easily explained.

There is a very important and useful operation, the **test cross,** that eliminates some of the complications of dominance. In this experiment, individuals of unknown genotype or with some particularly interesting genotype are crossed to individuals that are homozygous recessive for all loci being used. The latter can produce only gametes carrying recessive alleles, and so the phenotypes of all the progeny are effectively determined by gametes from the individuals being tested. In this case, we may cross the F_1 individuals with homozygous pr/pr flies; the latter can produce only pr gametes, while the former should produce $+$ gametes and pr gametes in equal numbers. Therefore, half of the progeny from this cross should be $+/pr$ red-eyed and half should be pr/pr purple-eyed. If approximately equal numbers of red- and purple-eyed flies are obtained, the genotype of the F_1 flies is confirmed. The beauty of the test cross is that it effectively reduces diploid problems to haploid problems.

Mutant flies can also be found that have *ebony* body color (*e*) instead of the normal gray color. Ebony is recessive to wild type, and a cross of +/+ by *e/e* individuals gives an F_1 generation that is phenotypically wild type. By appropriate crosses, we can make a strain of flies that is double mutant, *pr/pr e/e,* and cross the males to wild-type females. The F_1 generation, of course, is phenotypically wild type. Suppose we inbreed the F_1, collect the progeny, and find 172 wild-type, 57 purple-eyed, 63 ebony, and 18 purple-eyed ebony flies. The ratios of these four classes are very close to 9:3:3:1; let's see how we can account for this.

We assume that the two markers are on different chromosomes and assort independently. The P males can produce only *pr e* sperm and the P females can produce only + + eggs, so all of the F_1 must be heterozygous, +/*pr* +/*e*. However, the F_1 flies can make four kinds of gametes: + +, + *e, pr* +, and *pr e*. Therefore, there are 16 possible matings between these gametes; we could calculate the probability of each combination, but it is easier to tabulate the combinations as shown in Fig. 14-5. In nine combinations, there is at least one + allele at each locus, so these individuals will be phenotypically wild type. In three combinations, there are two *pr* alleles, making purple-eyed flies, and in another three there are two *e* alleles, making ebony flies. In the one remaining combination, all four recessive alleles are combined, making a purple-eyed ebony fly. Therefore, the observed ratio can be accounted for again on the basis of random events in meiosis and mating.

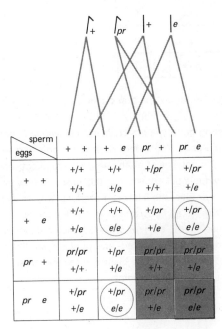

FIGURE 14-5
Tabulation of the 16 possible recombinant types in a two-factor cross. Notice where the four different phenotypes occur.

sperm / eggs	+ +	+ e	pr +	pr e
+ +	+/+ +/+	+/+ +/e	+/pr +/+	+/pr +/e
+ e	+/+ +/e	+/+ e/e	+/pr +/e	+/pr e/e
pr +	pr/pr +/+	+/pr +/e	pr/pr +/+	pr/pr +/e
pr e	+/pr +/e	+/pr e/e	pr/pr +/e	pr/pr e/e

A test cross in this case is also revealing. We cross the F_1 males to purple-eyed ebony females; the males can still produce four kinds of sperm, but they can fertilize only *pr e* eggs and there can be only four types of progeny: +/*pr* +/*e*; +/*pr e/e*; *pr/pr* +/*e*; and *pr/pr e/e*. These should occur in equal numbers. Since the parents were wild type and double mutant, the ebony flies and the purple-eyed flies are recombinants, and the frequency of recombination between *pr* and *e* is 50%. This supports the assumption that *pr* and *e* are unlinked. In fact, the test cross is the one normally used to establish a map in a diploid like *Drosophila*. In Section 14-7 we will see how it can be used for linked markers.

exercise 14-11 Use the χ^2 test to determine how well the numbers given above in the *pr e* cross fit the expected ratio of 9:3:3:1.

exercise 14-12 In peas, round seeds (*R*) are dominant over wrinkled seeds (*r*), yellow seeds (*Y*) are dominant over green (*y*) seeds, and colored flowers (*C*) are dominant over white flowers (*c*). In a cross between homozygous round, yellow peas and wrinkled, green peas, describe the F_1 and give their genotype. If the F_1 plants are inbred, what will be the appearance of the F_2 plants and what will be their genotypes?

exercise 14-13 Suppose a homozygous round, yellow, colored pea is bred to a wrinkled, green, white pea.
(a) Describe the F_1 and give their genotype.
(b) If we inbreed the F_1 and obtain an F_2, what is the probability that we will obtain plants with these phenotypes:

round, yellow, colored
round, green, colored
wrinkled, green, white
wrinkled, yellow, colored
wrinkled, green, colored

exercise 14-14 If the following genes are all unlinked to one another, how many kinds of gametes can be produced by an individual of genotype *A/a B/b C/c d/d E/e F/F*? If two individuals with this genotype mate, how many different genotypes could be found in their progeny? How many different phenotypes?

exercise 14-15 In man, the ability to taste phenylthiocarbamide (PTC) is due to a dominant gene *T*. Brown eyes (*B*) are dominant over blue eyes (*b*).
(a) What fraction of the offspring of two parents who have genotypes *T/t B/b* would be blue-eyed tasters; blue-eyed nontasters?
(b) What fraction of the children of two parents of genotypes *T/t B/B* and *t/t b/b* would be brown-eyed tasters; blue-eyed nontasters?

exercise 14-16 Mouse fur colors are determined by many genes. The basic dark color is due to a gene *E* for the eumelanin pigment; the recessive allele *e* produces pheomelanin, which has a red-yellow color. The factor *B* modifies *E*; *E/· B/·* is sepia and *E/· b/b* is brown. In a cross of *B/b E/e* individuals by themselves, what progeny do you expect and in what proportions?

exercise 14-17 Another factor *C* modifies the intensity of sepia or brown. *C/·* has 100% color; there are four other alleles that are recessive to *C* but codominant with one another. Assume that this gene makes an enzyme, that the four recessive alleles are mutants that make a bad enzyme with less activity, and that the activities in these mutants are all additive. From the following series, determine approximately what percent activity each allele has:

c^k/c^k	c^k/c^d	c^k/c^r	c^k/c^a c^d/c^d	c^d/c^r	c^d/c^a	c^r/c^r	c^r/c^a	c^a/c^a
80%	60%	45%	40%	25%	20%	10%	5%	0%

14-7 LINKAGE AND MAPPING

The test-cross technique can be easily applied to determine the linkage between any two markers, whether the alleles we use are dominant or recessive. In *Drosophila* and a few other organisms there is one unusual feature; crossing over does not occur in males, so we will always use homozygous recessive males as mates for the females of the strain we wish to test.

Drosophila mutants can be found that have *rough eyes* (*ro*) instead of the normal smooth eyes. If we cross purebred rough-eyed flies (*ro/ro*) to ebony flies (*e/e*), the F$_1$ flies are all wild type. If we mate these to get an F$_2$ generation, some rough-eyed, ebony flies appear which we can use for a test cross with the F$_1$. In this test cross, we obtain these numbers:

$\dfrac{ro\ +}{ro\ e}$	$\dfrac{+\ e}{ro\ e}$	$\dfrac{+\ +}{ro\ e}$	$\dfrac{ro\ e}{ro\ e}$	
168	176	41	45	total = 430

The frequency of recombination between *ro* and *e* is clearly 86/430 = 20%, so the two markers are 20 map units apart.

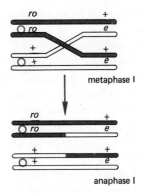

metaphase I

anaphase I

Again you can see how valuable the test cross is. The entire problem of mapping is effectively reduced to the haploid situation, since we can determine immediately the frequency of each combination of alleles in the gametes of the strain being tested and we can apply all the principles we have studied before in phage and similar haploid situations. Convince yourself that this is so by solving some of the following problems.

exercise 14-18 In a cross of *R b/r B* × *r b/r b*, the following results are obtained:

R b/r b	1040	*R B/r b*	6
r B/r b	1050	*r b/r b*	4

What is the frequency of recombination between these loci?

exercise 14-19 In a cross of *+ m +/w + p* × *w m p/w m p*, the following results are obtained:

+ m +/w m p	357	*w + +/w m p*	93
w m +/w m p	47	*w + p/w m p*	353
w m p/w m p	4	*+ + +/w m p*	6
+ m p/w m p	97	*+ + p/w m p*	43

Construct a map for these three loci.

exercise 14-20 The cross *Dd Ee Ff* × *dd ee ff* was performed and the following progeny were obtained:

Dd Ee Ff	29	*dd ee ff*	21
Dd Ee ff	235	*dd ee Ff*	216
Dd ee ff	210	*dd Ee Ff*	259
Dd ee Ff	27	*dd Ee ff*	23

Consider whether any of these loci are linked to one another and construct the best map you can.

14-8 SEX DETERMINATION

In the simplest eucaryotes, mating types appears to be determined by only a single locus that behaves like any other locus in crosses. The two mating types may differ only in some subtle factor, such as a chemical difference in their cell surfaces that permits gametes to fuse with each other. This simple system carries over through most of the plant kingdom. Generally each plant is **monoecious** (hermaphroditic) and produces both male and female gametes. In the smaller group of plants that are **dioecious**—one plant producing male gametes and another female gametes—the differences are almost always due to a single locus.

An interesting sexual cycle occurs in many of the fungi called **basidiomycetes,** which includes most of the "higher fungi" such as the mushrooms and shelf fungi we see so often in the woods and fields. A single, haploid mycelium that contains only one kind of nucleus is a **homocaryon;** when two homocaryons of the same species that differ in one or more loci grow together, their hyphae often fuse to form a **heterocaryon** that contains both kinds of haploid nuclei in the same cytosome. However, if the two homocaryons carry the appropriate mating factors (sometimes called incompatibility factors), they may form a special kind of heterocaryon called a **dicaryon,** in which the nuclei of the two parents associate in pairs and divide synchronously, so they remain in a one-to-one ratio. The dicaryon can grow vegetatively as a stable structure, and eventually it forms fruiting bodies in which caryogamy occurs, followed by meiosis and spore formation to complete the sexual cycle.

To maintain the dicaryon, the parental nuclei must undergo a conjugate division as shown in Fig. 14-6, with the formation of clamp connections. This requires that the parents differ at two loci, *A* and *B,* each of which seems to have many natural alleles in the species that have been studied. While the genetics of these systems is extremely complicated, we can generally say that the *A* factor controls nuclear pairing, conjugate division, and initiation of the clamp connection, while the *B* factor controls nuclear migration and the fusion of the hook cell of the clamp with the next-to-last cell of the mycelium. The *A* and *B* factors assort normally, so four types of haploid spores are formed after meiosis in the dicaryon and the system is termed **tetrapolar sexuality.** If we consider only two *A* alleles and two *B* alleles, you can see that there are 16 possible matings between different homocaryons, and only four of them can lead to stable dicaryons.

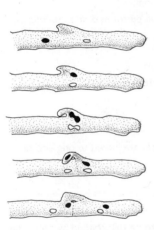

A clamp begins to form on the terminal cell of the hypha.

The two nuclei pair near this connection.

The nuclei divide; one of the black daughter nuclei moves into the clamp.

The nuclei migrate so the terminal cell retains one of each type,

and the clamp connection is finally completed so the subterminal cell retains one of each type.

Eventually the two nuclei in the terminal cell can fuse to make a diploid nucleus

that undergoes meiosis to make a tetrad of four basidiospores.

FIGURE 14-6

Conjugate division in a dicaryon and the mechanism of caryogamy and spore formation.

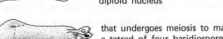

In most animals, there are two distinct sexes with clear morphological differences, and special mechanisms of sex determination have evolved. In the typical diploid animal, most of the chromosomes—the **autosomes**—can be paired with each other to make two sets of identical homologues, and these chromosomes carry most of the genes. There may then be one or more special **sex chromosomes** that determine the sex of the animal; there are two variants of this system. In an XY system (as in man and the other mammals) there are two extra chromosomes, X and Y, that are not homologous. A male carries X + Y, and since he produces two types of sperm, carrying either X or Y, he is **heterogametic.** A female carries two Xs, and since all of her eggs carry X, she is **homogametic.** There is a variant of this system in some insects; the male is heterogametic, but he carries only a single X chromosome and his sperm carry either the X chromosome or no sex chromosome; he is designated XO and the female is still XX.

In the ZW system, the females are heterogametic; there are two nonhomologous chromosomes, Z and W, the female being ZZ and the male ZW. This system occurs in moths and butterflies and in some birds. Other birds, like chickens, have a ZZ-ZO system, in which the female has only a Z chromosome and the male has two Zs.

Simply saying that males and females in each species are characterized by these combinations of chromosomes does not explain what determines maleness or female-ness. Now and then we find individuals with abnormal combinations of sex chromosomes that provide clues to the problem. It is clear that several different sex-determining mechanisms have evolved in different species. For example, in *Drosophila* the Y chromosome is essential for male fertility, but sex is determined by a balance between male factors in the autosomes and female factors in the X chromosome. One haploid set of autosomes contains 1 unit of male determinants and each X chromosome contains $1\frac{1}{2}$ units of female determinants. In a normal AAXY male, maleness outweighs femaleness by 2/1.5, and in a normal female, AAXX, femaleness outweighs maleness by 3/2. Any ratio between these two values, as in an AAAXX triploid, is a sterile intersex. More extreme ratios outside these values produce supermales or superfemales that are generally not viable.

In humans the situation is quite different. Maleness is determined by the Y chromo-some; anyone without a Y is female. People are occasionally found with abnormal numbers of X and Y chromosomes (Table 14-1). These occur because of an abnormality called **nondisjunction** in meiosis, in which a pair of chromosomes fail to separate at anaphase I. Nondisjunction of the X and Y chromosomes during spermatogenesis, for example, would produce a sperm with both X and Y plus a sperm with no sex chromo-somes. Some of the abnormal humans produced from such gametes have distinctive features. XO females exhibit **Turner's syndrome;** they are short, have impaired intelligence, and are sterile, among other features. Males with extra X chromosomes

TABLE 14-1

the range of known human sex aneuploids

NUMBER OF X	NUMBER OF Y CHROMOSOMES		
	0	1	2
1	monosomic XO Turner ♀	disomic XY normal ♂	trisomic XYY almost normal ♂
2	disomic XX normal ♀	trisomic XXY Klinefelter ♂	tetrasomic XXYY Klinefelter ♂
3	trisomic XXX normal or sterile ♀	tetrasomic XXXY extreme Klinefelter ♂	pentasomic XXXYY extreme Klinefelter ♂
4	tetrasomic XXXX super ♀	pentasomic XXXXY extreme Klinefelter ♂	hexasomic XXXXYY extreme Klinefelter ♂
5	pentasomic XXXXX super ♀	hexasomic XXXXXY (not found yet)	heptasomic XXXXXYY (not found yet)

exhibit **Klinefelter's syndrome;** among other things, they have long limbs and are sterile.

It is very easy to get some information about the number of X chromosomes in humans and other animals because all but one of the X chromosomes in each cell apparently becomes inactivated and appears as a dark Barr body in the nucleus. (The dark body was discovered by M. L. Barr.) Barr bodies can be seen easily by staining cells scraped off the inside of the cheek; naturally, every female should have one and every male should have none.

As this is written, the popular press has made a sensation of men who are XYY. These men are characterized by tallness and some other features, and it is claimed that they are abnormally aggressive and violent. Abnormally high percentages of them have been found among institutionalized criminals, and there may be some connection between their extra Y chromosome and their behavior. However, right now there is no firm clinical evidence that XYY men are unusually violent, and it is best to withhold judgement, particularly in view of the social importance of this problem.

exercise 14-21 The inhabitants of Tau Cygni E have a trisexual system, consisting of fatherkins, motherkins, and unclekins in the ratio of 2:5:2, respectively. The motherkins, who are heterogametic, mate first with a fatherkin and then with an unclekin. (To mate with an unclekin alone is considered a sinful perversion.) All individuals are triploid and have three sex chromosomes. What is the simplest system of sex chromosomes that will account for this mating system?

exercise 14-22 The inhabitants of Alpha Centauri B also have three sexes, but they are diploids and have two sets of autosomes plus two sex chromosomes. The sexes occur in the ratio of 1 male:2 females:1 shemale. The female produces ova; the male and shemale produce sperm and either of them can mate with a female. (The principal differences between males and shemales are social and psychological; the anthropologists have not worked this out yet.) What is the simplest system of sex chromosomes that will account for this? (Hint: This is more difficult because the system will be inherently unstable under some circumstances. Try a system of X, X', and Y chromosomes.)

exercise 14-23 Make a square of the 16 possible matings in a tetrapolar sexual system with two alleles at each locus. Which matings will be compatible?

exercise 14-24 What types of gametes and matings are required to produce the more unusual types in Table 14-1?

14-9 SEX LINKAGE

Many genes that have nothing in particular to do with sex are inherited in a rather unusual way because they are located on the sex chromosomes. A classical case in *Drosophila* was discovered by T. H. Morgan. Occasional mutants are found with *white eyes* (*w*). Morgan crossed a red-eyed female to a white-eyed male. All of the F_1 were red-eyed, and when these flies were inbred they produced an F_2 with a 3:1 ratio of red-to-white eyes. However, all of the F_2 females were red-eyed while the males were half red and half white. The reciprocal cross between a red-eyed male and a white-eyed female gave an F_1 generation in which all of the males had white eyes and all of the females had red eyes. When they were crossed to one another, they produced an F_2 with four equal classes of red- and white-eyed males, red- and white-eyed females.

This situation can be understood if the *w* marker is on the X chromosome. Figure 14-7 shows how this marker will be distributed in the two crosses. The important points are that (1) every male gets his X chromosome from his mother and (2) a single *w* allele in a female will be masked by the w^+ allele on the other X chromosome, but in a male a *w* allele will be expressed.

Many human characteristics, including red-green color blindness, are sex linked (which means X linked). A color-blind man obviously carries the *c* mutation on his single X chromosome; a woman can be color blind only if she is homozygous *c/c*, but a normal c^+/c heterozygote is a carrier for the condition and will pass the trait to half her sons and the gene to half her daughters.

The demonstration that the *w* marker for eye color in *Drosophila* is really on the

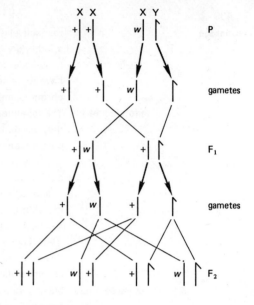

FIGURE 14-7

Sex-linked inheritance of white eye color (w) in Drosophila. Using this diagram as a guide, draw the corresponding pattern of inheritance when a white-eyed female (w/w) is crossed to a red-eyed male (+/Y).

X chromosome was a triumph of Calvin Bridges, one of Morgan's students. In a cross of red-eyed males by white-eyed females, which should have given an F_1 generation just the opposite of their parents, Bridges observed a few red-eyed males and white-eyed females. When he examined the females, he found that their X chromosomes were attached to one another; these **attached-X** chromosomes had not separated during anaphase I of meiosis (*primary* nondisjunction), and had both gone into the same egg. The white-eyed females therefore had the genotype XXY and the red-eyed males carried an X chromosome by itself. To carry the experiment through another generation, the exceptional females were mated to normal red-eyed males. As

$$X^w X^w \; ♀$$
primary nondisjunction

		$X^w X^w$	O
sperm	X^+	$X^+w^w X^{\,w}$ usually dies	X^+O red ♂
	Y	$X^w X^w Y$ white ♀	YO dies

Fig. 14-8 shows, the attached X chromosomes separate in about 92% of the eggs, but in about 8% *secondary* nondisjunction occurs and when these eggs are fertilized by normal sperm carrying X or Y chromosomes, the various classes shown are found. The ability to explain all of these results on the basis of the chromosomal theory of inheritance greatly strengthened that theory in its early days.

Sex-linked markers can be treated just like any others and they can be mapped by using ordinary test crosses. (Because of their unusual pattern of inheritance, the localization of genes on the X chromosome is often one of the simplest determinations that can be made.) You must always keep in mind that there is no homologous chromosome in the heterogamete, and it often helps to use a little symbol such as ⊦ for the Y chromosome (or W chromosome, if you are using that system). Furthermore,

	ova		
normal disjunction		secondary nondisjunction	
X^w Y	X^w	X^w X^w	Y
46%	46%	4%	4%
X^+ X^w Y	X^+ X^w	X^+ X^w X^w	X^+ Y
red ♀	red ♀	dies	red ♂
X^w Y Y	X^w Y	X^w X^w Y	Y Y
white ♂	white ♂	white ♀	dies

sperm: X^+, Y

FIGURE 14-8

The results of secondary nondisjunction in the attached-X females, with fertilization by normal X and Y sperm.

there must be an opportunity for crossing over to occur during meiosis and so the parent to be tested must be the homogamete and the recessive parent must be the heterogamete. Aside from these restrictions, you can prove to yourself that these situations can be analyzed easily by solving the following problems.

exercise 14-25 Human red-green color blindness is due to a sex-linked recessive, *c*. Unless you have reason to think otherwise, assume a probability of 1/100 that a normal woman is heterozygous for *c*.
(a) If a normal woman marries a color-blind man, what is the probability that their sons will be color blind? Their daughters?
(b) If one of their normal daughters marries a normal man, what is the probability that her sons will be color blind? Her daughters?
(c) If a normal son from the above marriage marries a color-blind woman, what is the probability that their sons will be color blind? Their daughters?

exercise 14-26 In cats, *black* fur is due to a sex-linked gene *B* and *yellow* fur to its allele *b*. Males can be either black or yellow; females can be heterozygous *B/b* and their fur is a mixture of black and yellow called *calico*.
(a) If a calico cat has a litter containing a black male, two yellow females, and a calico female, what color is their father?
(b) How many fathers must there be to produce a single litter containing a black male, two black females, and two calico females if the mother is black?

exercise 14-27 Normal *Drosophila* have oval eyes. The dominant sex-linked mutation *B* produces *bar* eyes. *B/b* females and *B/*⊦ males have wide bar eyes; *b/b* females have narrow bar eyes. The mutation *d* for *dark* eyes is also sex linked and is recessive to *D* for red eyes. A

normal

wide bar

narrow bar

heterozygous *B D/b d* female is mated to a *b d* male. Among their progeny are found 25 normal, red-eyed flies; 280 normal, dark-eyed flies; 290 wide-bar, red-eyed flies; and 35 wide-bar, dark-eyed flies. What is the map distance between the two loci? What fraction of each class of flies are males and females?

exercise 14-28 In chickens, which have a ZO sex system, *slow* growth of feathers (*Sl*) is dominant to *rapid* growth (*sl*) and *silver* feathers (*s*) is recessive to *golden* feathers (*S*). Purebred slow, golden males are bred to rapid, silver females; the F₁ males are then crossed to the parental females and the following progeny appear: 127 slow, golden; 118 rapid, silver; 11 slow, silver; and 14 rapid, golden. Both loci are on the Z chromosome; how close together are they?

exercise 14-29 Two normal people have a color-blind daughter with Turner's syndrome. Explain how this can happen.

14-10 PEDIGREE ANALYSIS

Unless you happen to be in the slave business, you cannot do breeding experiments with humans, so a great deal of human genetic analysis depends upon obtaining pedigrees of families with interesting traits to see how the characteristic is inherited. Pedigree data are summarized in the form of a chart that gives relevant information about the members of each generation, using some standard symbols. (These pedigrees can be drawn for any species, of course, but we'll confine ourselves to humans for simplicity.) A man is represented by a ☐ and a woman by a ◯. Phenotypically normal people are indicated by open symbols and those that are affected are indicated by some sort of shading. A marriage is shown by a horizontal line, ☐──◯ ; a double line, ☐══◯ , is used to emphasize consanguineous marriages. The children resulting from each marriage are connected by a horizontal line, joined by a vertical line to their parents, and they are generally listed in order of birth from left to right, since birth order may be important sometimes. Generations are given Roman numerals and individuals are given Arabic numerals. Other symbols are shown below.

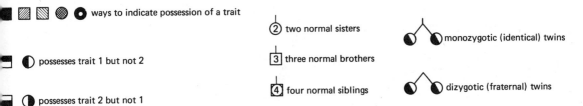

ways to indicate possession of a trait

possesses trait 1 but not 2

possesses trait 2 but not 1

② two normal sisters

③ three normal brothers

④ four normal siblings

monozygotic (identical) twins

dizygotic (fraternal) twins

In general, a study of pedigrees will reveal whether a trait is dominant or recessive and whether it is autosomal or sex linked. In the problems below, you will have to decide on the most likely mode of inheritance, but you cannot always be sure, even after studying many families. Remember that in humans you are dealing with a complex developmental system, and the relationship between a single gene that determines a single protein and some obvious feature, such as eye color, may be very obscure. Part of this obscurity is described by the term **penetrance;** an allele is said to have a penetrance of less than 100% if there is reason to think that it is not expressed in everyone who carries it. For example, polydactyly (more than five fingers on a hand) is thought to be the result of a dominant mutation, but it is apparently not manifested in everyone who carries the mutation. Furthermore, you cannot know a priori how common a recessive mutation is in the population; the best analyses now available of diabetes in humans suggest that the disease is due to a fairly common recessive with reduced penetrance, but the data are compatible with several different estimates of both frequency in the population and the degree of penetrance.

example 14-3   The classical case of an autosomal recessive, where the F_1 are unaffected and the F_2 show 3:1 segregation, could be diagrammed like this:

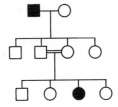

But, of course, you don't really expect to get perfect ratios in most cases. You must depend upon unique features of each mode of inheritance.

exercise 14-30   What is the most likely mode of inheritance in each of these cases?

(a)

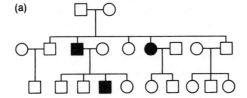

(b)

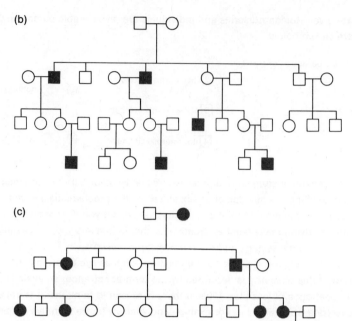

(c)

In some happy cases, one can get linkage data for human genes. Information about autosomal markers is relatively hard to obtain, but there are good linkage data for several mutations on the human X chromosome. The most generally useful method is to study the sons of women who are double heterozygotes for two sex-linked factors; since these males obtain their X chromosome from their mother, the genotype of their father is irrelevant and even illegitimacy can be ignored. However, there are two ways in which the alleles can be arranged in a doubly heterozygous woman, and the arrangement in each case can be determined from the genotype of her father, from whom she obtained one of her X chromosomes. (This method of analysis is therefore sometimes called the "grandfather method.")

Let's consider a simple, hypothetical case. The color-blindness locus, *c*, is fairly close to the *Xg* locus that determines a blood group (similar to ABO). In a population where both color blindness and the Xg^a blood type are common, we test a series of boys for both factors; if we find either one, we determine whether their mothers are heterozygous for these characteristics. (That is, we easily exclude homozygous women and look for women whose sons carry different blood type and color-blindness factors.) We then study every such family separately.

If we find that a mother is *C/c* and Xg/Xg^a, we are still uncertain about the arrangement of these four alleles on her chromosomes. However, if her father is color blind and Xg, her genotype must be *c Xg/C Xg^a*. Then all of her sons who are either *c Xg* or *C Xg^a* are noncrossover types, and those who are *c Xg^a* or *C Xg* are crossover types. The fraction of crossovers in a series of families is a measure of the distance between the two loci; a value of 38 map units has been obtained in some studies.

Suppose in all the cases in which the grandfather is Xg and not color blind, we find the following numbers of grandsons: 14 are Xg, normal; 10 are Xg^a, color blind; 5 are Xg, color blind; and 4 are Xg^a, normal. What map distance would you compute from these data?

14-11 FINE STRUCTURE ANALYSIS

A certain amount of detailed mapping has been done with some fungi because it is not hard to examine large numbers of spores, particularly when the factors in the cross are nutritional requirements and wild-type recombinants can be selected on minimal agar. There is often negative interference in these crosses and they generally give the same picture as crosses with bacteria and phage. Some interesting points that may indicate something about the mechanism of recombination will be discussed later.

Arthur Chovnick has shown that a region in *Drosophila* can be analyzed much like the *rII* region of phage T4 and that it has essentially the structure one would expect for a gene that is just a stretch of DNA. The *rosy (ry)* cistron carries the information for a xanthine dehydrogenase, and *ry* mutants have a brownish eye color. Chovnick found a set of recessive lethals that map on either side and quite close to the *ry* mutants; heterozygotes for any lethal are normal. He then set up crosses of the type

$$(1) \quad \frac{l_1\; l_2\; ry^x + + +}{+ + + ry^y\; l_3\; l_4} \; ♀$$
$$\times \frac{l_1 + ry\; l_3 +}{+ l_2\; ry + l_4} \; ♂$$
$$(2) \quad \frac{l_1\; l_2 + ry^y + +}{+ + ry^x + l_3\; l_4} \; ♀$$

If *ry^x* is to the left of *ry^y*, the survivors in the first cross should be almost entirely wild type due to a crossover between *x* and *y*; survivors in the second cross will be almost entirely rosy, and the frequency of survivors will be a function of the distance between *x* and *y*. About 95% of the flies are eliminated because they obtain two copies of the same lethal mutation; the remaining 5% can be counted and analyzed. By performing a series of these crosses, Chovnick has managed to arrange a number of mutants in an unambiguous sequence and he has established at least seven sites in the cistron. He can obtain frequencies of recombination as low as 0.25×10^{-3} and the greatest distance across the cistron is about 4.5×10^{-3}. There appears to be some negative interference in the data, but the region behaves in recombination like any similar region in phage or bacteria, except for the reduced resolution.

14-12 REGULATION OF HUMAN HEMOGLOBINS

The basic regulatory circuits elucidated in bacteria probably operate in all organisms, although, of course, they are expected to become more complicated as the organisms as a whole become more complex. One of the few systems in which repressor-operator circuits of limited complexity can be invoked as an explanation is the system of human hemoglobins.

All human hemoglobins (Hb) consist of two α chains and two chains of a slightly different type. The second chains may be either β, γ, or δ; the corresponding hemoglobins are $\alpha_2{}^A\beta_2{}^A$, Hb A, which constitutes about 98% of the normal adult hemoglobin; $\alpha_2{}^A\gamma_2{}^F$, Hb F (fetal hemoglobin), which is the normal hemoglobin of the fetus; and $\alpha_2{}^A\delta_2{}^A$, Hb A_2, which is the other 2% of normal adult hemoglobin. Hb F persists through most of fetal life, but near the time of birth Hb A and Hb A_2 begin to replace it, so Hb F has almost entirely disappeared by about eight months after birth.

A number of mutational defects are known in the hemoglobin system that can be interpreted as regulatory blocks; many of them are known clinically as thalassemias. For example, α-thalassemia is apparently an inability to make α chains; all reported cases of this mutation are stillborn, and their blood shows no trace of α chains. Instead, they have Hb H, $\beta_4{}^A$, and Hb Bart's, $\gamma_4{}^A$; clearly, the β and γ protomers have formed homotetramers, having no α chains to pair with. The existence of this defect indicates that there is only one locus for α chains and that it is expressed throughout life; hence, we will not have to assume any change in α chains at the time of birth.

The switchover from synthesis of γ chains to β and δ chains is very interesting. In the first place, it can be blocked by mutations, since there are people who have **high F character** in which fetal hemoglobin persists. Homozygotes for the high F factor never make adult hemoglobins, while heterozygotes make about half fetal and half adult. Since the two adult units, β and δ, appear to be regulated together, we are tempted to consider them a part of the same operon. Indeed, the β and δ genes appear to map very close together. Moreover, there are individuals who have Hb Lepore, which is $\alpha_2{}^A\text{Lepore}_2{}^A$. Amino acid analysis of the Lepore polypeptide shows that it carries the N-terminal sequence of the δ chain and the C-terminal sequence of the β chain; the easiest way to understand this mutation is to assume that a faulty pairing occurred once during diakinesis between the β gene of one chromosome and the δ gene of its homologue, with a crossover that created a hybrid gene (Fig. 14-9). So β and δ are probably close together, and considering that messengers are made from the 5′ end to the 3′ end and that the 5′ end codes the N-terminal portion of the protein, we will draw them in the sequence Operator$_A$-δ-β and call this the A operon. We will assume that the γ gene is in an operon that we will call F. Following a suggestion first made by Emil Zuckerkandl, we will assume that each operon contains the regulator for the other (Fig. 14-10).

The high-F mutation could clearly be at any of several points, such as R_F or O_F. However, we know that individuals who are transheterozygotes for sickle-cell anemia (Hb S or $\alpha_2{}^A\beta_2{}^S$) and for high-F, make half Hb S and half Hb F, with no Hb A. This means that the β gene cis to high-F is blocked, and the high-F mutation can only reside in O_A (or perhaps more precisely, in a promotor$_A$). Since the A operon cannot be expressed cis to this mutation, no β^A is made and the only F repressor made comes from the R_F gene cis to β^S.

Another thalassemia known as β-thalassemia, Cooley's anemia, or thalassemia major can also be understood within this framework. In this disease, there is no Hb A; Hb F persists and there are normal or increased amounts of Hb A_2. The mutation is probably a polar block of the β gene, which simultaneously eliminates the β polypeptide and F repressor. This disease, incidentally, is common among people of Italian and Greek ancestry. From the distribution of incidences in Italy, its origin—many generations ago—can practically be located.

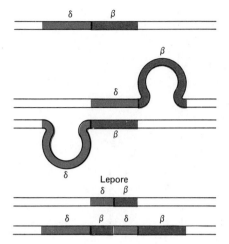

FIGURE 14-9

Probable origin of Hb Lepore through synapsis of a δ gene with a β gene.

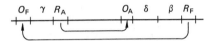

FIGURE 14-10

Theoretical structure of the Hb regulatory circuit. As long as the F operon is induced, it will make A repressor and repress the A operon. If an event near the time of birth blocks the A repressor, the A operon will become induced and make F repressor to turn off the F operon. (The two operons are not necessarily linked to each other.)

C. Nonchromosomal heredity

14-13 HETEROCARYOSIS AND THE PARASEXUAL CYCLE

When two different fungal hyphae fuse to make a heterocaryon, the nuclei do not usually undergo caryogamy; their products mix in a single cytosome, creating the perfect experimental situation for complementation analysis. Complementation tests are regularly carried out by spotting two mutant strains together on an agar plate and looking for the appearance of significant growth.

In *Aspergillus,* some of the nuclei in a haploid hypha may fuse to make diploid nuclei. If this happens in a heterocaryon, the resulting nuclei are heterozygotic. These diploid nuclei may continue to divide mitotically, but they may also exhibit recombination during mitosis and they occasionally haploidize. That is, in some way the diploid set is reduced to a haploid set, but through the loss of chromosomes in accidental processes rather than a regular meiosis. Bruno Pontecorvo has suggested that this cycle of diploidization and haploidization be called a *parasexual* cycle.

Pontecorvo has taken advantage of this system to map *Aspergillus.* Figure 14-11 shows how mitotic crossing over can result in the segregation of a homozygote from a heterozygous strain. If the mutant *a* is a recessive, the heterozygote will have the

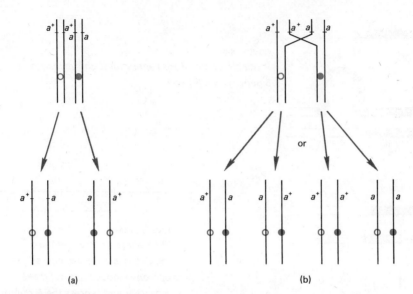

FIGURE 14-11

The consequences of a mitotic crossover. With no crossing over (a) *the diploid sets remain heterozygous for* a. *A crossover between* a *and its centromere* (b) *will produce the two homozygotes about half the time.*

wild-type phenotype. *Aspergillus* regularly produces mitospores (conidia) that can grow into new hyphae. Haploid hyphae, whether homocaryons or heterocaryons, produce small haploid mitospores and diploid hyphae produce larger, diploid mitospores. We can therefore look for diploid mitospores in which the *a/a* phenotype appears as a result of a crossover, and the frequency of such crossovers is a measure of the distance between *a* and its centromere.

14-14 NONCHROMOSOMAL FACTORS

One of the most significant additions to our conception of genetic organization has been the demonstration that part of the genome in eucaryotes consists of nonchromosomal (extranuclear) factors that are inherited in the course of cell proliferation in a somewhat different way from chromosomal genes. Most of the well-studied cases appear to involve the inheritance of plastids and mitochondria; the green chloroplasts involved in photosynthesis are only one of a class of simmilar vesicles found in plant cells, including amyloplasts, that carry starch, proteinoplasts that are full of protein crystals, and chromoplasts that contain characteristic pigments. DNA molecules have now been clearly demonstrated in both chloroplasts and mitochondria; the latter, at least, have circular molecules just like bacterial genophores and episomes, a fact that has some nontrivial implications for evolution. Mitochondria and chloroplasts contain ribosomes (which are rather more like bacterial than eucaryotic ribosomes), transfer RNAs, and activating enzymes, and can make some of their own proteins (Chapter 16). We can now give a few examples of clear cases of nonchromosomal inheritance.

Nonchromosomal factors are generally detected by their lack of Mendelian segregation. In general, the characteristic passed on to the next generation is the one carried by the gamete that contributes the largest part of the cell mass: the egg in higher organisms, the protoperithecium in fungi, and so on. This characteristic is called the *maternal effect* or *maternal inheritance*.

Some very clear cases of nonchromosomal inheritance have been found by Ruth Sager in *Chlamydomonas*. Mutants can be found that are resistant to streptomycin (sm^r), but the streptomycin-resistance marker is always inherited from the mt^+ parent. In a cross of $mt^+ sm^r \times mt^- sm^s$, all of the progeny are streptomycin resistant, but the mt marker segregates normally. When the mt^+ progeny from this cross are crossed to $mt^- sm^s$ cells, all of the progeny are again sm^r. However, if $mt^- sm^r$ cells are crossed to $mt^+ sm^s$ cells, all of the progeny are sm^s. A factor for acetate requirement (*ac*) is inherited in the same way, and this may well be a chloroplast gene, since the mutants are unable to photosynthesize.

John L. Jinks and his colleagues have described a cytosomal factor in the mold *Aspergillus*. Red or purple variants occur among the normally colorless hyphae, due to chromoplast pigments. Heterocaryons can be made between red and colorless hyphae carrying different chromosomal markers; the mitospores produced by these heterocaryons have haploid nuclei and small cytosomes. The red color always segregates among mitospores entirely independently of any other marker and its linkage to chromosomal markers cannot be established. Furthermore, a few colorless mitospores always appear among those collected from pure red hyphae; this is what we would expect for a population of chromoplasts that are distributed at random during spore formation, for there is always some chance of producing a spore with no chromoplasts.

Tracy M. Sonneborn studied some *killer* strains of *Paramecium* that release a substance called *paramecin* into the growth medium, which kills sensitive, nonkiller strains. To be a killer, a cell must carry particles called *kappa* in its cytosome, but it can maintain the kappa particles only if it also carries at least one dominant allele of several genes, of which we will consider only one, *K*. Killer strains will conjugate with sensitive strains, either *k/k* or *KK kappa⁻*; normally, the killer exconjugant remains a killer and the sensitive exconjugant remains sensitive. However, if conjugation is prolonged, more than just the micronuclei exchange between cells. As Fig. 14-12 shows, some kappa particles may be transferred to the sensitive strain, making it a killer. Thus, the killer trait clearly depends upon particles in the cytosome.

A related factor called *mu* has been investigated by G. H. Beale and his colleagues. Cells carrying mu are *mate killers;* they conjugate with other paramecia, but shortly thereafter the sensitive exconjugant dies, presumably because of some kind of cell surface reaction induced by contact with the mu cell. The mu is also a little particle in the cytosome, and its maintenance requires two chromosomal genes, M_1 and M_2, either of which is sufficient by itself. In a cross of $m_1 m_2/m_1 M_2 \times m_1 m_2/m_1 m_2$ cells, only half of the progeny are expected to be mate killers; however, all cells continue to be mate killers and contain large numbers of mu particles for seven generations.

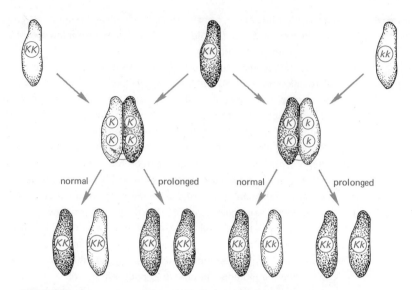

FIGURE 14-12

Conjugation between killer and nonkiller strains of Paramecium. *Kappa
particles are indicated by dots. During a brief conjugation with a
sensitive strain, either* KK *or* kk, *no particles are transferred. Transfer
does occur during a prolonged conjugation in either case.*

In the eighth generation, 5% of the cells become sensitive, and this fraction rises
steadily to 93% after 15 generations. When the remaining 7% of mate killers are
allowed to divide a few times more, most produce only one cell with mu particles out
of eight, a few produce two, and a very few produce three. This experiment is
interpreted as follows. Cells with at least one dominant *M* allele produce particles
called *metagons* that are required for maintenance of mu particles. After the loss
of the *M* gene by conjugation, no more metagons are made and those present
continue to be diluted out; there are evidently about 2^{15} such particles per cell
initially.

Further investigation reveals that the metagons are RNA particles. They are
apparently equivalent to messenger RNA molecules, except for their great stability,
and they presumably direct the synthesis of proteins necessary for mu maintenance,
since they are found attached to the ribosomes.

The mu and kappa particles, along with a host of similarly named cytosomic
elements in paramecia, have the appearance of small, degenerate bacteria. (They
seem to be closely related to the small bacteria called *rickettsias,* which are also
intracellular parasites and cause diseases in animals.) Sensitive cells can be infected
by free particles, and some of the particles can be maintained by themselves in
culture. Normally they maintain themselves very nicely in paramecia because the
relationship between the two cells seems to be mutualistic, with both partners
benefiting from this association.

The process of recombination is so important for our genetic analyses that it is a shame we don't understand it better. This is a matter of such great interest now that we will take a moment to see just how much we do understand.

Matthew Meselson and Jean Weigle performed an important experiment which proved that recombination in phage proceeds primarily by breakage and reunion of DNA molecules. First they had to find out how phage λ DNA is transferred from one generation of phage to the next, so they grew λ in a medium containing ^{13}C and ^{15}N to make their DNA dense. When they used these phage to infect ordinary bacteria in an ordinary medium, they found that the progeny phage fell into three classes in a CsCl gradient:

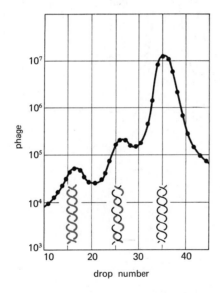

Most of the phage are made of completely light, new DNA. Some of them are semiconserved—that is, they have one strand of old, dense DNA combined with one of light DNA. And a very few are fully conserved, having both strands of dense DNA that has never replicated.

They then performed their critical experiment, by crossing wild-type phage with phage carrying two mutations: c, which makes clear plaques and maps near the middle of the genetic map, and mi, which makes minute plaques and lies near the right end of

the map. They grew the *c mi* in light medium, the + + phage
in $^{13}C^{15}N$ medium, and crossed them in bacteria growing in
ordinary, light medium. The graph below shows the distribution
of all the progeny phage in CsCl.

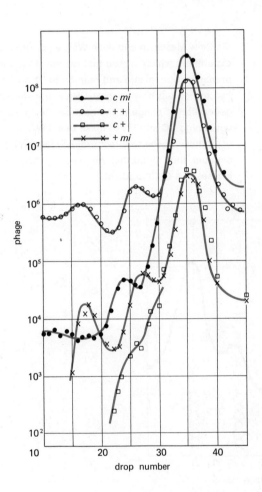

The parental + + phage are distributed about as before, but the
c mi and the recombinants fall into distinctly different bands.
For example, there is a peak of + *mi* phage that are a little lighter
than the conserved + + parents. Though their DNA has never
replicated it has recombined and presumably consists of a large,
dense piece carrying the c^+ marker and a smaller, light
piece carrying the *mi* marker. There are also peaks of both
recombinants that are a little lighter than the semiconserved + +
phage. There is a peak of *c mi* phage that are slightly denser
than the semiconserved + + phage; their DNA must consist of

a light half carrying both markers and a dense half from a + +
parent. This clearly shows that recombination can occur by
breakage and reunion of DNA molecules without any significant
replication. ("Significant" is used here as a hedge to cover some
later results.)

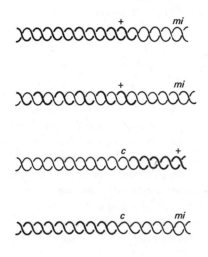

HETEROZYGOTES

When bacteria are infected simultaneously with an *r* and an *r*⁺
phage, about 1% of the progeny behave as if they are carrying
both alleles. They produce "mottled" plaques that look partially
like *r* and partially like wild type, and both types of phage can
be recovered from these plaques. They are called heterozygotes,
but to avoid confusion with legitimate Mendelian heterozygotes
we will adopt laboratory jargon and call them *hets*.

Cyrus Levinthal recognized that a het could have either this
structure:

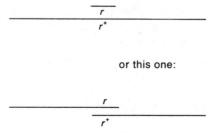

He suggested that they are really the second type because they
are intermediate stages in recombination. If this is true, a cross

of the phage $A^+ r B \times A r^+ B^+$ should give hets that are
recombinant for the outside markers A and B. Levinthal
performed this experiment and confirmed that the r/r^+ hets
are overwhelmingly $A^+ B^+$ or $A B$.

This discovery finally led to the suggestion that het genophores
are really *heteroduplexes* that have the information for r on one
strand and for r^+ on the other:

For example, if r^+ has an A-T pair in one position and r has a
G-C pair, the heteroduplex has either an A-C or a G-T pair, even
though these are "forbidden." Such heteroduplexes are formed
when genophores of opposite genotype suffer breaks close
to the same point and then recombine in the region where their
broken ends overlap. Now, suppose we perform a cross between
phage marked $a b c$ and $a^+ b^+ c^+$, where the markers lie within
a distance comparable to the length of the average overlap. A
heteroduplex structure might form like this:

It happens that bacteria contain a variety of enzymes that can
attack structures like this one. The principal function of these
enzymes seems to be in repairing damages to DNA from
ultraviolet light and various reagents, which are known to
produce abnormal base pairs and bases that are bonded
covalently to each other in ways that distort DNA structure.
These enzymes apparently search out these distortions and
destroy them by cutting out some of the damaged DNA and
recopying from an undamaged strand. Some enzymes attack
single-stranded DNA; others break DNA strands near damages
and excise part of one strand; and still others can recopy in
these regions and complete the covalent bonding. In concert,
they could change the heteroduplex as follows:

Now notice that in a single recombinational event—a single break and reunion—we have produced a strand that is doubly recombinant: *a b⁺ c.* The occurrence of an abnormally large number of double recombinants is what we have called negative interference; in fact, as the markers involved get very close we typically find *high* negative interference in which *S,* the coefficient of coincidence, reaches 20 to 50. In outline, high negative interference may be explained by the action of enzymes in repairing heteroduplexes, although there are many things we cannot explain yet. At least the role of enzymes in recombination seems clear; bacterial mutants that cannot recombine efficiently (*rec⁻* or "rec-less" mutants) are also defective in the repair of ultraviolet damages. Similar mutants that are deficient in various types of recombination have also been found in the phage λ system.

GENE CONVERSION

When crosses are performed between different mutants of *Neurospora* and other fungi, the alleles at all loci generally segregate in a 4:4 ratio among the spores of an ascus; we have seen many examples of this. However, Mary B. Mitchell performed some crosses between close markers in *Neurospora* in which she found the alleles at one site to segregate in a 6:2 ratio. She called this *gene conversion* because it looks as if the marker on one chromatid has been converted into the opposite allele. Conversion occurs much too frequently to be explained by mutation; furthermore, conversion always occurs from one allele in the cross to the other allele, not to some new type.

Margaret E. Case and Norman H. Giles studied gene conversion in the *pan-2* (pantothenic acid) region of *Neurospora;* their crosses always involved several markers within the locus and at least one marker on either side of it. Several asci showed 6:2 segregation for at least one of the *pan-2* markers (and sometimes for two or three of them together); the outside markers always showed 4:4 segregation, but half of the asci were recombinant for outside markers. This high frequency of recombination among asci showing conversion suggests that conversion occurs during a phase of the recombinational processes. Furthermore, all cases of conversion appear to affect only two of the four chromatids in a bivalent; this clearly implies that conversion occurs during a typical breakage and reunion process in a chiasma between two chromatids and that it may reflect strange events in the repair of chromosome breaks, just as high negative interference reflects these events in phage recombination.

Pascal Lissouba, Georges Rizet, Jean-Luc Rossignol, and their colleagues discovered a fascinating pattern of conversion in the fungus *Ascobolus,* which has brown ascospores. White-spored mutants are often found, and the mutations fall into groups that map close together. In a cross between two white mutants of the same group, asci often appear that have a pair of brown spores:

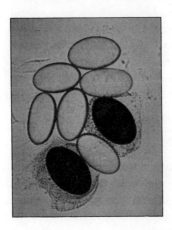

These could result from either a recombination or a conversion, so each spore must be grown to determine its genotype. Many of the asci turn out to be the result of a conversion. In a cross between $a +$ and $+ b$, which are both white mutants, two types of conversion tetrads can appear:

$$
\begin{array}{ccc}
a + & & a + \\
a + & & + + \\
+ + & \text{or} & + b \\
+ b & & + b
\end{array}
$$

Lissouba and his group defined the *majority parent* as $a +$ in the first case and as $+ b$ in the second case. They then made the remarkable observation that the white spore mutants fall into regions within which conversion is polarized, and they called these regions *polarons*. Within a polaron, the majority parent is almost invariably the one carrying the mutation on the left, and in a sense the mutation on the left can be said to "convert" the mutation on the right. However, if a cross is performed between two markers in neighboring polarons, wild-type spores almost invariably result from a recombination and the reciprocal double mutant is found in the ascus.

The construction of a model that will satisfy all of the available data is very difficult; no one has provided a fully satisfactory one yet. However, the most reasonable models to date all have one feature in common. They assume that a chromosome is divisible into short segments, which may be called regions of *effective pairing,* and that recombination occurs through the tight pairing of a single DNA strand from each parent in such a region. Remember that Huberman and Riggs' data show that chromosomes are divided into units of replication; J. Herbert Taylor has suggested that during the S period before meiosis some of these regions on each chromosome are only partially replicated, leaving single strands that can pair with each other. Franklin Stahl suggests that replication within such a region is stimulated by synapsis. In either case, the chromatids in early meiosis can interact over short segments where breaks, reunions, and repairs can occur. We cannot specify the events that must occur in one of these segments, but one possible sequence is suggested here.

We assume that in one region two of the four chromatids are complete and two have not replicated fully, so there are single strands with opposite polarities that can pair with each other:

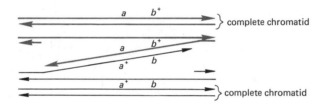

Since the newly hybridized strands are heteroduplex, they can be attacked by certain nucleases. However, not all of a strand will be digested, and therefore the probability that a marker will be removed increases as the marker gets closer to the end of the pairing region. This accounts partially for the polarity of conversion. As shown here, one strand will be partially degraded and then resynthesized near its end:

At this point, only one strand has been converted. If there were no further conversions and additional DNA strands were synthesized in this region to make all chromatids complete, there would be a 5:3 segregation of the b/b^+ markers and probably a recombination of outside markers on either side of this region. Such 5:3 segregations are also observed and they are associated with recombination. Although it is difficult to draw a simple, rational model, another conversion in the same direction might be necessary in order to straighten out all of the DNA strands, and this would produce a 6:2 segregation with recombination. We will not try to draw this because it would be too muddled and would have no rational basis.

One point at least seems clear: There is some additional DNA synthesis early in meiosis, and it is probably associated with breaks, reunions, and repairs that result in conversion and recombination. However, it should be obvious that a great deal of work will be required to sort out the details of the processes outlined here.

PART

15

the behavior
of proteins

Perhaps the most important thing we have learned about proteins recently is that they are very flexible, and this one concept provides the basis for understanding a great deal about biology. For our purposes, the trouble with the concept is that it is hard to document; that is, the evidence for conformational changes in proteins comes from a variety of techniques that we are hardly in a position to discuss in much detail, and so much of the information in this chapter will have to be taken on faith. These are the things that will require the careful, detailed explanation of a biochemistry course for a thorough understanding.

15-1 LIGAND-PROTEIN INTERACTIONS

Metabolism is largely a series of interactions between small molecules — ligands — and proteins. The ligands are primarily metabolites, but also various ions and coenzymes, and in some cases they are invaders that must be eliminated before they can do any damage. We can distinguish three major classes of interactions:

1. The protein is an *enzyme* and the ligand is a metabolite, its *substrate.* The substrate binds to the enzyme, generally covalently, and is changed through a series of reactions with side chains of the protein and released, leaving the protein as it was before.
2. The protein may be any of a large number of things, including an enzyme, and the ligand is an **effector** that binds to it noncovalently and changes its conformation so its activity is enhanced or diminished. The effector may be an **activator** or an **inhibitor.**
3. The protein is an **antibody** or **immunoglobulin** and the ligand is an **antigen,** generally a foreign substance. The antibody binds to the antigen and inactivates it. This reaction takes place primarily in birds and mammals, although even the lower fishes make some antibodies, and the general outcome is that many antibody and antigen molecules clump together into a mass large enough for a roving white blood cell to recognize, pick up by endocytosis, and digest.

We know about all of these interactions because we can measure enzyme activities, the abilities of various agents to inhibit or enhance these activities, the abilities of other compounds to induce or repress enzymes, and the amounts of specific antigens injected into animals. At the same time, we can measure changes in the surface charges of the proteins and changes in their shape as reflected by viscosity, sedimentation,

461

and diffusion measurements, and by various optical techniques. For present purposes, we will concentrate on the results of all these measurements and see what can be said about the changes in protein structure. We will begin with enzymes, since we know so much about them.

A. Enzymes

15-2 KINETICS OF ENZYME ACTION

If we consider a reaction of the type $A + B \rightleftharpoons C + D$, the rate at which one of the reactants disappears is a good measure of the overall rate of the reaction, so we define $v = -dA/dt = -dB/dt$. To achieve the correct kinetic expression for this reaction, we must consider the products and reactants to reside at certain levels on a free-energy scale separated by a free-energy barrier (Fig. 15-1). In order to pass the

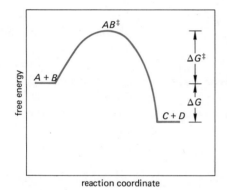

FIGURE 15-1
Relative energies of the molecular species in a chemical reaction.

barrier, the reactants A and B must be *activated* into a complex $AB^{\ddagger}$ through the free-energy difference $\Delta G^{\ddagger} = G^0_{AB^{\ddagger}} - G^0_{A + B}$. The complex $AB^{\ddagger}$ can decay back into A and B again, and for this phase of the reaction there is an equilibrium constant given by

$$K^{\ddagger} = \frac{[AB^{\ddagger}]}{[A][B]} \tag{15-1}$$

The complex can also decay into the products C and D with a velocity $v = k'[AB^{\ddagger}]$. As for any other chemical reaction, we can express $\Delta G^{\ddagger}$ as a function of $K^{\ddagger}$ by $\Delta G^{\ddagger} = -RT \ln K^{\ddagger}$. Combining all of these expressions into one, the forward velocity of the reaction is

$$v = k'e^{-\Delta G^{\ddagger}/RT}[A][B]$$

It can also be shown that the proper form for k' is kT/h, where k is Boltzmann's constant and h is Planck's constant. Just as we defined the height of the free-energy barrier as the difference between the standard free energies of the complex and the reactants, we can define an entropy and an enthalpy of activation which are related to $\Delta G^{\ddagger}$ by $\Delta G^{\ddagger} = \Delta H^{\ddagger} + T\,\Delta S^{\ddagger}$. Combining these results, the reaction rate is given by

$$v = \frac{kT}{h}\,e^{-\Delta H^{\ddagger}/RT}e^{\Delta S^{\ddagger}/R}[A][B] \qquad (15\text{-}2)$$

Examination of equation 15-2 shows that at constant temperature the reaction rate can be increased in two ways: by decreasing $\Delta H^{\ddagger}$ or increasing $\Delta S^{\ddagger}$. The first can be accomplished by activating the reactants, so their energy is closer to that of the complex, and the second by decreasing the entropies of the reactants relative to that of the complex. Enzymes do both of these things. They carry at least 20 different amino acid side chains, as well as special coenzymes, and they combine these groups appropriately to activate their substrates. Moreover, when the substrate is bound to these groups on the protein, it is held in a specific configuration and its entropy is decreased. This is why enzymes are such efficient catalysts; the reaction might be catalyzed by putting the same chemical groups into solution with the substrate, but an enzyme holds them in essentially a solid state in just the right configuration, so the reaction proceeds rapidly and efficiently. The **catalytic power** of an enzyme—the ratio of the catalyzed to the uncatalyzed rate of the reaction it performs—may be as high as 10^{12}.

If we define the rate of an enzymatic reaction as the rate at which the substrate S disappears, $v = -dS/dt$, a graph of v against $[S]$ looks like Fig. 15-2. The rising segment corresponds to a **first-order reaction,** with the velocity proportional to substrate concentration: $v = k[S]$. The horizontal part represents a **zero-order reaction,** independent of substrate concentration: $v = k'$. To explain these kinetics, Leonor Michaelis and Maud L. Menten developed a theory in 1913 that is still the basis for analysis of enzymatic reactions. We will discuss a slightly modified version developed by G. E. Briggs and J. B. S. Haldane.

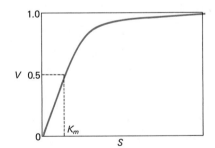

FIGURE 15-2

Velocity of a reaction showing simple Michaelis–Menten kinetics.

We assume that the active state of the substrate is a complex C in which the substrate is bound to the enzyme E, and that the complex decays into the product and the free enzyme again:

$$E + S \underset{k_2}{\overset{k_1}{\rightleftharpoons}} C \xrightarrow{k_3} E + P$$

Assume that the substrate concentration is much larger than that of the enzyme, so it can be considered constant. Furthermore, if $[E]$ is the total enzyme concentration, then the free enzyme is $[E - C]$. Then the rate of change in concentration of the activated complex is

$$\frac{d[C]}{dt} = k_1[E - C][S] - (k_2 + k_3)[C] = 0 \qquad (15\text{-}3)$$

which we set equal to zero at steady state. Equation 15-3 can be rewritten by combining the three rate constants into a single constant K_m, the *Michaelis constant*,

$$K_m = \frac{k_2 + k_3}{k_1} = \frac{[E - C][S]}{[C]}$$

The velocity of the reaction, in general, is the rate at which C decays into products: $v = k_3[C]$. The reaction proceeds at its maximum rate, V, when all of the enzyme is in the complex form; that is, $[E] = [C]$ so $V = k_3[E]$.

$$K_m = \frac{(V - v)[S]}{v}$$

or

$$v = \frac{V[S]}{K_m + [S]} \qquad \text{(Michaelis–Menten equation)} \qquad (15\text{-}4)$$

K_m and $[S]$ have the dimensions of concentration, generally moles per liter. If $K_m = [S]$, then $v = V/2$; that is, the Michaelis constant is just the substrate concentration that makes the reaction proceed at half its maximum value. This is the principal significance of K_m; it is generally said that K_m is the equilibrium constant of association between enzyme and substrate, so that $1/K_m$ is a measure of the degree to which a substrate is bound, but this is strictly true only if k_2 is much greater than k_3.

The Michaelis–Menten equation as written above is not easy to work with. H. Lineweaver and D. Burk showed that it is more useful to take its reciprocal and write

$$\frac{1}{v} = \frac{1}{[S]} \qquad (15\text{-}5)$$

Then, plotting $1/v$ against $1/[S]$ yields a straight line (Fig. 15-3) with slope K_m/V and intercept $1/V$; if the line is extended to the left, its intercept with the $1/[S]$ axis is $-1/K_m$. This graph makes it easier to determine the important parameters of the reaction.

It should be emphasized that this analysis is very superficial. The analysis of the kinetics of a given enzymatic reaction must be more complex and probe much deeper if it is to be useful. However, this is a matter that we leave for more advanced discussions and other textbooks.

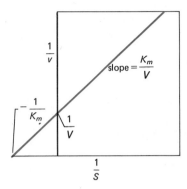

FIGURE 15-3
Lineweaver–Burk graph of kinetic data.

15-3 INHIBITION OF ENZYME ACTION

One of the classical observations of enzyme chemistry is that enzymes can be inhibited more or less specifically by various reagents. These inhibitors are often useful in studying the normal action of the enzyme, and we will briefly outline a kinetic analysis of their action.

Inhibitors are divided into two classes. *Competitive* inhibitors compete with the substrate for the enzyme; a competitive inhibitor must resemble the normal substrate. An example is the inhibition of succinic acid metabolism by malonic acid. Since the two molecules are so similar, it is presumed that malonate fits into the site on the enzyme that succinate would normally occupy and binds it so the enzyme is temporarily inactive.

COOH
|
CH_2
|
CH_2
|
COOH

succinic acid

COOH
|
CH_2
|
COOH

malonic acid

Noncompetitive inhibitors form a more diverse class. They can interact with the substrate-enzyme complex and inactivate it in some way. For example, heavy metals such as mercuric and silver ions are strong enzyme inhibitors; this is why it is dangerous for children to play with mercury and why one should not treat mercury spilled in the laboratory lightly. Some other inhibitors are cyanide (CN^-), hydrogen sulfide (H_2S), and iodoacetate (CH_2ICOOH).

The effect of an inhibitor can be analyzed as follows. If an inhibitor I competes with the substrate, then we have

$$E + S \rightleftharpoons C \rightarrow E + P$$
$$E + I \rightleftharpoons C_i$$

where C_i is the inhibited enzyme. Note that the free enzyme is $E - C - C_i$; we then define a constant

$$K_i = \frac{[E - C - C_i][I]}{[C_i]} \tag{15-6}$$

The rate at which $[C]$ changes is

$$\frac{d[C]}{dt} = k_1[E - C - C_i][S] - (k_2 + k_3)[C] = 0 \tag{15-7}$$

which we again set equal to zero at steady state. Then

$$C = \frac{[S][E - C_i]}{[S] + K_m} \tag{15-8}$$

where k_m is defined as before. Solving equation 15-6 for C_i and substituting into equation 15-8 gives

$$C = \frac{[E][S]K_i}{K_m K_i + K_m[I] + K_i[S]} \tag{15-9}$$

As before, we substitute v for C and V for E:

$$v = \frac{V[S]K_i}{K_m K_i + K_m[I] + K_i[S]} \tag{15-10}$$

which can also be transformed into an equation of Lineweaver–Burk type:

$$\frac{1}{v} = \frac{1}{V} + \frac{K_m}{V}\frac{1}{[S]}\left(1 + \frac{[I]}{K_i}\right) \tag{15-11}$$

If a graph is made of $1/v$ against $1/[S]$ as before (Fig. 15-4), the line has an intercept at $1/V$ and a slope $(K_m/V)(1 + [I]/K_i)$. Since K_m can be determined in the absence of inhibitor, K_i can be obtained from this graph.

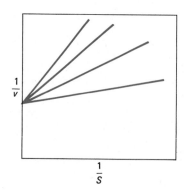

FIGURE 15-4
Kinetic data for varying concentrations of an inhibitor, showing competitive inhibition.

The derivation of an expression for noncompetitive inhibition is left as an exercise for the reader. The assumption must be made that the inhibitor can combine with all enzyme molecules, whether free or in an active complex. The resulting equation should be

$$\frac{1}{v} = \left(1 + \frac{[I]}{K_i}\right)\left(\frac{K_m}{V}\frac{1}{[S]} + \frac{1}{V}\right) \qquad (15\text{--}12)$$

Figure 15-5 shows the form of a Lineweaver–Burk plot for noncompetitive inhibition. This kind of graph enables one to distinguish easily between the two types of inhibitors, for in competitive inhibition the intercept of the line is independent of inhibitor concentration, while it increases with inhibitor concentration in the noncompetitive case.

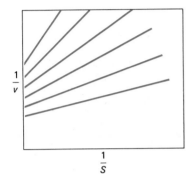

FIGURE 15-5
Kinetic data for varying concentrations of an inhibitor, showing noncompetitive inhibition.

15-4 MECHANISM OF ENZYME ACTION

The classical theory of enzyme action was formulated in the late nineteenth century by Emil Fischer, who conceived of an enzyme as having an **active site** in one place with just the right shape to accommodate the substrate and catalyze the reaction. Just as a key fits a lock, the substrate was supposed to fit the active site and be altered. This model explains a great deal, but it does not explain enough. More recently, Daniel Koshland and others have shown that the data can be accommodated by assuming that the active site is not a ready-made lock that the key slips into, but that the key **induces** the lock to fit around it. Specifically, it is assumed that the substrate binds to the active site and induces a conformational change in the protein which brings other reactive groups into position to catalyze the reaction.

Koshland gives the following example of a mechanism that cannot be explained by the lock-and-key model. The enzyme that splits the phosphate from AMP has a K_m of 10^{-4} and we may call its maximum velocity 100%. The enzyme will also react with another nucleotide in which the adenine is replaced by a smaller ring, nicotinamide. In the classical model, we expect this compound to bind more poorly because the nicotinamide ring does not fit so well into the space designed for adenine; the K_m for this compound is actually 2×10^{-3}, indicating somewhat weaker affinity for the enzyme. The reaction rate should not be drastically affected, since the ribose phosphate can still fit into its place in the active site; in fact, $V = 67\%$, which is not a great decrease. So far the model works. We now predict that ribose phosphate by itself should bind very weakly, because there is no ring at all to fit into the template, but its reaction rate should again not be affected much. But we find that the K_m is 3×10^{-3}, indicating no decrease in affinity, while V has dropped to 0.8%. This clearly violates the predictions of the classical model.

Figure 15-6 shows how the induced-fit model explains the action of adenosine phosphatase. The catalytic site binds a simple ribose phosphate just about as well as AMP, but the reaction proceeds much more slowly than with AMP because the adenosine is necessary to bring one reactive group of the protein into the proper configuration.

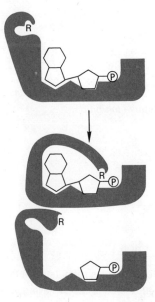

FIGURE 15-6
Induction of the proper enzyme conformation in adenosine phosphatase. The adenine brings the reactive R group into position to attack the phosphate bond. Ribose phosphate can bind but cannot induce this change in the protein.

The lock-and-key model can obviously explain competitive inhibition quite easily, for the competitor simply fits into the same site as the substrate. However, it is hard to see how it can explain noncompetitive inhibition, for if an enzyme is supposed to be a rigid structure with an active site of just the right shape for a substrate, one cannot easily explain how a molecule that acts somewhere outside the active site can temporarily inactivate the enzyme.

Noncompetitive inhibition is readily explained by the assumption that an enzyme is very flexible. Any group that reacts with a protein is potentially an inhibitor, since it may inactivate an essential part of the molecule remote from the active site. For example, heavy metals and iodoacetate react strongly with sulfhydryl groups, so it is not surprising that these reagents could upset a critical structure by breaking disulfide bonds or tying up a critical —SH group anywhere on the molecule.

The concept of flexibility is now well supported. We now consider its implications for metabolism as a whole when the enzymes are seen in their proper places inside cells.

exercise 15-1 A certain enzyme can perform a reaction with either glucose or mannose. The velocity of the reaction has been determined with these substrates at several concentrations:

CONCENTRATION MILLIMOLES/LITER	v, GLUCOSE	v, MANNOSE
0.01	20	25
0.02	30	35
0.05	42	46
0.10	49	51
0.20	52	54
0.30	54	55
0.50	55	55

Plot these data and estimate a K_m value for each substrate. Then make a Lineweaver–Burk plot and calculate K_m and V_{max}. Which substrate has a higher affiinity for the enzyme?

exercise 15-2 Ogden Edwards tried to correlate the maximum temperature at which certain bacteria (*Bacillus*) will grow with the minimum temperature at which their enzymes are inactivated (standing for 24 hours); he obtained results such as these:

STRAIN	MAXIMUM GROWTH TEMPER- ATURE	INDOPHENOL OXIDASE	CATALASE	SUCCINIC DEHYDROGENASE
		MINIMUM INACTIVATION TEMPERATURE (°C)		
B. mycoides	40°	41°	41°	40°
B. megatherium	46°	48°	50°	47°
B. alvei	46°	51°	50°	53°
B. vulgatus	55°	56°	56°	50°

Interpret these data in terms of adaptations of organisms to their environments and our knowledge of enzyme structure.

15-5 VECTORIAL ENZYME ACTION

When a chemical reaction takes place in solution, near-molar quantities of molecules may be bouncing around and interacting with one another. If we could watch this process, it would look completely chaotic; we might describe it by means of scalar quantities such as its equilibrium constant or activation energy, but the process has no spatial direction. This is still true even if we add enzyme molecules; the reaction goes faster than before, but the enzymes are moving about as randomly as their substrates.

Things are very different in a cell. A major share of the enzymes are bound to large structures such as membranes, ribosomes, and DNA molecules, and many enzymes are bound to other enzymes, with or without a larger support. The cell is not merely a bag of enzymes, as it was thought a few decades ago. A more realistic viewpoint comes from two principles of intracellular organization:

1. All molecules of a particular enzyme bound to the same support are bound in the same orientation. This is primarily a consequence of self-assembly; molecules bind to one another in their most stable conformations, and so, for example, all molecules of lactic dehydrogenase bound to an endoplasmic reticulum membrane must make the same bonds and they must all be sitting on (in?) the membrane in the same orientation.

2. All molecules of a particular enzyme are bound in a conformation where they can operate most efficiently with respect to other enzymes. This is just a restatement of our general belief that cellular systems are well designed by evolution.

When viewed in this way, it is easy to see that enzymes in a cell will not merely catalyze an undirected process. Most of them will carry out a *vectorial* process in which the movement of their substrates and products can be represented by a set of arrows. For example, Fig. 15-7 shows that a bank of identical enzymes situated identically on a membrane must all push their products along identical paths, because the substrates are all situated in the same positions and they all receive the same push from conformational changes in their enzymes during catalysis. This concept, which has been developed by Peter Mitchell, is basic to our understanding of many cellular processes, particularly membrane processes. Let's consider some additional examples.

Figure 15-8 shows how enzymes may be buried in a membrane so they can receive their substrates only from one side and can release their products only on the other side. This is a basic model for a *permease*. There is nothing mysterious here; translocation of a permeant from one side of the membrane to the other is a direct consequence of ordinary enzymatic catalysis, given only the restriction of a specific orientation and location. Thermodynamically, it may be necessary to use an ATP molecule or an equivalent amount of energy to drive this reaction, since in general the translocation will be against a concentration gradient, but even this is not absolutely necessary, since a great deal of free energy has been expended just in making the appropriate enzymes and placing them in position.

If translocation is simply the result of ordinary enzyme action, it is easy to understand how each permease can be specific for a small group of sterically related compounds (such as leucine, isoleucine, and valine). And of course it is just as easy to understand how these compounds can compete with one another for entry into the cell, just like any other enzymatic competition.

There are probably many different permease mechanisms. In some, there may have to be two enzymes whose active sites are close to one another (Fig. 15-9). There may then have to be a carrier molecule between them (Fig. 15-10), particularly if they are separated by lipid and the permeant is hydrophilic, or it may be sufficient just to have their two sites close together and the movement from one to the other—which may not be more than a few angstroms—may be diffusion, (although this is generally called a *facilitated diffusion*).

Notice that in these permeation reactions it is possible to have the same molecular

FIGURE 15-7
A bank of identical enzymes sitting in the same
position on a membrane will all push their products
in the same direction relative to the membrane.

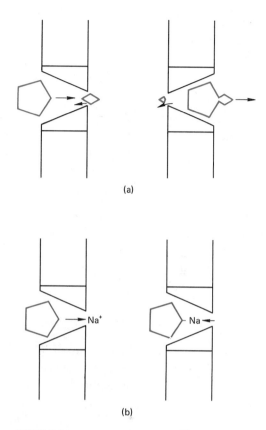

(a)

(b)

FIGURE 15-8
(a) A permease that receives "pentane"
from the left and "diamondic acid" from the
right and undergoes a conformational change
so it spits out "diamondyl pentane" on
the right and "water" on the left. (b) A
permease that attaches Na ions to a substrate, but can receive the substrate only
from the left and the ions only from the
right so it translocates ions from right to left.

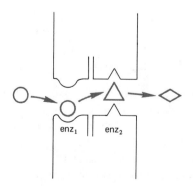

FIGURE 15-9
A permease system consisting of a set of
two enzymes. Enzyme 1 receives circles from
the left, makes them into triangles, and
pushes them into enzyme 2, which makes
them into diamonds and spits them out on
the right.

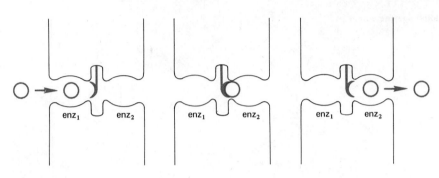

FIGURE 15-10

A permease system that requires an internal carrier. The ball by itself is highly insoluble in the internal environment of the membrane. It has to be bonded by enzyme 1 to a bound carrier crescent which rotates it into a second enzyme site, which breaks the crescent-ball bond and releases the ball on the right.

species appear on both sides of the membrane; but it is equally possible that the permeases are some of the early enzymes in a metabolic pathway, so the permeant is being transformed chemically while it is translocated. It is easy to conceive of an amino acid entering a bacterium through a series of activating enzymes, so it does not appear in the cytosol as a free amino acid but is always an amino acyl adenylate or an amino acyl tRNA.

We pointed out in Chapter 11 that secreted proteins appear in the lumen of the endoplasmic reticulum at the time they are synthesized. We can now understand this in terms of vectorial enzyme catalysis. For example, the ribosomes of a pancreatic secretory cell may be built into the ER membranes so the polypeptides are forced out and across the membrane during polymerization. The ribosomal complex could also include enzymes that attach the newly made polypeptides directly to carrier lipids in the ER membrane. It is too early to specify a detailed mechanism, but it should be easy now to see a general mechanism.

Another very general vectorial pattern is the transfer of a substrate from enzyme to enzyme in a metabolic pathway, just like on an assembly line, with the active sites of all the enzymes lined up so the substrate can move directly from one to another. We will see a number of examples of this as we proceed.

15-6 QUATERNARY STRUCTURE AND PROTEIN COMPLEXES

We have already looked at some structural protein complexes, such as the fibers and microtubules. It is now time to consider the role of quaternary structure in enzyme catalysis and control mechanisms. As a point of departure, we will consider the hemoglobin molecule.

We have already pointed out that normal, adult, mammalian hemoglobin consists of two α and two β chains. Both chains are clearly related very closely to myoglobin;

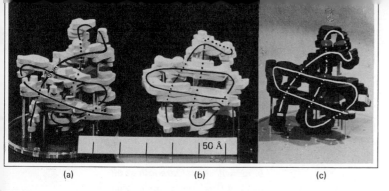

(a)   (b)   (c)

FIGURE 15-11

Models of myoglobin (a) compared with the α and β chains of hemoglobin, (b) and (c). The contours represent the distribution of electron density within the molecules; at this resolution, only the general shape of each molecule can be seen, but the chains can easily be traced. The N-terminal ends are at lower left. The great similarity between the three chains is obvious. (Courtesy of Max Perutz.)

FIGURE 15-12

Right, a view of the assembled hemoglobin molecule, consisting of two α chains (white) and two β chains (black). Notice that the four chains fit together very closely, and that there is a dyad axis (shown emerging at top, center) running through the tetramer. Two heme groups with their oxygens can be seen. (Courtesy of Max Perutz.)

their amino acid sequences are very similar and they are folded into the same conformations (Fig. 15-11). These similar or identical subunits are **protomers** and the whole hemoglobin molecule is a **multimer** (Fig. 15-12). Detailed examination shows that the surfaces of the protomers fit together very well and there are several points where bonds can occur between them; all these bonds are *noncovalent*. Notice that, in accordance with the principle of self-assembly, the molecule has an axis of symmetry.

    The interaction between the four protomers can be illustrated by the behavior of hemoglobin in an animal. The molecule, carried in red blood cells, picks up oxygen at the animal's surface (lungs, gills, etc.) and carries it to the muscles and internal organs, where it must be released. It is an advantage to have a molecule that binds oxygen strongly when the oxygen tension is high (at the surface) and weakly when it is low (in the muscles, etc.). Some species have hemoglobins that bind oxygen with different strengths under different conditions, while others bind equally well under all conditions,

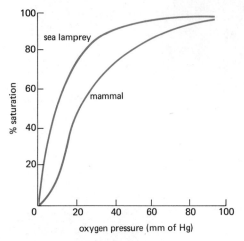

FIGURE 15-13
*Saturation curves of two hemoglobins.
The primitive type found in the lamprey
binds with equal strength at all
oxygen tensions (as myoglobin also
does), while the mammalian type
binds weakly at low tensions.*

as the curves in Fig. 15-13 indicate. The simple hyperbolic curve is characteristic of hemoglobins that have only a single polypeptide, like myoglobin. The sigmoid curve is typical of the mammalian hemoglobins whose four protomers, each with a heme, provide differential affinities for oxygen. At low oxygen tension, there is very little tendency to pick up oxygen; in other words, there is a great tendency to release oxygen. As the oxygen tension increases, there is a stronger tendency to bind oxygen.

The details of the interaction between the protomers are unknown. The hemes are too far apart to interact directly, so the effect must be transmitted through the protein. The sigmoid curve indicates that it is difficult to bind the first oxygen atom, but it is easier to bind the second, easier still to bind the third, and very easy to bind the fourth. Oxygenation of one protomer produces an effect that is transmitted to the other protomers, making them more likely to become oxygenated; this is therefore a *cooperative effect,* and because the same ligands are bound to all of the protomers, it is a **homotropic** effect.

The enzyme effects most closely related to cooperative binding in hemoglobins were discovered through some rather devious routes. Aaron Novick and Leo Szilard were growing a tryptophan auxotroph of *E. coli* in a chemostat with limiting tryptophan. The amount of tryptophan in the growth tube is negligible, of course, under these conditions and Novick and Szilard found that the bacteria were spewing out enormous amounts of an unidentified compound into the medium. Like tryptophan, the substance had a strong absorption maximum at 280 nm. The same bacteria grown in a medium with plenty of tryptophan did not pour out this substance.

Novick and Szilard concluded that the bacteria had a feedback-control mechanism for regulating tryptophan synthesis. We have already seen that excess amino acids will repress the synthesis of the enzymes of the pathway, but those that have already been made are quite stable and they would continue to operate if something did not turn them off; their end product is again the logical control device. Novick and Szilard therefore assumed that the bacteria can sense the amount of tryptophan in their intracellular pools and inhibit the enzymes of tryptophan synthesis in response to an excess. The chemostat is a highly unnatural condition for the tryptophan auxotrophs; they try to make tryptophan but they can only carry precursors up to the point of the mutational block. Since there is virtually no tryptophan to inhibit the enzyme system, they continue to make the last precursor and pour it into the medium.

Other studies led to the same conclusion. Richard Roberts and his group grew bacteria with radioactive glucose to study the pathways of biosynthesis from glucose. They found that if they added an unlabeled, exogenous amino acid, it appeared in the cell protein. This showed that the exogenous amino acid inhibited the biosynthetic enzymes and shut off the flow of radioactive glucose through the pathway. Following Edwin Umbarger, who has studied this phenomenon extensively, we call this **feedback inhibition** of enzyme action. Umbarger points out that feedback inhibition has at least three important features in all of the systems studied so far.

1. Only one enzyme in the pathway is generally inhibited and this is almost always *the first enzyme specific to the pathway.*

2. The inhibitor is the end product itself and not a special derivative. This is understandable; there is no particular reason to evolve a special enzyme to change the end product into some indicator substance.

3. The inhibitor seems to compete with the substrate of the control enzyme for the active site. However, the mechanism of inhibition turns out to be more complicated and interesting than a simple competition. Competitive inhibition occurs in cases where the substrate and the inhibitor look very much alike, but the end product of a pathway is generally very different from the first precursor to the pathway. The mechanism of inhibition turns out to have a great deal in common with the interaction between the hemoglobin protomers.

15-7 ALLOSTERIC INTERACTIONS

The enzyme L-threonine deaminase is the first enzyme specific to the biosynthesis of isoleucine and it is inhibited by its end product. Jean-Pierre Changeux found that the kinetics of inhibition (Fig. 15-14) are apparently competitive, even though threonine and isoleucine bear little resemblance to one another. However, the lines have some curvature near the $1/V$ intercept and if the inhibition were truly competitive they should be absolutely straight.

The enzyme can be made insensitive to isoleucine inhibition by heating it to 55°C for several minutes, but if isoleucine or threonine are present during the heat treatment, the enzyme is protected. Isoleucine and threonine also protect the enzymatic activity of the protein against inactivation by heating. This means that the substrate and inhibitor bind to sites on the enzyme and hold it in a conformation so it is somewhat protected against denaturation. However, Changeux showed that the two compounds bind to different sites; for example, isoleucine binding probably involves an —SH group because p-chloromercuribenzoic acid (PCMB), which attacks sulfhydryl groups, affects the inhibition but not the enzymatic activity. The enzyme can be protected against PCMB attack by isoleucine.

These studies indicate that there are two separate sites on the enzyme: A **catalytic**

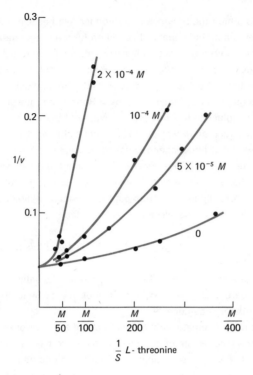

FIGURE 15-14
Kinetics of inhibition of L-threonine deaminase by varying concentrations of isoleucine.

site for threonine and an **inhibitory site** for isoleucine. There is clearly some interaction between them; the whole enzyme acts very much like hemoglobin, since the binding of one ligand apparently changes the conformation of the protein so the binding of the other ligand is affected. However, in contrast to hemoglobin, the effects are **heterotropic** because the two ligands have very different shapes and bind to different sites.

Both homotropic and heterotropic effects are examples of what Monod, Changeux, and Jacob call **allosteric effects.** They are all cases of indirect interaction between distinct binding sites on proteins. The ligands that interact with these sites are both substrates and **allosteric effectors,** either activators or inhibitors. At present, there are several alternative theories about allosteric effects and it is hard to distinguish between them experimentally; the differences can only be appreciated by real sophisticates, and for our purposes it will do to describe the theory developed by Monod and Changeux with Jeffries Wyman, since it fits so well with other concepts we wish to develop about protein structure.

The model emphasizes the fact that all allosteric proteins are small multimers — oligomers of about two to six subunits. It is assumed that each protomer of the oligomer has one distinct, specific binding site for each of the ligands that can interact with it; for example, there may be one site for the substrate and another for an inhibitor.

The whole oligomer must have at least one axis of symmetry. We assume that an allosteric oligomer can exist in at least two distinct conformations that differ in the domains of bonding between protomers, such as the distribution of bonds between them or the strength of their interaction; however, all of these conformations have the same symmetry. These two or more distinct states differ in the conformations of protomers, and the conformations are constrained by interactions with the other protomers. Therefore, in making the transition from one conformational state to another, each protomer is changed so its affinity for at least one of its ligands is changed. Finally, when the protein passes from one state to another its whole symmetry is conserved, so that all of its protomers impose the same conformational restraints on one another and all have the same affinity for a given ligand.

To fix these ideas a bit more, let's think about some of the interactions between protomers that we might observe. The tektin protomers we discussed in Chapter 11 were all assumed to associate **heterologously** through *different* domains of bonding. Figure 15-15 shows that we might also find heterologous associations within oligomers, but we would then expect the oligomer to have an axis of rotational symmetry as shown. (The degree of symmetry might be either odd or even, you notice.) We might

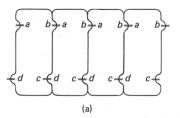

(a)

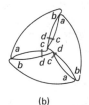

(b)

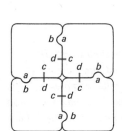

(c)

(d)

FIGURE 15-15
Types of bonding between protomers.
(a) In a heterologous association the domain of bonding is made of different binding sets (a-b, c-d). This can lead to an infinite, linear multimer (a), or to closed multimers with rotational symmetry, (b) and (c). In an isologous association (d), the domain of bonding is made of identical binding sets (a-b, b-a) and the domain of bonding itself is symmetrical. Only a finite multimer with an even rotational symmetry is possible.

also expect to find **isologous** associations as shown in the figure and these would always have only even symmetry numbers. The subunit structure of many enzymes has been investigated; the symmetry numbers are almost always even, suggesting that isologous bonding is more common than heterologous bonding.

The bonds between the protomers in an oligomer should be quite specific and stable, so any factor that causes a conformational change in one protomer will probably cause a change in structure in the other protomers. In isologous bonding, the domain of bonding always has a dyad symmetry axis, which means that any force that causes protomer 1 to push on protomer 2 will cause protomer 2 to push on protomer 1 in the same way. (If Tweedledee hits Tweedledum, Tweedledum will retaliate in kind.) Figure 15-16 shows a simple example of the way a change in the internal bonding of one

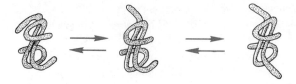

FIGURE 15-16

Above, transitions between two states of a dimeric protein. Because of the symmetry of bonding, any force that tends to change one protomer into a different conformation will also tend to change the other, so the mixed dimer in the middle will be almost nonexistent.

FIGURE 15-17

Right, the model applied to threonine deaminase. When the protomers are in their circular state, they have a high affinity for threonine (circles) and a low affinity for isoleucine (triangles). In their oval state, the situation is reversed. As the concentration of isoleucine rises, a larger number of molecules will be stabilized in the oval conformation, reducing the enzyme activity.

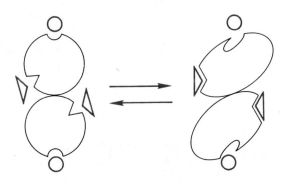

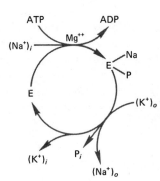

FIGURE 15-18

Left, the sodium-potassium pump. The enzyme (E) is activated by ATP and forms a complex with sodium ion. The activated complex is discharged by potassium ion, releasing the sodium and phosphate. The enzyme presumably undergoes conformational changes so these reactions are separated to different sides of the membrane.

protomer can make the other protomer undergo the same change. It thus becomes relatively easy to see how such a protein will exhibit cooperative effects; if the molecule were a two-protomer hemoglobin and the binding of oxygen to one protomer caused it to assume a different shape, the force would be transmitted to the other protomer, which would then assume a shape to which oxygen would bind more readily.

Now we can understand the allosteric inhibition of an enzyme with this kind of example. Suppose that threonine deaminase contains two protomers, each of which has one site for threonine and one for isoleucine (Fig. 15-17). Let the enzyme be capable of a reversible transition between two states, A and B, and suppose it is a good enzyme in state A and a poor enzyme in state B. We assume only that isoleucine has a greater affinity for state B. There will be a natural equilibrium between states A and B, but as the concentration of isoleucine increases the equilibrium will be driven farther in the direction of B, because proteins that assume the B conformation will tend to be stabilized by binding isoleucine.

By making essentially the opposite assumptions we could devise an enzyme system that was activated by an effector rather than inhibited; it is only necessary to assume that the effector tends to stabilize the system in the state where it is a *better* enzyme. Monod, Wyman, and Changeux have shown that the equations derived from their model will satisfy all of the available data about allosteric transitions. However, let us keep in mind that other, very similar models will also satisfy the data.

The important point that we will try to illustrate in later chapters is that multimeric proteins have just the properties required of allosteric systems and that they can be activated or inhibited by a wide range of ligands. In some cases, the transition from one state to another may actually be an association and dissociation of protomers. In others it is more subtle; for example, the actual conformational change in the protomers of hemoglobin is so slight that it has taken the highest resolution x-ray diffraction studies to detect it.

It should be pointed out that in one case we know something about the structure of an allosteric enzyme. John Gerhardt and Howard Schachman have shown that aspartic transcarbamylase—the first enzyme in the pathway of pyrimidine biosynthesis, which is inhibited by CTP—consists of six subunits. Four of them contain enzymatic sites and two contain inhibitory sites. Inhibitions must therefore occur through interactions between the protomers. It will be interesting to see whether other enzymes are built similarly.

We will leave this subject with one important point: Allosteric transitions are precisely what we expect of the repressors that regulate protein synthesis. We now know, primarily from the work of Walter Gilbert and Benno Muller-Hill, that the *lac* repressor is a protein that weighs about 150,000 daltons. It consists of four protomers. It binds inducers such as IPTG; it binds to DNA containing an intact *lac* operator, but not to the DNA of O^c mutants; and it does not bind well in the presence of IPTG. We therefore assume that IPTG and similar inducers produce an allosteric change in the repressor.

15-8 COUPLED TRANSLOCATION MECHANISMS

Many of the systems that translocate materials across membranes carry two molecules or ions together, either in the same direction (*symport*) or in the opposite direction (*antiport*). The best example of this kind of system is found in many animal and plant cells; it is an ATPase that simultaneously translocates K^+ ions inside and Na^+ ions outside. The rate of translocation of one ion is strongly dependent upon the concentration of the other, and there is approximately a one-for-one exchange of the two. Figure 15-18 shows how this ion pump apparently operates.

Halvor Christensen and his associates have shown that in many animal cells the system for accumulation of amino acids depends upon K^+ ions; the ion and the acid are apparently carried in opposite directions by a single permease. The uptake of sugars in some animal cells also depends upon a gradient of K$^+$ ions. On the other hand, William D. Stein has shown that the system for translocating glycerol and certain sugars into red blood cells behaves as if two molecules of the substrate must be carried simultaneously.

All of these systems are relatively easy to explain on the basis of allosteric transitions. Antiport systems presumably require the simultaneous bonding of two different ligands from opposite sides of the membrane, with a subsequent conformational change that deposits them in reverse on opposite sides. In a symport system, both ligands are bound from the same side. In either case, there is a cooperative effect exactly analogous to the cooperativity of hemoglobin or to the activation of enzyme action by an allosteric effector.

15-9 INTERALLELIC COMPLEMENTATION

Complementation was defined very simply in Chapter 5 for the *r*II mutants of phage T4. According to the simplest and most naive conception of genome-protein relationships, mutations that fall into different genes can be identified by complementation tests because each gene makes a single polypeptide which is itself an enzyme or other functional unit. But now we know that many — if not most — enzymes are multimers, and this fact permits us to understand some rather anomalous results from complementation experiments.

It did not take long after Benzer defined the cistron for a few people to find systems that did not conform to this simple concept. In the *r*II case, mutations could be classified into A and B cistrons by complementation tests, and all mutations in each cistron mapped in one locus. In other cases, however, mutations could be found that all produced a single phenotype and all mapped together, but they complemented one another in a strange pattern that was not simply related to the genetic map, and some failed to complement *any* of the others. The phenomenon is called **interallelic** or **intracistronic complementation.**

It is generally useful to establish a complementation map that looks superficially like a genetic map using deletions, but is really just a way of summarizing the data. We test all of the mutants against one another for complementation and summarize the results in a matrix, where + means that complementation occurs and 0 means that it does not. The mutants generally fall into classes with identical complementation properties; one of the simplest situations we could have would be three classes of mutants, A, B, and C, where all those in A complement all those in B, but those in C do not complement at all. This is shown in the following matrix:

	A	B	C
A	0	0	0
B	0	0	1
C	0	1	0

We then represent each class by a line and draw a map that is something like a deletion map, where two lines that overlap indicate no complementation and two nonoverlapping lines indicate complementation. (But you must not think of this as a deletion map!) The above matrix gives the map:

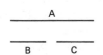

All cases of interallelic complementation have some interesting features in common. First, the complementation map is always linear; this in itself is rather remarkable, since there is no a priori reason why a map of function should be anything like a genetic map. Moreover, the complementation map tends to be collinear with the genetic map; this is even more amazing. In fact, the complementation map often assumes its simplest form when it and the genetic map are drawn as circles, spirals, and other simple loops. The significance of this, if any, still eludes us.

In several cases, the enzyme affected by the locus is known. The activity of enzymes made in complementation tests is never more than about a quarter of the wild type. When the protein can actually be examined, it is found to differ significantly from the wild type in heat stability, K_m, or some other characteristic. Perhaps the most significant point is that complementation within a cistron occurs only when the enzyme informed by the cistron is a multimer, and it is now clear that such complementation is a consequence of interaction between different protomers. This implies that one should be able to achieve *in vitro* complementation simply by isolating the proteins made by two complementing mutants, separating them into their protomers, and recombining them. This has been done now in a few cases. Our purpose here is to explain as far as possible just how two mutant protomers can correct one another's defects.

The general theory of interallelic complementation was developed by Francis Crick and Leslie Orgel. It is based on the premise that loci in which interallelic complementation occurs specify the structure of oligomeric proteins and that complementation results from the combination of protomers made by the two mutants. Furthermore, the mutations that complement one another are those that either affect the active sites of the enzyme or that affect amino acids in special positions relative to the symmetry axes of the molecule.

A very simple dimeric enzyme with the required properties is shown schematically in Fig. 15-19. We assume that the molecule has two active sites, formed jointly by the two

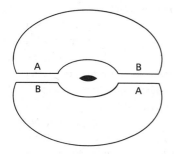

FIGURE 15-19
A protein with two active sites made from the A and B regions. A dyad symmetry axis runs through the center perpendicular to the paper.

protomers. A mutant that produces only protomers defective in region A or only those defective in region B will produce no active enzyme. However, if two genomes with separate mutations for A and B are combined they can produce dimers of nonidentical protomers that have one good site. There are four ways to combine the two protomers (AA, AB, BA, and BB); only half of these (AB and BA) are active and they have only half of their normal activity, so the maximum enzyme activity will be 25% of normal.

This model is probably too simple to be realistic. A better model is shown in Fig. 15-20. We again have a dimer of two identical polypeptides with a dyad axis running horizontally between them in the figure. There are two kinds of interactions between the protomers. First, there are points close to the dyad axis where like regions come together (B, D, F, and H); second, there are regions farther away from the axis where unlike regions come together (A with C, G with I).

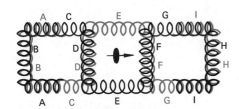

FIGURE 15-20

A more complex dimer, with a symmetry axis running in the direction of the arrow, that may exhibit interallelic complementation.

Now suppose a mutation occurs that affects the structure of region A. Since A comes in contact with C in the dimer, it is possible that the unmutated C region can hold the mutated A region in the proper shape; in this case, the mutation in A would not be detected ordinarily. However, if the C region cannot hold the molecule together properly, then the mutation will be detected but will probably be noncomplementing, since it is unlikely that a second defect, in C, can correct the first in A.

On the other hand, a mutation that affects region B will certainly appear as a mutant, since the two B regions cannot correct each other by coming into contact. But it is possible that a second mutation—for example one that affects region D or F—could complement the first because the good B region of one protomer can hold the defective B region of the other in the right shape, and similarly for the D or F region.

This leads us to expect that mutations that affect regions of the protein far from the axes of symmetry will either be undetected or noncomplementing and that those near the symmetry axes will often complement one another.

In general, a mutation can lead to a derangement of structure that will be transmitted to the active site, and this derangement can be corrected through contact with a normal protein segment on a different protomer. This contact can occur in many ways in a protein, but the contact regions will tend to be very close to the symmetry axes.

It is somewhat more difficult to make this theory account for the approximate collinearity between the complementation and linkage maps often observed. This arises in part from the collinearity of the nucleotide sequence in the genome and the amino acid sequence in the polypeptide, with the assumption that derangements of structure due to mutations will tend to spread along the polypeptide chain, rather than at some angle to it. However, the complexities of protein structure will tend to obscure collinearities. In considering applications of their theory to several specific enzyme models, Crick and Orgel conclude that there will tend to be collinearities, but there will be many exceptions.

B. Immunoglobulins

15-10 BIOLOGICAL RECOGNITION MECHANISMS

One of the most important things that a well-adapted organism must do is recognize certain features of its environment. Primitive organisms may have a very limited ability to do this, though even bacteria will swim toward the light or toward a source of carbon or oxygen. More sophisticated organisms have complicated sense organs and nervous systems that permit them to recognize many things and respond to them intelligently. Many also have chemical devices for detecting bacteria and other noxious foreigners and eliminating them; we will briefly describe this system here.

Chemical recognition can only take one form: chemical bonding. Two molecules can be said to recognize one another only if they bond to one another in some specific way. However, recognition of this type should also be spontaneous; the molecules should fall together by themselves, without an additional enzyme, and so in general, only weak, noncovalent bonds will be involved. In this sense, we can say that two DNA strands recognize one another through hydrogen bonding, or that a tRNA molecule recognizes a specific mRNA codon.

Healthy plants and animals have many mechanisms to protect themselves against pathogenic microorganisms. Some of these are purely mechanical or enzymatic; intact skin keeps them out and the mucus layers in the nose, throat, and other passages traps them from the air. Tears contain the enzyme lysozyme which attacks bacterial cell walls. However, if microorganisms get past these barriers into places where they can grow, there is generally a second line of defense, consisting of phagocytic white blood cells that pick up the invaders and digest them. In most vertebrates, particularly birds and mammals, the foreign material is picked up by large mononuclear leucocytes called **macrophages** (histiocytes). Rather than being completely digested, the foreign material is carried into a lymphocyte system in the spleen, lymph nodes, and other organs, where antibodies are made against it. These antibodies are released into the blood and carried back to the points of infection, where they combine with and inactivate the invading substances.

An enzymatic protein has one relatively small region that serves as an active site, but the rest of the protein also has a specific structure. Any small group of amino acids, monosaccharides, or other organic molecules of comparable size might, potentially, be recognized by an appropriate antibody; any such specific group is an **antigenic determinant** or **epitope.** In every case, we ask whether a structure is an epitope by injecting it into an immunologically competent animal and seeing whether the animal makes antibodies to it. In other words, an epitope is any structure that is **immunogenic** — capable of eliciting antibodies against itself — and the animal that can make those antibodies is **immune** to the epitope.

epitope binding
to a paratope

Since epitopes are relatively small structures, only a small part of the antibody molecule is required to bind with them; this portion of the antibody is a **paratope.**

An epitope and paratope that combine with one another specifically are **paratactic** and the act of combination is **parataxis.**

One peculiarity of the experimental system is that epitopes are not immunogenic unless they are carried on fairly large structures the size of proteins or bacterial cells. The term *antigen* is used for this whole structure regardless of how many distinct epitopes it carries. A **hapten** is an epitope that can combine with preformed antibodies, but is not immunogenic in itself.

Every immunologically competent animal must carry a **paratypic dictionary** consisting of a complete list of all the epitopes against which it can make antibodies. Each animal seems to begin life with a more extensive dictionary, but during its fetal life and shortly after birth it sorts out and discards paratopes against the epitopes of its own body—its own **idiotypes.** Some of these discarded epitopes may be common to every other individual in its species, while others may occur in no other individual, except an identical twin with exactly the same genotype. Babies are very naive, of course, and a fetal or neonatal animal can be fooled into accepting other epitopes as its own. For example, suppose you have mice of two strains, A and B, that make characteristic A and B epitopes. If A is injected with some cells from B, it will make antibodies against the B epitopes and will somehow reject or discard the foreign cells; B will do the same against A cells. However, if a very young A mouse is injected with B cells and then in later life it is challenged with another dose of B cells, it will not make antibodies against them; it has become **tolerant** of the foreign epitopes.

While immunological isolationism has its obvious advantages in fighting disease, it also has its drawbacks. It would be convenient if we could transplant organs willynilly into anyone who needs them, but our bodies tend to reject grafts of all kinds. If you need a new kidney, you have a fairly good chance of retaining one donated by a genetic relative, who is likely to have many of the same idiotypes, but if you have no generous siblings or cousins, you are in trouble. It has been suggested that there should be a systematic program of injecting infants with cells from other people to create groups of individuals who are mutually tolerant of one another and who can serve as donors for one another in emergencies; such a system would probably require a more altruistic society than we now live in—or would it help to create such a society?

The problems connected with blood transfusions are quite well known. The blood groups A, B, AB, and O are due to specific oligosaccharide epitopes on red blood cells and elsewhere. Everyone carries antibodies to the epitopes he does *not* have, so an A individual has anti-B antibodies and a B individual has anti-A antibodies. An AB individual has no antibodies at all in this class, but an O individual lacks both epitopes and has both antibodies. While everyone can accept blood from someone of the same type, there is a certain amount of leeway. Someone with type AB blood can accept a certain amount of blood from anyone, since the limited amount of antibody against his own blood cells will generally not hurt him; similarly, someone with type O blood can generally donate to anyone, since his blood cells carry no A or B epitopes. However, these transfusions are not always safe; death has sometimes resulted from them, and in practice a test mix is always done beforehand if this kind of transfusion is contemplated.

There are many other blood groups that make life complicated. The most serious difficulties arise from the **Rh (rhesus) factors.** At least three different genes seem to be involved in this system, but by way of example everyone's blood can be characterized as either Rh^+ or Rh^-; only about 15% of the population is Rh negative. A couple who are both Rh^+ or both Rh^- will have no problems with their children from the rhesus factors, nor will an Rh^- father and an Rh^+ mother. However, trouble can occur with an Rh^+ father and an Rh^- mother; the first one or two children from such a marriage will generally be normal, but during these first pregnancies rhesus antigens in the blood from the Rh^+ fetus may get across the placenta into the mother, inducing her to make anti-Rh antibodies, which will then get back into the blood of later babies and cause a severe anemia in the newborn child called **erythroblastosis fetalis.**

The immune system is often detrimental to an individual in another way. A person may acquire antibodies against a foreign substance upon first contact; then in subsequent contacts, the substance combines with these antibodies and produces a severe reaction, commonly called **hypersensitivity** or **allergy.** The reaction may be annoying and consist of itching or sneezing, or it may be deadly and consist of an **anaphylactic reaction** in which the person goes into shock, suffers asphyxia, and may die on the spot. It seems rather paradoxical that a system designed to protect an animal against pathogenic foreigners should contain elements that allow it to die when confronted by comparatively harmless substances such as bee venom, which is usually just painful, or pollen, which is normally quite innocuous.

15-11 ANTIGEN-ANTIBODY REACTIONS

It is important to understand just what an antibody preparation consists of. Remember that enzyme preparations are initially just broken cells that can catalyze some reaction; it takes a lot of purification to get a single protein with the enzyme activity. Similarly, antibody is initially just the blood serum from an animal that has been injected with an antigen; the serum itself contains an enormous variety of proteins, and immunization simply adds one more. At this stage, it is perhaps best to speak of an **immune serum** and not even dignify the crude preparation with a term that implies any degree of purity.

Suppose we inject rabbits with egg albumin (EA). The rabbits produce anti-EA antibodies, which we can detect with a number of tests, the simplest of which is perhaps the **precipitin reaction.** We mix a given amount of EA solution with the rabbit antiserum; the antigens and antibodies will interact and produce a precipitate, which can be centrifuged out of solution and analyzed. The amount of precipitate formed with an EA solution of known concentration can be a measure of the amount of antibody in the serum. However, things are not quite this simple. Figure 15-21 shows the amount of precipitate obtained by mixing different amounts of antigen and antiserum. There is a maximum at an optimal ratio of one reactant to the other, not a continuous increase as you might expect.

The reason for this curve is shown in Fig. 15-22. Each antibody molecule has two paratopes on it; this number emerges from a number of measurements, and it is best to simply accept it for the time being. Each antigen molecule is assumed to have at least two epitopes on it; the number is not important, so long as it is greater than one.

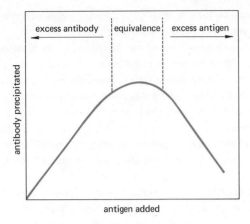

FIGURE 15-21

Precipitation in an antibody-antigen reaction.

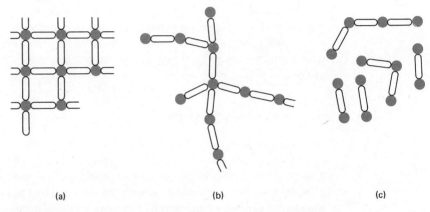

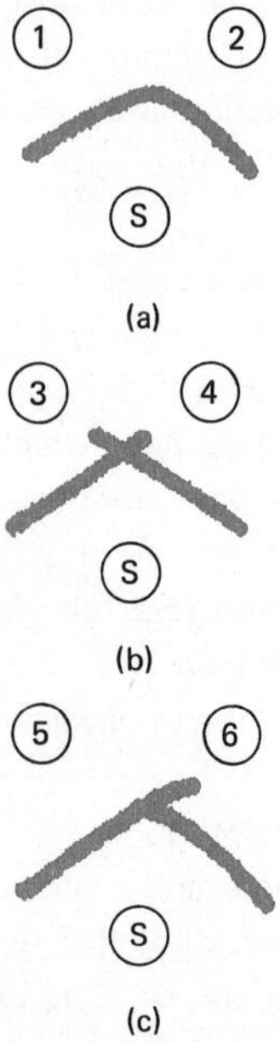

FIGURE 15-22

Antibody-antigen complexes. When the two components are mixed in the proper proportions (a) they form an extensive lattice. Even when there is some excess antigen (b) they will form chain-like precipitates. However, when there is a great excess of antigen (c) only soluble complexes will form.

FIGURE 15-23

Precipitin reactions in agar diffusion plates. In each case, an antiserum S, is placed in one well. If identical antibodies are placed in wells 1 and 2 (a), a single band will form. In (b), nonidentical antigens are placed in wells 3 and 4; two bands are formed. In (c), the antigens in wells 5 and 6 are partially identical, as indicated by the spur.

When there is exactly one paratope per epitope, the antibodies will react with the antigens to make a large, insoluble **lattice** that precipitates out of solution. If there is not enough antigen, there will simply be smaller lattices and less precipitate; the amount of precipitate will increase up to a point as more antigen is added, but eventually there will be an excess of antigen and then small, soluble complexes will form as shown, in which very few antibodies will be attached to two antigens simultaneously.

A similar reaction, called **agglutination,** occurs when the epitopes are part of a much larger structure, such as a bacterial or blood cell. Precipitation occurs here, also, but often much more dramatically because the components of the precipitate are much larger. Sometimes people take advantage of the strength of agglutination to increase the sensitivity of a precipitin test. If red blood cells are treated briefly with tannic acid, they become sticky and will adsorb small, soluble proteins. The coated cells can then be used as antigens against a serum containing antibodies for the protein itself and the reaction will be visible as a strong agglutination rather than a weak precipitation.

Precipitin reactions are often carried out in special agar plates (or for simplicity, on microscope slides covered with a thin layer of agar). An immune serum is placed in one well of the agar and various antigens in the other wells (Fig. 15-23). The components diffuse through the agar and form bands of precipitation where they meet. As the figure shows, various patterns are obtained. If the same antigen is put in each well, there will be a single curved band. If the antigens are different, they will form two distinct bands. However, sometimes two antigens are *partially identical* immunologically; an antiserum made against one will **cross react** with the other because they share some of the same epitopes, and on a diffusion plate the cross reaction will appear as a curve with a **spur** on it. The spur represents a region where antibodies are precipitating with the most reactive antigen—the one that has the most epitopes paratactic to the antiserum.

P. Grabar and Curtis A. Williams, Jr., developed an elegant extension of this method called **immunoelectrophoresis.** Figure 15-24 shows how a mixture of materials,

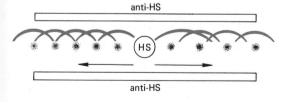

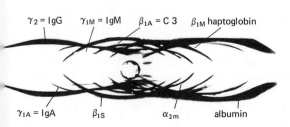

FIGURE 15-24

Immunoelectrophoresis. Normal human blood serum (HS) is placed in the center well and the current is turned on. After the components are separated, anti-HS is placed in the side wells and precipitin bands are allowed to form. A typical picture (courtesy of Nancy Holland) is shown below.

such as unfractionated human blood serum, can be separated by electrophoresis on a thin agar plate. After separation, the current is turned off and an antiserum against the mixture is placed in the long wells. When the components diffuse against one another, each antigen in the mixture forms a single arc. This is a sensitive way to identify components that might not be separable by diffusion alone; it is a standard way to examine components in blood, for example, and it makes it easy to detect abnormal amounts of various components.

The cross reaction is very important in genetic and metabolic studies. Mutants are often obtained that produce no traces of some enzymatic activity, but they can still be studied immunologically because the mutant enzyme has most of the epitopes of the wild type. It is then called a **cross-reacting material (CRM,** pronounced "crim" by the cognoscenti). Although we omitted the point from our discussion of genetic regulation, Jacob and Monod used Z^{CRM} mutants of the *lac* operon to confirm the predictions of the operon model; as you may expect, the CRM was induced and repressed just as it should have been if it were a normal enzyme.

15-12 KINETICS AND SPECIFICITY OF PARATAXIS

Since it involves only weak bonds, the combination of antigens with antibodies must be reversible. From equilibrium studies of the reaction, we can obtain some useful information about the structure of antibodies.

Suppose that every antibody molecule has n paratopes on it. If the antibodies are in solution with a concentration c of epitopes, they will bind an average of m epitopes each at equilibrium, leaving $(n - m)$ free paratopes. The equilibrium will be

$$\text{free paratopes} + \text{free epitopes} \rightleftharpoons \text{bound epitopes}$$
$$(n - m) \qquad\qquad c \qquad\qquad\qquad m$$

and we can define an equilibrium constant

$$K = \frac{m}{(n - m)\,c} \tag{15-13}$$

which can be rewritten

$$\frac{m}{c} = Kn - Km \tag{15-14}$$

The values of m and c can be obtained experimentally by **equilibrium dialysis.** We set up two cells separated by a semipermeable membrane through which soluble epitopes can diffuse freely (Fig. 15-25a); in one cell we put a salt solution (to negate possible Donnan effects later) and in the other the same salt solution with a known concentration of epitopes. At equilibrium there will obviously be equal amounts of epitopes in both cells. Now we add a known concentration of antibody specific for the epitopes to one cell. As Fig. 15-25b shows, the epitopes will bind to the antibodies on one side and the unbound epitopes will come to an equilibrium distribution. The difference in epitope concentration across the membrane will be due to antibody binding. By performing this experiment with different concentrations of epitopes, values for m and c can be obtained.

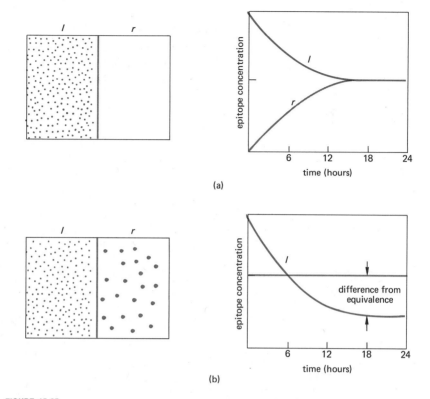

FIGURE 15-25

*Equilibrium dialysis. (a) In the absence of antibodies, the soluble epitopes equilibrate
across the membrane. (b) In the presence of antibody, there is an excess on the right
due to antigen-antibody binding.*

Equation 15-14 shows that values for m/c plotted against m will give a line with
slope $-K$. Furthermore, as c approaches infinity, m approaches n. Figure 15-26 shows
how data from an equilibrium dialysis experiment look when plotted in this way. The
intercept as m increases is 2. Data from many similar experiments give the same
result; the number of paratopes per antibody molecule (when the antibody is not of
the IgM type) always seems to be 2. This is why we are justified in drawing antibody
molecules as double-headed molecules.

Look at the graph again; if a single value of K could be defined, the line would be
straight, but it actually curves quite a bit, indicating that there are several different
equilibrium constants that contribute to the overall equilibrium. Now, remember that
the antibodies used for this kind of experiment were actually obtained by injecting
some antigen into a competent animal and then removing a blood sample. Even if the
epitope against which the antibodies are directed is very homogeneous, the paratopes
that combine with it can be quite heterogeneous, and the curvature of the graph
indicates they are. The meaning of experiments like this is simply that an immunogen

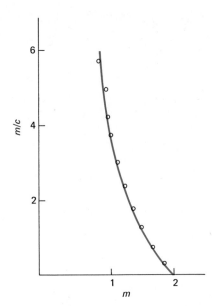

FIGURE 15-26

The results of several equilibrium-dialysis experiments graphed according to equation 15-14. All values are calculated for 160,000 molecular weight of antibody.

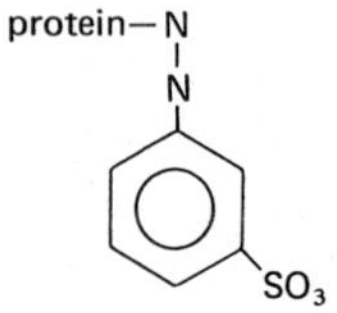

protein—N

Antibodies made against this epitope

were tested for precipitation against proteins bearing these epitopes:

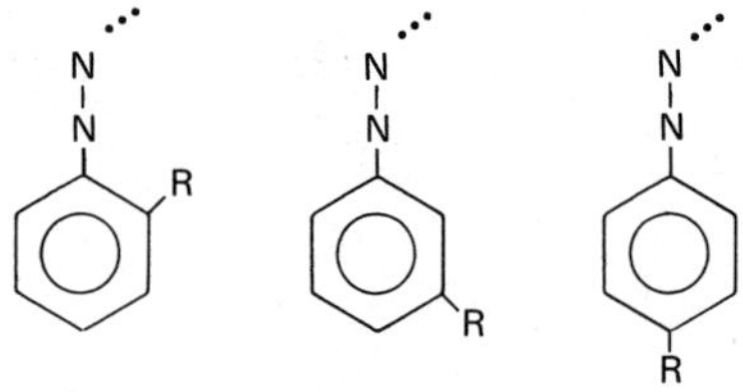

FIGURE 15-27

Antibody specificity.

	ortho	meta	para
R = —SO$_3$H	+ ±	+ +	±
R = —AsO$_3$H$^-$	0	+	0
R = —COO$^-$	0	±	0

Approximate amount of precipitation is indicated by +, ±, or 0 in descending order.

induces the formation of whatever antibodies happen to have paratopes capable of combining with it, and the immunized animal may be able to make a number of different paratopes with very different affinities for the same epitope.

The incredible specificity of antibodies is illustrated very well by a classical experiment performed by Karl Landsteiner and J. van der Scheer, shown in Fig. 15-27. They made antiserum against horse serum proteins bearing the metanilic acid epitope shown and then tested for precipitation with chicken serum proteins substituted with a series of related epitopes. Notice that the antibodies can distinguish between sulfate, arsenate, and carboxyl groups, and that they strongly distinguish the meta position, where the immunogen was substituted, from ortho or para.

This specificity enabled Linus Pauling, David Pressman, and Dan H. Campbell to perform a related experiment which proved that the two paratopes on each antibody molecule are identical. As Figure 15-28 shows, they immunized rabbits with proteins bearing either an R or an X epitope to get anti-R and anti-X serum. They tested this serum against a large ring compound bearing either X or R on its 1 and 6 positions.

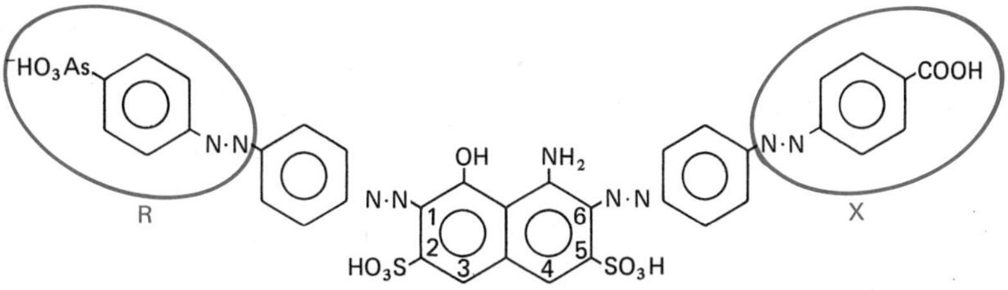

Groups Substituted	Antiserum		
	Anti-R	Anti-X	Anti-R + Anti-X
R R	+ +	0	±
X X	0	+ +	±
R X	0	0	+ +

The precipitate must have the structure:

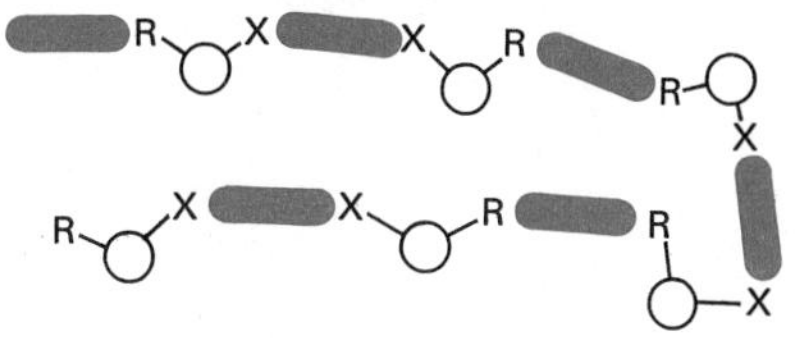

FIGURE 15-28

The R-X experiment that shows the identity of the paratopes on an antibody. The whole compound with R and X groups is shown above. Below is the kind of chain precipitate that R-X can form with anti-R and anti-X together.

Each serum by itself precipitates only the compound bearing a homologous epitope in both positions. A mixture of the two serums precipitates poorly with the compound bearing either two R groups or two X groups, but very well with the compound with one R and one X. In this case, a chain-like precipitate can form only if the epitopes on each antibody are the same.

15-13   BASIC STRUCTURE

The proteins in blood serum may be separated by electrophoresis into the fractions shown in Fig. 15-24, including a serum albumin and a series of serum globulins designated α_1, α_2, β_1, β_2, γ_1, and γ_2. The immunoglobulins are part of the γ-globulin fractions. There are several distinct classes, but on the basis of size (primarily) we can distinguish a type that sediments with a coefficient of about 7S and another of about 19S (the **macroglobulins**).

All of the immunoglobulins are built of two types of protomer—light (L) chains and heavy (H) chains. The immunoglobulin class is determined by the H chain. These classes are designated **IgG** (for **immunoglobulin G**), **IgA, IgM, IgD,** and **IgE.** The H chains of these immunoglobulins are designated γ, α, μ, δ, and ϵ, respectively. The L chains of all immunoglobulins in a particular species are essentially the same; these are designated κ and λ in man, and any individual may have some antibodies made of κ chains and some of λ chains, but these are never mixed in a given molecule. Thus, an IgG molecule has the formula $\kappa_2\gamma_2$ or $\lambda_2\gamma_2$, an IgA molecule is either $\kappa_2\alpha_2$ or $\lambda_2\alpha_2$, and similarly for IgD and IgE. The exceptions are the IgM macroglobulins, which have the structure $(\kappa_2\mu_2)_5$ or $(\lambda_2\mu_2)_5$.

Within each class, one can sometimes find subclasses. For example, in humans there are four types of γ chain and hence four types of IgG molecules (IgG^1, IgG^2, etc.).

The chains of immunoglobulins are held together primarily by disulfide bonds. Each H chain is bonded in this way to an L chain, and the two H chains are also bonded to one another. When an IgG molecule is subjected to mild proteolytic digestion, it separates into two fragments. One of these, Fab, has all of the antigen-binding ability, while the other, Fc, is crystallizable and has no affinity for antigen. In combination, these data lead to the model shown in Fig. 15-29, and this has been confirmed generally by high-resolution electron micrographs.

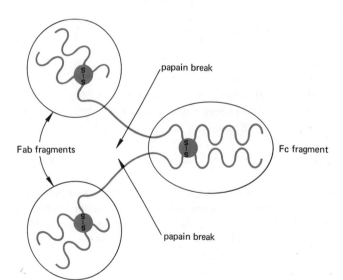

papain break

Fab fragments

Fc fragment

papain break

FIGURE 15-29
General structure of an IgG-type antibody.

The amino acid sequences for several immunoglobulin chains have now been determined. This work has been particularly aided by the discovery that humans and mice suffering from **myelomas,** which are a kind of cancer of antibody-producing cells, are pouring out enormous amounts of one specific immunoglobulin, directed against some specific (but unknown) antibody. The results of all these analyses can be summarized rather easily.

Making some allowance for hypothetical additions or deletions here and there, all of the mouse and human chains so far examined can be understood as multiples of a basic unit of about 114 amino acids. The L chains consist of two such units; there are many homologies between the two halves of an L chain. However, the C-terminal half of all L chains appears to be essentially the same in all human proteins and essentially the same in all mouse proteins. This half is the **constant** or **C portion.** The N-terminal half is the **variable (V) portion,** and here is where the differences between the different L chains lie.

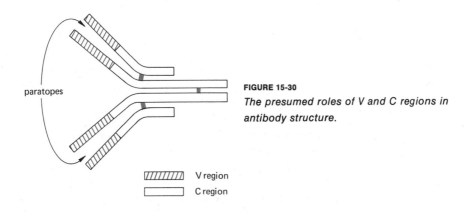

paratopes

FIGURE 15-30
The presumed roles of V and C regions in antibody structure.

////////// V region
☐ C region

Though the H chains have not been examined as thoroughly yet, the same thing can be said for them. A γ chain appears to be equivalent to four of the basic 114 amino acid sequences; the C-terminal $\frac{3}{4}$ of the chain is essentially constant from one H chain to another, but again the N-terminal $\frac{1}{4}$ is variable.

From these results and from the general picture of IgG structure, we can now picture the IgG molecule more or less as shown in Fig. 15-30. The V regions of the H and L chains are put together to make a paratope, while the C regions together make the rest of the molecule, including the whole Fc portion.

15-14 ANTIBODY SYNTHESIS

The whole matter of antibody synthesis and the specification of antibody structure is a long, complex story that we cannot go into deeply. However, it is important to see the story in outline.

The only way that antigen can get into the immune system and elicit antibody formation seems to be through the macrophages. If too large a dose of an antigen is given an animal, some of it may get into the immune system directly; in this form, it apparently kills those cells potentially capable of making antibody against it, producing **immune paralysis** and a subsequent inability of a later dose of antigen to elicit antibody formation. However, this does not happen often; the macrophages are voracious eaters of foreign material, and they quickly find antigens and carry them back to their target cells, which appear to be a type of white blood cell called a **small lymphocyte.** We assume that there is at least one cell—the **virgin lymphocyte**— that has never seen the particular antigen before, but is capable of producing antibodies against it. When stimulated by antigen carried by the macrophage, it begins to produce antibody, primarily IgM. The virgin lymphocyte now becomes an **experienced lymphocyte,** often called a **memory cell.** If the memory cell is challenged with a second dose of the same antigen, it begins an **anamnestic response** (literally, without amnesia) and begins to produce IgG as well as IgM. At the same time, the memory cell begins to proliferate, and the clone of cells **differentiates** at the same time; that is, their form changes in a characteristic way, and they eventually become **plasma cells.** These cells develop a lot of ER and become highly specialized for producing the particular antibody.

The obvious problem is to specify a way for cells to become committed to making specific antibodies—that is, to develop V regions that specify useful paratopes. The whole subject is in such a state of flux and controversy at this time that anything very specific will almost certainly be out of date before these words are set in type. However, we can make some general points.

It has been standard for years to distinguish between two possible ways of informing specific antibody structure—an *instructive* and a *selective* method. An instructive theory assumes that the polypeptides of immunoglobulins can be made to fold up into different conformations; the epitopes of a foreign antigen are presumably carried to the protein-synthesizing sites of lymphocytes, where they force the naive polypeptides to take on a shape that produces maximum parataxis. A selective theory, on the other hand, assumes that every competent animal can make many paratopes and that a foreign epitope somehow selects those cells that make paratopes against it, forcing them to make large amounts of antibody.

It is now clear that instructive theories are untenable. First, they would require at least one molecule of antigen to remain in each antibody-producing cell to act as a template, and it is very unlikely that antigen remains there as long as would be necessary. Antibody production can continue for long times. Antibodies against tetanus have been found 20 years after administration of the last dose of immunogen, and Sir Macfarlane Burnet found that yellow fever antibodies could be found in people who had had the disease 75 years earlier. In contrast, antigen is removed from the body very quickly. Labeled antigen has a half-life of about 6 weeks in rabbits and perhaps 100 days in man; at this rate, there might be one molecule left in the body out of the doses of immunogen usually given against tetanus or polio after 10 years or a little more.

One might be able to devise ways of accounting for these figures in an instructive theory, but the final blow to such a theory is the demonstration that immunoglobulins can be denatured and then carefully renatured in the absence of antigen, so they are once more good antibodies. Thus, immunoglobulins are no different from other proteins; their conformations are determined by their primary structures.

One therefore requires a theory to indicate how cells can be selected that produce specific antibodies and how those cells come to be specific in the first place. These theories can also be divided into two classes. *Germ-line theories* assume that the information resides in a large set of stable genes that are inherited like any other part of the genome. *Somatic theories* assume that there are relatively few genes and that the information carried within them can be "scrambled" to make a large number of possible paratactic polypeptides.

Germ-line theories may have an unreasonable sound to them. There seems to be no limit to the number of epitopes in the universe against which paratactic antibodies might be required. Even if the system was evolved over the history of our planet just to recognize the surfaces of bacteria, viruses, and other biological substances, it has the ability to recognize all kinds of strange rings that may never have existed before the advent of modern organic chemistry. Suppose there are a million different epitopes that can be recognized by the average competent animal; does that mean it has to have a million different genes for making paratypic polypeptides? The first answer we can give simplifies the problem by the square root of the number of epitopes; because two different chains cooperate to make a paratope, it would be sufficient in this case to have a thousand different L chains and a thousand different H chains, and the combination of the two would make a million paratopes. As Edward P. Cohen has pointed out, this also makes it easier to understand how all of this information can be retained in an organism in the absence of any selection. Suppose that 2000 years ago our ancestors found it necessary to make antibodies against some plague bacillus which has never been seen since. There is little doubt that we could make antibodies against the same organism if it ever appeared, but why should the information be retained for so long? The paratopes that were used two millenia ago might have been a combination of L_{57} with H_{734}, out of a possible 1000 of each. But L_{57} might be used by every generation as part of several other paratopes, and H_{734} might be a part of a common paratope needed every year against some common cold virus. Therefore, the information for the two will be retained, even though they might never be combined with each other again.

The situation becomes relatively simple if we assume that there is just one gene for each C region of an immunoglobulin, so there must be only about 10 such genes for the various classes. We then assume that there is a set of 1000 genes for each V region of a light chain and 1000 for each V region of a heavy chain. Two thousand genes, each coding for about 100 amino acids, requires $2000 \times 100 \times 3 = 6 \times 10^5$ nucleotide pairs. A whole haploid set of a mouse or a human may contain 10^{10} nucleotide pairs; the amount of information devoted to immunity would therefore be trivial in comparison with the whole, so the germ-line theory is not so unreasonable.

Theories of somatic diversity are also reasonable, although it is difficult to specify a mechanism. These theories assume that there is a gene for each C region, either combined with a corresponding gene for a V region or not; in any case, there are relatively few genes in any genome. There is then a **generator of diversity (GOD)** that is capable of scrambling the V region in many different ways; a mechanism for the GOD is unspecified, but it might cause repeated crossing over with some other DNA within the V region, or it might produce multiple mutations very close together. As the organism develops and produces a set of virgin lymphocytes, the GOD produces slight differences between their V regions and hence a population of lymphocytes potentially capable of making an enormous number of antibodies. It should be pointed out that experiments which ask about the capabilities of individual lymphocytes indicate that each one makes only a single species of antibody, so by the time the virgin lymphocyte appears, it is presumably programmed to make only one V region for each chain.

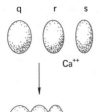

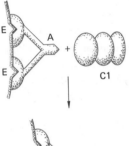

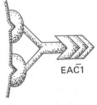

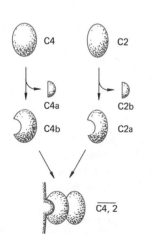

This is one of the most fascinating problems in contemporary biology; perhaps some of you will solve it.

15-15 COMPLEMENT: A STUDY IN QUATERNARY INTERACTIONS

One of the more obscure topics of immunology is becoming a prime example of the effects of proteins on one another in highly structured complexes. **Complement** is the name given to a series of protein components, designated C1 to C9, in normal blood serum. They are part of a complex system designed to lyse gram-negative bacteria (gram-positive bacteria are immune to the system) and foreign blood cells; complement exhibits a fascinating set of protein-protein interactions.

Complement is specific for IgG and IgM antibodies, but no complement component will bind to the antibodies by themselves. When IgG or IgM is bound to an appropriate antigen, it undergoes a change that creates a binding site for the first component of complement, C1. Just to keep this discussion reasonably specific and not lose sight of the purpose of this reaction, we will consider only an IgG antibody molecule (A) specific for certain surface epitopes of an invading bacterium. We will begin with the epitopes (E) bound to IgG, making an EA complex on the cell surface.

C1 is actually a set of three proteins, C1q, C1r, and C1s, which form a C1 complex with no known activity in the presence of Ca^{++} ions. When C1 binds to the EA complex, it undergoes a conformational change that passes from q through r to s, making C1s an active enzyme with esterolytic activity (i.e., it can split ester bonds). We'll draw a bar over any component that is enzymatically active, so the complex is now $EA\overline{C1}$.

C1 can now attack two components, C4 and C2. It splits C4 into C4a, a small piece with no activity, and C4b, which binds to certain cell surface sites. C1 also attacks C2, splitting it into C2b (which has no activity) and C2a, which binds to C4b in the presence of Mg^{++} ions and forms the tight complex $\overline{C4,2}$, which is enzymatically active and is often called C3 convertase. The enzyme $\overline{C4,2}$ also becomes bound to the cell; notice that all of its enzymatic and binding sites were apparently exposed by the C1 enzyme in attacking the inactive precursors C2 and C4.

C3 convertase ($\overline{C2,4}$) can attack C3 and cleave it into a large piece, C3b, and a small polypeptide, C3a, known as an *anaphylotoxin*. From the name, you can tell that this component is involved in the types of cellular reactions seen in allergy and anaphylactic shock. The anaphylotoxins cause contraction of the smooth muscles, including the muscles of the blood vessels; they induce increased permeability of the capillaries, and they also induce a type of white blood cell (polymorphonuclear leucocytes) to become more active and migrate toward the site where complement fixation is occurring.

Meanwhile, C3b combines with the $\overline{C4,2}$ complex to make an enzyme $\overline{C4,2,3}$ that has peptidase activity. This enzyme attacks C5 and again splits off a small polypeptide, C5a, which is an anaphylotoxin. The remainder, C5b, then binds to the cell surface, where it creates a site to which C6 and C7 can immediately bind. This complex has unknown enzyme activities that begin to attack the cell membrane. It serves also as a site to which C8 can bind, and once this has happened the cell starts to lyse, even if nothing else is added. However, C9 now binds to C8 and adds some cytolytic enzyme activities which result in destruction of the cell membrane and lysis.

The release of the anaphylotoxins has induced the appropriate leucocytes to migrate

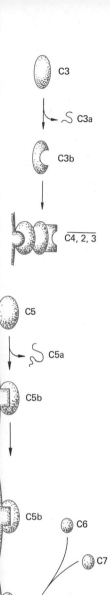

into the area and they now begin to clean up the bacterial debris by endocytosis. The same reaction occurs, however, if the foreign cell to which the antibody binds is a red blood cell, for example from an inappropriate donor, and so lysis of this cell, with all the deleterious effects it can bring, now results.

Complement fixation provides a very sensitive test for the presence of specific antibodies; it is used, for example, to test for antibodies against syphilis. A sample of serum from the patient is mixed with an appropriate antigen from the syphilis bacterium; if antibodies are present, they will complex with this antigen. Complement is then added, and this will react with an antigen-antibody complex if one has been formed. Finally, sheep red blood cells combined with antibodies against them are added; these sensitized cells will bind complement if it has not already bound to syphilis antigen-antibody complexes. If the sheep red blood cells lyse in a short time, they must have bound complement, indicating that there were no syphilis antibodies in the patient's serum.

SUPPLEMENTARY EXERCISES

exercise 15-3 When bacteria are grown on radioactive glucose, all their amino acids are radioactive. However, if unlabeled threonine is added to the medium, both threonine and isoleucine are unlabeled in the cells; and if unlabeled glutamate is added to the medium, glutamate, proline, and arginine are unlabeled in the cells. What do these results mean? (Hint: This has nothing to do with *repression* of enzyme synthesis.)

exercise 15-4 Werner Kundig, Saul Roseman, and Sudhamoy Ghosh have described a general system for translocation of at least nine sugars in several bacteria. It involves two enzymes, designated I and II, and a small (9400-dalton) protein containing histidine, designated HPr:

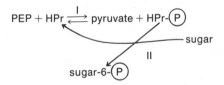

(The histidine of HPr is phosphorylated.) In several different bacteria, single mutants are known that have no enzyme I activity at all; other mutants have no HPr at all; but other mutants have lost the ability to translocate glucose alone, or mannitol alone, or mannose alone, or some other specific sugar by itself. From this information, you should be able to say how specific each part of the permease system is.

exercise 15-5 Shugi Tanaka and E. C. C. Lin have examined this sugar permease system in more detail. They find that, in contrast to the wild type, a mutant for the mannitol enzyme II cannot accumulate mannitol-6-P; they suggest that the enzyme II phosphorylates and translocates simultaneously. Why is it advantageous for cells to use one enzyme for both operations?

exercise 15-6 R. Damadian and A. K. Solomon collected *E. coli* mutants that cannot accumulate K^+ ions. By selecting for amino acid auxotrophy simultaneously, they found one mutant that is a methionine auxotroph and also cannot accumulate K^+. Thirteen independent revertants of this mutant are methionine prototrophs and have also regained the ability to translocate K^+, suggesting that a single mutation can abolish both activities. What do these data suggest about the K^+ permease?

16

**membranes
and energy-
conservation
mechanisms**

We have considered a number of structures built of protein subunits that can contract, bend, carry gas molecules, regulate metabolism, and do other complicated jobs. We will now consider how extensive sheets of proteins, in the form of cellular membranes, can perform some other tasks, principally the transportation of electrons and the regulation of energy-yielding reactions. In order to understand membrane structure in general we will first consider the electron-transport systems of mitochondria, chloroplasts, and bacterial cell membranes.

A. Electron-transport components

16-1 QUINONES

A number of ring structures with ortho- and para-substituted hydroxyl groups can be reduced to diketo forms called **quinones.** For example,

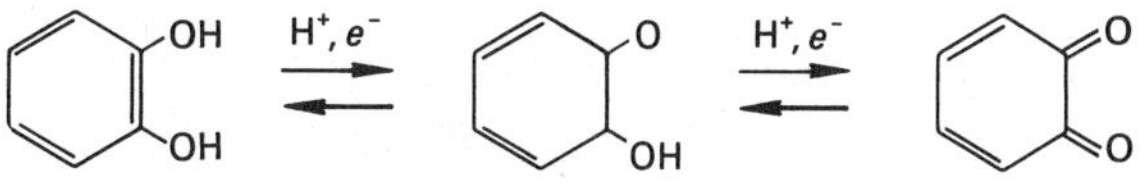

| quinol | semiquinone | quinone |

Because the reaction is reversible, quinones can act as electron carriers. All the known quinones of ET systems are characterized by a side chain made of varying numbers of *isoprenoid* units; they may be designated by the number of units, n, as a subscript, or by the number of carbon atoms in the side chain, $5n$, in parentheses. The following four series are the most important.

1. *Ubiquinone* or *coenzyme Q* is the most common and important quinone; it is really almost ubiquitous. The various members of the series are designated either ubiquinone ($5n$) or coenzyme Q_n, abbreviated $UQ(5n)$ or CoQ_n. The value of n varies from 6 (yeast) to 10 (most plants and animals).

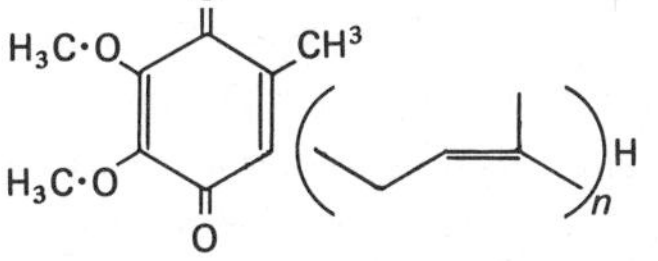

ubiquinone

2. *Plastoquinone* replaces ubiquinone in chloroplasts, where it is probably the chief quinone of photosynthetic electron transport.

501

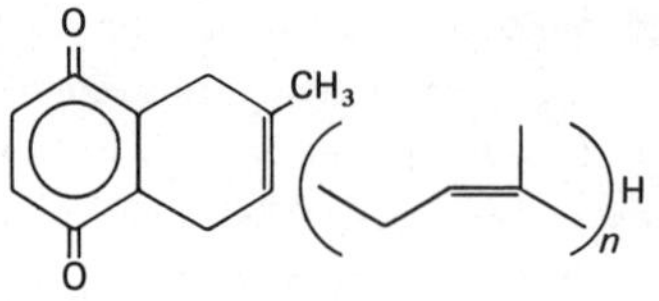

vitamin K$_2$

3. *Farnoquinone* (vitamin K$_2$) has been isolated from a number of bacteria, where it probably replaces ubiquinone.

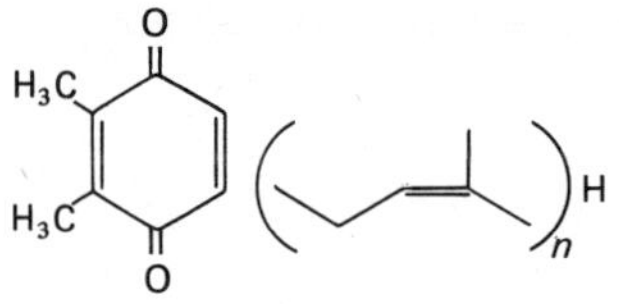

plastoquinone

4. *2-demethyl vitamin K$_2$* [DMK$_2$(5n)] has been isolated from bacteria. Members of the series with $n = 5$, 6, and 7 have been found.

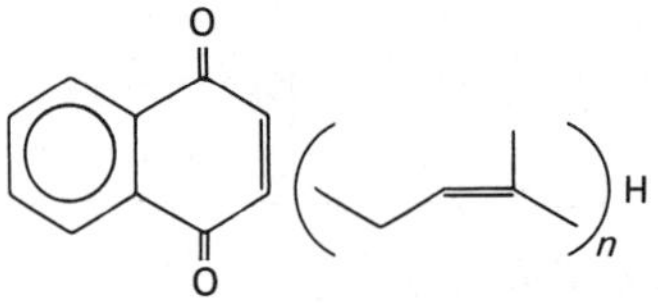

2-demethyl vitamin K$_2$

16-2 THE CYTOCHROMES

Electron-transport systems are built primarily of cytochromes, which are probably arranged so that electrons can pass quite directly from the heme of one to the heme of another with minimal movement and conformational change. Cytochromes were first identified by David Keilin, who showed that characteristic absorption bands in the spectra of animal tissue suspensions could be attributed to colored substances, which he called cytochromes *a, b,* and *c.* A typical cytochrome spectrum is shown in Fig. 16-1. Each protein has three strong absorption maxima: an α band in the region from 550 to 600 nm, a β band at a lower wavelength, and a very strong γ or Soret band close to 400 nm. The shape of the spectrum changes with oxidation and reduction, and most studies of cytochrome action depend upon spectrophotometric observations of reaction mixtures during electron transport.

The many cytochromes that are now known are divided into groups *a, b, c,* and *d,* and are further identified by subscripts. The major characteristics of these groups are given in Table 16-1. When a new cytochrome is discovered, it can generally be

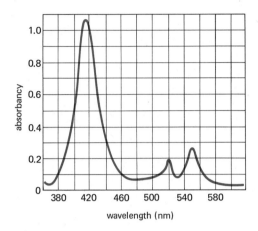

FIGURE 16-1

Absorption spectrum of cytochrome c.

assigned to one of these classes on the basis of these criteria. Until it is finally characterized, it is designated by the name of the organism from which it was obtained and the position of its γ band. For example, cytochrome c(551, *Chromatium*) is a cytochrome of c type derived from *Chromatium* whose γ band is at 551 nm. There are also some cytochromes with two distinct heme groups per molecule, and in some of these one heme may not be a hemochrome. These are named for both hemes (as, cytochrome *cd*) and a nonhemochrome heme is designated by an attached "prime" mark (as, cytochrome *aa'*).

TABLE 16-1

major characteristics of the cytochromes

CLASS	PROSTHETIC GROUP	α-BAND (nm)*
a	heme with formyl side chain	580–590
b	protoheme	556–558
c	heme with covalent bonds to protein	549–551
d	iron chlorin	600–620

*Absorption band of pyridine ferrohemochrome in alkaline solution.

The best-known cytochromes are of the c type, whose hemes are bonded covalently to the protein as shown in Fig. 16-2. The amino acid sequences around the attachment

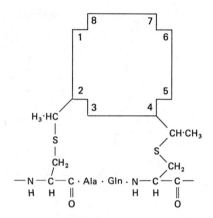

pig, rabbit, beef, horse, kangaroo	Val — Gln — Lys—CYS— Ala —Gln—CYS—His—Thr— Val — Glu
human	Ilu — Met— Lys—CYS— Ser —Gln—CYS—His—Thr— Val — Glu
chicken	Val — Gln — Lys—CYS— Ser —Gln—CYS—His—Thr— Val — Glu
tuna, salmon	Val — Gln — Lys—CYS— Ala —Gln—CYS—His—Thr— Val — Glu
moth, silkworm	Val — Gln — Arg—CYS— Ala —Gln—CYS—His—Thr— Val — Glu
yeast	Lys— Thr — Arg—CYS— Glu—Leu—CYS—His—Thr— Val — Glu
Rhodospirillum cytochrome c_2	? — Ser — Lys—CYS— Leu—Ala —CYS—His—Thr— Phe— Asp
Chromatium cytochrome *c'*	Ala — Gly — Lys Ser —Gln—CYS—His—Thr— Leu— Val

FIGURE 16-2

Heme attachment and amino acid sequences in c-type cytochromes. The close evolutionary relationships between the various proteins is obvious; the more closely related organisms, in general, have the most similar sequences. Notice that the residue before the first cysteine is always basic — either lysine or arginine — and that histidine and threonine always follow the second cysteine. The Chromatium *protein is not a hemochrome type and has only one covalent linkage.*

point are known for several *c* cytochromes, as shown in the figure. The evidence for their evolutionary relationship is obvious.

The number of cytochromes in an ET system varies considerably. A few species, such as the *Clostridium* group of bacteria, have none at all and depend upon electron transport only as far as the flavoprotein level, along with energy released in the amphibolic reactions themselves. Some bacteria have as many as five cytochromes, forming a sequence from the FP-quinone level to the terminal electron acceptor. The general pattern in chemotrophs is

$$\text{cytochrome } b \rightarrow \text{cytochrome } c \rightarrow \text{cytochrome } a$$

There are often several different cytochromes, often of the *a* type, which oxidize cytochrome *c* and in turn reduce a terminal electron acceptor.

16-3 NONHEME IRON PROTEINS

A series of small proteins are known with several iron atoms in them that are *not* part of heme groups. The best known is **ferredoxin,** whose name is meant to remind you of both its structure (*ferrum* = iron) and its function (*redox* = oxidoreduction in laboratory jargon). The ferredoxins have now been found in several different ET systems; they all have extremely low potentials ($E_0' = -425$ mV) and they are used at the most highly reducing ends of ET chains or in special systems where electrons must be transported to acceptors with comparably low potentials, such as hydrogen.

Ferredoxins have evolved from a primitive molecule of about 55 amino acids to more advanced ones, such as the ferredoxin in spinach chloroplasts, which have about 97. Their iron atoms are bound between a series of sulfur atoms, in an arrangement rather like that shown in Fig. 16-3. (Ferredoxin will sometimes be abbreviated **Fd,** and we will sometimes use the subscript "NH" to indicate a nonheme iron protein that may be different from the ferredoxin type).

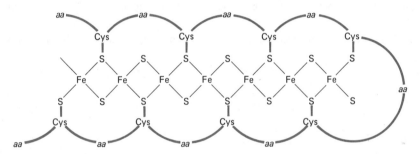

FIGURE 16-3
Presumed structure of ferredoxin, where aa *represents an indefinite number of inter-vening amino acids.*

16-4 STRUCTURAL PROTEINS

It might seem strange to include a purely structural material in the list of components of the ET systems, but we will see very shortly that the subunit structure of these systems is critical for understanding how they operate and that a purely structural protein is an important part of them—as well as all of the other functional membranes that have been studied carefully.

The first structural protein of a membrane was isolated by David E. Green, H. D. Tisdale, R. S. Criddle, and R. M. Bock by treating mitochondria with several detergents. They found a fraction amounting to about a third of the total protein that is readily soluble in 8-*M* urea, strong acetic acid, and other solutions that attack hydrophobic bonds, but not in water. Similar proteins were later found in the membranes of erythrocytes and other cells, and they are so similar in molecular weight and amino acid composition to the proteins of microtubules and the mitotic apparatus that Mazia and Ruby included them among the *tektins,* in their definition of the class. Once again, we find these important proteins forming the base of one of the most essential of all cellular structures, the membrane. We will now try to see how these proteins fit into the total picture of membrane architecture.

B. Chemosynthetic systems

16-5 MITOCHONDRIA

The mitochondria of readily available materials like beef heart and yeast have been studied more thoroughly than any other ET systems. They are easily obtained by differential centrifugation and can be handled like any other enzyme preparation. They can also be broken into smaller particles, but many operations that separate components from one another destroy the ability to translocate electrons, indicating that a relatively complex, intact structure is necessary for full activity. We are coming to view the mitochondrion as a whole as an integrated machine for conserving energy; the enzymes and electron carriers built into it must all be quite highly integrated with one another.

The sequence of electron carriers is now quite firmly established; the mitochondrial

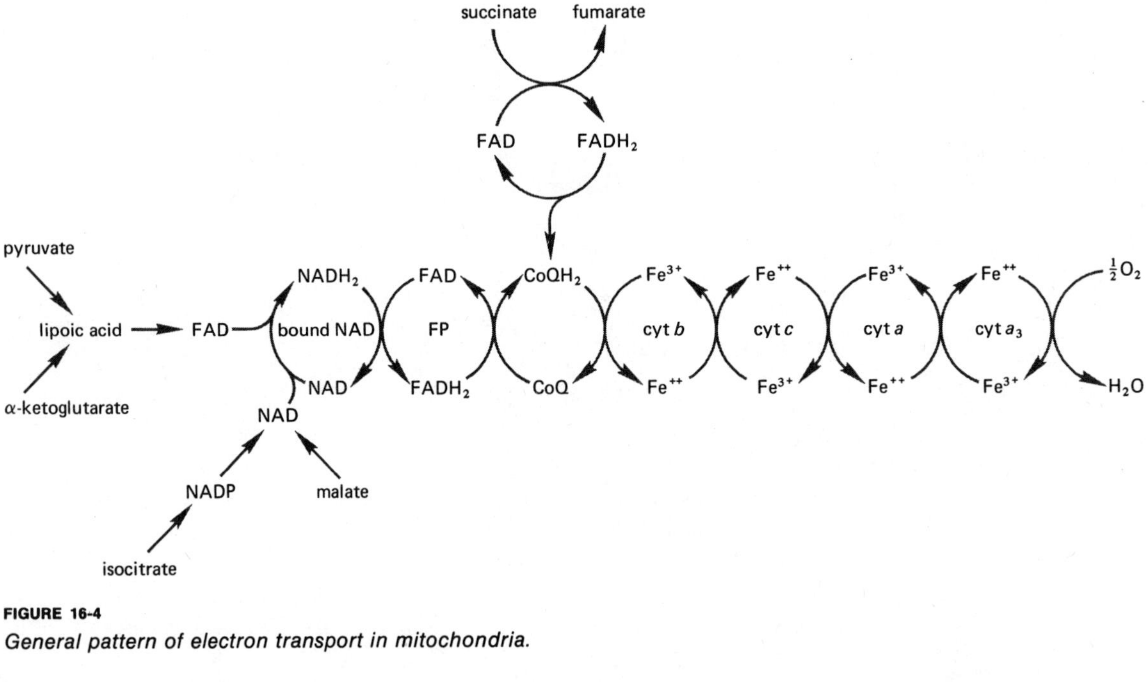

FIGURE 16-4
General pattern of electron transport in mitochondria.

sequence is shown in Fig. 16-4. An important factor in establishing this seqence was Britton Chance's development of a double-beam spectrophotometer that compares the light passing through two cuvettes simultaneously and records a **difference spectrum** giving the absorption in one relative to the other. If the two solutions were identical, the difference spectrum would be a flat, horizontal line. However, if a component is more oxidized in one cuvette its absorption spectrum will generally be different from that of the reduced component in the other cuvette and this difference will appear as one or more peaks. In this way, Chance could identify each component and study the kinetics of its oxidation and reduction.

There are many reagents that will inhibit one phase or another of mitochondrial action; many of these are antibiotics. They have been very useful in dissecting ET systems because they generally block at only a single point. For example, antimycin seems to inhibit the transfer from ubiquinone to cytochrome c_1. If antimycin is added to one cuvette of mitochondria and the mitochondria in another cuvette are allowed to respire normally, a difference spectrum will show that cytochromes c_1, c, a, and a_3 are oxidized in the poisoned preparation while the other components are reduced.

Let's remember where the electrons come from that enter the ET system. The reactions of the TCA cycle and some of the associated pathways—such as the glycolytic pathway from glucose to pyruvate—are continually forming NADPH and NADH; these are probably the chief sources of electrons. The transformation from succinate to fumarate within the cycle is also an oxidation step that yields electrons for the ET system directly. There are a few other reactions in which various dehydrogenases remove electrons from their substrates and reduce FAD or FMN, which in turn contribute to the ET system.

ATP is somehow generated as electrons traverse this series of carriers. It takes a pair of electrons to reduce $\frac{1}{2}O_2$ to H_2O, nitrate to nitrate, or any comparable terminal reaction, and it is customary to specify the number of ATPs generated per $2e^-$ (or per oxygen atom, when that is the terminal acceptor and is easily measured). This is expressed as the $P/2e^-$ or P/O ratio, where P represents phosphate esterified as ATP. The classical numbers, which seem well established, are $P/2e^- = 3$ when a molecule of NADH is the electron donor and $P/2e^- = 2$ when succinate is the electron donor. There are apparently specific points within the ET chain where ATP synthesis can occur. One is at the level of the NAD/FAD couple; the second is at the level of cytochrome b; and the third is at the level of cytochrome a-a_3. An electron pair traversing any of these points can generate an ATP molecule, even if artificial inhibitors are added that prevent its traversing some other part of the chain.

Figure 16-5 shows the general structure of a mitochondrion. David Green and his colleagues have broken mitochondria as carefully and specifically as possible by using reagents such as detergents and physical methods such as sonication. They find that the outer membrane contains all the enzymes of the TCA cycle plus some accessory biosynthetic activities, such as heme biosynthesis. The inner membrane, including the cristae membranes, contain the cytochromes and other ET elements. These can be broken into several functional units, shown in Fig. 16-6, which have well-defined protein and phospholipid compositions; the lipid is essential and its removal inhibits electron transport. It is clear that these are multimeric complexes that differ from those we have considered before chiefly in their lipid content. Each complex is a functional unit that can accept electrons from an appropriate donor, can be inhibited as a whole by characteristic reagents, and can transfer electrons between its components. (Notice that the reductant of each complex is not itself a part of the

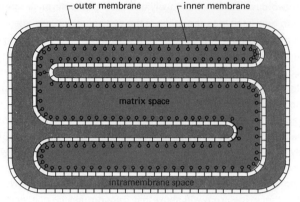

FIGURE 16-5

Topology of the mitochondrion. Notice the little globular subunits that protrude into the matrix space; these are the ATPase units.

FIGURE 16-6

The four complexes of electron carriers that can be isolated from mitochondria.

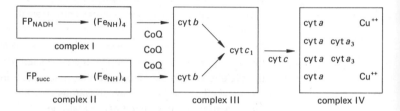

complex, although it is an important part of the ET system as a whole). Furthermore, all four complexes have approximately the same molecular weight, about 300,000 daltons.

In examining the cristae membranes by negative staining and high-resolution electron microscopy, Humberto Fernandez-Moran discovered that they are made of many small subunits (Fig. 16-7). Each subunit looks somewhat like a little phage particle; it has a round head, a short stalk, and a basepiece. These can be separated from one another. The basepieces can now be identified with the four functional ET complexes; the dissociated complexes by themselves can be reassociated into membranes that look just like the membranes formed by the basepieces.

Each complex by itself will form a membrane; however, a membrane formed by complex II will not transfer electrons to one formed by complex III. Electron transport occurs only when the complexes are mixed together and allowed to form a single, mixed membrane. The relative numbers of each complex in the mixture and their exact positions in the reconstituted membrane do not seem to be important, but the concentrations of the mobile carriers—coenzyme Q and cytochrome *c*—are very important. Cytochrome *c* is a fairly small protein, and it apparently has many lysine residues exposed on its surface that are bonded to an equal number of phosphatidyl ethanolamines, so the whole complex has a lipoidal exterior. Coenzyme Q is a lipid, and the phospholipids of the complexes are also important for electron transport; it is safe to say that the interactions between complexes in the membrane occur largely through lipid interactions.

A negatively stained beef heart mito-chondrion whose cristae are covered with repeating globular units. The outer mem-brane clearly has a different structure. (Courtesy of D. E. Green.)

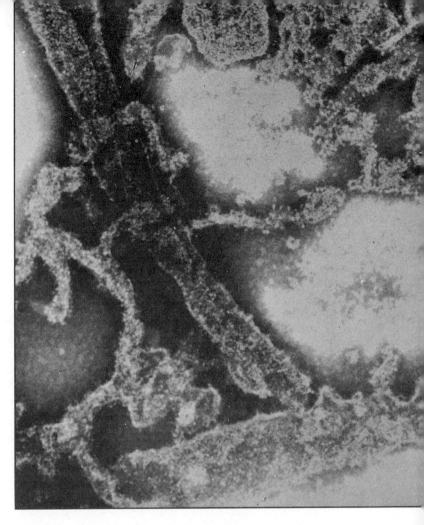

The function of the stalk and headpiece have also been established. Intact mito-chondria can make ATP from ADP and inorganic phosphate and they have an ATPase activity that reverses this reaction; these activities are inhibited by oligomycin. Efraim Racker has now shown that the headpiece is the ATPase, which by itself is not sensitive to oligomycin. When the headpiece is attached to the stalk, the oligomycin sensitivity returns. So now the point at which phosphorylation coupled to electron transport occurs seems to be established. The problem remains of specifying how the energy of electron transport is converted into the energy of the activated phosphoryl group of the ATP. We will take up this problem after we have examined some other ET systems.

16-6 BACTERIAL ELECTRON-TRANSPORT SYSTEMS

Figure 16-8 is a summary of the principal patterns of electron transport found in various chemotrophs, mostly bacteria. The variety is really quite fantastic; it looks like organisms have found every possible inorganic energy source during their evolution on earth. The organics are all lumped together, but you could probably find some bacterium that would grow on just about any organic compound on the shelf; and if

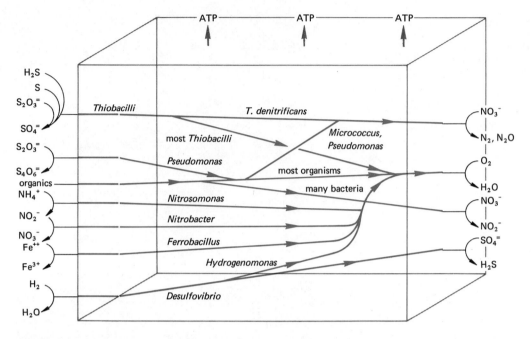

FIGURE 16-8

*A summary of the ET systems found in various chemotrophs. Electron-donor reactions are listed
on the left and electron-acceptor reactions on the right. The internal arrows show most of
the paths used by various bacteria.*

not, you could select a mutant that could do so by adding the compound to an
appropriate chemostat. There is no point in belaboring the subject, particularly because
the important points about all of these systems can be summarized very simply.

First, the ET systems are always basically the same, and their sequence is almost
always flavoprotein → quinone → b-type cytochrome → c-type cytochrome →
a-type cytochrome. The variations in cytochromes can be summarized by

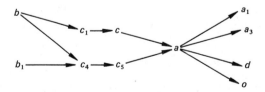

Second, the enzymes that reduce these ET systems are more loosely attached
to the membranes. In the cells that oxidize iron, sulfides, etc., the initial reactions

occur on the membrane surfaces, frequently out in the growth medium, where the byproducts (oxidized iron, sulfates, etc.) are deposited as wastes. When the electron donor is an organic compound, the dehydrogenases that oxidize it may be inside the cell, but still attached to the membrane much more tenuously than the components of the ET system themselves.

Third, the electron acceptor is usually oxygen (aerobic respiration), more rarely nitrate (nitrate respiration), and very rarely sulfate (sulfate respiration). Quite a few bacteria can reduce nitrate to nitrite or even to nitrogen and ammonia. In combination with the bacteria that can oxidize nitrogen compounds (*Nitrobacter, Nitrosomonas,* etc.) these organisms contribute to a major flow of nitrogen through the biosphere.

Sulfate respiration is rather remarkable. It requires a strange cytochrome, c_3, that has an E_0' of only -205 mV. However, the sulfate/sulfite couple has an E_0' of -486 mV, and it could not possibly be driven by cytochrome c_3. Harry D. Peck and M. Ishimoto have shown that the bacteria actually use one of their precious ATP molecules, which they combine with sulfate to make adenosine phosphosulfate (APS; see Section 8-6d). APS accepts electrons from cytochrome c_3 and is split into sulfite and AMP.

You must understand that there are good reasons why organic compounds and oxygen are the most commonly used electron donors and acceptors. The potential difference between, say, glucose and oxygen is much greater than the potentials between the various inorganic compounds that are used by chemolithotrophs, sulfate reducers, and so on. Electrons falling through this great potential drop can be made to yield much more ATP than those in the inorganic systems, so the aerobic chemorganotrophs dominate the earth. The other organisms have found appropriate ecological niches in the remote past and they stick with them, but they do not grow as rapidly or efficiently as organisms that make their living by burning carbohydrates. Most of them, however, do have the advantage of being able to switch from one mode to another in response to environmental conditions, and this adaptability is one of the features that makes them interesting tools for studying the structure of membranes.

The $P/2e^-$ ratios obtained for bacteria are generally much lower than those for mitochondria. Typical experimental values are less than one; it is hard to know at this time whether these low values reflect damage to the systems incurred during their extraction and preparation or some intrinsic inefficiencies in coupling electron flow to phosphorylation. In view of the efficiency that bacteria exhibit in their other reactions, the former seems much more likely than the latter.

David C. White and Lucille Smith have studied induction and repression in the cytochromes of some strains of *Haemophilus* that are capable of making cytochromes b_1, c_1, a_1, d, and o. Instead of ubiquinone, they have DMK_2; from White's work, it is clear that this quinone is the oxidant of a series of dehydrogenases and the reductant of cytochrome b_1. The molar ratio of DMK_2 to cyt b_1 remains about 14:1 under a variety of growth conditions while the ratios of other components vary over a wide range. This suggests that these two components are part of a single regulatory unit, such as an operon.

The bacteria change their terminal oxidases to conform with environmental conditions. Cytochrome a_1 is a nitrate reductase, while d and o use oxygen as an electron acceptor. These strains are all heme auxotrophs, so heme must be supplied in the medium for cytochrome synthesis. If the bacteria are grown with heme and with both oxygen and nitrate, they make a complete set of cytochromes and reduce both electron acceptors. As with the other cellular components that we discussed in Chapter 8, the amount of cytochrome made is a function of growth rate; in a rich medium where they can grow rapidly they make a lot of cytochrome, while they make relatively little in a poor medium. However, if they are grown with only nitrate as an electron acceptor, the synthesis of cytochromes d and o is repressed and only a_1 is made. If the bacteria are grown without heme, they make no cytochromes at all; their NADH dehydrogenase becomes an aerobic dehydrogenase that reduces oxygen to peroxide.

The point of all this is that the bacterial membrane must be extremely flexible. Many industrial and educational architects have learned recently that it is valuable to make rooms into which many different modules can be fitted for different purposes. When the function of the room must be changed, you simply roll out the old units and put new ones in, because they are all basically the same size and shape. The bacterial membrane must be very much the same. The various cytochrome and enzyme systems that it has to accomodate must be built into functional complexes with similar dimensions, much like the basepiece complexes of mitochondria, so they can all be fitted together into a close-packed array and so they can be put together in different proportions under different growth conditions. Here again, it is necessary to picture the membrane as a functional set of subunits; the fact that the subunits of membranes contain a great deal of lipid does not alter the basic picture, but it places some restrictions on the type of bonding that may exist between the subunits. The bonds that are principally responsible for the integrity of the membrane are expected to be hydrophobic, and all of the relevant experimental data indicate that they are. We will now consider the structure of the photosynthetic ET systems, to see what information they contribute to this general picture of energy-conserving membrane systems.

exercise 16-1 Using equation 3-21, you should now be able to calculate the minimum potential drop that can generate an ATP molecule with a phosphoryl transfer potential of 9 kcal. Remember: *Two* electrons are transferred for each ATP, and 1 cal $= 4.182$ joules.

exercise 16-2 Consult Table 16-2 for emf values. Consult Fig. 16-8 for specific bacteria. In each case, calculate the potential drop and the maximum number of ATPs that could be made in each of the following ET systems:

(a) *Nitrosomonas* using oxygen as an acceptor.

(b) *Ferrobacillus* using oxygen as an acceptor.

(c) *E. coli* growing with succinate and oxygen.

(d) *Thiobacillus* oxidizing sulfide and using oxygen as an acceptor.

exercise 16-3 Calculate the potential drop from NADH to oxygen. What is the efficiency of energy conservation if the $P/2e^-$ ratio is 3.0 for this span?

exercise 16-4 Calculate the potential drop from succinate to oxygen. What is the efficiency of conservation if the $P/2e^-$ ratio is 2.0 for this span?

TABLE 16-2

standard potentials of oxidoreductions

SYSTEM	E'_0 (mV)
acetaldehyde + H_2O/acetate + H^+	−598
acetaldehyde + CoA/acetyl-CoA	−412
NH_4^+ + H_2O/$NH_2OH \cdot H^+$	+562
cytochromes Fe^{++}/Fe^{3+}	
a	+290
b	+ 70
b_2	+120
b_5	−120
b_6	+ 50
c	+250
c_6	+365
ethanol/acetaldehyde	−197
FADH + H^+/FAD	−200
ferredoxin red/ox	−413
glutamate + H_2O/α-ketoglutarate + NH_4^+	−133
glycollate/glyoxylate	−90
HS^- + H^+/S (rhombic)	−272
lactate/pyruvate	−190
malate/oxalacetate	−166
NADH + H^+/HAD$^+$	−320
NADPH + H^+/NADP$^+$	−324
NO_2^- + H_2O/NO_3^- + H_2	+421
pyruvate + H_2O/acetate + CO_2	−699
succinate/fumarate	+31
$SO_3^=$ + H_2O/$SO_4^=$	−454
$0.5S_2O_4^=$ + H_2O/$SO_3^=$ + H^+	−471
$S_2O_3^=$ + H_2O/$S_2O_4^=$	−484
ubiquinone red/ox	+100
H_2O/$0.5O_2$	+816
H_2O_2/$0.5O_2$ + H_2O	+295

C. Photosynthetic electron-transport systems

16-7 GREEN PLANT PHOTOSYNTHESIS

We will first examine the photochemical process in green plants and then compare it to the bacterial photochemical systems. In 1943, Robert A. Emerson and his associates measured the efficiency of photosynthesis using monochromatic light; they found that the efficiency fell off more sharply than they expected as they moved toward the red end of the spectrum and that the reddest light was almost totally ineffective. However, if chloroplasts irradiated with light that is longer than about 680 nm are also given light of shorter wavelength, full photosynthesis is restored. This is called the **Emerson enhancement effect.**

The spectrum of the enhancement by shorter-wavelength light corresponds to the absorption spectrum for known accessory chloroplast pigments. From Emerson's experiments and others it is now clear that there are two pigment systems, designated **photosystem I** and **photosystem II.** Photosystem I absorbs most of the far red light, and seems to depend primarily upon absorption by Chl *a,* while photosystem II absorbs light of shorter wavelength, largely through accessory pigments such as Chl *b* and carotenoids. J. Myers and C. Stacy French have shown that the two systems do not have to absorb light simultaneously; they achieve maximum efficiency with alternating flashes of light of two different wavelengths, proving that absorption of light produces intermediate products that may endure for as much as a few seconds.

We must also say now that 4 quanta of light are not enough to make 1 molecule of oxygen. Since 2 quanta of light must cooperate to mobilize each electron, chloroplasts require at least 8 quanta per oxygen. Measurements from several laboratories now indicate that 8 to 10 quanta are required, with 8 being the most likely value at present.

16-8 THE PRIMARY PHOTOCHEMICAL EVENT

When Robert A. Emerson and William Arnold were studying the relationship between light intensity and photosynthetic activity, they observed that one molecule of oxygen was evolved for every 2400 chlorophyll molecules. If the quantum requirement for oxygen evolution is 8, then every electron that goes into the evolution of oxygen is generated by about 300 chlorophyll molecules. This figure of 200–400 chlorophylls appears repeatedly in photosynthesis research. Green plants appear to have a **photosynthetic unit** that consists of about 300 chlorophyll molecules that cooperate to trap light and emit electrons. They are presumably built into a compact, semi-crystalline array in the chloroplast thylakoids so they can share electrons with one another. Each electron must then be considered a part of the chlorophyll unit as a whole, not of any single molecule; and if one molecule absorbs a photon and becomes excited, its excitation energy must be considered to belong to the whole array of molecules. Each unit is therefore a group of pigment molecules that presents a large surface for trapping light and can transfer the light energy quickly to points some distance from the absorbing molecule, where it can be used for chemical processes.

A photosynthetic unit can also be identified in photosynthetic bacteria; it contains about 40–50 bacteriochlorophyll molecules. In the following discussion we will assume that the primary events in a photosynthetic unit are essentially the same, regardless of the organism.

The primary event in photosynthesis is the absorption of one light quantum by one photosynthetic unit; we will write this as chlorophyll going from its ground state to an excited state: $Chl^0 \rightarrow Chl^*$. The nature of the excitation and transfer of the excitation energy are matters of controversy at present, but we can afford to ignore these details anyway. It is sufficient to say that an electron is excited from a low-energy orbital to a high-energy orbital and that this energy is transferred through the unit to a **photochemical reaction center** where it is used to excite other chemical compounds. Each reaction center contains a unique pigment designated by the wavelength of light that it absorbs strongly (as P700 in green plants, P870, or P890 in some bacteria); it is probably always a chlorophyll molecule in a special environment where its absorption spectrum is changed. This pigment accepts the energy of the chlorophyll, so the sequence of events may be described:

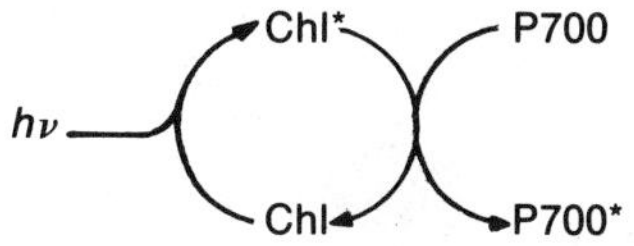

$$h\nu \underset{\text{Chl}}{\overset{\text{Chl}^*}{\diagdown}} \underset{\text{P700}^*}{\overset{\text{P700}}{\diagup}} \qquad \text{(for example)}$$

Roderick Clayton has shown that the proper kinetics of transfer can be obtained only by assuming that any photosynthetic unit can excite any reaction center that is not already excited. We should therefore picture the membranes containing these photosystems as extensive arrays of chlorophyll molecules with reaction centers scattered throughout so *on the average* there are a few hundred chlorophylls associated with each one.

   In addition, every reaction center must contain an electron acceptor, which we will designate Z, that can oxidize the excited pigment and an electron donor that can reduce it. We do not know what Z is — and it probably is different in each system — but the donor is always a c-type cytochrome, and since its prosthetic group is also a tetrapyrrole (heme) it could be conjugated to the chlorophyll so a direct electron transfer can occur. In any case, there must now be two reactions that reduce the reaction center pigment to its ground state again:

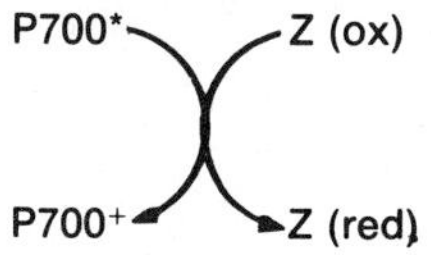

$$\underset{\text{P700}^+}{\overset{\text{P700}^*}{\diagdown}} \underset{\text{Z (red)}}{\overset{\text{Z (ox)}}{\diagup}}$$

and

$$\text{cyt } c \quad \underset{\text{Fe}^{++}}{\overset{\text{Fe}^{3+}}{\diagdown}} \underset{\text{P700}}{\overset{\text{P700}^+}{\diagup}}$$

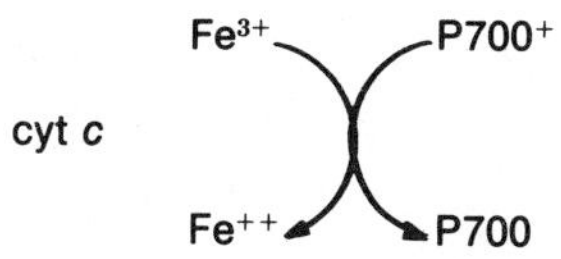

This leaves Z in reduced form at a high reducing potential (perhaps -600 mV) and oxidized cytochrome at a moderate oxidizing potential. Therefore Z is in a position to reduce something else and the cytochrome is in a position to be reduced. We will now see how the various electron-transfer systems can effect both reactions and produce the compounds necessary for further metabolism.

16-9 THE PHOTOCHEMICAL ELECTRON-TRANSPORT SYSTEMS

In all phototrophs there are apparently two pathways of electron transfer: a **cyclic pathway** from Z back to the cytochrome *c* and a **noncyclic pathway** in which Z reduces a stable compound such as NADP$^+$ and the cytochrome is reduced by an accessory electron donor. ATP synthesis (sometimes called **photophosphorylation**) can occur in either pathway, but there is good evidence that in the photosynthetic bacteria, at least, the pathways are essentially specialized and separated. That is, the cyclic pathway produces primarily ATP and the noncyclic one produces primarily NADPH.

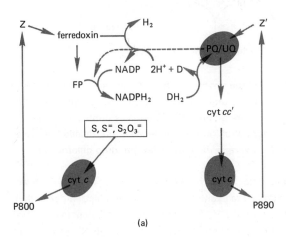

(a)

FIGURE 16-9

A summary of the ET systems found in photo-trophs: (a) photosynthetic bacteria and (b) algae and plants. Two separate light reactions are assumed in each case—one associated primarily with a cyclic pathway and one with a noncyclic pathway. Dashed lines indicate possible alternate routes, and colored areas are the most likely points of phosphorylation.

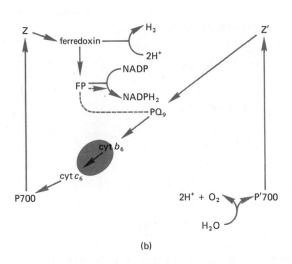

(b)

You will recognize, of course, that both are necessary for the major biosynthetic processes.

Figure 16-9a shows the general patterns inferred from bacterial studies. Although the carriers differ somewhat from one species to another, the patterns seem to be common to all. There is always a quinone and at least one cytochrome in addition to the cytochrome c. The noncyclic pathway always contains ferredoxin in its usual role as a reducing agent for $NADP^+$ or other stable compounds; in some cases, you notice, hydrogen ions may be reduced to hydrogen gas. The various sulfur compounds serve as sources of electrons for the noncyclic system; there is also the possibility of a reverse electron transfer out of the cyclic system to produce NADPH, and in this case organic compounds such as fumarate may serve as electron donors. This whole scheme should be considered rather tentative; there are other possible ways to draw the paths of electron transfer.

In the algae and green plants, the two separate photosystems have apparently evolved into a more concerted system. Photosystem I produces an energized electron that can travel the cyclic path to produce ATP or the noncyclic path to produce NADPH. In the second case, the photochemical system is left oxidized and must be reduced by some accessory electron donor; this is where photosystem II enters. By using another quantum photosystem II oxidizes water into free oxygen and produces electrons that pass into photosystem I (Fig. 16-9b). This noncyclic pathway also generates ATP. Again, other pathways are possible, and there may be a reverse electron flow out of the cyclic system to generate NADPH.

Somewhat quixotically, the systems that oxidize the accessory electron donors are among the most poorly understood parts of the photosynthetic apparatus, in spite of their importance. The system that oxidizes water in plants may be considered the reverse of the cytochrome oxidase system in mitochondria, except that it uses Mn^{++} ions instead of Cu^{++}. Oxidation of the sulfur compounds must require a series of specific enzymes. It should be pointed out that the organic electron donors used by some of the bacteria also function as major carbon sources, and their role in photosynthesis *per se* may be somewhat secondary.

Some separation of chloroplast membranes into subunits has been achieved with detergents and other reagents. Fractions that correspond approximately to photosystem I and photosystem II have been separated, but the separation is not complete. In any case, it is clear from electron micrographs that the membranes contain substantial subunits (see Fig. 7-19). There are several sizes of subunit in these membranes, but so far it is not possible to identify any of them with specific chloroplast components.

D. Three accessory reactions

The energy-conserving ET reactions of mitochondria, chloroplasts, and bacterial membranes are accompanied by many side reactions, which either contribute electrons or use some of the energy for important biosyntheses. Any attempt to list these would produce a rather boring encyclopedia with a few highlights; here we will pick out three of the highlights that illustrate interesting points.

16-10 PYRUVIC DEHYDROGENASE: A COMPLEX MULTIMER

We have already outlined the reactions that lead from pyruvate to acetyl-CoA, but let's look at them again from the organizational point of view, because the reaction sequence actually takes place in a large multimer of enzymes. Richard Schweet first found the complex in mitochondria, but the one in *E. coli* has been studied very extensively by Lester J. Reed and his associates. Reed's analysis of its structure is instructive because it depends upon some fine electron microscopy and considerations of symmetry in self-assembly processes, as well as a knowledge of the reaction mechanism.

The multimer can be broken down by various treatments (such as dilute acid) into three separate enzymes, each of which can be dissociated into protomers. The protomers reassociate spontaneously when the treatments are reversed and normal pH, buffering, Mg^{++} concentrations, and so on are restored; thus the multimer is a self-assembling unit held together by noncovalent bonds. The enzymes are as follows:

1. The pyruvic decarboxylase (C), which catalyzes the conversion of pyruvate into CO_2 + hydroxyethyl-TPP. Each multimer probably consists of 24 protomers of about 90,000 daltons each which have covalently bound TPP prosthetic groups.

2. The lipoic transacetylase (T), which transfers the acetyl group from lipoic acid to coenzyme A. It forms a unit by itself with a total molecular weight of about 980,000; this can be broken into eight enzymatically active units of 120,000 daltons each, and each of these, in turn, is three protomers of 40,000 daltons. Each protomer contains one bound lipoic acid, as a long lipoyl-lysine chain.

3. The lipoyl dehydrogenase (D) is a flavoprotein containing bound FAD which oxidizes the lipoic acid. The multimer contains 24 protomers of the enzyme which weigh 55,000 daltons each.

Reed's analysis indicates that the multimer is built as shown in Fig. 16-10. The transacetylase clearly forms the core since the T protomers associate with one another and make a large unit and since protomers of the other two enzymes bind to this core but will not bind to each other. The eight active transacetylase units apparently form a cubical packet which can be seen in electron micrographs of the isolated transacetylase and in the whole multimer. The 24 C units can fit on the edges of the cube and the 24 D units can fit on its faces. This model fits perfectly with electron micrographs, chemical data, and considerations of symmetry.

The model has one other interesting feature. The lipoyl-lysine of the transacetylase is about 1.4 nm long, and Reed suggests that it is located centrally between the active sites of the three enzymes as shown in Fig. 16-11, waving back and forth from one to the other to participate in the cycle of reactions. An enzyme complex like this would certainly operate very efficiently.

The dehydrogenase system that converts α-ketoglutarate into succinyl-CoA is exactly analogous to this one and Reed's studies indicate that it is built very much like the pyruvic dehydrogenase. In fact, the flavoproteins of the two multimers are similar, if not identical, since they can be exchanged from one complex to the other with full activity.

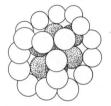

The core consists of eight T units,
each of which is a trimer.

  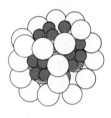

On each edge of the core there are two C units.

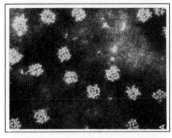

Four units of F will fit on each face of the core.

FIGURE 16-10

*Some of Reed's pictures of the pyruvic dehydrogenase
complex and the model based upon them. (Courtesy
of Lester J. Reed.)*

FIGURE 16-11

*A model for the role of lipoic acid as a
carrier between the active sites of the three
enzymes.*

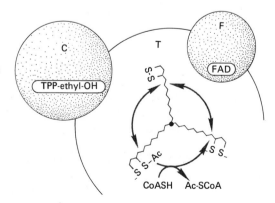

Reed has also taken some pictures of the purified pyruvic dehydrogenase complex from mitochondria. Figure 16-12 shows that it is different from the bacterial complex; it probably has elements of five-fold rotational symmetry, very much like the viruses we shall discuss in Chapter 20, whose capsids are closed shells. The molar ratios of the enzymes in this complex are consistent with a structure that is basically a dodecahedron. The decarboxylase molecules can fit on the 30 edges of this figure and the flavoproteins on its 12 faces.

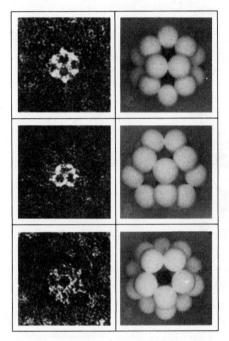

FIGURE 16-12
*Three views of the transacetylase core from the mito-
chondrial pyruvic dehydrogenase, with corresponding
views of a model consistent with these pictures.
(Courtesy of Lester J. Reed.)*

16-11 FATTY ACID OXIDATION

A great deal of energy can be stored in the form of fat. The excess carbohydrate a cell takes in is often converted into fatty acids; when the need arises, the energy of these fats can be released through an interesting cycle of reactions known as **β-oxidation.** β-oxidation yields electrons that go directly into the ET chain, and the fatty acids are broken down into C_2 units of acetyl-CoA, which go directly into the TCA cycle where they are oxidized more completely.

The cycle is very simple; it starts with formation of a CoA derivative of the fatty acid, and this is then oxidized in four steps:

1. Dehydrogenation across the α-β bond:

$$RCH_2CH_2CH_2CO-SCoA + FAD \rightarrow RCH_2CH=CHCO-SCoA + FADH_2$$

2. The double bond is hydrated:

$$RCH_2CH{=}CH_2CO—SCoA + H_2O \rightarrow RCH_2CHOH—CH_2CO—SCoA$$

3. The β-carbon is reduced:

$$RCH_2CHOH—CH_2CO—SCoA + NAD^+ \rightarrow RCH_2CO—CH_2CO—SCoA + NADH + H^+$$

4. A terminal acetyl-CoA molecule is removed, leaving an acyl-CoA shorter by two carbons:

$$RCH_2CO—CH_2CO—SCoA + HSCoA \rightarrow RCH_2CO—SCoA + CH_3COSCoA$$

The cycle then repeats to remove another molecule of acetyl-CoA.

Notice that the cycle results in reduced FAD, whose oxidation provides two molecules of ATP, and NADH. Oxidation of NADH provides three more molecules of ATP. Thus each cycle produces five ATP molecules directly, plus a molecule of acetyl-CoA whose oxidation through the TCA cycle produces 12 more ATPs per mole. This is therefore an extremely efficient cycle which conserves a great deal of energy in usable form.

16-12 THE CALVIN CYCLE IN PHOTOSYNTHESIS

We have outlined the reactions that lead to ATP and NADPH production in photosynthesis, but we have not indicated how these are used to reduce carbon dioxide to the level of intermediate metabolites. The actual reaction sequence was worked out by a group of workers led by Melvin Calvin, James Bassham, and Andrew Benson. In fact, we now know that the sequence is used by all autotrophs for incorporation of CO_2; it is not unique to phototrophs at all.

The sequence of reactions forms a very complicated cycle. It is interesting biochemically because most of the reactions, when run in reverse, can be made into a cycle for the oxidation of glucose through a pathway different from the usual Embden–Meyerhof glycolytic pathway. However, at the level of this course, most of these reactions would be quite meaningless, so we will just outline some of the major features.

The first reaction is a phosphorylation of ribulose-5-phosphate to ribulose-1, 5-diphosphate:

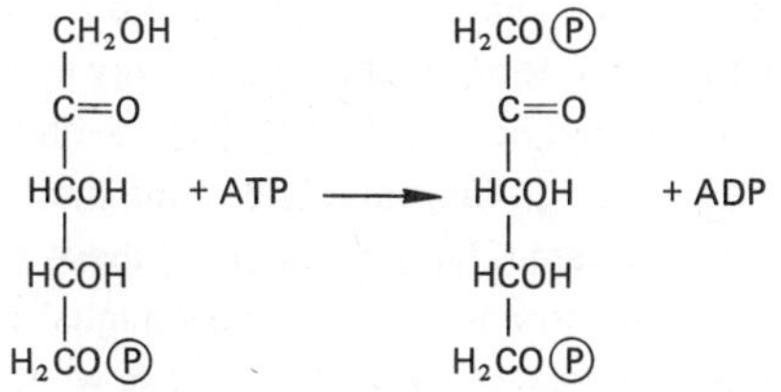

Carbon dioxide is then combined into ribulose diphosphate in a sequence that requires two more molecules of ATP and two of NADPH:

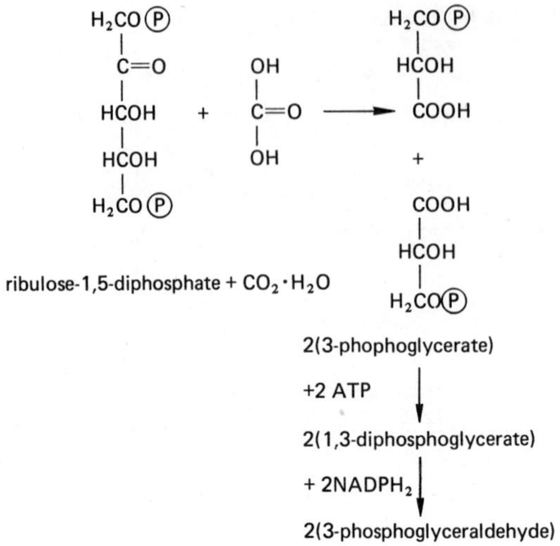

$$
\begin{array}{ccc}
\begin{array}{c}
H_2CO\,\textcircled{P} \\
| \\
C{=}O \\
| \\
HCOH \\
| \\
HCOH \\
| \\
H_2CO\,\textcircled{P}
\end{array}
& + \;
\begin{array}{c}
OH \\
| \\
C{=}O \\
| \\
OH
\end{array}
& \longrightarrow \;
\begin{array}{c}
H_2CO\,\textcircled{P} \\
| \\
HCOH \\
| \\
COOH
\end{array} \\[2em]
& & + \\[1em]
& &
\begin{array}{c}
COOH \\
| \\
HCOH \\
| \\
H_2CO\,\textcircled{P}
\end{array}
\end{array}
$$

ribulose-1,5-diphosphate + $CO_2 \cdot H_2O$

2(3-phophoglycerate)

+2 ATP $\Big\downarrow$

2(1,3-diphosphoglycerate)

+ 2NADPH$_2$ $\Big\downarrow$

2(3-phosphoglyceraldehyde)

After the formation of these C_3 compounds, there is a cycle of reactions that regenerates longer carbohydrates. In balance, if three molecules of ribulose phosphate are condensed with three CO_2 molecules, the cycle regenerates the ribulose phosphates and leaves one molecule of phosphoglyceraldehyde which can go into the TCA cycle through pyruvate. The reduction of each CO_2 is not exactly cheap, since it requires three ATP molecules and two reducing equivalents of NADPH, but as we saw in Chapter 1, phototrophs still manage to reduce an enormous amount each year.

E. Membranes as transducers

16-13 GENERAL PROPERTIES OF MEMBRANES

The membranes we have been discussing are all designed for one major purpose: to conserve the energy of oxidation in the form of ATP. If we loosen a term from engineering slightly, we can call these membranes a kind of *transducer* (Section 3-10). To the physicist, a machine is a transducer if it converts one form of energy into another. Motors convert electrical energy into mechanical energy and dynamos do just the opposite. A battery converts chemical energy into electrical energy. A chloroplast converts light energy into chemical energy, and the ET systems of chemosynthetic organisms convert one form of chemical energy into another. The important point is that each object must be designed through evolution for its particular task, and what is a good design for a mitochondrion may not necessarily be good for a chloroplast. The remarkable thing is that all of these membranes really are so much alike in their basic structure. The thing we must be careful of is allowing ourselves to pretend they are exactly alike, and this is what too simple a model can suggest.

Look back at Fig. 7-18 and 7-19 again. They emphasize the great differences that must exist between different kinds of membranes and the pitfalls involved in assuming that a very simple model can accomodate all of these types. There are probably some, like the myelin sheath membranes, that are little more than bimolecular layers extending *ad infinitum,* with some protein attached on the outside. The purpose of the myelin sheath around a nerve axon is, essentially, to insulate the nerve and help the nervous impulse move along with less energy loss; layers of lipid make very good insulators, so this is a good design for a myelin sheath membrane. However, most cellular membranes are complexes of proteins such as enzymes and electron carriers; they cannot be simply lipid layers even though their lipids are essential to their operation.

Are you disappointed to find the simple unit-membrane model destroyed and yet not replaced with something equally simple? Well, it is true that we are no longer pretending we can tell how all the molecules of phosphatidyl choline are lined up in relation to a layer of protein, but this was really just a pretense, and the truth, once it is faced, is much prettier. The subunit membrane can *do* things, not just enclose things. It can contain a variety of functional units in whatever patterns are necessary and change their relative numbers just like any other inducible or repressible proteins. The model is consistent with the subassembly and self-assembly principles that seem to apply to all other large cellular structures.

Studies of membrane structures, including those using optical techniques and the methods of nuclear magnetic resonance, proton magnetic resonance, etc. of organic chemistry, indicate that all membranes have certain common features. (1) They are built of many lipid-protein subunits that associate with one another primarily through hydrophobic bonding. (2) The subunits generally contain significant percentages of structural, tektin proteins. (3) The protein has little pleated-sheet structure, but it does have a great deal of helical structure. The helical regions tend to penetrate the membrane, while the more random regions are closer to the membrane surface. (4) Lipid molecules seem to sit between protein molecules in positions similar to those depicted in the old unit-membrane pictures, so their hydrocarbon chains can act as a hydrophobic cement between proteins. The hydrophilic groups of lipids and proteins will tend to be at the membrane surfaces.

Now, given this subunit picture of the membrane, we can extend our discussion of the relationships between protomers from Chapter 15. The protomers of the membrane can again be bonded to one another isologously or heterologously (Fig. 16-13). Depending upon the type of bonding, the membrane can possess a transverse polarity or not. Depending upon the structure of the subunits, the membrane surface is also expected to display some kind of regular, close-packed structure, either hexagonal or rectangular. Regular surface structures in membranes have now been observed by several investigators; one of these is shown in Fig. 16-14.

The membrane as a whole can also be considered a collection of interacting protomers that may exhibit allosteric or cooperative effects just like multimeric enzymes or any of the other structures we have considered. Changeux and his colleagues have developed equations to describe these changes in membranes by extending their treatment of allosteric oligomers. It is again assumed that the protomers contain sites to which various ligands can bind and that the protomers are capable of undergoing at least one change between different conformations. One of these states is assumed to have greater affinity for the ligand than the other. Figure

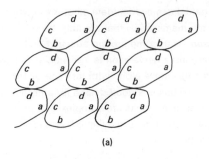

(a)

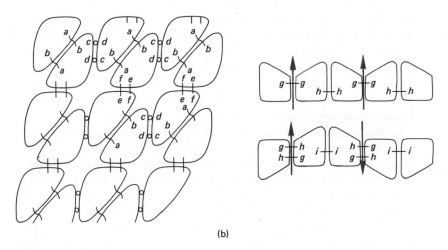

(b)

FIGURE 16-13

Types of bonding in an infinite sheet of protomers making a membrane. (a) Heterologous. (b) Isologous, with cross sections of a membrane that is anisotropic (with a transverse polarity) or isotropic (with no transverse polarity).

FIGURE 16-14

A surface view of an isolated rat liver cell membrane, negatively stained with phosphotungstate, to show its subunit structure with hexagonal packing. (Courtesy of P. Emmelot and E. L. Benedetti.)

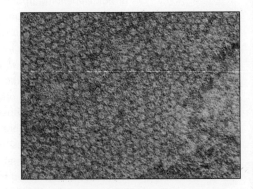

16-15 shows how a change in conformation can pass along the entire membrane from one point; this change may be induced by contact with a single molecule of the ligand and it will be stabilized by further binding of ligand molecules.

Changeux's equations, which we will not attempt to reproduce here, show that membranes should be able to respond to ligands with either a graded, continuous shift into a new conformation or with a sharp, all-or-none response. The former will occur when the protomers interact weakly and the latter when they interact strongly. Both kinds of response have been observed in membranes under different circumstances, and it is comforting to know that such membrane responses can be derived

FIGURE 16-15
The spreading of a conformational transition in a membrane from a single point. The white conformation has a greater affinity for ligand than the gray conformation. Interaction with a single ligand molecule can initiate a wave that spreads from one unit to the next in all directions. (After Changeux.)

from a simple theory that conforms with the successful theories of allosteric effects in enzyme systems with which we have already dealt. We will now see how this concept of membrane behavior can explain certain interesting phenomena, and we will begin with some possible explanations of phosphorylation in ET membrane systems.

16-14 CONFORMATIONAL CHANGES IN ELECTRON TRANSPORT

Most pictures of mitochondria, either in cells or isolated, show them to have the internal structure we have shown in several pictures, with a number of separated, parallel cristae. This has come to represent the "orthodox" concept of mitochondrial structure, but a few years ago Charles Hackenbrock observed that mitochondria in different metabolic states exist in very different configurations. David Green and his group have taken up this observation and pursued it toward its logical limit. They have assumed that the conformational changes observed are the result of different states of the electron-carrier complex in the membranes and that they are, in turn, the causes of such active processes as phosphorylation. In other words, they assumed that the energy derived from electron transport is conserved in the form of membrane changes.

Green and his group have identified at least three states of the mitochondria, which they call **nonenergized, energized,** and **energized-twisted;** mitochondria in these conditions are shown in Fig. 16-16. Cycles between these states can be demonstrated in different energy-exchange processes. Suppose we start with nonenergized mitochondria; the addition of either substrate (i.e., a metabolite that will reduce the ET chain) by itself or ATP by itself causes the mitochondria to take on their

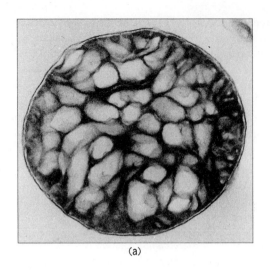

(a)

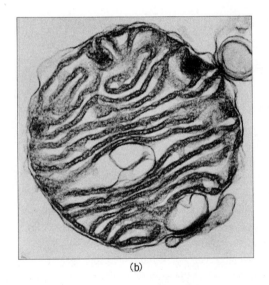

(b)

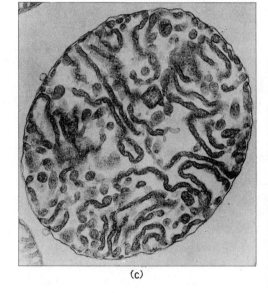

(c)

FIGURE 16-16

Configurations of isolated mitochondria. (a) Nonenergized, (b) energized, and (c) energized-twisted. The matrix space (in which some subunit structure can occasionally be seen) is always gray, the intramitochondrial space is always light. (Courtesy of D. E. Green.)

energized state. (This might seem somewhat paradoxical, but it has been known for some time that the ET system can be activated either by an electron donor or by ATP; the argument has been over the form the activation takes.) The addition of inorganic phosphate to the preparation will then throw the mitochondria into their energized-twisted state. This state can be discharged to the nonenergized state by adding ADP, which is phosphorylated to make ATP.

One of the complicating and puzzling observations on mitochondria has been their ability to translocate specific ions under various conditions. The translocation of divalent cations, particularly Ca^{++}, is intimately related to the active state of the mitochondria. When calcium ions are added to a mitochondrial preparation, there is a burst of electron transport, several of the components of the ET system are transiently oxidized and reduced, and the Ca^{++} ions are taken up by the mitochondria, while at the same time H^+ ions are ejected, mole for mole with the calcium. The Ca^{++} has exactly the same effect on the conformation of the mitochondria as ADP; it also leads to a discharge of the energized-twisted state. Thus the mitochondrion can either conserve its energy by forming ATP or it can use it by translocating an ion.

The behavior of various poisons and antagonists is also consistent with this model. Reagents that prevent electron transport prevent the transformation into the energized conformation. Uncoupling reagents, which permit electron transport but uncouple it from phosphorylation, transform the mitochondrion from the energized to the nonenergized state.

The picture of energy conservation in mitochondria can therefore be stated essentially as follows (Fig. 16-17). Each of the ET complexes of the inner membranes has at least two conformations; it is thrown into its energized conformation by electron transport. When this happens in a whole membrane full of units, the conformational

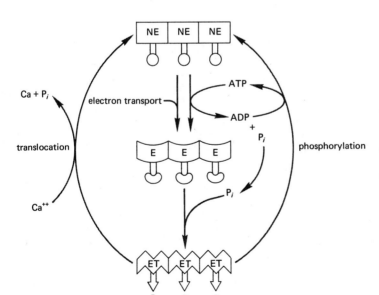

FIGURE 16-17

A schematic interpretation of conformational changes in mitochondrial subunits from nonenergized (NE) to energized (E) to energized-twisted (ET), associated with different reactions.

change can be seen as a change in the membrane; however, it is only the interaction between subunits that matters, not the gross, visible changes. When a complex is thrown into its energized state, it causes a conformational change in the mito-chondrial ATPase system, allowing it to bind a free phosphate as a phosphoryl group with enough energy to form ATP. This may be reflected as a further conforma-tional change, seen as the energized-twisted state of the membranes. The energy of this state may be discharged either by ADP, which becomes phosphorylated, or by Ca^{++} ions, which are translocated.

There is a significant alternative view of phosphorylation that we have not considered yet, even though it is generally consistent with the model we have been developing. The traditional theory of phosphorylation assumed that electron transport creates a series of high-potential intermediates in the ET membranes which effect a phosphoryl-ation essentially like the substrate-level phosphorylations that occur during glycolysis. However, with all the skill of modern biochemistry, these compounds escaped detection. Peter Mitchell then suggested a rather elegant alternative. In his model,

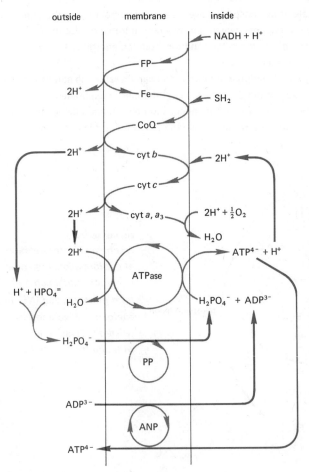

FIGURE 16-18
One possible version of Mitchell's model for a membrane which concentrates hydrogen ions on one side during electron transport and thus drives the synthesis of ATP. PP is a phosphate permease and ANP is an adenine nucleotide permease. (After G. D. Greville.)

the electron carriers are arranged in the ET membrane so they translocate hydrogen ions during electron transport and create a charge and/or a pH gradient across the membrane (Fig. 16-18). Now, if you write a properly balanced equation for ATP synthesis, you can see that hydrogen or hydroxyl ions are involved, depending upon the pH and the way you choose to describe the process:

$$\text{adenosine-O}-\overset{\overset{\text{O}}{\|}}{\underset{\underset{\text{O}^-}{|}}{\text{P}}}-\text{O}-\overset{\overset{\text{O}}{\|}}{\underset{\underset{\text{O}^-}{|}}{\text{P}}}-\text{O}^- + \text{HO}-\overset{\overset{\text{O}}{\|}}{\underset{\underset{\text{O}^-}{|}}{\text{P}}}-\text{O}^- \longrightarrow \text{adenosine-O}-\overset{\overset{\text{O}}{\|}}{\underset{\underset{\text{O}^-}{|}}{\text{P}}}-\text{O}-\overset{\overset{\text{O}}{\|}}{\underset{\underset{\text{O}^-}{|}}{\text{P}}}-\text{O}-\overset{\overset{\text{O}}{\|}}{\underset{\underset{\text{O}^-}{|}}{\text{P}}}-\text{O}^- + \text{OH}^-$$

$$\text{ADP}^{3-} \quad + \quad \text{HPO}_4^= \quad \longrightarrow \quad \text{ATP}^{4-} \quad\quad + \text{OH}^-$$

In this equation, ATP formation is accompanied by the release of a hydroxyl ion. An equilibrium constant for the reaction,

$$K_p = \frac{[\text{ATP}^{4-}][\text{OH}^-]}{[\text{ADP}^{3-}][\text{P}_i^{2-}]}$$

has a value of about $10^{-5.8}$ for $\Delta G^0 = -9$ kcal/mole.

Mitchell assumed that the membrane is impermeable to H^+ ions and that it contains an ATPase that is activated by H^+ ions from the outside and has access to adenosine phosphates and inorganic phosphate from the inside, more or less as the figure shows. (ATPases with essentially the right properties do exist in mitochondria.) If electron transport creates a charge separation and a pH gradient it can drive the ATPase in the direction of synthesis, even though the equilibrium constant strongly favors ATP hydrolysis. Mitchell calculates that a pH difference of 3.5 units or a potential of 210 mV across the membrane is sufficient.

Although there are many difficulties with Mitchell's theory (and the model can be written in different ways, some of which may be more realistic than others), there is really a great deal to be said for it. The expulsion of H^+ ions from mitochondria during electron transport can be demonstrated; however, some investigators believe that this ion movement is an immediate consequence of electron transport and a direct cause of phosphorylation, while others believe it occurs later as a consequence of other processes. It is interesting to notice two facts from studies with chloroplasts. First, in chloroplasts hydrogen ions are taken up as a consequence of photoinduced electron transport; this is the opposite of the mitochondrial system, and the little knobs on the chloroplast membrane subunits—which, you recall, are the ATPase molecules—are also oriented in the opposite direction. Second, Andre Jagendorf and his associates have demonstrated that phosphorylation can be induced in intact chloroplasts by creating an artificial pH gradient, by incubating the chloroplasts in an acid medium and then suddenly shifting them to a more basic medium.

Information about the action of uncoupling agents is also consistent with this theory. Mitchell proposes that compounds which uncouple phosphorylation from electron transport do so by disrupting the membrane structure and making it more permeable to the various ions responsible for the potential. It has been shown that dinitrophenol and other uncouplers have this effect.

The matter is extremely controversial at present. It is quite likely that the truth lies in a marriage of Mitchell's so-called "chemiosmotic" viewpoint with information about changes in membrane conformations. It is quite possible that potentials are created across the ET membranes that have critical effects on the conformations of the ATPase system and the various permease mechanisms of the membranes. For the present, it is important to focus on the major ideas we have stressed here. The membrane is made of functional subunits; these subunits and the membranes as a whole clearly undergo extensive conformational changes in response to electron transport; and the energy conserved in membrane conformation may be used either for ATP formation or for translocation. Furthermore, the membranes have a transverse polarity, and this in itself may allow them to perform a considerable amount of work through the creation of electrochemical potentials.

16-15 NERVOUS-SIGNAL CONDUCTION

The whole vast subject of neurophysiology is one that you will have to learn about in a future course, but it would be wrong not to indicate here, in a general way, how nervous conduction can be understood in terms of the cell-membrane structure.

We have already seen that the properties of every cell membrane give the cell a transmembrane potential of approximately -70 mV. For this reason, every cell is capable, in principle, of responding to stimuli from the outside by suddenly changing its potential; a long string of these cells put together could carry a signal from one end of a large animal to another if each one could stimulate the next in line. However, this kind of transmission would be very slow and inefficient. Instead, most animals have developed specialized nerve cells with extremely long processes (axons) that can run for several feet. Then the stimulation of the cell in one part of the body can produce a signal that travels along the axon to the other end of the nerve cell, without having to jump a million gaps.

Most of the significant work on membrane potentials during nervous conduction is now done by inserting very fine glass electrodes filled with KCl into nerve fibers. One electrode is set up to deliver a stimulus to the nerve and another is connected to an oscilloscope to record changes in membrane potentials. The following basic picture of electrical events during the nerve impulse comes primarily from studies by A. L. Hodgkin and A. G. Huxley, who measured the movements of various ions in and about the nerve membrane.

Because of its sodium-potassium permease system, the nerve cell has about 150 mM K^+ inside and 5.5 mM K^+ outside, 150 mM Na^+ outside and 15 mM Na^+ inside, and Cl^- ions distribute themselves across the membrane freely so their concentration is 125 mM outside and 9 mM inside. The nerve has a net charge that is positive outside and negative inside (Fig. 16-19). When the neuron is stimulated with one electrode, changes in potential can be recorded with the second. The potential can rise significantly above the "resting" potential of -70 mV during a brief stimulation without any other effect; the potential simply reverts to its resting level. However, if the neuron is given enough of a stimulus to make the voltage rise above a critical **threshold level**—

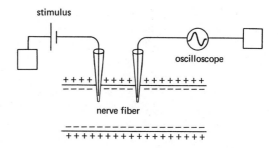

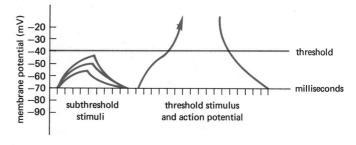

FIGURE 16-19

Measurements of nerve potentials. (a) The arrangement of microelectrodes for stimulating a fiber and measuring potentials. (b) Potential changes recorded after subthreshold stimuli and after a stimulus great enough to reach the threshold and elicit an action potential.

FIGURE 16-20

The action potential is passed along the nerve fiber as an inward flow of Na^+ followed by an outward flow of K^+.

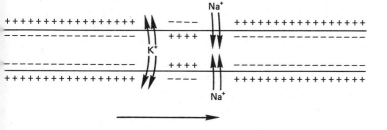

generally around −50 mV — the potential suddenly rises, not just to zero, but to as much as +40 mV. This **action potential** lasts for a few milliseconds, and the potential then drops back to about −70 mV again. If we place a series of recording electrodes all along the neuron, each of them will record an action potential in turn. This means that a wave of depolarization is running along the neuron (Fig. 16-20) usually at a rate of about 5–10 m/second, but in large myelinated nerves sometimes as much as 100 m/second.

From measurements of ion movements, we know that the principal reason for this potential is a reversal of the distribution of cations compared to the resting neuron. Potassium ions begin to flow out of the neuron and sodium ions flow in, and the action potential is apparently due to the great influx of sodium. The nerve is restored to its resting potential first by an active expulsion of potassium, to balance the intake of sodium, and then by restoration of the sodium-potassium pump which removes the sodium and pulls in potassium.

We have some way to go before we will have a completely satisfying molecular picture of this process, but some notable features have been outlined by Julian Tobias and his students. It is worth considering this model in some detail, for even if parts of it are uncertain it indicates the general direction in which research on the problem will be aimed in the future.

The earliest measurable event in excitation is an outward flow of cations; K^+ is most likely to be responsible for this flow, because of its high intracellular concentration and high mobility. Tobias and his colleagues have shown that there is a considerable amount of Ca^{++} in the membrane bound to phosphatidyl serine. If phosphatidyl serine is bound in between the protein units of membranes, where it serves as a hydrophobic glue along with other phospholipids, then the bound calcium simply increases the strength of this glue. A number of studies on lipid structures indicate that divalent cations tend to bind components together more strongly while monovalent cations bind them more weakly. This is easy to understand if you picture the divalent ion sitting between two bound anions and holding them close together, while the anions bound to two separate cations can still drift apart. Tobias therefore proposes that the flux of K^+ toward the membrane tends to displace Ca^{++} ions on phosphatidyl serine; this loosens the membrane structure, allows water molecules to move in, and makes the membrane suddenly permeable to both Na^+ and K^+.

When the membrane is viewed as a large multimer capable of allosteric transitions, this general model becomes even easier to understand. The membrane subunits may be capable of existing in two conformations. In state A, the sodium-potassium permeases are active and the resting potential is maintained; in state B, these permeases are inactive, but another set of permeases that can pump sodium inward are activated. The transition from one state to another can easily be effected by changes from K^+ to Ca^{++} ions in the membrane phospholipids. When the membrane is stimulated locally to shift from state A to state B, its potential changes locally, but at the same time a wave of depolarization is initiated that passes along the entire neuron as each subunit pushes its neighbor into state B. Behind this wave, another wave occurs as the subunits revert to state A.

Again, it must be emphasized that this is a sketchy and provisional model, but it conforms with everything else we know about membranes and it illustrates the point that rather complex phenomena at a gross observational level may have explanations that are relatively simple when viewed within the developing framework of molecular biology.

16-16 SYNAPTIC TRANSMISSION

The transmission of an impulse from one neuron to another can be understood in chemical terms even more easily than the nerve impulse within a single neuron. Allosteric structures like membranes can be shifted into different conformations through the stimulus of specific ligands, and we know quite a bit about the ligands that are active between neurons.

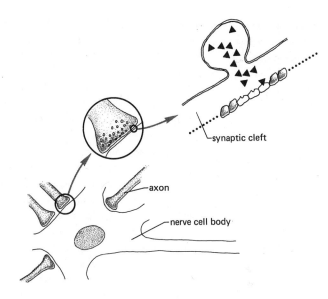

FIGURE 16-21

Structure of a synaptic cleft, enlarged to show one vesicle releasing its transmitter substance (triangles) into the cleft where they stimulate receptors on the opposite membrane.

synaptic cleft

axon

nerve cell body

Figure 16-21 shows the general structure of a **synapse,** where a process (such as an axon) from one neuron joins the body of another. The synapse is very tight; the two membranes are held together very strongly with a narrow **synaptic cleft** between them. At the tip of the axon, a number of tiny vesicles can usually be seen and it is now clear that these vesicles are filled with one or another **transmitter substance.**

All neurons that have been investigated can be classified as either **adrenergic** or **cholinergic,** depending upon the transmitter they release, which is either **norepinephrine (noradrenalin)** or **acetylcholine.** For acetylcholine, at least, the sequence of events is reasonably clear. A stimulus passing through a cholinergic nerve causes the presynaptic vesicles of acetylcholine (just *before* the synapse) to move to the cell membrane and eject their contents by exocytosis. The acetylcholine diffuses across the synaptic cleft and combines with specific receptors on the postsynaptic membrane. This stimulates the postsynaptic neuron to produce an action potential, which runs all the way to its other end, where the process may be repeated at another synapse. Acetylcholine therefore acts as an allosteric effector on the membrane permease systems. Meanwhile, the acetylcholine bound to the postsynaptic membrane is destroyed by the enzyme **cholinesterase,** which splits it into choline plus acetate and thus removes the ligand from its site.

The electric eel, *Electrophorus,* possesses **electroplax cells** that develop a considerable potential and enable the animal to give a painful shock. Changeux and his associates have studied the electroplax system as a model for allosteric transitions in membranes. The electroplax is induced to discharge by acetylcholine and its analogues, such as carbamylcholine, which presumably all bind at a common site. The discharge is inhibited by a variety of other analogues, including *d*-tubocurarine. The competition between these compounds is precisely what is predicted for effectors that bind to the same site. Acetylcholine apparently stabilizes the membrane subunits

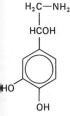

noradrenalin

acetylcholine

in their discharged state (until the effector is destroyed by cholinesterase); the inhibitors stabilize the subunits in their charged state and prevent the transition associated with a discharge. These inhibitors, of course, are not inactivated by the cholinesterase.

We must expect to find this kind of specific, internal ligand system operating generally within a nervous system. Free-living cells can presumably undergo transitions in response to many external ligands; in the evolution of organisms with nervous systems, one or two of these ligands have been selected as specific, internal effectors. However, other cells remain that are sensitive to a greater variety of external stimuli. Thus, the cells of taste buds and the smell receptor cells must have allosteric sites into which certain specific ligands can fit and produce a discharge. When you taste something sweet, for example, it is because a sugar or a very similar compound has fit into a site on a receptor cell in your tongue and started a series of discharges all along a specialized part of your nervous system, which your brain interprets as "sweet." The enormous problem of describing the events within your nervous system still remains, but it is easy to see generally what molecular events must be occurring at the point where your system meets the chemical environment in your mouth.

From what we have said before about the possible transitions within a membrane, you can also see that one possible state for a neuron is a permanent transition into a distinctive membrane configuration—for example, a configuration in which a much smaller stimulus is required for a membrane discharge than before. What we call memory may very well depend upon the establishment of this kind of state within specific neural pathways, so a slight stimulus to the pathway will always be followed by a further discharge of the entire pathway.

16-17 THE REGULATION OF MUSCLE CONTRACTION

The general outline of events in muscle contraction now seems quite clear; it fits perfectly with the picture of membrane structure and function we have been discussing.

The system of membranes in a muscle fiber is shown in Fig. 16-22. There are clearly two separate systems: a **sarcoplasmic reticulum** that lies more or less parallel to the bundles of myofibrils and a **transverse** or **T system** that runs radially through the reticulum. There are now many good pictures which show that the tubules of the T system are continuous with the limiting cell membrane (sarcolemma) so they are just deep invaginations of extracellular space that get very close to the contractile elements of the muscle. The sarcoplasmic reticulum seems to be an internal system that is isolated from the sarcolemma and T system; the reticulum membranes terminate at the Z line, so they are confined to single sarcomeres.

A. F. Huxley and his colleagues demonstrated that muscles could be forced to contract locally by applying micropipettes to their surfaces and depolarizing the sarcolemma with a short pulse of current. This shock was only effective when applied at particular regions of the sarcomere, particularly near the Z line. It now turns out that the transverse canals of the T system are always at those points where a local depolarization is effective. The T system is therefore the mechanism for transmitting an external stimulus into the heart of the muscle fiber.

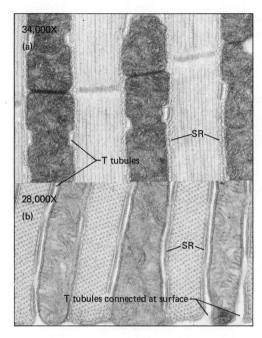

34,000X
(a)

T tubules

SR

28,000X
(b)

SR

T tubules connected at surface

FIGURE 16-22

The internal membrane systems of dragonfly flight muscle in (a) longitudinal and (b) cross section. Sarcoplasmic reticulum (SR) surrounds the myofibrils; and T tubules are long invaginations of the cell surface that pass through the SR but do not connect. (From David S. Smith, Insect Cells, *Oliver and Boyd, Edinburgh, 1968, reprinted by permission.)*

The immediate stimulus of contraction at the myosin-actin level is clearly Ca^{++} ion; the evidence for this has been growing for a long time, and there are now many experiments which show that Ca^{++} stimulates the myosin ATPase and causes the myofilaments to contract, both in intact muscle and in isolated preparations. (The interaction with calcium apparently depends in a complex way upon additional proteins, particularly tropomyosin.) Calcium ions are clearly bound in the membranes of the myofibril, and here again phospholipids—presumably including phosphatidyl serine—are involved. Thus contraction occurs when a muscle fiber receives a stimulus from a nerve, in the form of a depolarizing action potential; the nerve endings are on the plasmalemma, but the stimulus passes through the plasmalemma and down the T system, where Ca^{++} ions are ejected into the myofibrils.

As soon as the contraction occurs, there must be an opposing process to relax the muscle. The relaxing factor of muscle has now been identified as a large, insoluble material, which is the sarcoplasmic reticulum itself, and a poorly characterized soluble factor produced by the reticulum. The sarcoplasmic reticulum has an enormous affinity for Ca^{++}, which it picks up very actively after a contraction, to cause relaxation of the muscle. The soluble factor may interact with the contractile proteins even more directly; it inactivates the myosin ATPase and is not affected by Ca^{++} ions, but it is strongly inhibited by norepinephrine, which is the mediator between nerve and muscle. The inactivation of this factor by norepinephrine and the subsequent removal of ATPase inhibition may therefore be the first events in contraction, even preceding the direct action of calcium.

Similar events probably occur in other contractile systems. It is quite likely that ciliary contraction is initiated by processes in the ciliary membrane, and the same can probably be said for contraction in the cytosol of cells that move by means of a myxomyosin-actin system.

SUPPLEMENTARY EXERCISES

exercise 16-5 R. P. Levine isolated acetate requiring (*ac*) mutants of *Chlamydomonas reinhardi.* Several of them are unable to photosynthesize because of a block between photosystem I and photosystem II. Here are several mutants, with the components they are missing and the condition of two cytochromes in their chloroplasts:

MUTANT	MISSING FACTOR	cyt *b*-559	cyt *c*-553
ac-21	"M-component"	reduced	oxidized
ac-80a	P700	reduced	reduced
ac-115	cyt *b*-559	absent	oxidized
ac-206	cyt *c*-553	reduced	absent
ac-208	plastocyanin	reduced	reduced

(Plastocyanin is a copper-protein whose function is unknown.) The sequence of carriers can be determined from these data, if you remember that photosystem I oxidizes carriers and photosystem II reduces them. If a component is absent, everything between it and photosystem II will be reduced and all the carriers between it and photosystem I will be oxidized. P700, of course, comes just before photosystem I. What is the sequence of carriers?

17

biosynthesis and its regulation

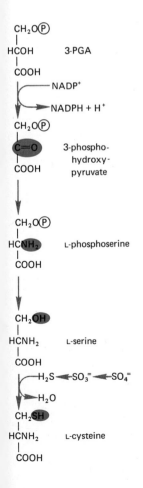

CH$_2$O(P)
|
HCOH   3-PGA
|
COOH

    — NADP$^+$

    → NADPH + H$^+$

CH$_2$O(P)
|
C=O   3-phospho-
| hydroxy-
COOH   pyruvate

CH$_2$O(P)
|
HCNH$_2$ L-phosphoserine
|
COOH

CH$_2$OH
|
HCNH$_2$ L-serine
|
COOH

   —H$_2$S ◄— SO$_3^=$ ◄— SO$_4^=$

   —H$_2$O

CH$_2$SH
|
HCNH$_2$ L-cysteine
|
COOH

In Chapter 4 we noted that the various transformations of carbon compounds can all be seen as parts of a carbon cycle in the biosphere. In Chapter 8 we looked more closely at some parts of the cycle and saw that a great deal of metabolism can be related to the reactions in and around the tricarboxylic acid (TCA) cycle. We have looked briefly at other phases of metabolism and we have seen how energy is obtained to drive biosynthesis. Now we can go back and look at some aspects of biosynthesis in a little more detail. In particular, we can look at many of the pathways that lead out of the TCA cycle and at some other pathways of biosynthesis. This will give us an opportunity to look at a variety of interesting things that illustrate points that have been made before; we can see some of the mechanisms used in transforming various compounds in metabolism, some of the regulatory devices superimposed on these pathways, and some of the consequences of genetic mistakes in these pathways, for some diseases are simply the result of a single enzymatic deficiency. But these examples are *not* presented with the expectation that you will memorize them; even though everyone eventually gets a general map of metabolic pathways in his head if he does serious biological research, no one is foolish enough to try to learn all of this.

One thing you should notice is the relatively small number of reaction mechanisms employed in these pathways; the same processes are employed over and over again. However, please try to keep your head above water and don't lose sight of the whole picture; it will help to refer back to Figure 8-11 and see how these pathways relate to one another. You will see that the various pathways all branch out from one another in a small set of families, and you should try to see the general structure of each family.

A. The pyruvate and oxalacetate family

The largest biosynthetic pathway begins with the transamination of oxalacetate to aspartate and leads to the pyrimidines and several amino acids. However, at several points it becomes involved with pyruvate or amino acids derived from pyruvate, so we will briefly dispose of these and then examine the main pathway.

17-1 ALANINE, SERINE, AND CYSTEINE

Alanine, of course, is derived directly from pyruvate by transamination. Serine comes from a precursor of pyruvate, 3-phosphoglycerate, which is oxidized and transaminated according to the pathway shown here.

539

COOH
|
CH$_2$
| aspartate
HCNH$_2$
|
COOH

⟍ ATP
⟍ ADP

COO(P)
|
CH$_2$ β-aspartyl
| phosphate
HCNH$_2$
|
COOH

⟍ NADPH

P$_i$ ⟵ ⟶ NADP

HC=O
|
CH$_2$ β-aspartic
| semialdehyde
HCNH$_2$
|
COOH

⟍ NADPH
⟍ NADP$^+$

H$_2$COH
|
CH$_2$
| homoserine
HCNH$_2$
|
COOH

As we noted in Chapter 8, sulfate is carried into the cell and reduced through PAPS to sulfite, which then is reduced to sulfide. This sulfide is incorporated directly into serine to make cysteine. Once in this organic form, the sulfur atom can easily be carried into other compounds.

17-2 THE MAIN PATHWAY

The step from oxalacetate to aspartate is another transamination. There is then a set of three reactions whose pattern is repeated so often that we will simply call it **phosphorylation-reduction.** In this case, it is phosphorylation by ATP, reduction to the semialdehyde, and reduction again to the alcohol (homoserine), with the release of inorganic phosphate.

Homoserine can be transformed in two ways; let's first follow its conversion to threonine. It is first phosphorylated into O-phosphohomoserine and then converted into threonine through the mechanism we have seen before of removing water and adding it back in reverse:

CH$_2$OH   CH$_2$O(P)   CH$_2$ CH$_3$
| ATP | H$_2$O || H$_2$O   HCOH
CH$_2$ ⟶   CH$_2$ ⟶   CH   ⟶ |
| | | HCNH$_2$
HCNH$_2$ HCNH$_2$ HCNH$_2$ |
| | | COOH
COOH   COOH   COOH

homoserine   O-phosphohomoserine   L-threonine

The other fate of homoserine is to make methionine. It is first made into O-acetyl- or O-succinyl-homoserine (depending on the organism); the sulfur is added from cysteine through the simple mechanism of condensing the two molecules and then splitting on the other side of the sulfur atom:

H$_2$CSH
| S
L-cysteine   HCNH$_2$ CH$_2$ CH$_2$ CH$_3$
| | | |
COOH   HCNH$_2$ CH$_2$ HCNH$_2$ L-alanine
                                COOH   HCNH$_2$ |
CH$_2$O—CH$_2$ succinate | COOH
| | COOH
CH$_2$ CH$_2$ SH
O-succinyl- | | cystathionine |
homoserine HCNH$_2$ COOH   CH$_2$
| |
COOH   CH$_2$
                                                     |
                                                     HCNH$_2$
                                                     |
                                                     COOH

L-homocysteine

The resulting homocysteine is methylated to make methionine.

The transfer of C$_1$ units, including methyl groups, is a rather complicated subject we will not discuss here. Methionine itself is a methylating agent, and other compounds are used occasionally as methyl donors, but most of the C$_1$ transfer reactions require

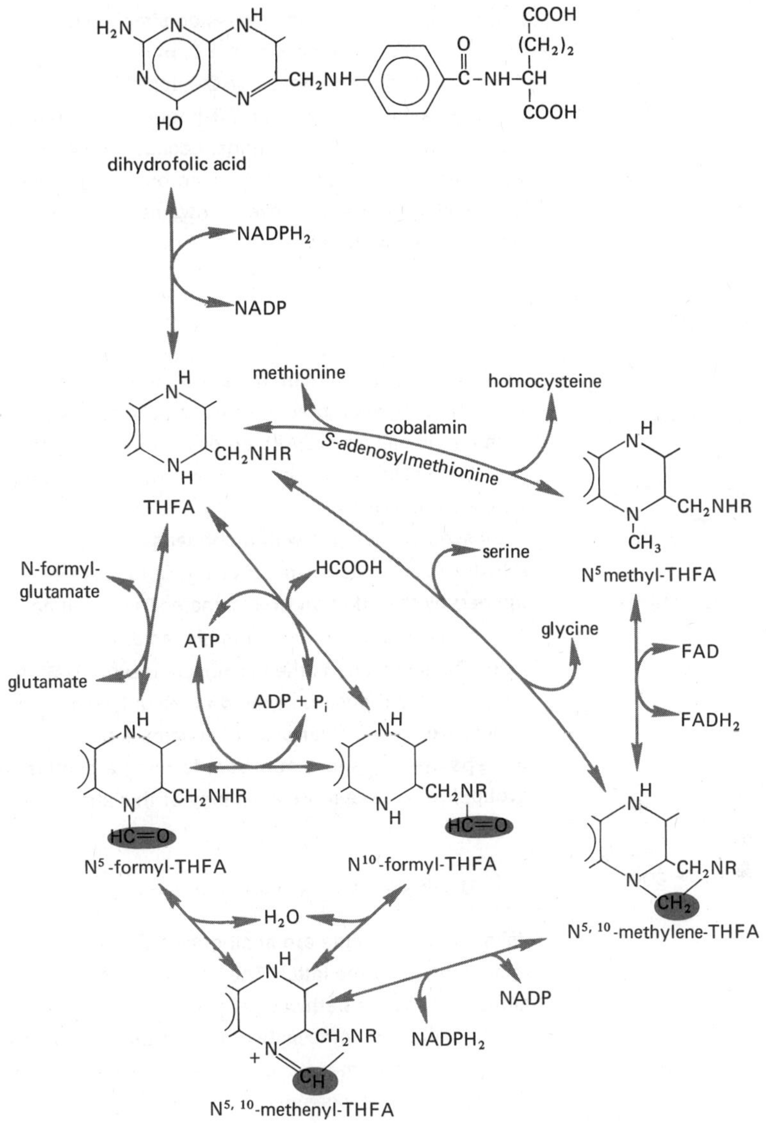

FIGURE 17-1

The C₁ pool: Transformations between the forms of tetrahydrofolic acid (THFA).

some form of **tetrahydrofolic acid (THFA),** whose reactions are shown in Fig. 17-1. Several different kinds of C_1 groups are carried about by these coenzymes; to make matters worse, some of the reactions involve a large, complex cofactor, cobalamin or

HC=O
|
CH₂
|
HCNH₂   +
|
COOH

β-aspartic
semialdehyde

CH₃
|
C=O
|
COOH

pyruvate

→ H₂O

NADP →

dihydropicolinate   tetrahydropicolinate

COOH
|
C=O
|
(CH₂)₃   O
| ‖
HC—NH—C—(CH₂)₂—COOH
|
COOH

—succinyl CoA
— CoA

COOH
|
H₂N—CH
|
(CH₂)₃   O
| ‖
HC—NH—C—(CH₂)₂—COOH
|
COOH

→ succinate

COOH
|
H₂N—CH
|
(CH₂)₃
|
HCNH₂
|
COOH   CO₂

L-DAP

NH₂
|
(CH₂)₄
|
HCNH₂
|
COOH

L-lysine

---

vitamin B_{12}. To keep the discussion simple, we'll simply refer to the whole set of possible C_1 donors as the **C_1 pool.**

Now, why is the O-acetyl or O-succinyl added in this pathway if it is removed almost immediately? The question is not so easy to answer here as it will be in similar reactions, because the reactions around cystathionine are not entirely clear, but groups like this are often used as **blocking agents** to prevent unwanted side reactions. We will see a better example shortly.

17-3 THE LYSINE-DAP BRANCH

Most organisms that can make their own lysine (humans, for example, cannot) do so via a branch of the oxalacetate family. This branch includes diaminopimelic acid (DAP), an essential wall component of many bacteria, and these cells obviously get a dividend from making both compounds by the same route.

Biosynthesis begins with a condensation to make a ring compound, dihydropicolinic acid, and after this is reduced once more, the ring is opened by the addition of succinate. The resulting compound, N-succinyl-ε-keto-L-α-aminopimelic acid, is transaminated to the diamino acid. The succinate is then removed, leaving DAP, which can be isomerized and decarboxylated to make lysine.

Here we have a clear case of succinate being used as a blocking agent; it keeps the α-amino group from forming a peptide bond with the ε-keto group. We will see more examples of this strategy later.

17-4 THE BRANCH-CHAIN AMINO ACIDS

Valine and isoleucine are analogues that differ only in the addition of a methyl group to the latter. They are synthesized by a single set of enzymes along most of the pathway.

Isoleucine biosynthesis begins with threonine, which is dehydrated and transaminated to make α-ketobutyrate. α-Ketobutyrate and pyruvate are the precursors of isoleucine and valine. In both cases, the first reaction

CH₃
|
HCOH
|
HCNH₂
|
COOH

L-threonine

H₂O →

CH₃
|
CH
‖
CNH₂
|
COOH

NADPH₂ → NADP →

CH₃
|
CH₂
|
HCNH₂
|
COOH

α-ke

is a condensation with acetyl-TPP, and the two sequences then proceed as indicated.

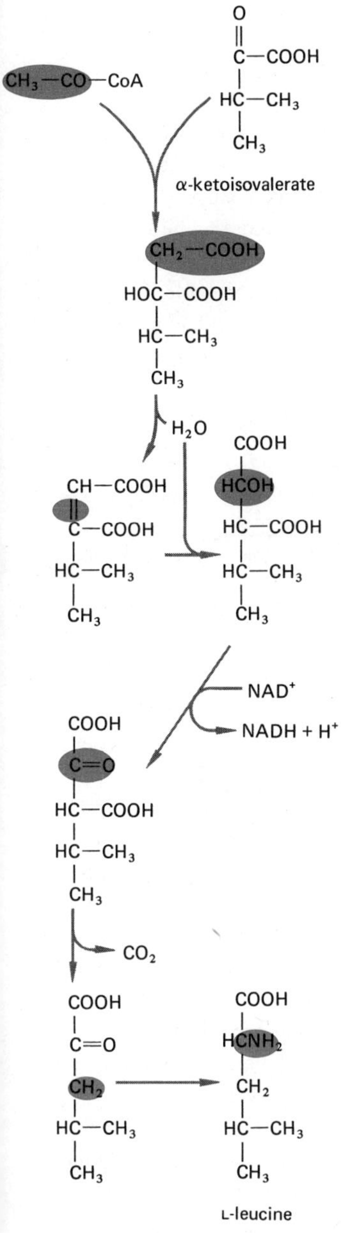

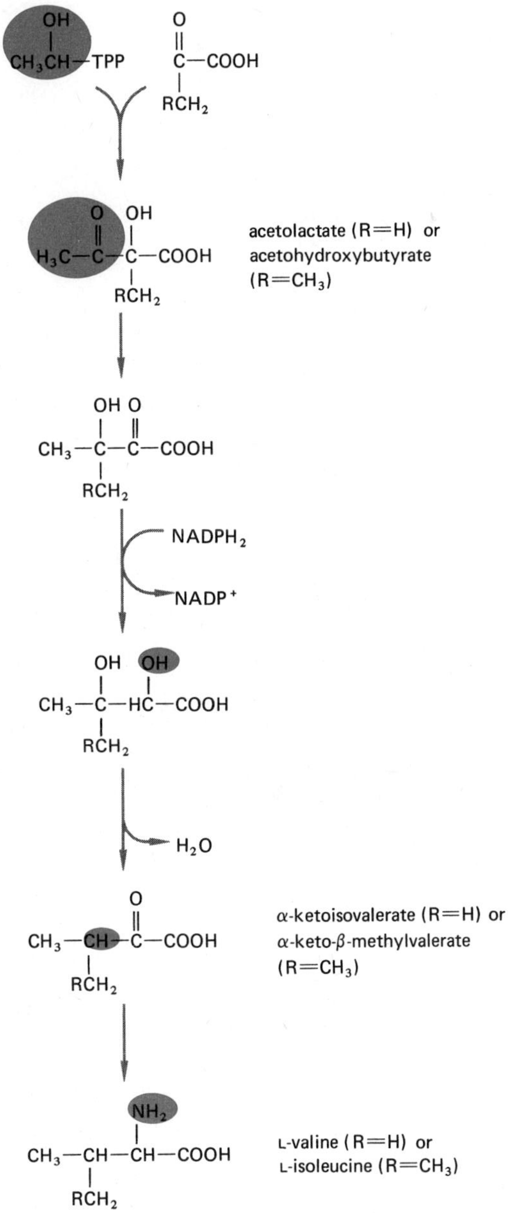

$CH_3-CO-CoA$

α-ketoisovalerate

$$\begin{array}{c} O \\ \parallel \\ C-COOH \\ | \\ HC-CH_3 \\ | \\ CH_3 \end{array}$$

$$\begin{array}{c} CH_2-COOH \\ | \\ HOC-COOH \\ | \\ HC-CH_3 \\ | \\ CH_3 \end{array}$$

H_2O

$$\begin{array}{c} CH-COOH \\ \parallel \\ C-COOH \\ | \\ HC-CH_3 \\ | \\ CH_3 \end{array}$$

$$\begin{array}{c} COOH \\ | \\ HCOH \\ | \\ HC-COOH \\ | \\ HC-CH_3 \\ | \\ CH_3 \end{array}$$

NAD^+

$NADH + H^+$

$$\begin{array}{c} COOH \\ | \\ C=O \\ | \\ HC-COOH \\ | \\ HC-CH_3 \\ | \\ CH_3 \end{array}$$

CO_2

$$\begin{array}{c} COOH \\ | \\ C=O \\ | \\ CH_2 \\ | \\ HC-CH_3 \\ | \\ CH_3 \end{array}$$

$$\begin{array}{c} COOH \\ | \\ HCNH_2 \\ | \\ CH_2 \\ | \\ HC-CH_3 \\ | \\ CH_3 \end{array}$$

L-leucine

$$\begin{array}{c} OH \\ | \\ CH_3CH-TPP \end{array} \quad \begin{array}{c} O \\ \parallel \\ C-COOH \\ | \\ RCH_2 \end{array}$$

$$\begin{array}{c} O \quad OH \\ \parallel \quad | \\ H_3C-C-C-COOH \\ | \\ RCH_2 \end{array}$$

acetolactate (R=H) or
acetohydroxybutyrate
(R=CH$_3$)

$$\begin{array}{c} OH \quad O \\ | \quad \parallel \\ CH_3-C-C-COOH \\ | \\ RCH_2 \end{array}$$

$NADPH_2$

$NADP^+$

$$\begin{array}{c} OH \quad OH \\ | \quad | \\ CH_3-C-HC-COOH \\ | \\ RCH_2 \end{array}$$

H_2O

$$\begin{array}{c} O \\ \parallel \\ CH_3-CH-C-COOH \\ | \\ RCH_2 \end{array}$$

α-ketoisovalerate (R=H) or
α-keto-β-methylvalerate
(R=CH$_3$)

$$\begin{array}{c} NH_2 \\ | \\ CH_3-CH-CH-COOH \\ | \\ RCH_2 \end{array}$$

L-valine (R=H) or
L-isoleucine (R=CH$_3$)

Leucine biosynthesis branches off from α-ketoisovalerate through the series of reactions shown here. We will not go into some other reactions of this family, but you should be aware that coenzyme A is made mostly from these compounds.

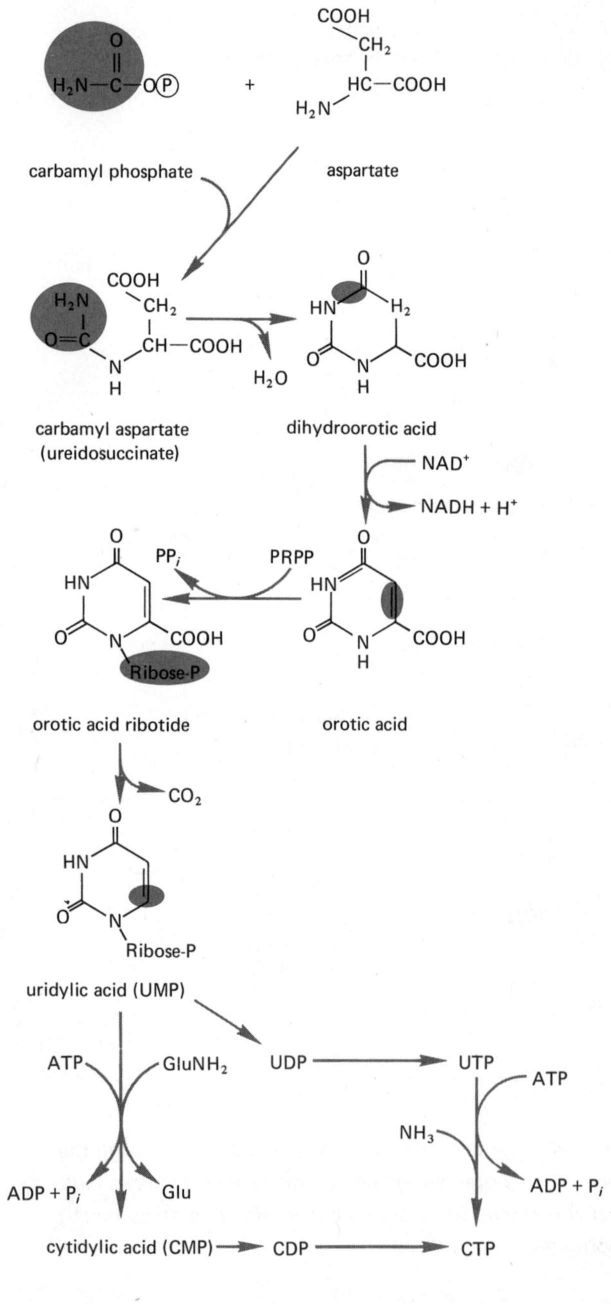

carbamyl phosphate   aspartate

carbamyl aspartate
(ureidosuccinate)   dihydroorotic acid

orotic acid ribotide   orotic acid

uridylic acid (UMP)

cytidylic acid (CMP) → CDP ——→ CTP

17-5 THE PYRIMIDINES

Because they constitute such a large part of the cell mass, the pyrimidine branch is a major drain on the oxalacetate family. The first reaction in the pathway is a condensation of aspartate with carbamyl phosphate, a molecule formed by the combination of ammonia and carbon dioxide:

$$NH_4^+ + CO_2 \rightarrow H_3N\text{—}COOH$$
$$H_3N\text{—}COOH + ATP \rightarrow H_3N\text{—}COO\,\text{\textcircled{P}} + ADP$$

The first reaction occurs spontaneously only in bacteria; in mammalian cells, it requires another ATP molecule. Carbamyl phosphate is used in various pathways to add a nitrogen and a carbon atom simultaneously. In this case, the product is carbamyl aspartate. Removal of a water molecule closes the ring. The dihydroorotic acid is reduced to orotic acid and the ribose base is added. This is an important reaction; the activated form of the sugar is phosphoribosyl-1-pyrophosphate (PRPP), which is made via

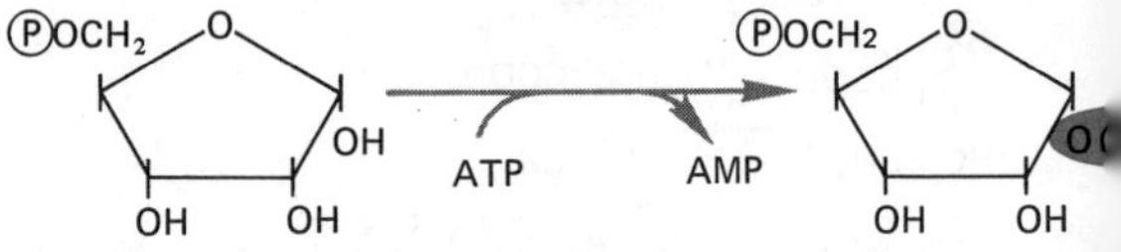

Decarboxylation yields uridylic acid, which may be transaminated to cytidylic acid. The other nucleotides are made by phosphorylation with ATP.

17-6 REGULATION OF THE OXALACETATE FAMILY

The interrelationships between the various end products of this family are so complex that a number of novel mechanisms have had to evolve to ensure that there will be a balanced flow of material into all branches of the system and that the cells will be able to grow under many conditions. The points of feedback inhibition are shown in Fig. 17-2. While this is generally what one would expect a priori, there are some really interesting points that should be explained.

One general solution to regulatory problems has been achieved in *E. coli* by means of multiple enzymes for a single reaction. There are three

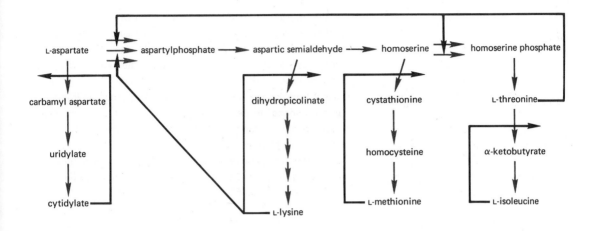

FIGURE 17-2

Regulation in the oxalacetate family. The colored arrows indicate points of feedback inhibition. Notice the separate enzymes from aspartate to aspartyl phosphate, inhibited by different products.

aspartokinases; type I is inhibited by threonine but repressed only by threonine and isoleucine together, II is repressed by methionine, and III is both repressed and inhibited by lysine. The two inhibitions are classical examples of allosteric mechanisms. You can easily prove to yourself that with this combination of controls the cells can grow efficiently with any combination of amino acids in the medium.

The next set of controls are exerted at homoserine dehydrogenase, which has two forms. Type I is repressed by threonine and isoleucine and allosterically inhibited by threonine; type II is repressed by methionine, but not inhibited. But J.-C. Patte, Georges N. Cohen, and their associates have shown that there is one protein complex that carries both homoserine dehydrogenase I and aspartokinase I activities and another that carries the corresponding type II activities of both enzymes. This explains several points about the regulation of these enzyme systems and shows what a remarkable and intricate set of controls these cells have evolved.

Martin Freundlich, R. O. Burns, and Edwin Umbarger have demonstrated a complex repression mechanism for the four enzymes that synthesize valine and isoleucine. They find that these enzymes are repressed only if valine, leucine, and isoleucine are all present simultaneously, and they propose that this phenomenon be called **multivalent repression.** In other words, the repressor in this system must have sites for all three amino acids and must be inactive if any site is empty. Since the same enzymes are required for all three amino acids, it is hard to imagine another way in which their synthesis could be regulated.

exercise 17-1   The enzyme complex I, with both aspartokinase and homoserine
                    dehydrogenase activities, is repressed by threonine and isoleucine

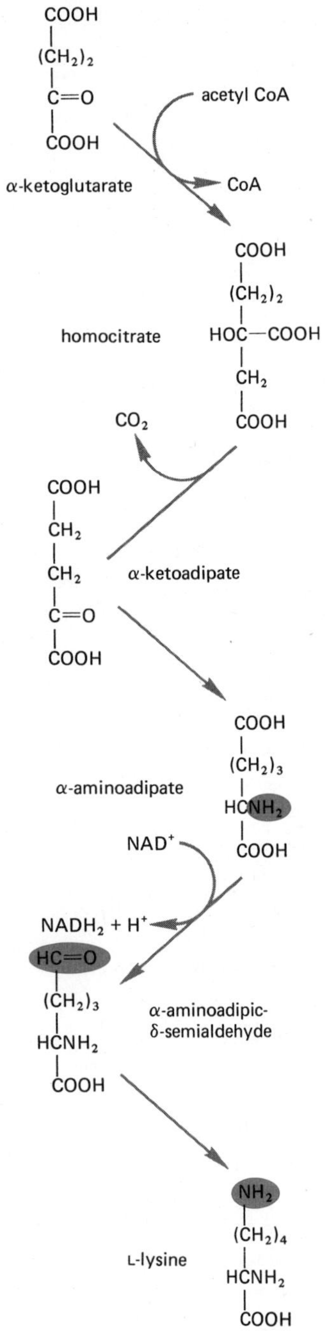

COOH
|
$(CH_2)_2$
|
C=O
|
COOH

α-ketoglutarate

acetyl CoA

CoA

COOH
|
$(CH_2)_2$
|
HOC—COOH
|
CH_2
|
COOH

homocitrate

CO_2

COOH
|
CH_2
|
CH_2
|
C=O
|
COOH

α-ketoadipate

COOH
|
$(CH_2)_3$
|
HCNH₂
|
COOH

α-aminoadipate

NAD^+

$NADH_2 + H^+$

HC=O
|
$(CH_2)_3$
|
HCNH₂
|
COOH

α-aminoadipic-
δ-semialdehyde

NH₂
|
$(CH_2)_4$
|
HCNH₂
|
COOH

L-lysine

together. A *thr ilu* double auxotroph is grown in a chemostat with either limiting threonine and excess isoleucine or limiting isoleucine and excess threonine; in both cases, the complex I enzymes are highly derepressed to the same extent. What does this indicate about the repression mechanism?

exercise 17-2 The aspartokinase activity in other bacteria (such as *Rhodopseudomonas*) resides in a single enzyme that is not inhibited allosterically by threonine or lysine by themselves, but is strongly inhibited by the two of them together. This is called **concerted inhibition;** what must be the general mechanism of concerted inhibition? Do you expect the inhibition to be complete in this case?

B. The α-ketoglutarate family

17-7 LYSINE, GLUTAMATE, AND PROLINE

The true fungi and the euglenoids (and who knows what else) synthesize lysine through a pathway from α-ketoglutarate. The first step is like the condensation of acetyl-CoA and oxalacetate in the TCA cycle, but there is an extra carbon atom and the product is homocitrate. There then follows a series of decarboxylation, transamination, reduction, and transamination again to produce lysine.

As we have already seen, α-ketoglutarate is transaminated directly into glutamate. Proline is then synthesized via a simple pathway using the familiar mechanism of phosphorylation-reduction again. The ring apparently cyclizes spontaneously, and a final reduction produces proline.

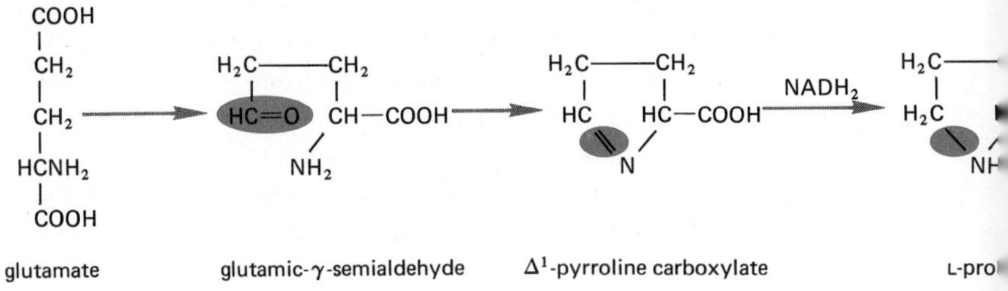

COOH
|
CH_2
|
CH_2
|
HCNH₂
|
COOH

glutamate

H_2C——CH_2
| |
HC=O   CH—COOH
        /
      NH₂

glutamic-γ-semialdehyde

H_2C——CH_2
| |
HC   HC—COOH
   \   /
    N

Δ¹-pyrroline carboxylate

$NADH_2$

H_2C——
|
H_2C
    \
    NH

L-pro[line]

17-8 ARGININE AND THE UREA CYCLE

The biosynthesis of arginine begins with reactions similar to those leading to proline. But we have just seen that the ring closes spontaneously in proline synthesis, and this is precisely what must not happen in arginine synthesis, so a blocking group is used again to prevent this reaction. Bacteria acetylate the α-amino group and other organisms use a succinyl group.

N-acetyl glutamate is phosphorylated and reduced to its semialdehyde, and when this is transaminated and deacylated it becomes ornithine, which is an important product in itself. Ornithine is not found in any large proteins, but it is used in several polypeptide hormones. Carbamyl phosphate is also used in this pathway; the product of its condensation with ornithine is citrulline.

Citrulline then undergoes a very interesting condensation with aspartate, which results in deamination of the aspartate and the formation of arginine. Thus far, the pathway is common to all organisms, for it produces arginine for protein synthesis. However, plants and most animals have an additional enzyme that removes the two terminal nitrogens and a carbon from arginine, producing urea and a new molecule of ornithine. This creates a closed pathway, the **urea cycle,** which continually picks up carbamyl phosphate and produces urea, which has one more amino group. Each turn of the cycle therefore eliminates an ammonia molecule; this is one of the principal mechanisms that higher organisms have for getting rid of their ammonia, which comes chiefly from amino acids. In animals, this cycle takes place primarily in the liver, and deamination is one of the chief functions of that organ.

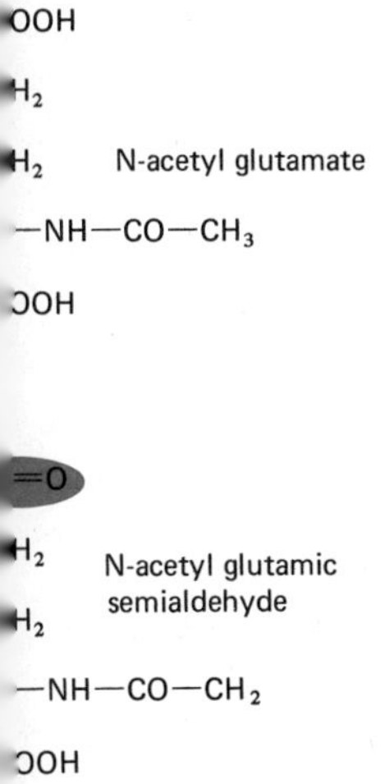

17-9 CONTROL IN THE α-KETOGLUTARATE FAMILY

The points of feedback inhibition in this family are shown in Fig. 17-3. Note that proline inhibits the first enzyme specific to its synthesis, the glutamate kinase, but arginine for some reason inhibits the second one in its pathway, the *N*-acetyl glutamate kinase. This has only been demonstrated in one case and it may not be universal.

OOH
H$_2$
H$_2$ N-acetyl glutamate
—NH—CO—CH$_3$
OOH

=O
H$_2$ N-acetyl glutamic
H$_2$ semialdehyde
—NH—CO—CH$_2$
OOH

H$_2$
H$_2$)$_3$ N-acetyl ornithine
—NH—CO—CH$_3$
OOH

H$_2$
℗—O—C—NH$_2$
H$_2$
H$_2$)$_3$
NH$_2$
OOH
hine

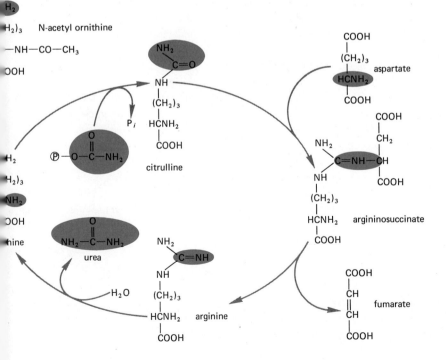

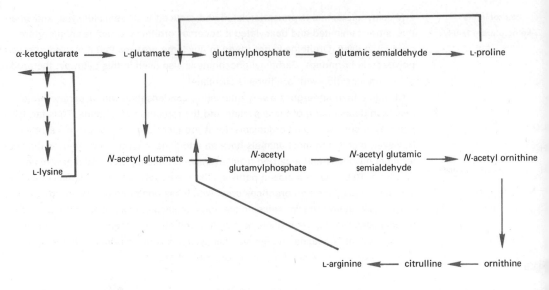

FIGURE 17-3

Regulation in the α-ketoglutarate family.

    Notice also that arginine synthesis involves just those precursors—carbamyl phosphate and aspartate—that are combined to make the pyrimidines. This provides a point of coupling between an amino acid—a protein precursor—and the pyrimidines—nucleic acid precursors. While protein and nucleic acid synthesis are intimately intertwined with one another in direct heterocatalytic mechanisms, this may be another point where their relative rates of synthesis could be controlled.

    The synthesis of the arginine enzymes in *E. coli* presents a very interesting elaboration of the operon model proposed by Werner Maas. The eight enzymes of the pathway are informed by eight genes that map in four different regions, and yet the enzymes are all coordinately repressed by arginine. A single locus, *argR*, has the properties of a regulator and mutations for *argR* can lead to constitutivity for all eight enzymes together. Maas proposes that there are four operons, all regulated by a single aporepressor informed by *argR*; he calls the whole unit a **regulon.** This adds another dimension to the possibilities of regulation. One could easily devise a system in which the entire regulon is coordinated by one mechanism, through the aporepressor, and where individual operons within it are coordinated by other mechanisms acting at the operator level.

FIGURE 17-4

Biosynthesis of inosinic acid (IMP) and its conversion to the other purine necleotides. Other conversions, some of which are salvage routes for bases and nucleosides, are shown below; the dashed lines indicate purely catabolic pathways.

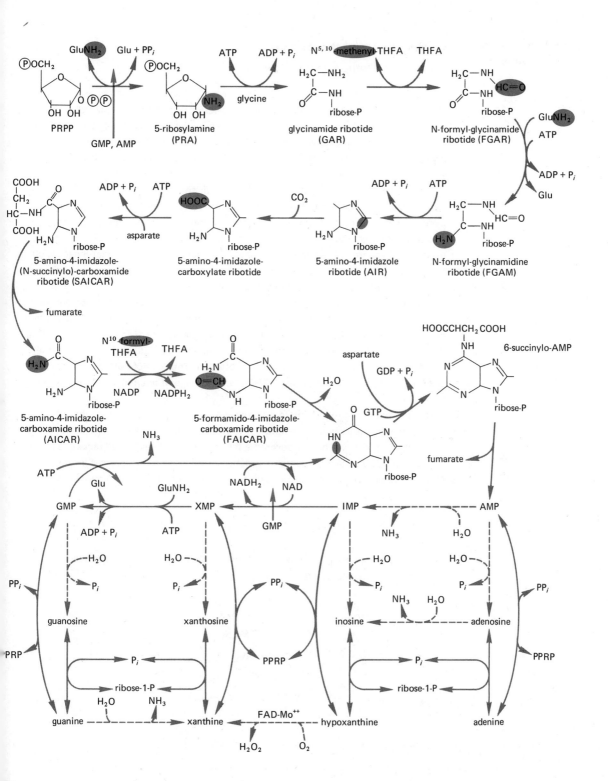

C. Histidine and the purines

17-10 PURINE SYNTHESIS AND ITS CONTROL

The pathway of histidine biosynthesis was shown in Fig. 9-11 to illustrate the intimate connection between the genetic map and the series of enzymes. If you looked at that pathway carefully, you may have been disturbed by the fact that a very large molecule, amino imidazole carboxamide ribotide (AICAR) was apparently thrown away. In fact, it isn't thrown away, for AICAR is one intermediate in the pathway of purine biosynthesis shown in Fig. 17-4.

There is nothing unusual about this pathway, in spite of its length. Notice that the purines are made in their ribotide form while the ribose molecule is attached late in pyrimidine biosynthesis. A series of reactions at the end of the pathway convert one nucleotide into another; these reactions are partly degradative, but they can also be used to salvage free bases and make them into nucleotides. Notice that these reactions lead to xanthine; xanthine can be oxidized to uric acid, which is excreted by many animals as a way of getting rid of excess nitrogen.

Figure 17-5 shows the interrelationships between histidine and the major purine nucleotides. There are evidently several possible control points; IMP itself stands in a central position and its concentration is critical. GMP and AMP inhibit PRA synthesis, but AICAR—and therefore ATP for histidine synthesis—can still be formed from the histidine pathway even if PRA synthesis is blocked. The path of IMP resynthesis from GMP is useful for AMP synthesis if GMP is in excess, and this reaction is blocked by ATP.

The importance of regulation in a pathway of this kind has been brought out most forcefully in a study reported recently by J. E. Seegmiller, Frederick M. Rosenbloom, and William N. Kelley. They discovered a family carrying a sex-linked recessive mutation resulting in cerebral palsy, mental retardation, aggressive behavior, and a variety of other neurological and behavioral disorders. Two brothers who were studied had an

FIGURE 17-5

Relationships between the intermediates of histidine and purine biosynthesis. Notice the complex pattern of feedback inhibition.

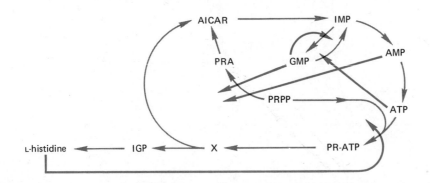

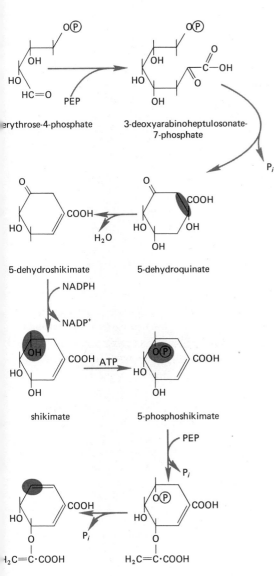

erythrose-4-phosphate

3-deoxyarabinoheptulosonate-
7-phosphate

P_i

5-dehydroshikimate

H_2O

5-dehydroquinate

NADPH

NADP⁺

shikimate

ATP

5-phosphoshikimate

PEP

P_i

chorismate

P_i

3-enolpyruvyl-5-phosphoshikimate

excess of uric acid, a condition also associated with the painful disease gout, which does not, however, produce similar neurological disorders. This suggested a defect of purine biosynthesis, but the patients did not respond to drugs, such as azathioprine, which inhibit uric acid synthesis in gouty conditions. Seegmiller and his coworkers then found that the patients lack the PRPP transferase enzyme that converts free guanine and hypoxanthine back into GMP and IMP, which in turn feed back to inhibit the first step in purine biosynthesis. This enzyme is also necessary for converting drugs used in the treatment of gout into their ribotide forms, so they can inhibit purine biosynthesis. The PRPP transferase appears to be critical for regulation of the pathway, and somehow the lack of regulation leads to the neurological syndrome. Discoveries of this kind, along with carefully controlled observations on people under the influence of LSD and other hallucinogenic agents, could lead to a reevaluation of abnormal human behavior in terms of metabolic disturbances.

D. The aromatic family

17-11 THE SHIKIMIC ACID PATHWAY

In addition to its oxidative function, the pentose phosphate cycle produces compounds that are precursors to the biosynthesis of the aromatics. The pathway begins with the condensation of erythrose-4-phosphate and PEP to make a compound that cyclizes and is then reduced and made into another PEP derivative, chorismic acid.

Chorismate is the branch point for a large number of important pathways. For example, the benzene ring of folic acid is made in this way. The pyruvate group of chorismate is lost, and the ring is then aminated and joined to glutamate. The pteroyl ring is made elsewhere.

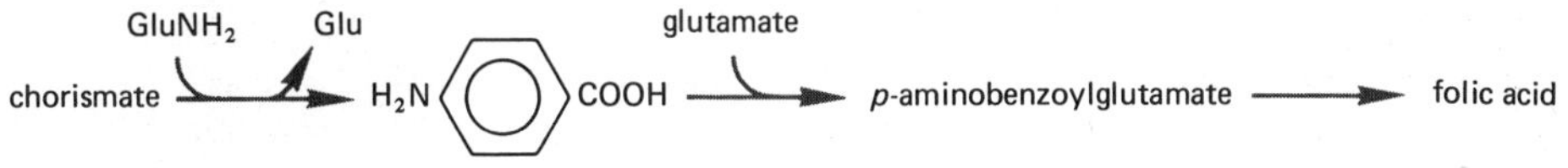

chorismate — GluNH₂ / Glu → H_2N⟨◯⟩COOH — glutamate → p-aminobenzoylglutamate ⟶ folic acid

A shift of the pyruvyl group of chorismic acid yields prephenic acid. This compound can be aromatized if its carboxyl group is eliminated; this apparently can be done either by the simultaneous elimination of the hydroxyl group to yield phenylalanine, or by its retention to yield tyrosine:

hydroxyphenylpyruvate   chorismate   phenylpyruvate

tyrosine   prephenate   phenylalanine

In mammals, however, tyrosine is made directly from phenylalanine by hydroxylation. A number of human mutants have been found with defects in parts of this complex. For example, **phenylketonuria** is the result of a defect in the hydroxylase that converts phenylalanine to tyrosine; people with this defect accumulate large amounts of phenylalanine in their blood and phenylpyruvate in their cerebrospinal fluid, and they are mentally defective. This disorder can be controlled by restricting their intake of phenylalanine.

FIGURE 17-6

The tryptophan system of E. coli. *The sequence of genes and enzymatic steps is essentially collinear, but the enzymes are formed by a set of protein-protein interactions.*

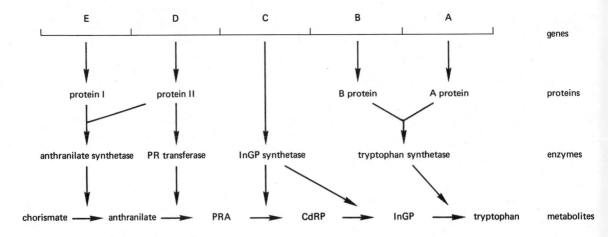

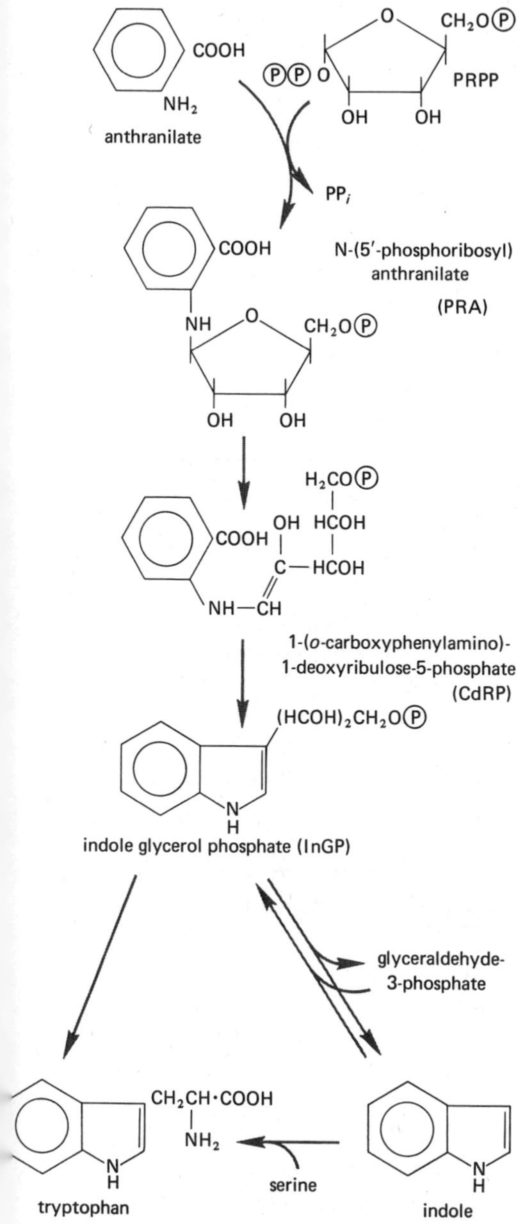

anthranilate

PRPP

PP$_i$

N-(5'-phosphoribosyl) anthranilate (PRA)

1-(o-carboxyphenylamino)-1-deoxyribulose-5-phosphate (CdRP)

indole glycerol phosphate (InGP)

glyceraldehyde-3-phosphate

tryptophan

serine

indole

Alkaptonuria is a disorder in which the person excretes large amounts of homogentistic acid, a breakdown produce of p-hydroxyphenylpyruvate, in his urine. The urine turns black due to oxidation of this product on exposure to air. The defect is in a normal oxidase that catalyzes one step in the catabolism of tyrosine to TCA cycle intermediates.

17-12 TRYPTOPHAN

Tryptophan is also made from chorismic acid; in the first reactions, the hydroxyl and pyruvyl groups are eliminated and an amino group is added to make anthranilic acid, which is then condensed with PRPP and transformed as shown here.

The tryptophan pathway provides a wonderful example of a multienzyme complex and its regulation. In both *E. coli* and *Salmonella typhimurium,* the genes for the five enzymes of the pathway all map together (Fig. 17-6) and it is clear that the enzymes are associated with one another. Ronald Bauerle and Paul Margolin have found that the proteins with anthranilate synthetase and PRPP transferase activities, informed by genes E and D, are bound together and the first enzyme is active only when bound to the second. They have found interesting polarity mutants that have a mutation in the end of the *trpD* gene nearest the operator and yet have no anthranilate synthetase activity; the protein made by the mutant *trpD* gene has a little PRPP transferase activity, but it cannot bind the first enzyme. Furthermore, the activities of the two enzymes are inhibited by tryptophan, and tryptophan clearly causes the two proteins to become more closely associated. This is a perfect case of an allosteric transition involving a change in quaternary structure, where tightening the complex inhibits the enzymes.

The tryptophan synthetase itself is a complex of two polypeptides, A and B. The complete enzyme can catalyze all three reactions between indole glycerol phosphate, indole, and tryptophan. The A chain alone can catalyze the forward and reverse transformations of indole and indole glycerol phosphate, while the B enzyme alone can convert indole into tryptophan.

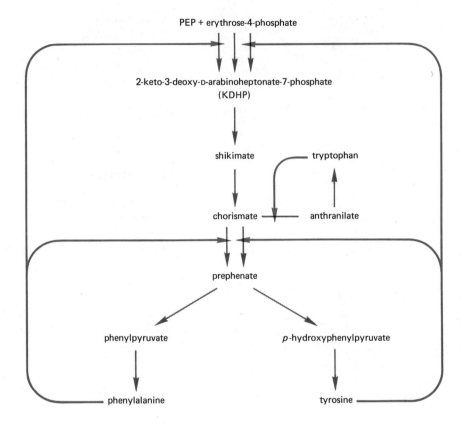

PEP + erythrose-4-phosphate

2-keto-3-deoxy-ᴅ-arabinoheptonate-7-phosphate
(KDHP)

shikimate

chorismate —————— anthranilate

tryptophan

prephenate

phenylpyruvate

p-hydroxyphenylpyruvate

phenylalanine

tyrosine

FIGURE 17-7

Regulation in the aromatic pathways in E. coli. *Notice that here again there are points where different enzymes which respond to different end products catalyze the same reaction.*

Figure 17-7 shows that there are some very interesting feedback inhibition mechanisms in this family. The system drawn here is for *E. coli,* which has three distinct enzymes that condense PEP with erythrose-4-phosphate. One is inhibited by phenylalanine, one by tyrosine, and the third is not sensitive to any end products known so far. There are also two chorismic mutases that make prephenic acid. One of them is part of an enzyme complex that makes phenylpyruvate and it is inhibited by phenylalanine, while the other is part of a complex that makes *p*-hydroxyphenylpyruvate and it is inhibited by tyrosine.

  Bacillus subtilis has a different system. There is only a single PKDH synthetase at the beginning of the pathway, and it is inhibited by prephenic acid and by high concentrations of chorismic acid. Tyrosine and phenylalanine inhibit the first steps in their conversion from prephenate. You can see that if both tyrosine and phenylalanine are at high enough concentrations, prephenate will accumulate and inhibit the PKDH synthetase. And if there is also enough tryptophan, the chorismic acid pool will build up and contribute to this inhibition.

17-14 A COMPLEX CASE: FLOWER COLORS

Many of the pigments in flowers are flavonoid compounds that are synthesized by a branch from the shikimic acid pathway. There is an enormous variety and only a few of them will be shown here as an example of an apparently complicated situation that turns out to have a fairly simple basis. The work we will discuss is based on the snapdragon, *Antirrhinum majus.*

This pathway is not known in detail, but analysis of four genes that affect it has allowed its outlines to be sketched in. A gene *N* affects some early stage of biosynthesis, for *n/n* homozygotes are dead white. In the presence of at least one *N* allele, the other genes, *P* (pink), *Y* (yellow), and *M* (modifier) produce the series of pigments and colors shown in Table 17-1. Together with biochemical data, the pathway shown in Fig. 17-8 is suggested. Homozygous *n/n* plants do not contain flavonoids; they have compounds such as *p*-coumaric and caffeic acids, which arise from the shikimic pathway. Either these two or some very similar compounds are the precursors of the flavonoid B ring. The enzyme informed by the *N* gene adds the A ring, apparently from three acetyl-CoA molecules. The resulting compound is a chalkone, with an open chroman ring.

TABLE 17-1

pigments of Antirrhinum majus mutants*

GENOTYPE	PHENOTYPE	PIGMENTS			
		CYANIDIN QUERCITIN	PELARGONIDIN KAEMPFEROL	LUTEOLIN	APEGENIN
PPMMYY	magenta	+		+	+
PPMMyy	red-orange	+		+	+
PPmmYY	pink		+		+
PPmmyy	yellow-orange		+		+
ppMMYY	ivory			+	+
ppMMyy	yellow			+	+
ppmmYY	ivory				+
ppmmyy	ivory				+

*All strains are genotype *NN.*

Gene *Y* probably directs an enzyme that closes the chroman ring and makes a flavonoid, though the block is not complete even in *y/y* plants. Gene *M* enzyme probably adds the 3'-hydroxyl group and gene *P* enzyme makes flavonols by adding the

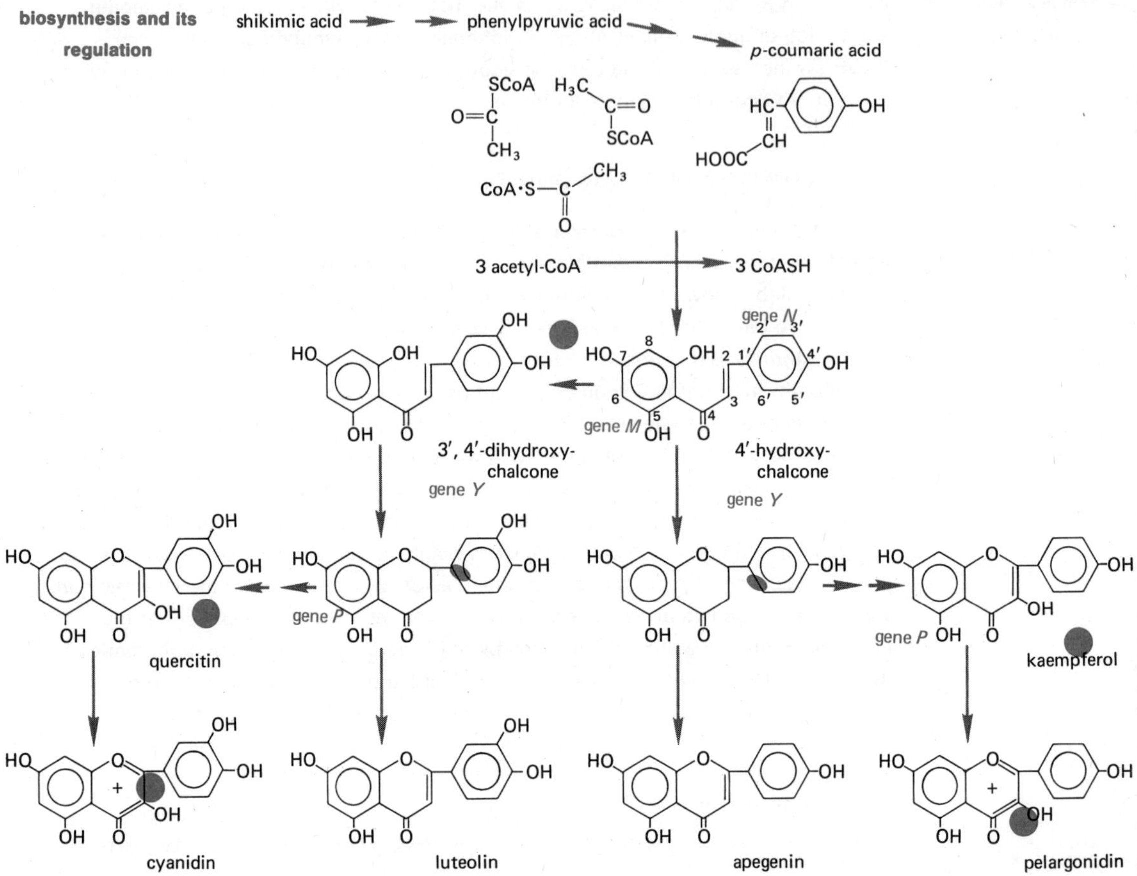

FIGURE 17-8

Probable biosynthetic pathway of the flavonoid pigments of Antirrhinum majus.

3-hydroxyl group. There are other factors yet to be discovered, but in its outlines this scheme is essentially correct. It is interesting to see how a phenotype like flower color can be analyzed to relatively simple components by a combination of genetic and biochemical techniques.

E. Lipids and related compounds

17-15 FATTY ACID BIOSYNTHESIS

The formation of fatty acids begins with acetyl-CoA. Since this is the chief product of carbohydrate metabolism, it is evident that excess carbohydrates can easily be converted to fats if the acetyl-CoA is not burned up in the tricarboxylic acid cycle through exercise and activity, or if there are other defects that tend to inhibit respiration. Hogs grow fat on a carbohydrate diet, and many people tend to emulate them.

The first reaction is CO_2 fixation by a biotin enzyme and condensation with acetyl-CoA to form malonyl-CoA:

$$CO_2 + ATP + biotin{-}enzyme \rightarrow CO_2{-}biotin{-}enzyme + ADP + P_i$$
$$CO_2{-}biotin{-}enzyme + CH_3CO{-}SCoA \rightarrow biotin{-}enzyme + COOH{-}CH_2{-}CO{-}SCoA$$

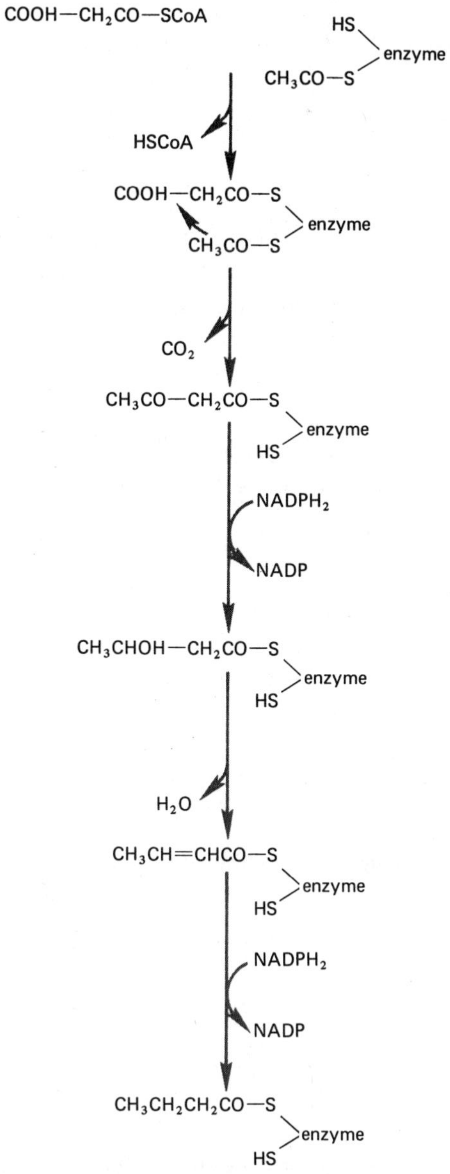

$COOH{-}CH_2CO{-}SCoA$

$$\begin{array}{c} HS \\ \diagdown \\ \quad > enzyme \\ CH_3CO{-}S \diagup \end{array}$$

HSCoA

$$\begin{array}{c} COOH{-}CH_2CO{-}S \\ \qquad \diagdown \\ \qquad\quad > enzyme \\ CH_3CO{-}S \diagup \end{array}$$

CO_2

$$\begin{array}{c} CH_3CO{-}CH_2CO{-}S \\ \qquad\quad \diagdown \\ \qquad\qquad > enzyme \\ HS \diagup \end{array}$$

$NADPH_2$

$NADP$

$$\begin{array}{c} CH_3CHOH{-}CH_2CO{-}S \\ \qquad\qquad \diagdown \\ \qquad\qquad\quad > enzyme \\ HS \diagup \end{array}$$

H_2O

$$\begin{array}{c} CH_3CH{=}CHCO{-}S \\ \qquad\qquad \diagdown \\ \qquad\qquad\quad > enzyme \\ HS \diagup \end{array}$$

$NADPH_2$

$NADP$

$$\begin{array}{c} CH_3CH_2CH_2CO{-}S \\ \qquad\qquad \diagdown \\ \qquad\qquad\quad > enzyme \\ HS \diagup \end{array}$$

The malonate molecule and a second molecule of acetyl-CoA then bind to the thiol groups of an enzyme and all further reactions occur in the enzyme-bound form. The acetyl group is transferred to the malonyl, with the release of the CO_2 that was fixed in forming the malonate, and there is a round of reduction, dehydration, and reduction again that is essentially the reverse of β-oxidation. The addition of a second malonyl-CoA then lengthens the chain by two more carbons and this cycle continues until stearyl and palmityl molecules are synthesized. The fatty acids are released as CoA derivatives, which can be used directly for synthesis of triglycerides and phospholipids.

It is particularly significant that this biosynthetic pathway is almost the exact reverse of β-oxidation, but that different enzymes and cofactors are used. Oxidation requires FAD and NAD, while biosynthesis requires NADPH. Biosynthesis also proceeds on a large enzyme complex that effects all of the transformations shown above on a bound substrate. The obvious reason for making the two processes so different is that it allows them to be controlled separately. The same thing is true of many other pathways, in which degradation and biosynthesis are almost exact opposites except for the coenzymes. There is very little information about control in this system. The enzyme that catalyzes malonyl-CoA formation is inhibited by the long-chain fatty acids that are the end products of the pathway, and it is also activated in a complex way by inter-mediates of the TCA cycle, including citrate.

The straight, saturated chains that come from this pathway can be oxidized and branched by many different enzymes. They can also be lengthened, particularly by a system of enzymes that has been identified in mitochondria which adds acetyl groups to fatty acids already formed in other systems.

The acyl-CoA molecules that are synthesized as shown

above are incorporated directly into triglycerides or phospholipids by combination with glycerol. The glycerol itself comes already phosphorylated through reduction of the dihydroxyacetone-phosphate of the EMP pathway:

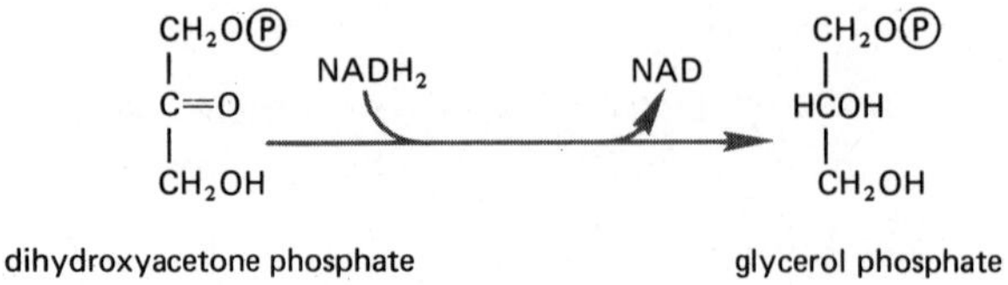

$$
\begin{array}{ccc}
\text{CH}_2\text{O}\textcircled{P} & & \text{CH}_2\text{O}\textcircled{P} \\
| & \text{NADH}_2 \qquad \text{NAD} & | \\
\text{C}=\text{O} & \xrightarrow{\hspace{3cm}} & \text{HCOH} \\
| & & | \\
\text{CH}_2\text{OH} & & \text{CH}_2\text{OH} \\
\end{array}
$$

dihydroxyacetone phosphate   glycerol phosphate

Two fatty acids added to this yield phosphatidic acid. If the phosphoryl group is removed, a third molecule can be added to make a triglyceride, for fat storage. Alternatively, a base may be added to the phosphate to make a phospholipid.

17-16 PHOSPHOLIPID BIOSYNTHESIS

The various phospholipids can be synthesized from phosphatidic acid and one of the free bases through one of two patterns, each of which requires CTP as a cofactor to make an activated form of one component. The first pattern, which is used for choline and ethanolamine, may be written as follows (for choline):

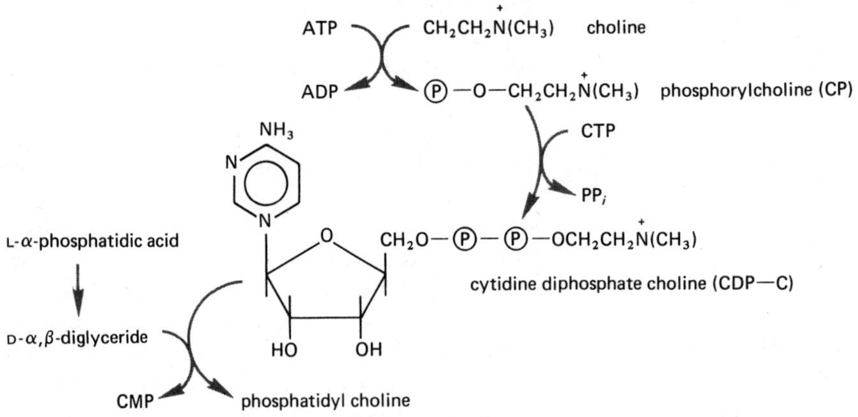

The second pattern is similar, except that the diglyceride is activated by CTP instead of the base. This may be written for inositol as follows:

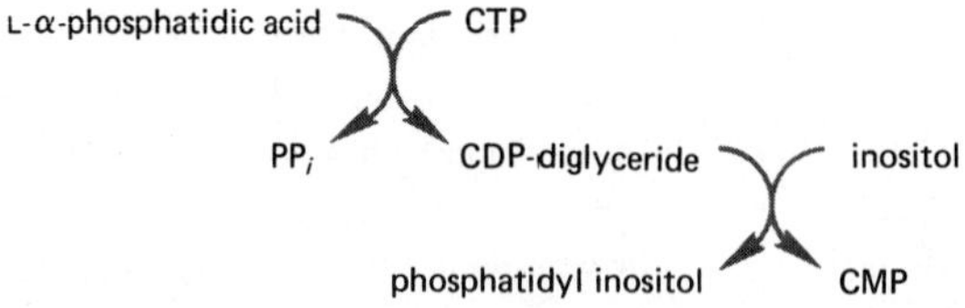

The bases serine, ethanolamine, and choline can also be converted into one another. The reactions require considerable shifting of C_1 groups, so it is easiest to show them

as exchanges with a C_1 pool. All of the reactions probably occur in the phosphatidyl form. The whole series makes a cycle that may be regarded as partially biosynthetic and partially degradative:

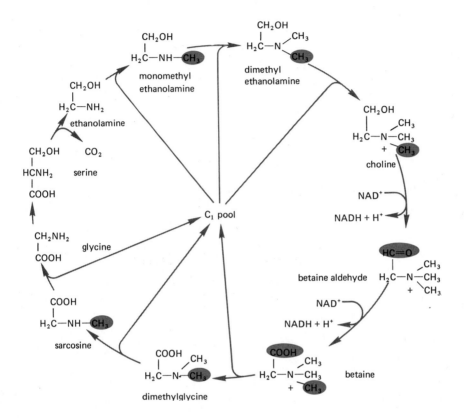

F. Carbohydrate metabolism

17-17 MONOSACCHARIDE SYNTHESIS

The Embden–Meyerhof pathway in reverse provides a mechanism for getting from the central pathways to the hexoses. There are only two points where biosynthesis takes a different course from catabolism. First, there is a major barrier in getting up to the PEP level, but as we have already seen this is overcome in one of the Wood–Werkman reactions, where oxalacetate is combined with CO_2. The second barrier is at the phosphofructokinase level. In glycolysis, this enzyme makes fructose-1,6-diphosphate from fructose-6-phosphate and ATP, and the reaction cannot be reversed easily. In biosynthesis, the enzyme diphosphofructose phosphatase is substituted to remove the 1-phosphate.

This pathway therefore leads directly to fructose-6-phosphate. Remember that almost all of the important monosaccharides are related to D-glucose, D-galactose, or D-mannose. All of them are formed in a simple pattern of reactions from glucose-,

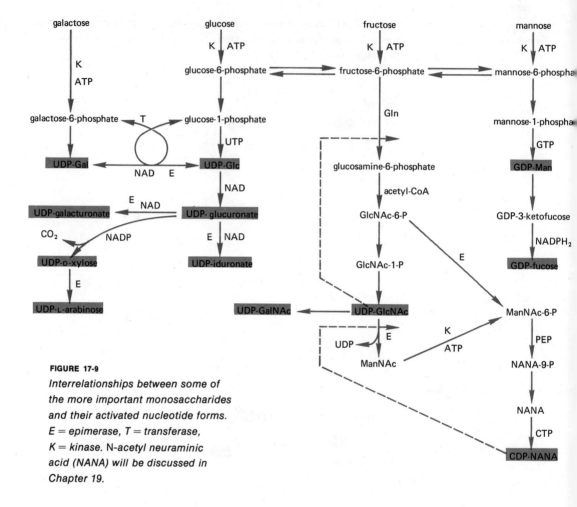

FIGURE 17-9

Interrelationships between some of the more important monosaccharides and their activated nucleotide forms. E = epimerase, T = transferase, K = kinase. N-acetyl neuraminic acid (NANA) will be discussed in Chapter 19.

mannose-, or fructose-6-phosphate, as shown in Fig. 17-9. These reactions have a few key features in common.

1. There are at least two isomerases—one for phosphoglucose and one for phosphomannose—that catalyze the conversions with fructose-6-phosphate:

mannose-6-phosphate   fructose-6-phosphate   glucose-6-phosphate

2. There are mutases that transfer a phosphate group from the 6 position to the 1 position and vice versa.

3. There are kinases that form phosphorylated sugars from free sugars and ATP.

4. There are epimerases that reversibly catalyze the epimerization of glucose to galactose or to mannose by changing the configuration at one carbon atom.

5. There are several systems that activate the monosaccharides so they can be transferred to make polysaccharides and other glycosides. This always requires a conversion to a nucleoside diphosphate ester form. Uridine diphosphate esters are most common, but esters of CDP, GDP, ADP, and deoxy-TDP have also been found. These compounds are formed by nucleoside triphosphate and combined with the sugar as a nucleoside diphosphate:

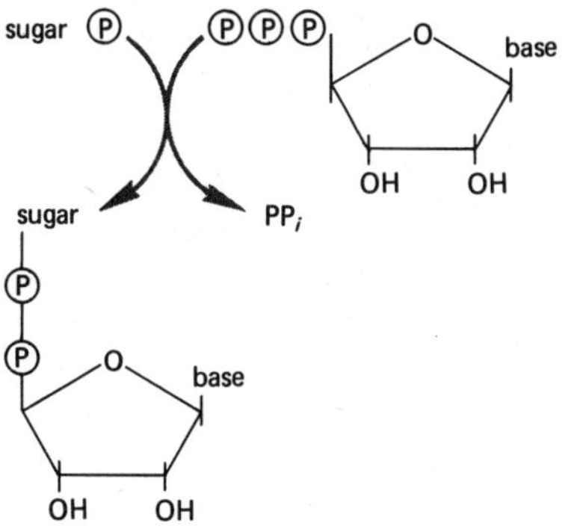

You should now be able to trace the various transformations from one hexose to another. Notice the interesting transformation in galactose metabolism. There is a transfer reaction, marked T in Fig. 17-9, where galactose-1-phosphate becomes UDPGal while UDPG becomes glucose-1-phosphate. UDPGal is then epimerized to UDPG so that galactose is continuously converted to glucose by repetition of this cycle.

17-18 GLYCOSIDE BIOSYNTHESIS

The nucleotide sugar esters are the immediate donors of sugars for oligo- and polysaccharide biosynthesis; this is their primary importance in metabolism. The most common biosynthetic pattern is very simple, and may be illustrated by lactose formation:

$$\text{UDP-galactose} + \text{glucose} \rightarrow \text{lactose} + \text{UDP}$$

An alternative synthesis leaves disaccharide phosphates that must be dephosphorylated:

$$\text{UDPG} + \text{fructose-6-phosphate} \rightarrow \text{sucrose phosphate} + \text{UDP}$$
$$\text{sucrose phosphate} \rightarrow \text{sucrose} + P_i$$

The simplest homoglycans are synthesized by continuous repetition of this process. Branched homoglycans are made from linear molecules by the action of transglucosylases, commonly called branching enzymes. For example, the branched homoglucans amylopectin and glycogen are built primarily of $\alpha(1 \rightarrow 4)$ bonds. Their

occasional $\alpha(1 \rightarrow 6)$ bonds are made by branching enzymes that break off short
glucosides and reattach them in $\alpha(1 \rightarrow 6)$ linkages:

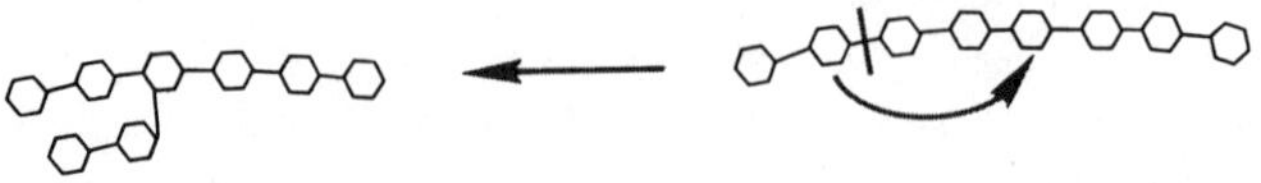

G. A wider view of metabolism

It is so easy to get bogged down in details and lose sight of the broad perspective
that we had better go back and see the significance of all these transformations for
the biosphere as a whole. The mechanisms we have been discussing are strategies
that cells have evolved over billions of years for transforming molecules and making
their own structures out of the available raw materials. These are merely small pieces
in the whole pattern of biosynthesis; and the biosynthesis of each organism is merely
a small piece in the whole, enormous transformation of material through the food
chains of the ecosystem. We might represent the whole pattern like this:

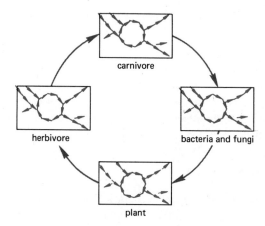

Just as we can identify a carbon cycle in the biosphere by concentrating on the
organic compounds, we can identify interesting cycles of other elements. As a simple
case, let's consider sulfur. Sulfur can be oxidized and reduced through several steps
from an oxidation state of -2 in H_2S and organic sulfur compounds, through 0 in
elemental sulfur, to $+2$ in thiosulfate ($S_2O_3^=$), $+4$ in sulfite, and $+6$ in sulfate.
Sulfates are very common in nature, but sulfides are unstable in aerobic environments
and are rapidly oxidized to sulfur and sulfates. Many organisms can pick up sulfate
and incorporate it into cysteine and other organic compounds through the PAPS
route; this is biosynthetic or assimilatory sulfate reduction. A large group of microorganisms involved in decay can convert the $-SH$ groups of proteins back into H_2S,
which may be oxidized spontaneously in the air or by the bacteria that get their energy

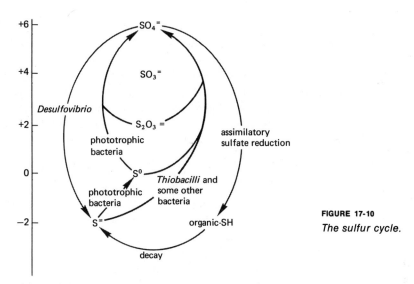

FIGURE 17-10
The sulfur cycle.

in this way—principally the *Thiobacilli* and the photosynthetic sulfur bacteria. A minor pathway involves *Desulfovibrio*, which reduces sulfate back to sulfite in electron transport (dissimilatory sulfate reduction). The general pattern of the sulfur cycle is summarized in Fig. 17-10.

Another important cycle can be drawn for nitrogen, as shown in Fig. 17-11. Most of the individual steps of the cycle have been discussed before. One of the major steps

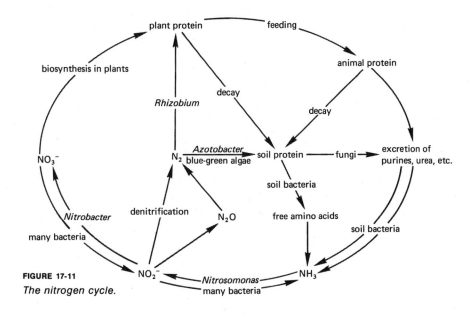

FIGURE 17-11
The nitrogen cycle.

that we have not discussed is the reduction from nitrate to the amino group of protein. In the plants and microorganisms where this has been studied, there is always a molybdoflavoprotein that can reduce NO_3^- to NO_2^-. Other metalloflavoproteins then reduce nitrite to the level of HNO and reduce hydroxylamine, NH_2OH, to NH_3. These are biosynthetic or assimilatory nitrate reductions; remember that many bacteria also reduce nitrate to nitrite as a sink for electrons in their electron-transport systems.

An important and dramatic reaction is the reduction of free atmospheric nitrogen to ammonia; since nitrogen constitutes at least three-quarters of our atmosphere, transformations that substantially increase or decrease the amount of N_2 are important. Nitrogen fixation is carried out by many organisms; but the process requires a great deal of energy and a system that transports electrons at a very low oxidation potential. As in so many other systems that have similar requirements, ferredoxin is used as an electron donor; the electrons may come from a variety of sources, but the reaction is always essentially the same (Fig. 17-12).

Many soil bacteria are nitrogen fixers; nitrogen fixation can also occur in bacterial photosynthesis (as part of another system that uses ferredoxin, you will notice) and some of the blue-green algae are very important in reducing nitrogen. However, the most interesting and important nitrogen-fixation apparatus is probably the symbiotic association formed between some bacteria of the genus *Rhizobium* (rhizobia) and the roots of legumes (peas, clover, and related plants). The rhizobia burrow into

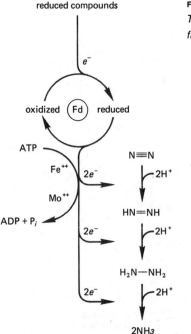

reduced compounds

FIGURE 17-12
*The reactions in nitrogen
fixation. Fd is ferredoxin.*

the growing roots of these plants by breaking down the root cell walls enzymatically. As they invade the roots and proliferate, the root cells are stimulated to grow; eventually the two organisms produce large nodules consisting of enlarged cells that are full of bacteria. Neither rhizobia nor legumes can fix nitrogen alone, but they combine forces to make a powerful system which wise farmers have depended on for generations to enrich their soils periodically (by rotating their main crops with clover, alfalfa, etc.). At this time, we cannot tell what components of the nitrogenase system come from each organism.

SUPPLEMENTARY EXERCISES

exercise 17-3 By now you have seen that glutamine is a supplier of amino groups in many reactions. It therefore makes sense that glutamine synthetase, which catalyzes the reaction

$$\text{glutamate} + NH_4^+ + ATP \rightarrow \text{glutamine} + ADP + P_i$$

is subject to concerted inhibition by at least eight end products of pathways in which glutamine is required. In this case, each inhibitor reduces the activity of the enzyme by a certain fraction, independently of the other inhibitors. Excess tryptophan reduces the activity to 84% of maximum, CMP to 86%, carbamyl phosphate to 87%, and AMP to 59%. What percent activity would you expect from all four of them together?

exercise 17-4 In *Neurospora,* there are apparently two enzymes that synthesize carbamyl phosphate — one to make precursors for arginine biosynthesis and the other for pyrimidine biosynthesis. A mutation in one enzyme creates an arginine auxotroph and a mutation in the other creates a pyrimidine auxotroph. What does this imply about the pools in this eucaryote?

exercise 17-5 2-Thiazole-alanine is an analogue of histidine that inhibits growth of bacteria by acting as a false feedback inhibitor. What kind of mutants would be resistant to 2-thiazole-alanine? In what gene should they map?

some general aspects of regulation

Throughout this book, we have tried to view biotic systems in the simplest possible terms and to dissect them into elementary components that can be understood by themselves. We would negate this whole approach to biology if we suddenly tried to view an organism in all its complexity, for we must maintain that the whole is the sum of these elements. In this chapter we will try to achieve a somewhat more integrated view by looking at some aspects of metabolism that we have glossed over, but we will still try to confine ourselves to things that are reasonably well understood.

18-1 REGULATION OF THE AMPHIBOLIC PATHWAYS

There are so many reactions that lead into the TCA cycle reactions and those associated with the TCA cycle, and so many paths that drain from the cycle, that it is virtually impossible to understand all of these factors at once. Some of the most interesting work in this field is being done with computer programs; if the factors that regulate the levels of enzyme activity can be identified, then a computer program can often be evolved to show how certain enzymes will respond to a variety of changes in substrate and inhibitor concentrations. However, the things that a computer can digest and understand are not the same as those a human can understand, so we will try to see how a few of these factors, by themselves, will affect metabolism.

The general direction of flow around the TCA cycle is determined primarily by a few key reactions that are essentially irreversible. The oxidations of pyruvate and α-ketoglutarate are perhaps the most significant; the condensation of acetyl-CoA and oxalacetate to form citrate is also essentially irreversible. These features are made most conspicuous by a case in which the cycle *is* reversed—in the phototrophic bacteria *Chlorobium* and *Chromatium*. M. C. W. Evans and Bob B. Buchanan have shown that the cycle can be driven in reverse to fix CO_2 by inserting ferredoxin at the most critical points (Fig. 18-1). Since ferredoxin is reduced directly in photosynthetic electron transport, the light energy can be used very efficiently for reduction, without the synthesis of NADPH.

In the usual TCA cycle, the concentration of oxalacetate is probably the most critical regulatory device. The oxidation of malate to oxalacetate is not favorable when coupled to NAD^+; since the equilibrium lies in favor of malate formation, this tends to create a pool of malate that serves as a reserve against the possibility that oxalacetate will be depleted. Oxalacetate is also

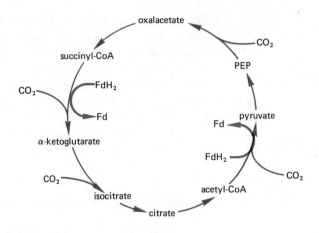

FIGURE 18-1
A reductive TCA cycle powered by ferredoxin (Fd).

a competitive inhibitor of succinic dehydrogenase; this may not be an allosteric effect, but it *is* a case of an end product inhibiting an early reaction in its pathway.

The isocitric dehydrogenase is also subject to control from within the cycle. In *Neurospora*, it has been shown that this enzyme is activated by citrate and inhibited by α-ketoglutarate. These effects do have the properties of allosteric regulation.

exercise 18-1 The following diagram shows the points of regulation in the glyoxalate cycle. Isocitratase formation is repressed by PEP. Suppose that bacteria in which the enzyme is repressed are given acetate. Describe a short sequence of events that should result in derepression, assuming that oxalacetate and PEP are in equilibrium with one another.

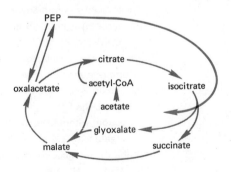

exercise 18-2 Now explain why isocitratase is repressed when the TCA cycle is operating without exogenous acetate.

exercise 18-3   Look at the diagram again and think about an organism that is given
fatty acids as a carbon source. Show how the glyoxalate cycle will aid
in converting the fat to carbohydrate.

exercise 18-4   You might find it amusing to place oxalacetate at the center of a piece
of paper and draw as many reactions as you can leading to or from it.
This will help to solidify your conception of its role in metabolism.

18-2   NICOTINAMIDE NUCLEOTIDE LEVELS

The nicotinamide nucleotides are involved so intimately in metabolism that the relative
amounts of their reduced and oxidized forms must be quite critical. NADH and NADPH
are being produced continually during oxidation via the glycolytic and TCA pathways
and they must be oxidized somehow. NADH clearly is a major reductant of the
mitochondrial ET chain, but, paradoxically, the mitochondrial membranes are virtually
impermeable to NADH. Therefore, the reduction of NADH by mitochondria must occur
in a roundabout way; the mechanisms are called **shuttles,** and the one that is most
clearly demonstrated is the α-glycerophosphate-dihydroxyacetone cycle.

Figure 18-2 shows how this cycle operates. External NADH reduces dihydroxyacetone
phosphate to glycerol phosphate. The mitochondria are freely permeable to the latter,
so it diffuses in and is oxidized by an α-glycerophosphate dehydrogenase that
contains FAD and reacts directly with cytochrome b. The dihydroxyacetone phosphate
that is formed there diffuses out again to complete the cycle.

18-3   REGULATION BY ADENOSINE NUCLEOTIDES

A large number of reactions in and around the amphibolic pathways are regulated
in one way or another by ADP or ATP. Taking the broadest view of energy metabolism,

FIGURE 18-2

The glycerophosphate-dihydroxyacetone phosphate shuttle for oxidation of NADH.

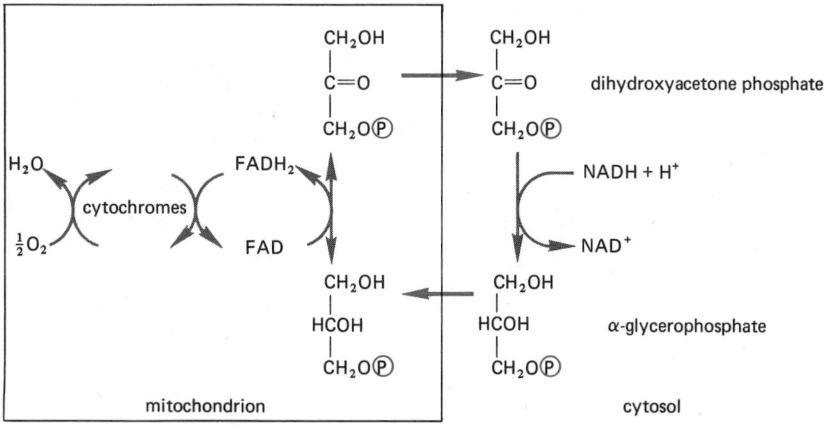

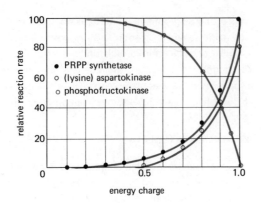

FIGURE 18-3

Effects of energy charge on two enzymes. The biosynthetic enzymes PRPP synthetase and aspartokinase are activated as the energy charge increases while the catabolic enzyme phosphofructokinase is inhibited.

FIGURE 18-4

Points of regulation in glycolysis and the TCA cycle. Inhibitions are indicated by × and activations by supplementary arrows.

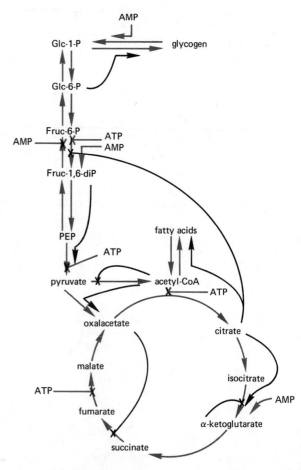

this is a very useful kind of regulatory mechanism, since the purpose of these reactions is ATP synthesis and the relative amounts of the adenylate coenzymes is a good measure of the amount of free energy available to the cell or required by it.

Daniel Atkinson has assembled the information about this level of regulation into a rather nice picture. He proposes as a measure of ATP/ADP level the number of phosphoryl groups in adenosine nucleotides that can be transferred with a large change in free energy, divided by the total number. Since ATP has two, ADP has one, and AMP has none, the **energy charge** is defined as $(ATP + \frac{1}{2}ADP)/(ATP + ADP + AMP)$. If an enzyme is at all sensitive to the energy charge, it should respond in one of two directions. If it is in a pathway that generates ATP, it should be inhibited by a high charge; if it is in a pathway that uses ATP, it should be inhibited by a low charge. Atkinson and his group have examined the activities of several key enzymes as a function of energy charge, and they find that the enzymes respond just as would be expected (Fig. 18-3).

Figure 18-4 summarizes the results of Atkinson and many other investigators. The points where adenylates activate or inhibit are shown, along with the regulation by several other key intermediates. It is not clear in every case just how the activation or inhibition is effected. Most of these situations probably come under the heading of allosteric interactions. When two or more effectors produce the same effect, they amplify one another. For example, phosphofructokinase is inhibited by both ATP and citrate. At a given energy charge, the affinity of the enzyme for its substrate, fructose-6-phosphate, decreases with increasing citrate concentration; of course, this does not indicate how many distinct sites of interaction there are and how the effectors interact with one another.

The regulation by acetyl-CoA of the two main paths from pyruvate is particularly interesting. When acetyl-CoA is low, there is a deficiency of TCA-cycle intermediates leading to ATP production; when it is high, there is enough material running through the TCA cycle so the cell can afford to recycle pyruvate back to glucose and other carbohydrates. High concentrations of acetyl-CoA inhibit pyruvic dehydrogenase and stimulate pyruvic carboxylase, which makes oxalacetate, which can in turn be converted into PEP. The stimulation of pyruvic carboxylase may also help to ensure an adequate supply of oxalacetate for citrate formation.

The important thing to see in Figure 18-4 is the multiplicity of controls. As always, the controls make sense functionally, and most of them are feedback inhibitions. However, the first enzyme system represented in the figure, the phosphorylase which splits glycogen into glucose-1-phosphate, is different. As Fig. 18-5 shows, the active enzyme is a tetramer, phosphorylase *a;* each of its protomers has an active site containing phosphoserine. When the phosphates are removed, the enzyme dissociates into two dimers, phosphorylase *b,* and this dissociation can be reversed by ATP. However, both phosphorylase *a* and *b* can be activated, apparently in an allosteric transition, by AMP; phosphorylase *b* is completely inactive unless bound to AMP, while phosphorylase *a* is only slightly more active when this effector is bound. This is a clear case of positive feedback that will promote glycolysis when the AMP level is high and thus increase the energy charge.

The kinase that converts phosphorylase *b* to phosphorylase *a* also exists in an active and an inactive form. We will see shortly that the conversion from one to the other is another important control point.

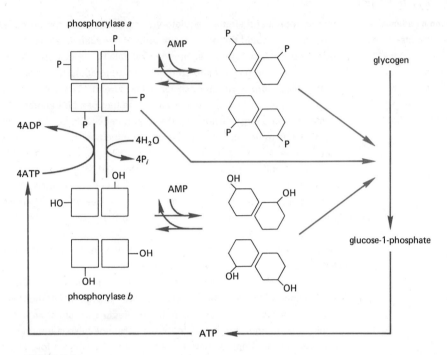

FIGURE 18-5
Interconversions between the forms of phosphorylase.

Gordon M. Tompkins and K. L. Yielding have studied another enzyme that is similar to phosphorylase—the ill-named enzyme glutamic dehydrogenase, which primarily catalyzes the amination α-ketoglutarate $+ NH_4^+ + NAD(P)H + H^+ \rightarrow$ glutamate $+ H_2O + NAD(P)^+$. The whole enzyme is a hexamer of six protomers, each of which weighs about 53,000 daltons, but it can also form large multimers of about 2×10^6 daltons, which is much more than the individual hexamer should weigh. Both the molecular weight of the enzyme and its activity are influenced by several ligands, as shown in Fig. 18-6. It now appears that the hexamer can undergo allosteric transitions between two forms; in one state it is a good alanine dehydrogenase and in the other state it is a good glutamic dehydrogenase. Only the latter is able to polymerize still more, so ligands which promote conversion to the alanine dehydrogenase form also inhibit polymerization. The controls are not easy to understand when one considers that the conversion of alanine to pyruvate supplies additional energy while the synthesis of glutamate is a drain on the available energy. This suggests that there are some important aspects of this regulatory system that we do not understand yet.

18-4 CATABOLITE REPRESSION

When Jacques Monod first started to do experiments on bacterial growth around 1940, he used a mixed carbon source of glucose plus some other carbohydrate, such

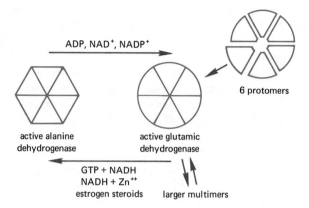

FIGURE 18-6

Association and dissociation of the glutamic dehydrogenase complex.

as sorbitol or lactose. Rather than the simple exponential growth curve we discussed in Chapter 2, he got the complex curve shown in Fig. 18-7. It took some time to appreciate the full significance of this pattern of **diauxic growth.** It is now clear that the glucose represses the formation of inducible enzymes, such as β-galactosidase, needed to catabolize the second carbon source. In Monod's experiments the bacteria grew on glucose until they exhausted it; they went into a temporary lag until the second set of enzymes could be induced and then grew on the second carbon source until it or some other factor became limiting.

This phenomenon was once called the "glucose effect"; now it is usually called

FIGURE 18-7

Diauxic growth of a bacterial culture on lactose and glucose.

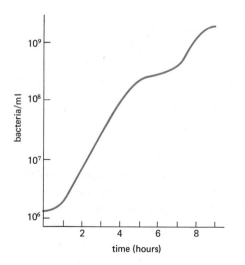

catabolite repression, because it is induced by other carbon sources that are close to the central pathways, such as glycerol. Catabolite repression is generally exerted over the enzymes necessary for the initial catabolism of amino acids, carbohydrates, and other compounds rather remote from the central pathways. It remains generally puzzling. The suggestion has been made that it is an inhibition of the permease systems for the secondary carbon source, but it can be shown that the permeases are fully functional in several cases where these are constitutive. Catabolite repression cannot be due simply to competition for amino acids, etc. for protein synthesis, since it is not relieved by adding these substances. Other obvious explanations also do not work.

Frederick Neidhardt and Boris Magasanik have pointed out that the repressed enzymes are always those that catabolize a carbon source to intermediate metabolites the cell can obtain more easily from glucose itself. They proposed that the catabolism of glucose, glycerol, and other compounds close to the central pathways fills an intracellular pool of some intermediate which represses the synthesis of other enzymes that would tend to fill the pool more indirectly. However, there must be distinct repression systems for each enzyme, since William F. Loomis and Magasanik have found a mutant in which the *lac* operon alone is insensitive to repression.

Catabolite repression is not affected by operator or regulator mutations; it is a control superimposed over the Jacob–Monod system. Loomis and Magasanik also discovered a related but quite distinct control system that produces **transient repression.** If a culture grown on a carbon source such as glycerol or lactate is making β-galactosidase, either inducibly or constitutively, the sudden addition of glucose will produce a severe, temporary repression of enzyme synthesis. Transient repression occurs in the mutant that is not subject to catabolite repression and it is induced by glucose analogues that cannot be metabolized.

Catabolite repression is still not understood; but we are closer to an understanding of transient repression because the phenomenon has now been linked with a very important aspect of cellular regulation, which we will now discuss.

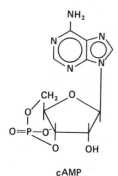

cAMP

18-5 CYCLIC AMP

Investigations of certain hormones in mammalian cells by E. W. Sutherland, G. A. Robison, and their associates revealed that the hormones affected the intracellular concentrations of the nucleotide **3′, 5′-cyclic AMP (cAMP).** As we shall see shortly, cAMP not only has been implicated in the action of many hormones, but it has also been found in bacteria and other unicells in a particularly interesting connection.

Cyclic AMP is made from ATP by **adenyl cyclase:**

$$ATP \rightarrow cAMP + PP_i$$

It is destroyed by a specific phosphodiesterase, which converts it into AMP. Cyclic AMP is an important effector in the phosphorylase system shown in Fig. 18-5, because the phosphorylase kinase which converts phosphorylase *b* into phosphorylase *a* is itself inactive unless it is bound to cAMP. Thus we have the sequence of events:

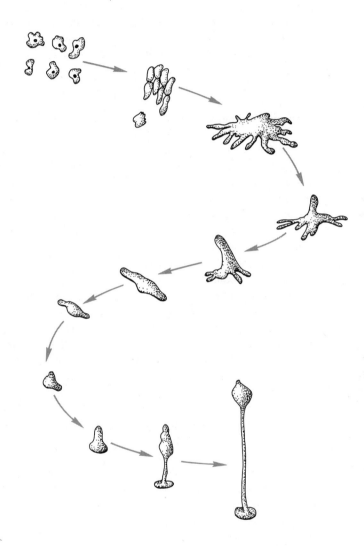

FIGURE 18-8
Aggregation and fruiting-body formation in the cellular slime mold Dictyostelium.
*Separate amebas form a slug that migrates for a few hours and then settles down and
produces a stalk and fruiting body in which spores are formed. (From J. T. Bonner,
The Cellular Slime Molds, 2nd ed., Princeton University Press, Princeton, N.J., 1967,
by permission of Princeton University Press.)*

activation of adenyl cyclase → cAMP → activation of phosphorylase kinase → activation of phosphorylase → breakdown of glycogen.

In bacteria, the adenyl cyclase activity is increased by pyruvate, α-ketoglutarate, and other α-keto acids. The level of cAMP is very closely correlated with the amount of glucose present in the medium. It is very low in the presence of glucose, but rises immediately and sharply when the glucose is exhausted. This change can be understood if the bacteria have a phosphorylase system that is anything like the well-studied animal system; the increase in cAMP will then stimulate phosphorylase to increase the flow of glucose from glycogen.

This relationship between glucose and cAMP is apparently also closely related to transient repression. Ira Pastan and Robert L. Perlman have shown that transient repression is abolished by cAMP. Cyclic AMP is effective in a variety of mutants, including those for the *lac* operator and regulator and the Loomis–Magasanik mutant that is not sensitive to catabolite repression. Furthermore, cAMP acts at the level of transcription, since it is not effective when mRNA synthesis is blocked. However, cAMP is not effective in a *promotor* mutant of the *lac* operon; therefore Pastan and Perlman believe that cAMP increases the rate of induced enzyme synthesis by acting on the promotor, and that transient repression is caused by the decrease in cAMP level produced by glucose.

Cyclic AMP is also known as an allosteric activator of phosphofructokinase. It is beginning to emerge as a major intracellular effector for control systems at several levels. One of the most interesting cases is that of the cellular slime mold *Dictyostelium.* This organism reproduces vegetatively in the form of little ameboid cells; after a time of proliferation, many of these amebas respond to a chemical stimulus that has been called *acrasin,* which makes them aggregate into a large mass or "slug" (Fig. 18-8). Eventually the "slug" settles down and sends up a little stalk, which forms spores at its tip. It has now been shown that acrasin is cyclic AMP.

A bacterium, as we have emphasized before, is an organism in itself. It does not have to respond particularly to other cells and its regulatory problems are internal. However, when multicellular organisms began to evolve they had to find mechanisms for regulating the activities of all cells in concert with one another; the slime mold is a very simple case in which an internal effector, cAMP, was turned into an external effector for regulating a multicellular activity. At a later stage in evolution, multicellular organisms began to emerge that were considerably more complicated because they contained several types of specialized, differentiated cells performing their particular tasks; it then became necessary to coordinate these different activities into a whole. These organisms therefore began to develop intercellular effectors that could be used by one cell type to communicate with other cell types; many of these effectors are what we call *hormones.* The messages that these hormones carry would translate into English: "I have a need for glucose; start excreting glucose into the blood." "There is too much Na$^+$ in the blood; the kidneys must excrete more sodium." "A fresh supply of amino acids has just arrived; all growing tissues must start to make more protein."

In evolving these hormonal controls, it was only natural for organisms to superimpose this level of intercellular regulation on an intracellular regulatory system that was already acting efficiently. Since cAMP was already being used as a major

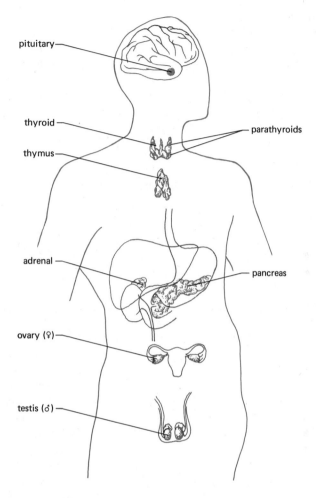

FIGURE 18-9
Locations of the major human endocrine glands.

internal effector, many of these hormonal systems were evolved as activators or inhibitors of cAMP synthesis. In the last few years, we have begun to realize just how vast and general this control system is. Many hormones can now be seen as primary intercellular messengers which act on the surfaces of specific **target cells.** An adenylic cyclase system is then built into these cells which is activated (allosterically?) by the hormone, and this in turn increases the intracellular cAMP level and causes specific internal metabolic changes.

There are probably other internal effectors besides cAMP that are affected by hormones. In some cases, the hormones are known to act specifically as inducers at the level of DNA transcription. Without trying to give a complete account of all the known hormonal systems, we will discuss a few to show how some of these systems operate.

18-6 A BRIEF MAP OF THE ENDOCRINES

Figure 18-9 shows the location of the chief glands in the human body that have an endocrine hormone secretion. There is little point in trying to memorize all this, for we will concentrate on selected parts of the system.

The hormones secreted by these tissues fall generally into three classes: peptides and small proteins, sterols, and amino acid derivatives.

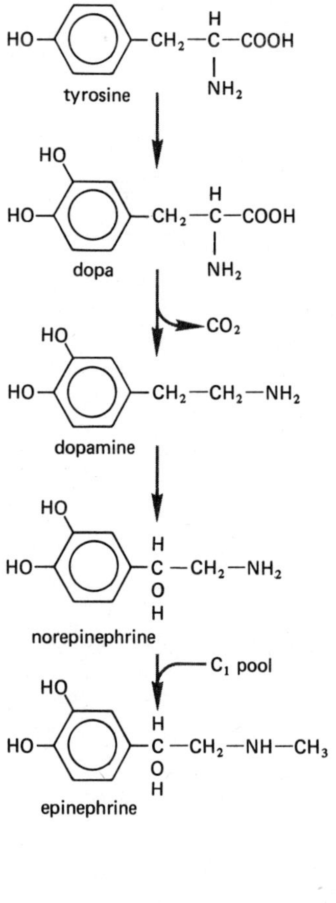

18-7 REGULATION OF CARBOHYDRATE METABOLISM

An average, healthy person has about 50–60 mg of glucose in 100 ml of his blood. This level is kept quite constant by a series of hormones. When the blood glucose level rises to about 70 mg/100 ml, beta pancreatic cells, which synthesize **insulin,** are stimulated to release this hormone. Glucose itself seems to be an effector that raises the intracellular cAMP level and stimulates this process. The insulin released into the blood has a general effect on many tissues; principally, it stimulates the glucose permeases of muscle cells and makes them take up the excess blood glucose. Whether these cells convert the glucose to glycogen or metabolize it will depend upon a variety of factors.

Suppose, on the other hand, that the blood glucose level falls toward 40 mg/100 ml, a state of **hypoglycemia.** Then the alpha pancreatic cells are stimulated to release **glucagon,** which passes to the liver and stimulates the formation of cAMP. This activates liver phosphorylase, which begins the conversion of glycogen into glucose and results in an increase in blood glucose. **Epinephrine,** produced by the adrenal medulla, has an effect similar to glucagon. The epinephrine system is under nervous control; in times of stress, the medulla is stimulated to produce epinephrine to raise the body's supply of mobile glucose.

At the same time, epinephrine, norepinephrine, and glucagon stimulate the breakdown of fats by acting on adipose (fatty) tissues where the neutral fats are stored. This effect is also

tyrosine

dopa

dopamine

norepinephrine

epinephrine

HO—⟨benzene⟩—CH$_2$—$\overset{\text{H}}{\underset{\text{NH}_2}{\text{C}}}$—COOH

tyrosine

↓ I$_2$

HO—⟨benzene, positions 3, 5, I⟩—CH$_2$—$\overset{\text{H}}{\underset{\text{NH}_2}{\text{C}}}$—COOH

diiodotyrosine

↓ → alanine

HO—⟨benzene, I⟩—O—⟨benzene, I⟩—CH$_2$—$\overset{\text{H}}{\underset{\text{NH}_2}{\text{C}}}$—COOH

thyroxine

↓ → I

HO—⟨benzene, I⟩—O—⟨benzene, I⟩—CH$_2$—$\overset{\text{H}}{\underset{\text{NH}_2}{\text{C}}}$—COOH

triiodothyronine
(five times as active
as thyroxine)

mediated by cyclic AMP. Lipolysis can also supply the metabolic intermediates leading to glucose synthesis and eventually to an increase in blood glucose. Insulin, on the other hand, stimulates lipid synthesis in the adipose tissues, so it has just the opposite effect.

18-8 PITUITARY CONTROL CIRCUITS

The anterior lobe of the pituitary gland is one of the most important of all endocrine organs because it produces hormones which act, in turn, on other endocrine glands. For example, it produces hormones which act on the various sex glands, particularly in females, where it controls the functions of ovulation, pregnancy, milk production, and so on.

Among its hormones is the **adrenocortical tropic hormone (ACTH)** that promotes the formation of the steroid hormones of the adrenal cortex. Cyclic AMP is clearly involved in this process. Although it is too soon to be sure, it appears that the increased level of cAMP in the adrenal tissue stimulates specific enzymes in the pathways of steroid biosynthesis. The reason for hedging on this point is that ACTH may stimulate an increase in protein synthesis that can be blocked by actinomycin D, but we cannot tell if this is a direct or an indirect effect.

The adrenal corticoids, in turn, have many hormonal activities that are far from being understood. One of these is a **glucocorticoid** effect—the stimulation of glycogen synthesis. Some investigators believe that these steroids act directly at the DNA level to induce the synthesis of certain enzymes in the pathway of glucose synthesis, while others believe the effects are more indirect.

The adrenal corticoids also have a **mineralcorticoid** effect on the kidneys, where they stimulate the retention of sodium ions. Another pituitary hormone, **vasopressin,** stimulates the kidneys to retain water. Both of these effects are probably mediated through cAMP. Hormones which have specific effects on the cells of the kidney and bladder of various animals stimulate an increase in cAMP, while those that do not have these effects stimulate no increase.

Another interesting system involves the thyroid gland, which is stimulated by **thyrotropin (TSH)** from the pituitary. TSH is clearly an activator of adenyl cyclase in the thyroid tissue; the increase in cAMP then promotes iodine uptake and the synthesis of thyroxine from tyrosine. Thyroxine is released into the blood, and it has a feedback inhibitory effect on the synthesis of TSH.

The site of thyroxine action has been a mystery for a long time. It stimulates metabolism in general by speeding up respiration and uncoupling it from phosphorylation. Its most probable primary effect is on the mitochondria, for it clearly stimulates mitochondrial swelling and permeability changes. In view of what we have said about the mitochondrion as a machine for conserving energy through changes in membrane conformation, it is not hard to conceive of thyroxine as an allosteric effector that can change these membranes so that phosphorylation is inhibited.

Certainly one of the most dramatic and familiar examples of a hormonal regulatory

FIGURE 18-10
Summary of the major events in the human menstrual cycle.

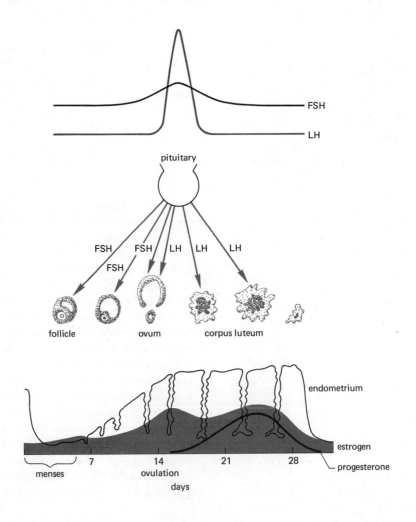

circuit is seen in the estrous cycle of female mammals; estrous is a time when females are in heat and ready to accept a mate. The closely related menstrual cycle in humans is shown in Fig. 18-10. This 28-day cycle is counted from the beginning of the menses (menstruation), which lasts about five days. Under the influence of **follicle stimulating hormone (FSH)** from the anterior pituitary, the ovum begins to mature inside a **Graafian follicle;** on the fourteenth day, with a sudden burst of **luteinizing hormone (LH)** the ovum is released from the follicle and begins to migrate down the fallopian tube toward the uterus. The empty follicle develops into a **corpus luteum.** Meanwhile, the lining of the uterus **(endometrium)** has been thickening in response to estrogens secreted by the follicle and progesterone from the corpus luteum. If the ovum is fertilized during its migration, it implants itself in the endometrium around the twenty-

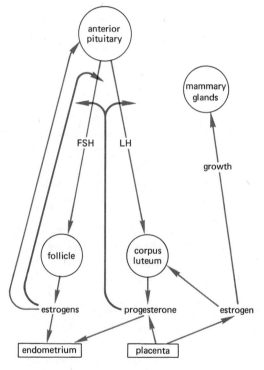

FIGURE 18-11
The pattern of stimulation and inhibition
of human female sex hormones.

first day and begins to develop into an embryo. The embryo produces a placenta through which it exchanges gases and food with the mother's blood; the placenta secretes additional hormones that help to maintain the corpus luteum, which in turn maintains the endometrium and the embryo. If fertilization does not occur, the corpus luteum is not maintained and around the twenty-eighth day the endometrium begins to break down and menstruation begins. The feedback-regulation system is shown in Fig. 18-11.

In spite of a great deal of research, the cycle is not really understood well enough, particularly in humans. Thus the cycle can be governed artificially with birth-control pills and ovulation can be prevented by using appropriate analogues of these steroid hormones, but no one is sure how these analogues work. The most likely route is through inhibition of FSH production, so that growth of the follicle is inhibited, but there may be other important effects as well.

After all we have said about the interactions between small molecules and the sites on proteins, it should hardly be necessary to point out the possible complications of introducing a hormone analogue into the body once a day. The birth-control pills do have side effects; while some women may experience nothing unusual, others may have headaches, visual problems, weight gain, psychological effects, and so on. The effects will be different in different women because each of them has a different set of receptor sites on her various proteins and cells. It is simply impossible to predict the effects of a molecule that looks like a natural estrogen but has an additional side chain. Until more effective means are available, every individual will have to weigh the possible side effects against the benefits of family planning and the impending disaster of overpopulation.

18-9 GENETIC REGULATION BY HORMONES

Studies on simple organisms such as bacteria have revealed many principles of genetic regulation of protein synthesis. We are now entering a period in which similar phenomena are being investigated in more complex organisms, and here we expect to see these principles amplified and expanded. The discoveries have already started; if this were being written even a year or two later, there would probably be a great deal that could be said with some certainty. At this time, however, it is best to speak in generalities.

A plant or animal cell may have 100 times as much DNA as a bacterium, yet it cannot carry out 100 times as many distinct enzymatic reactions. Each cell can do some things that the other cannot, and on balance their total metabolic capacities are probably about the same. It is very reasonable to assume that most of the extra DNA has a regulatory function.

A multicellular organism really needs two distinct levels of regulation. First, it needs a more or less permanent set of controls that can be applied to each cell as it becomes specialized. Every cell that retains its genome during development has the same information in it as every other cell in the body, yet one becomes a liver cell and another a muscle cell; they make very different proteins and respond in very different ways to stimuli. The process of differentiation is primarily a matter of repressing a large part of the genome so it can never be expressed under normal circumstances, and leaving a rather limited part exposed so its genes can be expressed in varying degrees. The second control system must regulate this expression in response to specific stimuli.

We are beginning to get some insight into the nature of the first system, but we cannot spell it out in detail yet. There are clearly a variety of RNA molecules in nuclei that are distinct from tRNA, rRNA, cytoplasmic mRNA, and any of their precursors. Some of them have a very distinctive base composition (for example a high content of uracil or dihydrouracil) and some of them have a high affinity (or even a covalent linkage?) for nuclear proteins — either distinct histones or distinct acidic proteins. Furthermore, these RNA molecules or RNA-protein complexes have a high affinity for some of the DNA of these cells. All this suggests that a differentiated cell has a considerable amount of DNA covered up with RNA-protein complexes, which bury certain sequences so they cannot be transcribed and do not have to be regulated.

The second type of control system must respond to various stimuli and make significant changes in a period of minutes or a few hours. Certain hormones are almost certainly among the effectors at this level. One excellent illustration of this point appears in certain two-winged (dipteran) flies such as *Drosophila* and *Chironomus* which have giant **polytene chromosomes** (Fig. 18-12) in the cells of their salivary

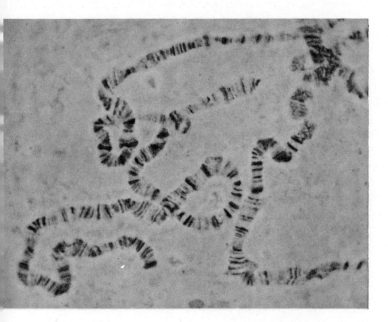

FIGURE 18-12
Phase-contrast micrograph of giant polytene chromosomes of Drosophila. *(Courtesy of Sara E. Pollard.)*

glands and other tissues. Each chromosome must be about 50 chromatids lined up in tandem and stretched out so they can be seen easily with a light microscope; the dark and light bands, which must reflect the binding of RNA and protein as well as the folding of DNA, are very characteristic, and the genetic map of *Drosophila* can be aligned with them to some extent. Thus each band must mark the position of a few genes. The bands change their appearance in characteristic ways as the fly develops from a larva to an adult; Wolfgang Beerman and his associates have described regular sequences during development in which one band after another enlarges greatly and forms a *puff*, which disappears after a time. Puffs have always looked like places where the DNA is stretching out so it can be transcribed, and now it is clear that there is a very active temporary synthesis of RNA at these points. A larva will not molt if it lacks the steroid hormone **ecdysone,** and Peter Karlson has now shown that when ecdysone is injected into a larva, its first visible effect is to induce certain puffs. In fact, there is a regular pattern of puffing as the level of ecdysone increase.

There is now good evidence that other hormones, particularly other steroids such as hydrocortisone and testosterone, act as inducers at the DNA level; a major part of this work depends on the observation that increases in protein synthesis effected by the hormone are blocked by actinomycin D and therefore involve new RNA synthesis. Some of the pituitary hormones may also be inducers of genetic regulatory units in their target cells. Plants also contain a variety of hormones and growth factors which promote cell division and many other processes; little is known about their immediate effects, but they might be specific inducers. We will not spell out these phenomena in more detail because so little is really known. Notice also that we have to speak of "genetic regulatory units." We do not dare to describe any of these systems in terms of operons and regulons, and in addition we must anticipate many more complex levels of regulation. In Chapter 14 we examined some evidence for a relatively simple regulatory system for human hemoglobins; it would not be hard to draw infinitely more complicated regulatory networks, with a variety of regulatory elements that can switch one another on and off in a system as complicated as a computer circuit.

A developing multicellular organism must have something like a computer program built into it. According to a defined timetable, it must produce a series of components, and this requires that one set of genes after another be turned on to make their products and then turned off again. As each set is turned on, it may well make a product that induces the next set, which in turn will repress the set that is no longer needed. There is little point in elaborating on this theme; there are endless possibilities. The point to remember is that we can expect to find that these circuits and sequences are simply more elaborate combinations of the basic regulatory elements we have already described. What we need now are experiments that can reveal specific examples.

cell surfaces and supports

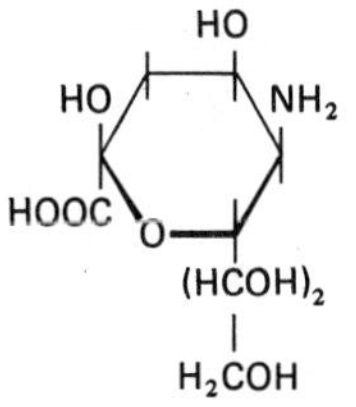

19

HO

HO — NH₂

HOOC — O

(HCOH)₂

H₂COH

NA

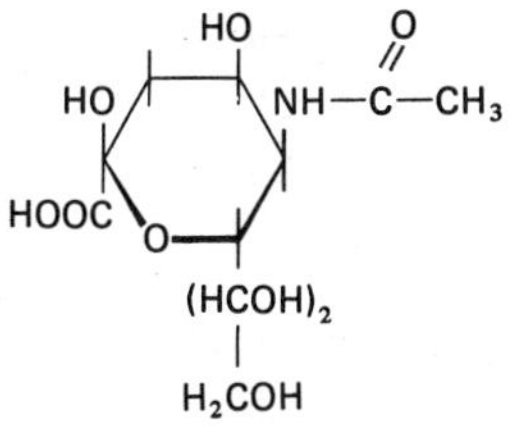

HO — O
         ‖
HO — NH—C—CH₃

HOOC — O

(HCOH)₂

H₂COH

NANA

HO — O
         ‖
HO — NH—C—CH₂OH

HOOC — O

(HCOH)₂

H₂COH

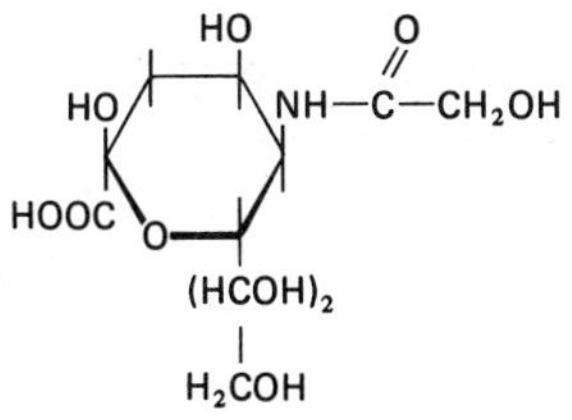

N-glycolyl-NA

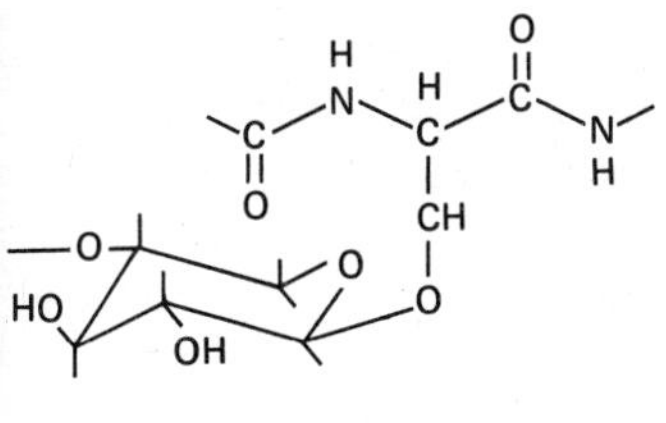

The surface of a cell is its means of communication with the rest of the world. Surfaces have very specific structures, and this specificity allows them to recognize and be recognized. The surface of the cell membrane may also produce a heavy wall around itself that provides strength and protection. We will briefly consider some of these features here because of their general importance.

19-1 GLYCOLIPIDS AND GLYCOPROTEINS

The surfaces of eucaryotic cells are always covered with oligo- or polysaccharides. These polymers are attached to the base membrane either by lipids or proteins. The entire molecule is therefore either a **glycolipid** or **glycoprotein.** Some of these structures have also been found in bacteria but we do not know very much about them as yet.

The oligosaccharide chains that have been identified can probably be bound either to lipids or proteins, so we can consider these independently and then see what kind of linkages there are between them and the membrane. While some chains have been isolated with only a very few monosaccharides, most chains are longer and contain one of two unusual residues on their ends: either L-fucose or one of the **sialic acids.** These acids are all based on **neuraminic acid** (NA), which is acylated with either acetate or glycollate. The most common sialic acid is N-acetyl-neuraminic acid (NANA); N-glycollyl-neuraminic acid and O,N-diacetyl-neuraminic acid are also known.

The oligosaccharide may be very short, as NANA-(1 → 4)- Gal-, or very complex, as

NANA-(1 → 4)-Gal-(1 → 4)-GNAc-GNAc-GNAc-(4 ← 1)-Gal-(4 ← 1)-Fuc

|
Man
|
GNAc
|
Man
|
protein

These chains are always attached to proteins through a glycosidic linkage from the first monosaccharide to the hydroxyl group of *serine, glutamate,* or *aspartate,* as shown here.

The compounds formed by linking these oligosaccharides to lipids have been given many names, but they are really all of a kind (see Table 19-1). Just as the simple phospholipids can be based on either glycerol or sphingosine, the glycolipids can also contain either of these. The oligosaccharide is usually bonded directly to the glycerol or sphingosine, but it may also

587

be bonded to an *inositol phosphate.* Inositol has also been found in a similar position on proteins, so it may be quite a common "linker" in cell surfaces.

TABLE 19-1

some glycolipids

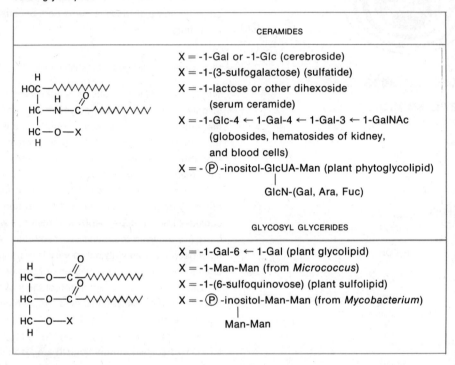

	CERAMIDES
H HOC —\/\/\/\/\/\ | H O HC—N—C—\/\/\/\/\/\ | HC—O—X H	X = -1-Gal or -1-Glc (cerebroside) X = -1-(3-sulfogalactose) (sulfatide) X = -1-lactose or other dihexoside (serum ceramide) X = -1-Glc-4 ← 1-Gal-4 ← 1-Gal-3 ← 1-GalNAc (globosides, hematosides of kidney, and blood cells) X = - (P)-inositol-GlcUA-Man (plant phytoglycolipid) | GlcN-(Gal, Ara, Fuc)
	GLYCOSYL GLYCERIDES
H O HC—O—C—\/\/\/\/\/\ | O HC—O—C—\/\/\/\/\/\ | HC—O—X H	X = -1-Gal-6 ← 1-Gal (plant glycolipid) X = -1-Man-Man (from *Micrococcus*) X = -1-(6-sulfoquinovose) (plant sulfolipid) X = - (P)-inositol-Man-Man (from *Mycobacterium*) | Man-Man

Given the enormous variety of oligosaccharides, it should be easy now to see how every cell can have its own specific surface structure. Perhaps the best examples we can give are the blood type antigens, such as A-B-O, Rhesus, and so on; these are due to specific oligosaccharides, primarily on red blood cells. We have already discussed the possible consequences of mixing the wrong blood types; every person's paratypic dictionary tells him which cells are his and which are not, and trespassers are dealt with ruthlessly (to the detriment of the landowner, not the invader).

19-2 GENETICS AND BIOSYNTHESIS OF BLOOD-TYPE EPITOPES

The A-B-O system of human blood types has been investigated very thoroughly by W. T. J. Morgan, Winifred M. Watkins, and Elton A. Kabat, and from their work we know a great deal about the biosynthesis of the oligosaccharides in the system and the relationships between the different genes responsible for these structures.

The primary oligosaccharide is made through an unknown pathway; we will simply call it precursor substance (P). Its structure is

$$Gal-(1 \rightarrow 3)-\beta-GNAc-(1 \rightarrow 3)-\beta-Gal-(1 \rightarrow 3)-GalNAc$$

which we will abbreviate slightly as

$$Gal_1\text{-}GNAc\text{-}Gal_2\text{-}GalNAc \qquad (P)$$

The *Lewis* gene, *Le,* informs an enzyme that attaches a fucose residue to GNAc by a $1 \rightarrow 4$ bond, to make the Lewis[a] substance:

$$Gal_1\text{-}GNAc\text{-}Gal_2\text{-}GalNAc \qquad (Le^a)$$
$$| $$
$$Fuc$$

The A-B-O epitopes are carried on many places other than the surfaces of red blood cells; people who carry at least one dominant allele of the *secretor* gene, *Se,* secrete them in their saliva. (Most people carry at least one *Se* allele and are therefore secretors.) *Se* informs an enzyme that attaches a second fucose to Gal_1 in a $1 \rightarrow 2$ bond; the resulting compound, if Lewis[a] is the substrate, is called Lewis[b]:

$$Gal_1\text{-}GNAc\text{-}Gal_2\text{-}GalNAc \qquad (Le^b)$$
$$|\quad| $$
$$Fuc\ Fuc$$

However, in a recessive homozygote *le/le,* there is no Lewis enzyme and no Le[a] substrate. The *Se* enzyme attaches a fucose residue to the precursor and makes a compound called H:

$$Gal_1\text{-}GNAc\text{-}Gal_2\text{-}GalNAc \qquad (H)$$
$$| $$
$$Fuc$$

Finally, an enzyme directed by the *I* gene acts on this complex; this enzyme requires the terminal fucosyl residue, so it cannot act on the Le[a] substrate. The enzyme informed by I^A attaches another molecule of GalNAc to make A substance:

$$GalNAc\text{-}Gal_1\text{-}GNAc\text{-}Gal_2\text{-}GalNAc \qquad (A)$$
$$|\quad| $$
$$Fuc\ (Fuc)$$

The enzyme informed by I^B attaches another molecule of Gal instead, to make the B substance:

$$Gal_3\text{-}Gal_1\text{-}GNAc\text{-}Gal_2\text{-}GalNAc \qquad (B)$$
$$|\quad| $$
$$Fuc\ (Fuc)$$

People who have the genotype I^A/I^A or I^A/i^0 are therefore type A; those with the genotype I^B/I^B or I^B/i^0 are type B; and those who are I^A/I^B have a mixture of the two compounds and are type AB. The recessive heterozygote i^0/i^0 has no active enzyme, and therefore only Lewis and H substances.

exercise 19-1 What compounds would you find on the blood cells of people with the following genotypes:

(a) *Le/le, se/se,* I^A/I^B.

(b) *le/le, Se/se,* I^A/I^A.

(c) *Le/Le, Se/Se,* I^B/i^0.

exercise 19-2 Give the possible blood-group phenotypes of children from the following matings:

(a) $I^A I^B \times I^A i^O$.

(b) $I^A i^O \times I^B i^O$.

(c) $I^B i^O \times I^B i^O$.

(d) $i^O i^O \times I^A I^B$.

exercise 19-3 A woman with type A blood bears an illegitimate child with type AB blood and accuses a man of type O of being its father. If you were the judge, would you make the accused pay for support of the child?

The specific oligosaccharide patterns appear to be very important in producing functional organs in animals. The nervous system, for example, depends upon an arrangement of neurons with the proper surface structures. One of the normal brain glycolipids (so-called ganglioside) has the structure

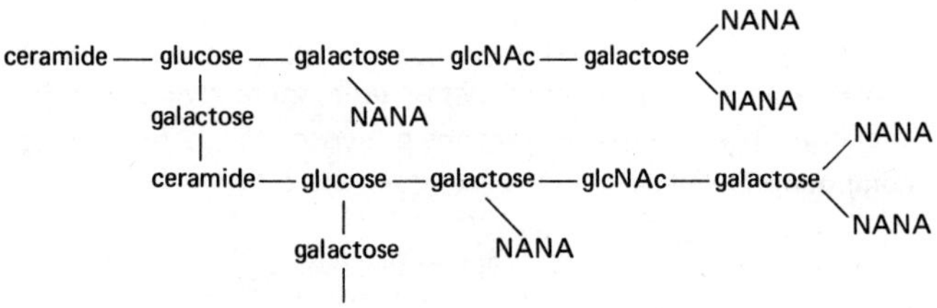

There are a whole series of congenital brain malfunctions (infantile idiocies) associated with the lack of one enzyme for making this structure. For example, Tay-Sachs disease is characterized by the fragment

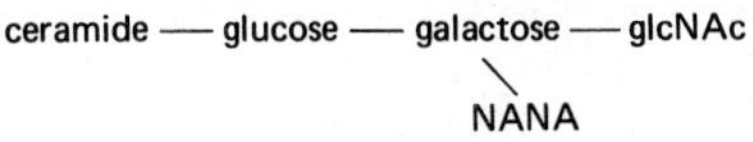

This kind of defect can be understood if neurons have to recognize one another by their surfaces and make specific connections—to form appropriate circuits in the computer, if you will. A simple enzymatic defect can be as serious in the brain as a lack of solder at an IBM plant.

19-3 BACTERIAL CELL WALLS

We noted in Chapter 8 that bacteria can be divided into gram-positive and gram-negative groups by a simple staining test, and that the difference resides in the walls, which are much thicker and more complex in gram negatives. In both cases, the underlying structure seems to amount to a thickening of the cell membrane; this layer has a very distinctive structure, which we will discuss here.

The thickened surface of the gram-positive cell is probably the same as the corresponding layer in gram-negative cells. It contains very few amino acids; these are alanine, glutamate, either lysine or diaminopimelic acid (DAP), glycine, and, more rarely, serine or aspartate. DAP is unique to these peptides; furthermore, they contain

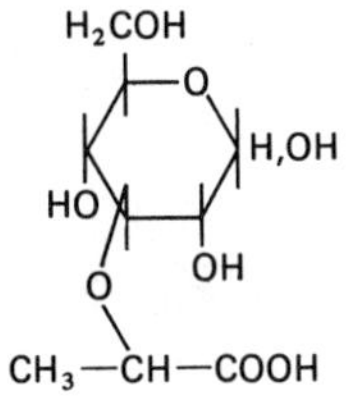

H$_2$COH

H,OH

HO

OH

CH$_3$—CH—COOH

muramic acid

the D isomers of glutamate, aspartate, and alanine, which are found nowhere else and often outweigh the corresponding L isomer in this material. Short peptides of these acids are connected to a unique sugar, **muramic acid** (*Mur:* 3-O-lactyl-glucosamine).

J. T. Park and others have shown that cell walls from both gram-negative and gram-positive bacteria can be broken down into an elementary **muropeptide** whose repeating disaccharide is *N*-acetyl-glucosamine-β(1 → 4)-*N*-acetyl-muramic acid, with the lactyl moiety of the latter bonded in turn to an oligopeptide. The composition of this oligopeptide varies from one species to another, but a typical muropeptide might have the structure shown in Fig. 19-1. M. R. J. Salton and J. M. Ghuysen have shown that these disaccharides are bonded to one another with additional β(1 → 4) bonds to form polysaccharides. Furthermore, Ghuysen and others have shown that the peptides can be interconnected, for they have isolated dimers of the muropeptide with peptide bonds between the terminal alanine of one oligopeptide and the lysine or DAP of the other.

Wolfhard Weidel and H. Pelzer have evolved a model for the basic structure of the cell wall from these data. They point out that the muropeptide can form bonds in two directions: laterally through glycosidic linkages and vertically through peptide linkages. Therefore, the muropeptides can form a perfect network polymer (Fig. 19-2). Weidel and Pelzer suggest that the polymer be called a **murein,** by analogy with a protein. A murein can be extended indefinitely; it can close on itself to form a tube and the ends

FIGURE 19-1
Structure of a muropeptide unit.

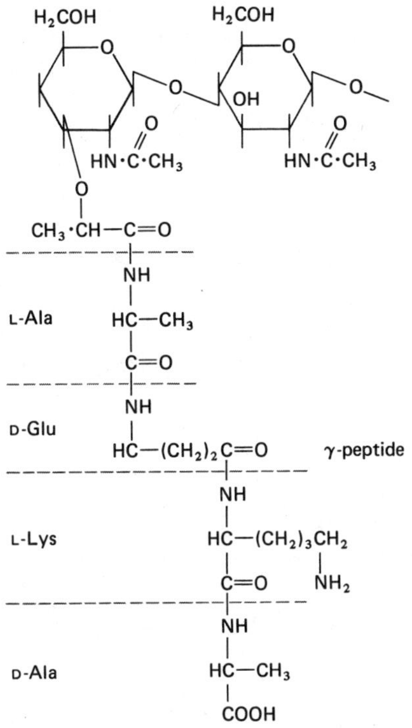

GlcNAc—MurNAc
|
L-Ala
|
D-Glu
|
DAP—NH—C=O
| |
D-Ala   D-Ala
| |
HOOC   DAP—NH₂
|
D-Glu
|
L-Ala
|
Glc—NAc—MurNAc

$$=$$

(a)

(b)

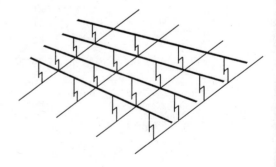

(c)

FIGURE 19-2

*Structure of a murein. (a) The
elementary dimer made of two
muropeptides, with an abbreviated
representation. (b) The general pattern
of bonding within the murein. (c) A
possible three-dimensional arrangement
of polysaccharide chains in two layers
at right angles to one another, connected by vertical peptide links.*

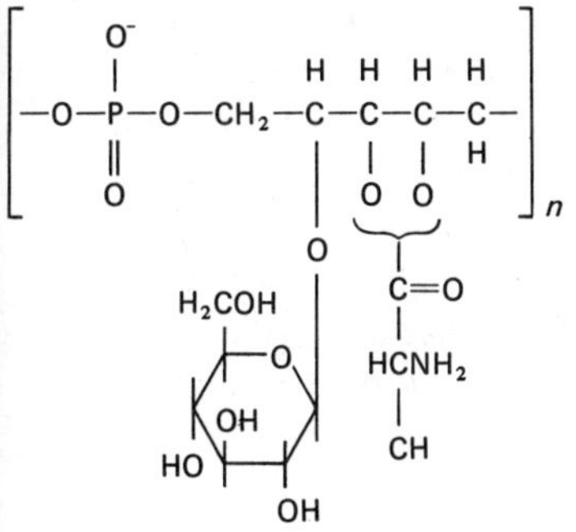

ribitol teichoic acid repeating unit

of the tube can close to form a sack. It is therefore proposed that this closed murein sack forms the basic framework of the bacterial cell wall. It is a framework upon which more elaborate structures can be built, including functional enzymes and special polysaccharides.

In addition to muropeptides, many gram-positive bacteria contain interesting polymers called **teichoic acids;** one common type is a glucose-ribitol-phosphate-alanine polymer. This is interesting primarily because of the unusual ester attachment of the alanine, although it is not clear to which ribitol carbon it is linked. This alanine may be identical with the terminal alanine of the muropeptide, so that some walls may be built of a teichomurein, as shown in Fig. 19-3.

FIGURE 19-3

Two possible bonding patterns in murein-teichoic acid complexes.

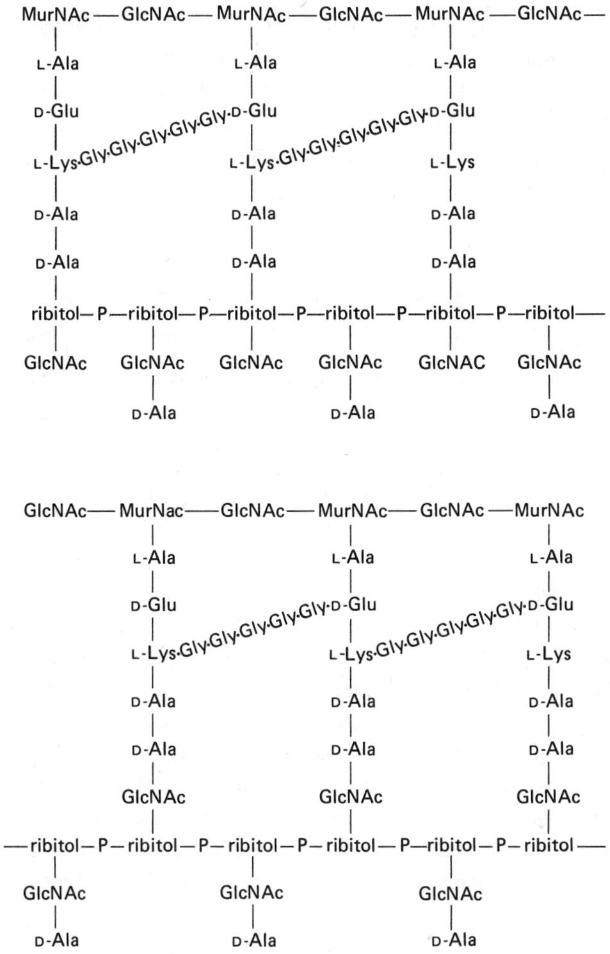

Gram-positive cell walls may be entirely murein or they may contain up to 60% teichoic acid, but they probably contain little else. However, in gram-negative cells the basic murein accounts for no more than 20% of the total mass, and it is overlayed with the extensive polysaccharide and lipoprotein coats we have already noted. The structure of these layers is largely unknown; some very complex glycolipids have been isolated from them, as well as polysaccharides and mucopolysaccharides. It is certain that they contribute to the intricate surface patterns of the cells; Weidel has demonstrated that some phages attach specifically to lipopolysaccharide structures and others to lipoproteins. This aspect of the wall structure remains a major problem for future research.

19-4 BACTERIAL CELL WALL BIOSYNTHESIS

The biosynthesis of the murein layer has been studied in detail by Jack L. Strominger and his colleagues. His results are worth discussing here because mureins are still among the best-known cell surface structures and because the pathway illustrates several mechanisms cells must use to synthesize very large structures in specific locations.

The initial stages of biosynthesis leading to UDP-N-acetyl-glucosamine have already been described. UDP-N-acetyl muramic acid (UDP-MurNAc) is synthesized from the glucosamine by addition to pyruvate as PEP and its oxidation to lactate. In the bacterium Strominger uses (*Staphylococcus aureus*), the muramic acid carries a pentapeptide: L-Ala-D-Glu-L-Lys-D-Ala-D-Ala. Each of these amino acids is activated by ATP and added in turn:

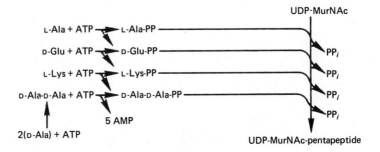

All of these reactions proceed inside the cell, presumably at an enzyme complex on the inner surface of the membrane. This complex UDP-MurNAc-pentapeptide must now be transported across the membrane to its outer surface; this is done by temporarily bonding the complex nucleoside to a phospholipid. The repeating disaccharide is then formed by adding a molecule of GlcNAc, and the entire complex crosses the lipid to be bonded into the growing murein sheet (Fig. 19-4).

Notice that this process conforms with the general model for translocation that we developed earlier. It uses a set of oriented enzymes on both sides of the membrane and a lipid carrier that can traverse the hydrophobic barrier. A completely analogous biosynthetic mechanism has been demonstrated for one of the specific polysaccharides

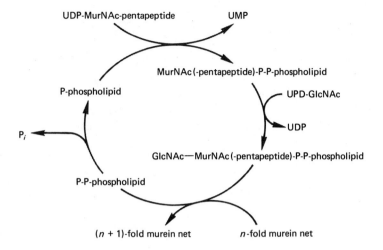

FIGURE 19-4

*A cycle of murein synthesis in which a muropeptide unit is made intact and then trans-
located across the membrane by a phospholipid carrier.*

on the surface of the bacterium *Salmonella* (one of the so-called *O*-antigens). This
molecule is

$$\text{Gal-(Man-Rham-Gal)}_n\text{-GNAc-Glc-Gal-Glc-O}\overset{\displaystyle Gal}{\underset{|}{-}}\text{P}\overset{\displaystyle O^-}{\underset{\|}{O}}\text{O-ethanolamine}$$

Its biosynthetic pathway is shown in Fig. 19-5. Notice that this pathway uses bonding
to a lipid carrier, sequential donation from nucleoside diphosphate sugars, and initial
formation of the repeating subunit. In this case, the abequose branches are added
before complete polymerization, but the other mechanisms are in accord with the
general pattern for synthesis of large molecules.

19-5 PLANT CELL WALLS

Plant cells of various kinds lay down an incredible variety of polysaccharide coverings.
These may become just strong enough to support a fragile stem or they may be thick,
tough trunks that support gigantic trees and can be used to hold tons of weight. They
may also become the soft, sweet flesh of fruits.

   Aside from cellulose itself, there are two major polysaccharides in plant cell walls.
The **hemicelluloses** are somewhat more soluble than cellulose but less soluble than
starches; they are often made of a repeating disaccharide called an **aldobiuronic**
acid—an aldose plus a uronic acid. A few of these, out of the many now known, are
shown in Fig. 19-6.

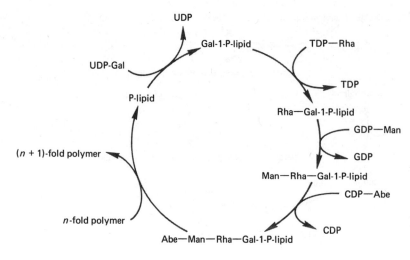

FIGURE 19-5

A cycle of surface polysaccharide synthesis, in which the tetrasaccharide unit is made intact and then translocated across the membrane by a phospholipid carrier.

FIGURE 19-6

Three representative hemicelluloses.
(a) Arabino-4-O-methylglucurono-xylan.
(b) An arabinogalactan from larch.
(c) A type of polymer found in bark.

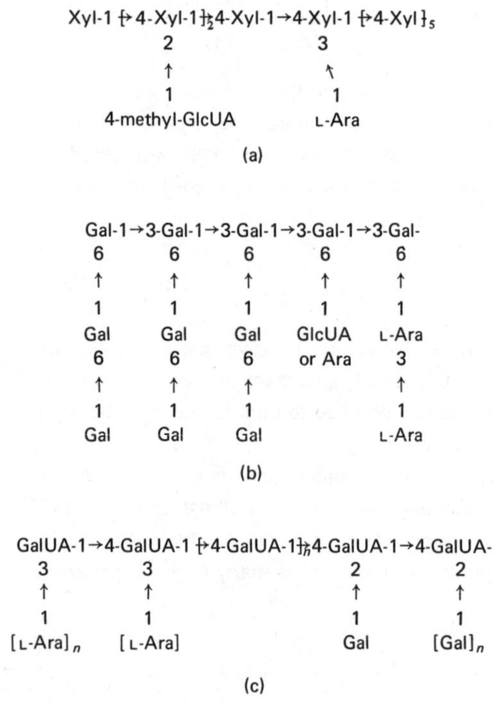

Xyl-1 $\{$→4-Xyl-1$\}_{\frac{1}{2}}$4-Xyl-1→4-Xyl-1 $\{$4-Xyl$\}_5$
            2   3
            ↑   ↖
            1   1
      4-methyl-GlcUA   L-Ara

(a)

Gal-1→3-Gal-1→3-Gal-1→3-Gal-1→3-Gal-
     6   6   6   6   6
     ↑   ↑   ↑   ↑   ↑
     1   1   1   1   1
    Gal   Gal   Gal   GlcUA   L-Ara
     6   6   6   or Ara   3
     ↑   ↑   ↑   ↑
     1   1   1   1
    Gal   Gal   Gal   L-Ara

(b)

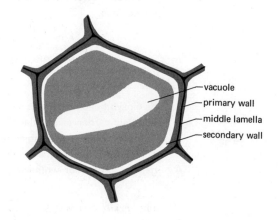

FIGURE 19-7

The principal layers of a generalized plant cell wall.

- vacuole
- primary wall
- middle lamella
- secondary wall

GalUA-1→4-GalUA-1 $\{$4-GalUA-1$\}_{\frac{1}{n}}$4-GalUA-1→4-GalUA-
       3   3   2   2
       ↑   ↑   ↑   ↑
       1   1   1   1
   [L-Ara]$_n$ [L-Ara]   Gal   [Gal]$_n$

(c)

The **polyuronides** are uronic acid glycans. The principal polyuronide of green plants is **pectic acid,** a poly-D-galacturonic acid; the brown algae have alginic acid (poly-D-mannuronic) instead. The native large molecules are partially methylated and are called protopectin. The extracted small molecules called pectins are used to form jams and jellies.

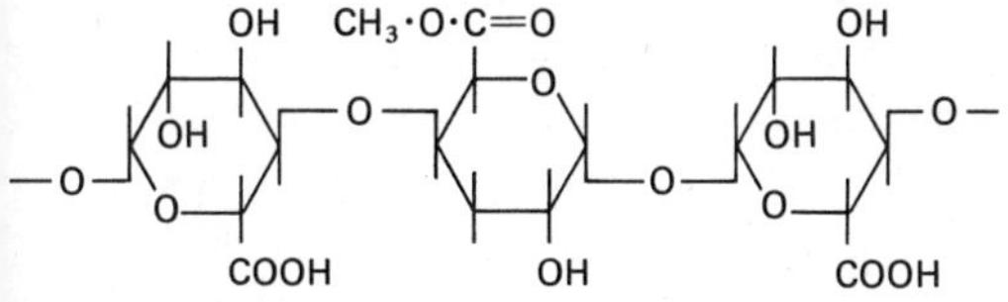

A plant cell wall may have three layers (Fig. 19-7). The outermost part is the **middle lamella,** which is a layer of cement — mostly polyuronides — between cells. Inside this is the **primary cell wall;** nearly half of this layer is regular, crystalline fibrils of cellulose that are built up into larger and larger fibers, as in the fibrous proteins. This makes the basic network, which generally encloses the cell in a regular sheath; hemicelluloses and pectic substances are woven into this network more irregularly. Many proteins, some with enzymatic activity, are also woven into the wall structure.

pectic acid

The **secondary cell wall** is deposited inside the primary wall in cells that are developing into the strong fibers of wood, where the wall is strengthened and cytosol disappears. It is basically built of the same materials as the primary wall, but in addition it is **lignified** by deposits of **lignin,** a complex polymer of coniferyl alcohol (Fig. 19-8). Lignin sheets and masses intertwine with the polysaccharides to make very tough, hard walls.

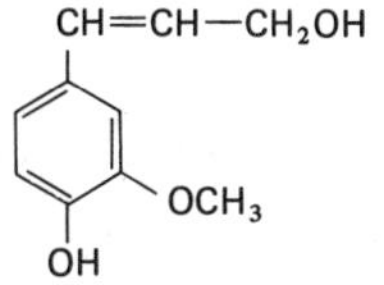

coniferyl alcohol

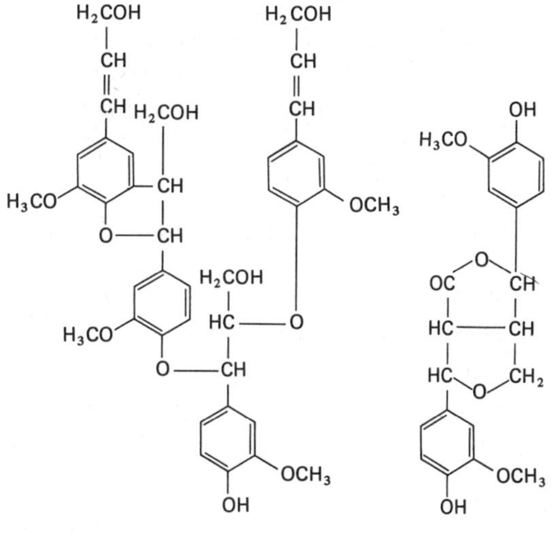

FIGURE 19-8
Types of bonding in lignin polymers.

19-6   THE MUCOPOLYSACCHARIDES

Some of the most interesting materials found in animals are secreted polysaccharides that are important components of *mucins,* the slippery, viscous fluids that lubricate and protect many organs. Hence, these substances are called **mucopolysaccharides.** They are closely associated with large-scale body movements. They form the synovial fluids in the joints between bones, the mucus of the respiratory tract and birth canal, and the mucins that snails and other invertebrates glide over. They are also a part of saliva and the fluids of the eye cavities, and they make up a major part of bones and cartilage.

As Table 19-2 shows, most of them can be built from a repeating disaccharide unit usually consisting of a uronic acid and an amino sugar. They are generally highly sulfurated, although this is typical of many structural polysaccharides. Muco-polysaccharides often form large complexes with proteins. Hyaluronic acid-protein complexes may weigh up to 10 million daltons and chondroitin sulfate-protein complexes generally range around 100,000–200,000 daltons. They are bonded to their proteins just like the oligosaccharides we discussed earlier—through glutamate, serine, or aspartate. When several chains are bound to the protein close to one another, the monosaccharides may be in a position to chelate calcium ions between their sulfate groups (Fig. 19-9). These calcium salts may give great strength and stability to the tissues they form, as in cartilage and bones.

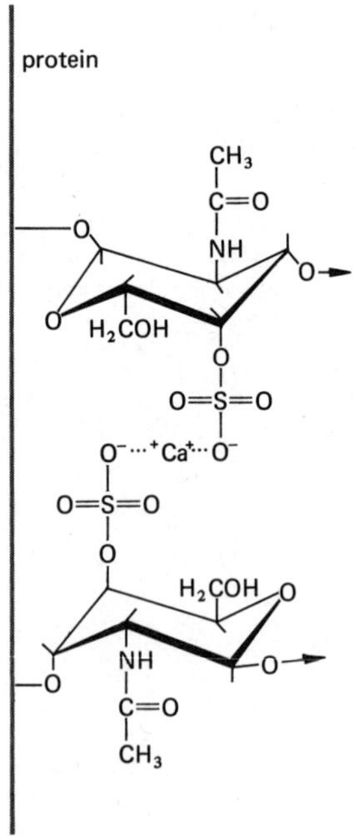

FIGURE 19-9

Chelation of a calcium ion between two sulfated residues of a mucopolysaccharide complex.

TABLE 19-2

some mucopolysaccharides

NAME AND OCCURRENCE	REPEATING DISACCHARIDE
Chitin (poly-N-acetyl-glucosamine) Structural component in many invertebrates and in cell walls of fungi and algae.	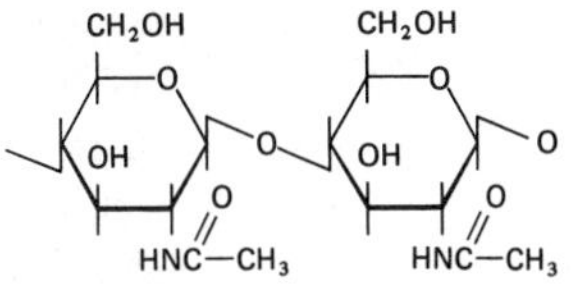
Hyaluronic acid (GUA + GNAc)$_n$ Synovial fluid (lubricant of joints); skin; jelly of umbilical cord; vitreous humor of eye.	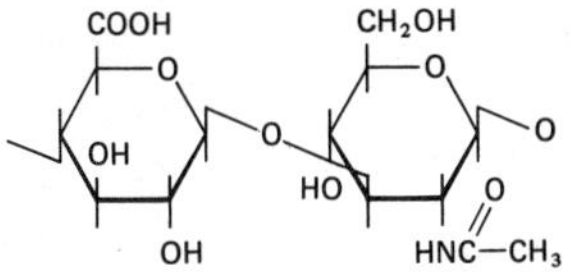
Chondroitin 4-sulfate (GUA + GalNAc-4-SO$_4$)$_n$ 40% of some cartilage; skin; cornea of eye; bones; umbilical cord jelly.	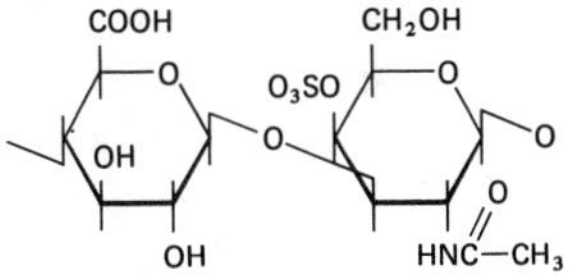
Chondroitin 6-sulfate (GUA + GalNAc-6-SO$_4$)$_n$ Tendon; cartilage; skin; umbilical cord; saliva.	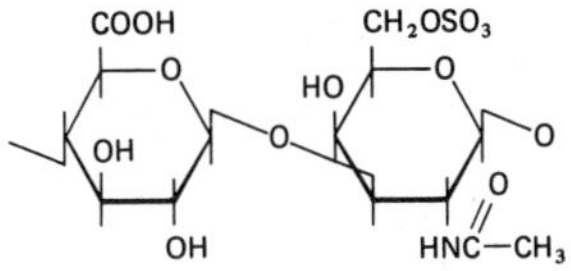
Dermatan sulfate (IdUA + GalNAc-4-SO$_4$)$_n$ Skin; tendon; heart valves and blood vessels.	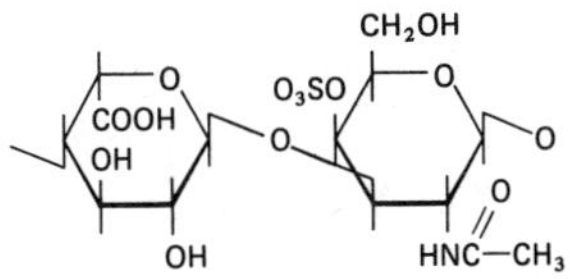
Mucoitin sulfate (GUA + GNAc-4-SO$_4$)$_n$ Mucus.	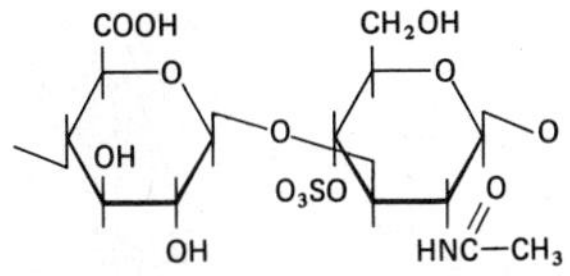
Keratan sulfate (Gal + GNAc-6-SO$_4$)$_n$ Cornea of eye; cartilage.	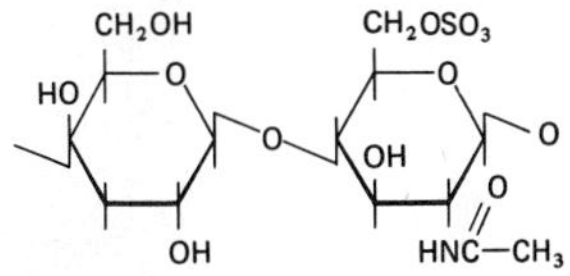
Heparin (GUA + GlcN-N-SO$_4$-6-SO$_4$)$_n$ Prevents coagulation of blood.	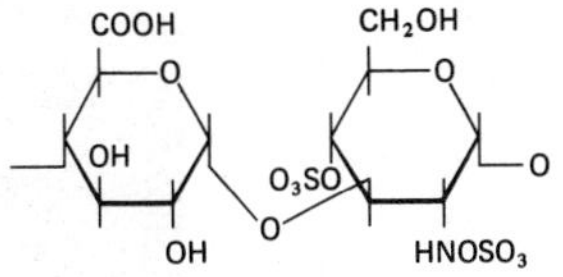

Much of the mass of connective tissue that is not mucopolysaccharide consists of the interesting protein **collagen;** for example, it is 35% of the dry weight of bone. The protein occurs in long fibers that have a striking pattern of bands (Fig. 19-10) with a repeat period of 690 ± 10 Å that presumably arises from the alignment of characteristic amino acid side chains from different molecules. Most work on collagen structure has been directed toward understanding what makes this pattern.

Collagens have a very unusual amino acid composition; a third of the residues are glycines and another quarter are proline or hydroxyproline. (The latter is found only in collagens and plant cell wall proteins; another rare amino acid, hydroxylysine, is confined to the collagens.) The primary structure of much of the molecule consists of sequences like -Gly-Pro-X, where X may be any amino acid, and -Gly-Pro-Hyp-, which make very regular stretches; more irregular stretches fit between these. The more regular regions contribute to a characteristic x-ray diffraction pattern, including a 2.86-Å meridional reflection and equatorial reflections at 4.5 and 11 Å. There were many attempts to explain this pattern, beginning with Astbury, but the first really successful model was proposed by G. N. Ramachandran and his colleagues. Their model consisted of three hydrogen-bonded helices wound around each other, with three residues per turn and a pitch of 9.5 Å. One of the three residues had to be Gly, the second either Pro or Hyp, and the third any except Pro or Hyp. This first approximation did not explain the 2.86-Å meridional reflection, however, and it did not entirely fit the amino acid sequence data.

L-hydroxyproline

L-hydroxylysine

FIGURE 19-10

Fibrils of bovine collagen that show the typical banding pattern. (Courtesy of Jerome Gross.)

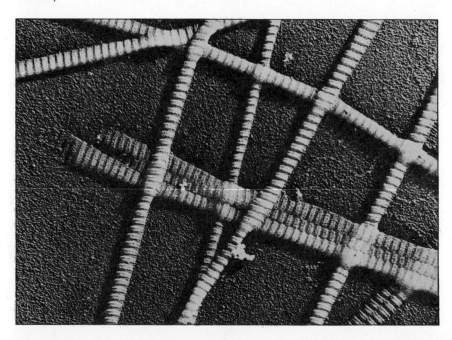

Francis Crick and Alexander Rich then proposed a modification based on the polyglycine structure (Fig. 19-11). Notice that each chain has a three-fold screw symmetry and forms hydrogen bonds with six neighbors. The distance between residues is 3.1 Å, making each turn 9.3 Å long—approximately the dimensions of the Ramachandran model. Rich and Crick then showed that two structures for natural collagens could be derived from polyglycine by taking three left-handed strands and giving them a slight right-hand twist to make a coiled coil that repeats every 10 elementary coils (30 residues), like the coiled coil of keratin. It is now agreed that this structure gives a satisfactory reflection at 2.86 Å and that it will easily allow the sequence -Gly-Pro-X-. It will probably allow Hyp in the third position without too much strain. The principal points of disagreement now are minor features, so collagen should be thought of as a coil of three strands with rather gentle pitches.

Francis Schmitt, Jerome Gross, and Cecil Hall have shown that in 1% NaCl collagen dissociates into a basic molecule called **tropocollagen,** which weighs 300,000 daltons and is 3020 ± 45 Å long and about 15 Å wide. It consists of three strands: two of them (α_1 strands) are identical and the third (α_2 strand) has a different composition. John Petruska and Alan Hodge have recently proposed a model in which each α_1 strand consists of five elementary polypeptides, designated σ_1, and each α_2 strand consists of seven elementary σ_2 polypeptides. However, the σ_1 polypeptides are seven units long and the σ_2 polypeptides are five units long (where one unit is the length of a turn of the coiled coil, about 30 residues). If the tropocollagen molecule starts to assemble itself from one end, then it determines its own length by the vernier principle; five molecules that are seven units long equal seven molecules that are five units long, so the tropocollagen unit stops growing when it is 35 units long, where the three strands come out even. In essence, the α_2 strand measures the α_1 strands. This arrangement and the higher-order organization of collagen is shown in Fig. 19-12.

A variety of unusual linkages are found in collagens. The strands may be linked to one another by ester or amide bonds, so when tropocollagen is dispersed in acid or high salt it forms not only α strands, but also β units that consist of two α units joined together and some γ units that consist of all three strands joined together. The number of these linkages increases with time and is apparently a normal part of the aging process.

The higher-level organization of collagen parallels that of α-keratin. The tropocollagen molecule of three α strands is analogous to a protofibril of three α-helices. A large number of tropocollagen molecules make a microfibril. If the repeating distance of 690 ± 10 Å found in native collagen is called a D unit, the length of each tropocollagen molecule is L = 4.375 D units. They associate in the microfibril with a staggering of 1 D unit, leaving overlaps of 0.375 D units between molecules that are displaced by 4 D units from one another and gaps of 5 D − L = 0.625 D units between molecules. These gaps are spaces that may be filled with calcium or other hardening agents. This model gives a good agreement between the calculated and observed patterns of banding.

At still higher levels, microfibrils of collagen make macrofibrils and these make fibers, but all of this can be understood in terms of the structure of the elementary units and their modes of association. The regular pattern of rather large, visible structures is therefore due entirely to the regularities in the small molecules of which

FIGURE 19-11

Structures of polyglycine I and II (a) viewed from above and (b) viewed from the side.

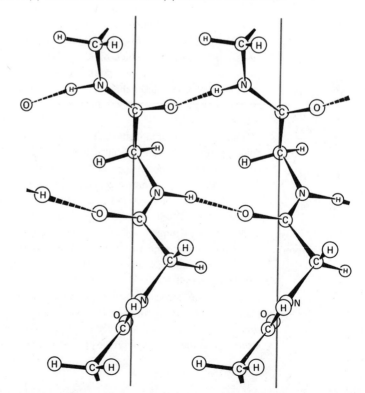

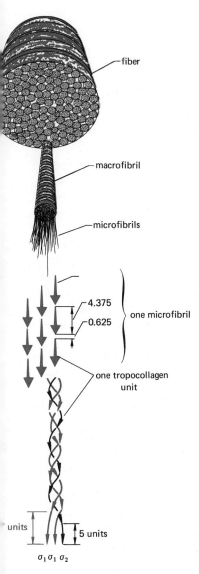

fiber

macrofibril

microfibrils

4.375
0.625 } one microfibril

one tropocollagen
unit

units 5 units

$\sigma_1\,\sigma_1\,\sigma_2$

FIGURE 19-12
Levels of collagen organization.

they are built. Few molecules have been analyzed as carefully as collagen, but we may predict that many other macroscopic patterns will be found to have a structural basis similar to that of its banding pattern. In other words, "large-scale" order is probably always a reflection of "small-scale" order in single molecules, along with specific modes of aggregation.

The mode of tropocollagen aggregation conforms to both the subassembly and self-assembly principles and probably illustrates a general method for limiting the growth of linear structures that assemble themselves. It would be hard to imagine a very long enzyme molecule that can act like a measuring rod by stopping assembly after a certain length is achieved. However, the Petruska–Hodge model for tropo-collagen employs a very simple self-limiting element, for the molecule may be expected to stop growing when it reaches a "beat" point between five-unit and seven-unit elements. It will be interesting to see whether other large structures employ this mechanism.

19-8 THE SIGNIFICANCE OF SURFACE CONTACTS

The child who gets a minor cut and loses a few drops of blood is terrified at the thought that his whole insides seem to be coming out (like the teddy bear's stuffing); he soon learns that the blood stops flowing, a scab develops, and—if he is patient enough—he is soon left with a patch of new skin indistinguishable from the old one, down to all of its tiny creases and ridges. While we get used to seeing this repair process time after time, most of us remain rather awed by it and astonished that it can occur so precisely. We still cannot explain the precision of wound repair, but we have at least learned that part of the process depends upon interactions between the cells of the wound area—interactions that are mediated by their surface structures.

During the few days after a wound occurs, several different types of cells are proliferating under the scab and building up new tissue. For example, fibrocytes have migrated into the wound area and are secreting collagen and mucopolysaccharides which orient themselves into connective fibers along the lines of stress in the tissue, to hold the other cells in strong, connected masses. At the surface, the epidermal cells around the wound area have also migrated in and started to proliferate; they form a new layer of epidermis just under the scab, and this eventually grows until it pushes the scab off. We may ask two questions about these cells, that are really part of the same question: If the epidermal cells were able to proliferate, why didn't they do so before the skin was cut? And after they start to repair the wound, what makes them stop when they have formed a new layer of skin—why don't they keep growing and make a ball of cells on top of the wound area?

This same inhibition of growth is seen in many tissue culture cells. A monolayer of cells in a culture generally proliferates very well until the cells start to come into close contact with one another, and then they stop dividing. This phenomenon, which is really the same one seen in wound repair, is called **contact inhibition.** It remains one of the major puzzles of cell growth, but we can begin to understand this phenomenon within a framework of cell-surface interactions.

To understand the action of hormones, we developed a picture of two effectors operating in series; first, the hormone itself, which is carried to a specific cell, and second, an internal effector, such as cyclic AMP, which is stimulated by the hormone. In this model the hormone must interact with surface receptors on its target cells, and the stimulation of the internal effector may occur through an allosteric stimulation of enzymes located in the cell membrane. Contact inhibition may be generally understood in the same way, simply by saying that the inhibition of growth processes inside a cell may be produced by contact with inhibitory ligands carried on the surface of another cell.

Figure 19-13 shows how this can occur. The cells are growing and proliferating because an internal effector, X, is being produced continually by enzymes in the cell membrane. However, the cells carry surface structures that can induce a general change in one another's membranes. This occurs only when they are pressed together, inhibiting the enzyme system producing X, and thus inhibiting growth.

This picture is supported by two types of information. First, there are proteins called **mitogens** that stimulate the proliferation of various animal cells. One of them, phytohemagglutinin, promotes the proliferation of animal leucocytes; one of its effects is to stimulate the synthesis of inositol phospholipids in the cell membrane, which may be quite a direct effect or the indirect result of a general growth stimulation. Since the mitogens are apparently glycoproteins, it is unlikely that they get inside cells, but they could easily interact with specific cell surface structures.

The other line of evidence comes from studies on antibody production. The best way to understand proliferation of lymphocytes in response to the introduction of specific immunogens is to assume that each lymphocyte—which, you recall, is already determined for the production of a single antibody—contains surface receptors that are identical to the paratope they can produce, and thus paratactic to one specific epitope. The virgin lymphocyte is stimulated to produce antibody by a contact with the appropriate epitope on its surface (Fig. 19-14); the experienced lymphocyte is stimulated by this epitope to proliferate. In either case, it is easiest to conceive of these effects being mediated through the cell surface.

Occasionally a cell loses some phase of this contact-inhibition system. It begins to proliferate even when other cells are packed tightly around it and becomes a **tumor.** The cells of some tumors remain compacted together, so, even though they make a potentially dangerous mass, they can be removed with careful surgery; however, the cells of other tumors may get free and migrate throughout the body or **metastasize.** New tumors then sprout up at any point where one of these migrant cells settles down, so there is little possibility of permanent control. In this case, the tumor is **malignant.** Much of the energy of contemporary medical research is going into efforts to understand tumors and find ways to control them. It should be obvious that we cannot really hope to understand them unless we can identify the controls in normal cells that keep them from proliferating wildly. Within the framework outlined here, you can see a number of points that can be investigated—the allosteric regulation of enzymes, the

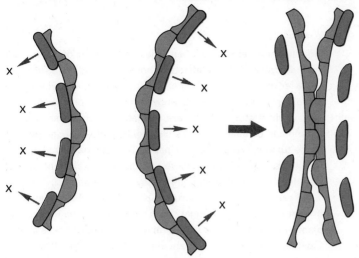

FIGURE 19-13

One possible explanation of contact inhibition. Each cell contains enzyme systems that manufacture compound X, which induces continued proliferation. Each surface contains allosteric sites and complementary inhibitors; when they are pressed against one another, they mutually inhibit each other's enzymes.

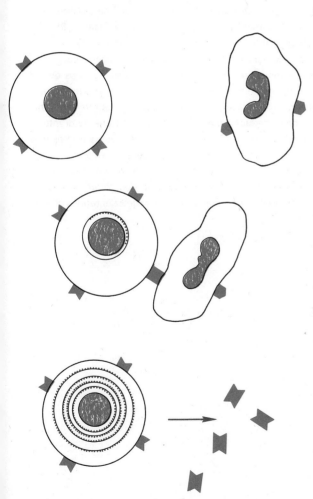

FIGURE 19-14

Contact stimulation of antibody synthesis. The small lymphocyte contains its specific paratopes on its surface. When complementary epitopes are carried to it by a macrophage, the lymphocyte is induced to make more endoplasmic reticulum and to start producing antibody. Presumably a later contact of the same kind will induce proliferation of the lymphocyte.

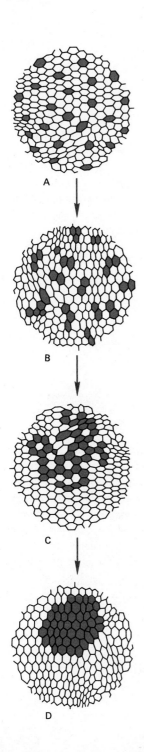

A

B

C

D

induction and repression of enzymes, the systems of general effectors such as cyclic AMP, the effects of hormones, and the structure and behavior of the cell membrane. Carcinogenesis can involve any of these control points; a mutation within a line of proliferating tissue cells could destroy any of these regulatory mechanisms, and any of them could be affected by the chemical agents—**carcinogens**—that cause tumors. We will see in Chapter 20 that viruses and episome-like factors can cause tumors; these foreign genetic elements may also introduce information that affects any of these points.

We can also see the importance of specific cell-surface structures in another context in normal organisms. Each tissue in a complex plant or animal consists of several different types of cells compacted together in just the right arrangement. It is becoming obvious that the precision of these arrangements depends upon surface-surface interactions between the cells of the tissue. In general, each type of cell recognizes others like itself, and these cells tend to stick to one another; at the same time, different cells are excluded and they tend to associate nearby in their own distinctive block. For example, in a developing embryo the cells that are destined to become the muscles of the limb remain together inside the limb and those destined to become the skin and other external structures remain together outside. Why don't these cells get mixed up? How do they recognize each other? We assume they make contacts between specific groups on their surfaces, although we don't know which groups are involved in the system. However, Malcolm S. Steinberg has demonstrated that the association can be understood merely in terms of the energy of interaction between cell surfaces. He can dissociate embryonic tissues into separate cells with mild trypsin digestion. The cells migrate by ameboid movement and gradually sort themselves out into tissues (Fig. 19-15). The arrangement they achieve is an *equilibrium* arrangement, since it occurs even when intact tissue fragments (undissociated cells) are put together.

FIGURE 19-15
Segregation of a mixture of cells into two distinct tissues.

Furthermore, there is a *hierarchical ordering* of cell types (Fig. 19-16). That is, if cells of type A are mixed with those of type B, A will segregate internal to B; and if cells of types B and C are mixed, B will segregate internal to C; then if cells of types A and C are mixed, A will segregate internal to C. No explanation is needed for this ordering other than differences in the strengths of bonding between surfaces; this sequence means that bonds between A cells are most stable, bonds between B cells are weaker, and those between C cells are still weaker.

Investigations of this type are just beginning; they are designed to answer some fundamental questions which are part of the much larger question of how a developing, differentiating organism achieves its final form—a question that we haven't tried to attack directly in this book. However, experiments like Steinberg's suggest that the same principles of self-assembly we established for molecular interactions at the subcellular level will also apply to cell-cell interactions and may be much more general than we have thought until now. This is another of the exciting areas of biology that is still wide open and waiting to be investigated.

(a)

FIGURE 19-16
The hierarchical ordering of cell types.
(a) *Cells of limb bud precartilage from a four-*
day chick embryo are surrounded by
five-day heart ventricle cells.
(b) *Five-day heart ventricle cells are surrounded*
by five-day liver cells.
(c) *Four-day limb bud precartilage cells are*
surrounded by five-day liver cells.
(Courtesy of Malcolm S. Steinberg, reprinted by
permission from Cellular Membranes in Development,
Michael Locke, ed., Academic Press, New York, 1964.)

(b)

(c)

20

the viruses

Now that the major patterns of macromolecular synthesis and regulation in a normal cell have been outlined, it becomes interesting to see how these patterns are altered when the cell's metabolism is taken over by the foreign nucleic acid of a virus. The result of a viral infection is generally an obvious destruction of host structure—a disease—and the disease itself has long been the principal focus of attention. Viruses have been studied primarily in medicine or agriculture by people whose primary concern has been to identify diseases and cure them. However, even though we still have to touch on the diseases themselves (if only because viruses are generally named for the diseases they cause) we will be concerned primarily with viruses as biochemical agents.

A. Structure and classification

20-1 VIRUS ARCHITECTURE

The principles of subassembly and self-assembly were first brought to light in connection with virus structure, and the two concepts are illustrated by viruses more clearly than by any other biotic systems. Every virus consists of a protein capsid with a nucleic acid genome; the combination of the two makes a **nucleocapsid,** and the architecture of this structure can be understood very easily.

The structural units of the capsid can associate in only two ways, so we observe viruses with two general forms: helical or spherical (Fig. 20-1). We have seen plenty of examples of helical structure by now; the structural units in this case assemble themselves into a tube of indefinite length by repeating a given bonding pattern over and over. The nucleic acid winds around inside this tube as shown in Fig. 20-2.

The closed shell is a slightly more complex problem in architecture; a general solution for its design was given, appropriately, by an architect, R. Buckminster Fuller, who showed that the shell can be built of quasiequivalent subunits arranged in a hexagonal pattern that contains 12 points of five-fold symmetry. Fuller used this principle to design geodesic domes, such as the one shown in Fig. 20-3. Each structure unit in this shell is an equilateral triangle; a polyhedron made entirely with equilateral triangular faces is called a **deltahedron.** The triangles all look alike, although in fact there are 12 different shapes and they are only quasiequivalent. The pattern is primarily hexagonal, but a few pentagonal points can be seen where only five triangles are arranged around a vertex.

The 12 pentad vertices permit the formation of a closed shell

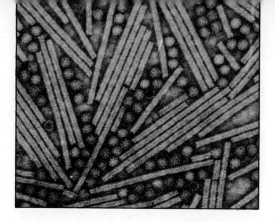

FIGURE 20-1

The two major types of virus: long helical rods of tobacco mosaic virus (TMV) mixed with spheres of turnip yellow mosaic virus (TYMV). (Courtesy of Robert W. Horne.)

FIGURE 20-2

Structure of the TMV nucleocapsid. (After a drawing by D. L. D. Caspar.)

from a plane. As Fig. 20-4 shows, a hexagonally packed plane can be folded if some subunits are removed. The minimum distortion is achieved by removing one subunit from a group of six to make a pentad vertex; 12 such vertices are required to make a closed polyhedron. The resulting polyhedron has 20 faces and is called an **icosahedron;** since its faces are equilateral triangles, it is an **icosadeltahedron.** The removal of two or three subunits at each vertex would lead to the formation of an octahedron or a cube, respectively, but these would require much greater distortions of the bonding between the structure units, and the icosadeltahedra are clearly the most likely shapes for biological structures that must form symmetrically and spontaneously.

All icosadeltahedra have $20T$ faces, where T, the **triangulation number,** is given by the rule $T = Pf^2$, where f is any integer and P is any number in the series 1, 3, 4, 7, 9, 13, ... The number of structure units required to build a shell is always $S = 60T$. In all real capsids, the structure units bond rather tightly into units equivalent to the pentamers and hexamers visible in the geodesic dome. While the structure units themselves are too small to be resolved clearly by electron microscopy, the hexamers and pentamers are readily seen with negative staining. These are called the **morpho-**

FIGURE 20-3
*One of Fuller's geodesic domes; two pentad units are indicated. (Courtesy of
R. Buckminster Fuller.)*

FIGURE 20-4
*Removal of a 60° sector from a hexagonally packed sheet (a) makes a pentad vertex
with minimum distortion and less strain on interprotein bonds than the removal of a
90° sector from a square-packed sheet (b).*

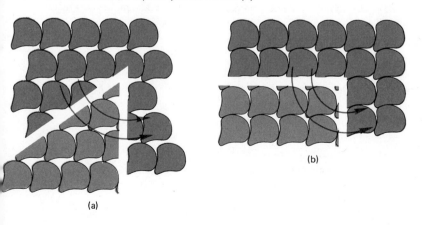

(a)

(b)

logical units or capsomeres. Every capsid necessarily has 12 pentamers and as many hexamers as are necessary to fill in between them, so the larger capsids are all made by adding more hexamers. The number of capsomeres, which is determined by electron microscopy associated with model building, is a useful means of identifying and classifying viruses. In some of the following figures, a number of examples are shown of real viruses with models that conform to these rules and fit the pictures perfectly.

exercise 20-1 P is actually given by $h^2 + hk + k^2$, where h and k are any integers that have no common factor. Calculate the values of P up to 40.

exercise 20-2   Write down all of the possible triangulation numbers through 25.

exercise 20-3   Determine a formula for the number of capsomeres in a capsid with a given triangulation number.

exercise 20-4   How are the structure units clustered (in hexamers and pentamers) for $T = 1$; $T = 3$; $T = 4$; $T = 9$?

20-2   VIRUS CLASSIFICATION

With this information about the architecture of virions, it is very simple to classify viruses into conceptually useful categories. We will use a system that was first devised by Andre Lwoff, Robert Horne, and Paul Tournier; it provides a very useful framework for much of virology.

The first criterion for classification is the nucleic acid of the virus. This is always either DNA or RNA (but never both), so there are deoxyviruses (DNA) and riboviruses (RNA). The second criterion is the symmetry of the nucleocapsid; this is either helical (H) or cubic (C), and so there are two classes under each major group. However, within the deoxyviruses we must add a third class for the large bacteriophages with tails. The tails apparently have hexad symmetry, but they are attached to the pentad vertices of an icosahedral head. This creates a dyad symmetry for the whole virion and it is necessary to separate these viruses as a special class. They are best described as urophages (Greek: oura = tail).

In some cases, the nucleocapsid by itself is the whole virion. However, other viruses consist of the nucleocapsid surrounded by a mantle or peplos, which is always a bit of membrane taken from the host cell, perhaps with some additional virus-specific materials added. Whether the virion is naked (N) or enveloped (E) makes a third criterion for classification. These criteria establish eight orders of viruses, and the urophages, which are always naked, make a ninth group.

Within each of these groups, families of similar viruses can be defined, primarily on the basis of size: for cubic viruses, by the triangulation number of the capsid, and for helical viruses by the diameter of the helix. These families are convenient conceptual groups, for the viruses in each family have very similar nucleocapsids, nucleic acids, and patterns of biosynthesis. However, they do not necessarily have the same host; the viruses of plants, animals, and bacteria may all be lumped in the same group. Most of these families, with their chief characteristics, are listed in Table 20-1. Some of these families are so important and familiar that it is worthwhile to single them out.

TABLE 20-1

major groups of viruses

Capsid dimensions in nanometers. Nucleic acid is represented by color: lines for open molecules and circles for ring molecules, either double- or single-stranded. Molecular weights in millions of daltons.

DEOXYHELICAL VIRUSES

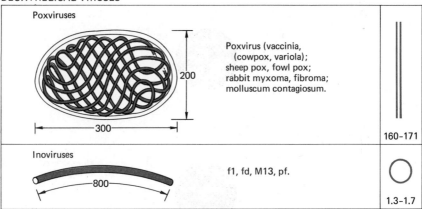

| Poxviruses | Poxvirus (vaccinia, (cowpox, variola); sheep pox, fowl pox; rabbit myxoma, fibroma; molluscum contagiosum. | 160–171 |
| Inoviruses | f1, fd, M13, pf. | 1.3–1.7 |

DEOXYCUBICAL VIRUSES

Microviruses $T = 1$	ΦX 174, S13.	1.7
Parvoviruses $T = 3$	Kilham rat virus (KRV); minute virus of mice (MVM); H1, H3, X14.	1.5
Papillomaviruses $T = 7$	Shope papilloma; human papilloma (warts); polyoma; SV 40.	2.5–4.2
Adenoviruses $T = 25$	Human adenoviruses 1–28; simian, bovine, canine, murine, and avian adenoviruses.	10.0
Iridoviruses $T = 81$	Insect iridescence viruses: Tipula (TIV), Chilo (CIV), mosquito (MIV), Sericesthis (SIV).	156
Herpesviruses $T = 16$	Herpes simplex, H. zoster (varicella); pseudorabies; equine abortion virus (EAV); equine herpes 2 (LK).	54–93

RIBOHELICAL VIRUSES

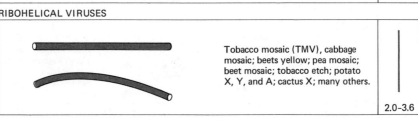

| | Tobacco mosaic (TMV), cabbage mosaic; beets yellow; pea mosaic; beet mosaic; tobacco etch; potato X, Y, and A; cactus X; many others. | 2.0–3.6 |

TABLE 20-1
(*continued*)

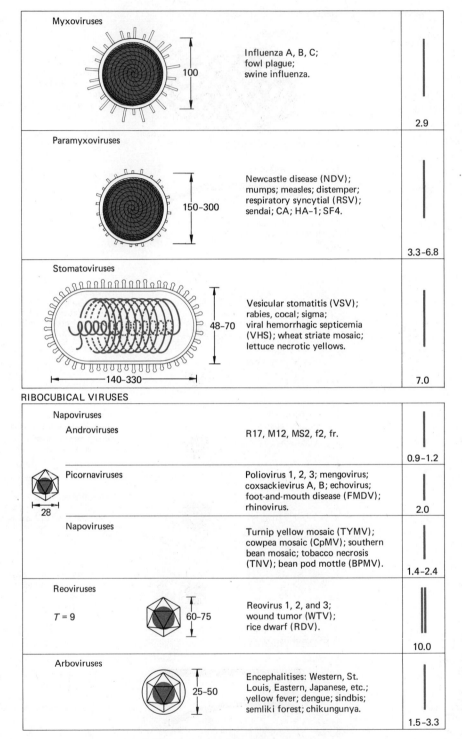

Myxoviruses	Influenza A, B, C; fowl plague; swine influenza.	2.9
100		
Paramyxoviruses	Newcastle disease (NDV); mumps; measles; distemper; respiratory syncytial (RSV); sendai; CA; HA–1; SF4.	3.3–6.8
150–300		
Stomatoviruses	Vesicular stomatitis (VSV); rabies, cocal; sigma; viral hemorrhagic septicemia (VHS); wheat striate mosaic; lettuce necrotic yellows.	7.0
48–70 / 140–330		

RIBOCUBICAL VIRUSES

Napoviruses		
Androviruses	R17, M12, MS2, f2, fr.	0.9–1.2
Picornaviruses	Poliovirus 1, 2, 3; mengovirus; coxsackievirus A, B; echovirus; foot-and-mouth disease (FMDV); rhinovirus.	2.0
28		
Napoviruses	Turnip yellow mosaic (TYMV); cowpea mosaic (CpMV); southern bean mosaic; tobacco necrosis (TNV); bean pod mottle (BPMV).	1.4–2.4
Reoviruses $T = 9$	Reovirus 1, 2, and 3; wound tumor (WTV); rice dwarf (RDV).	10.0
60–75		
Arboviruses	Encephalitises: Western, St. Louis, Eastern, Japanese, etc.; yellow fever; dengue; sindbis; semliki forest; chikungunya.	1.5–3.3
25–50		

The **poxviruses** have large, oval virions in which a double-stranded DNA in a very flexible nucleocapsid is wound in a very characteristic pattern, as shown in the table and Fig. 20-5. They generally cause lesions (pocks) on the epidermis of their animal hosts. The most famous of them is smallpox (variola). Edward Jenner's observation that milkmaids who had had the related disease cowpox did not develop smallpox laid the foundations for the modern technique of vaccination, in which a mild, local infection is used to induce antibodies against the more serious disease. *Vaccinia,* the virus now used for vaccinations, is still immunologically related to the smallpox virus, but it has been derived from cowpox by a gradual process of evolution over the years.

The **herpesviruses** (Fig. 20-6) include herpes simplex, which causes fever blisters and cold sores around the mouth, as well as more serious lesions. Herpes zoster is the cause of chickenpox (mostly in children) and shingles (mostly in adults), a painful, incapacitating disease of the skin and nerves.

The **microviruses** (Fig. 20-7) have the smallest known virions ($T = 1$). This group includes the bacteriophage ΦX 174 and S13, which have a single-stranded, circular DNA and have been studied very intensively.

The **papillomaviruses,** which have a double-stranded, circular DNA, all cause some kind of tumor, and for this reason they are very interesting as models for the involvement of viruses in the induction of cancer. One group consists of the papilloma or wart viruses, including human warts virus, which causes the common skin growths in man, and Shope papilloma, which produces dark skin growths in the common cottontail rabbit. The second group includes the polyoma viruses that produce widespread, generally fatal tumors in mice and hamsters.

FIGURE 20-5

Orf virus, one of the poxviruses, negatively stained to reveal the arrangement of the nucleocapsid inside the peplos. The interpretation of this picture is shown in Table 20-1. (Courtesy of Robert W. Horne, reprinted from Virology, **23,** *1964, by permission of Academic Press.)*

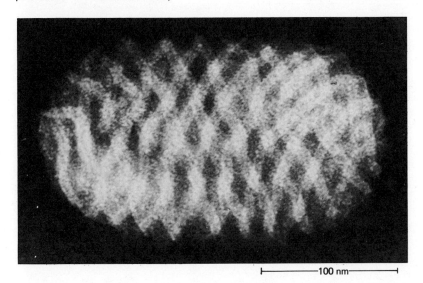

├───────100 nm───────┤

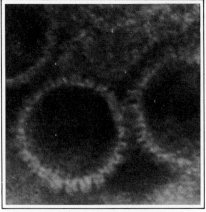

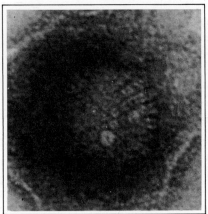

(a)

(b)

FIGURE 20-6

Herpes simplex virus. (a) Negative staining reveals the arrangement of capsomeres and shows that they are hollow. (b) A nucleocapsid inside its irregular peplos. (c) Model of a T = 16 capsid for comparison. (Courtesy of Robert W. Horne, reprinted from Virology, **12,** *1960, by permission of Academic Press.)*

FIGURE 20-7

Electron micrograph of phage ΦX 174, compared with a model of a T = 1 capsid. (Courtesy of Robert W. Horne, reprinted from Virology, **15,** *1961, by permission of Academic Press.)*

├─10nm─┤

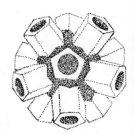

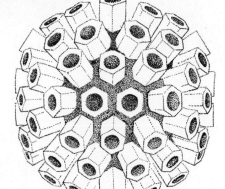

(c)

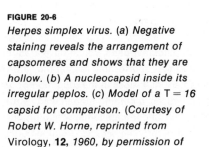

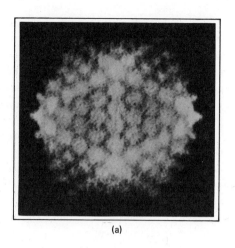

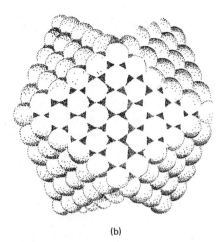

(a)

(b)

FIGURE 20-8
*Structure of adenovirus. (a) A negatively stained
virion that compares perfectly with the T = 25
model in (b). (c) Another preparation shows
that each pentad unit is connected to a rod and
ball. (Photographs by R. C. Valentine, courtesy
of H. G. Pereira.)*

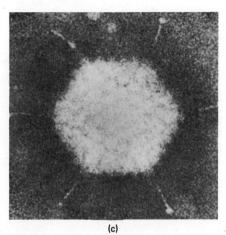

(c)

The **adenoviruses** (Fig. 20-8) are all too familiar; they are responsible for a variety of respiratory diseases that would generally be classified as common colds, with their miserable fevers, catarrhs, etc. Many of them also produce conjunctivitis of the eye, and at least three types can cause tumors in hamsters.

The **iridoviruses** are extremely large viruses ($T = 81$) that produce insect diseases known as iridescences, because the body fluids of the insects become so filled with mature virions that they become milky and iridescent.

The naked ribohelical viruses are all plant parasites that produce an enormous variety of diseases. For example, there are *mosaics,* in which the leaves of the plant take on a mosaic pattern of coloring and the whole plant becomes stunted. *Vein diseases* appear primarily as discolorations about the leaf veins, through which the viruses pass very easily, while in *ringspot diseases* each focus of infection appears as a little discolored ring.

The **myxoviruses** are also familiar enemies; the family includes the influenza viruses whose effects are all too well known to almost everyone. There are several groups of these viruses, each characterized by a different pattern of antigens. Every few years, a new mutant appears with antigens against which no one has any antibodies; the new strain spreads rapidly and causes a severe epidemic. For example, one type of influenza A virus was common from 1934 until about 1946. Then an A_1 strain appeared and spread rapidly through the population; this remained dominant until about 1957, when the A_2 ("Asian flu") strain appeared. New mutants and new epidemics can be expected at any time.

The related **paramyxoviruses** are larger; the group includes viruses of measles, mumps, dog distemper, and a variety of pneumonias. The **stomatoviruses** (Fig. 20-9) have an interesting bullet-shaped peplos with the nucleocapsid wound up inside. This family includes the viruses of rabies, vesicular stomatitis of horses and cattle, and several plant diseases.

The **arboviruses** were originally characterized as those that were *arthropod borne*,

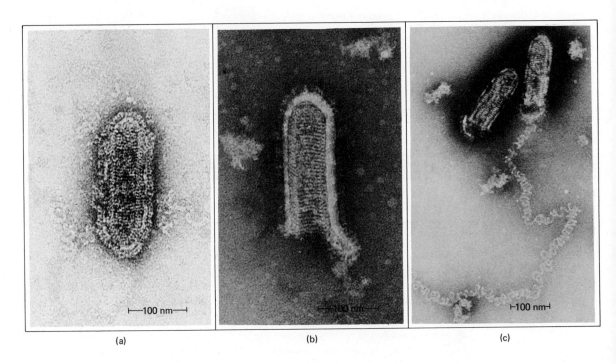

|———100 nm———| |———100 nm———| |—100 nm—|

(a)   (b)   (c)

FIGURE 20-9

Structure of vesicular stomatitis virus (VSV). (a) An intact virion strongly penetrated by phosphotungstate stain to reveal the fine striations. (b) A virion that has broken open and released its nucleocapsid, but that shows internal striations and surface projections. (c) A virion releasing its long, flexible, and highly coiled nucleocapsid. (Courtesy of Robert W. Simpson, reprinted from Virology, **29**, 1966, *by permission of Academic Press.)*

because they are carried from one host to another by arthropods such as mites, ticks, and mosquitoes; however, this is no longer a definitive characteristic, since many viruses are carried in this way. The group includes yellow fever, equine and human encephalitis, and dengue fever.

The **napoviruses** include many plant parasites and many small bacteriophage. The **picornaviruses** include poliovirus and its relatives, which also produce paralyses, as well as foot-and-mouth disease of cattle and the rhinoviruses, which cause many human colds. The **reoviruses** are unique in having double-stranded RNA; they produce respiratory and enteric diseases of mammals. The group also includes wound tumor virus, which produces tumors on plants at places where they have been injured.

B. Patterns of biosynthesis

20-3 INITIATION OF INFECTION

Since the nucleic acid of a virus is its most important component, the initial stages of infection must be directed toward introducing a naked nucleic acid molecule into a cell where it can replicate and direct the synthesis of all the proteins needed for multiplication. The infection process can be understood for each virus in relation to the characteristics of the host cell.

Plant cell walls are heavy, rigid layers of cellulose and other polysaccharides. It is difficult to break through them, and all plant viruses depend on some other agent to damage the cell wall enough so they can enter the wound. The damage may be as simple as a little scratch induced by a nearby branch blowing in the wind. One standard method of infecting plants in the laboratory is to simply rub a solution of viruses over the leaf with a slightly abrasive rod, which causes irritations and allows viruses to enter and initiate local lesions.

However, other plant viruses have more complex life cycles that depend upon insects or other animals that act as **vectors**—that is, intermediate organisms that carry the virus from one host to another. Many insects feed by sucking plant juices; viruses may establish themselves in the insect's mouth parts and be introduced directly into the plant tissue while the animal feeds. Aphids are probably the most important single group of virus-transmitting insects. Nematodes, the tiny round worms so common in soils, are also very important vectors.

Bacteria also have rigid walls, but they are not as heavy and impervious as plant cell walls. As we have already noted for phage T4, the uroviruses have special mechanisms in their tails for breaking through the cell wall and injecting their DNA directly. The sheath proteins of the T4 tail are contractile; when the phage is properly adsorbed, the sheath contracts, and this somehow initiates DNA injection, although the forces at work are not really known.

Two other groups of phage use a different entrance; they can only infect male bacteria that have sex pili, such as F^+, Hfr, and some colicinogenic strains. The little androviruses, like f2, adsorb all along the sides of the pili, while the inoviruses, like fd, adsorb to their ends—they are so thin that they look like extensions of the pili (Fig. 20-10). These viruses are very specific, and pili can be separated on the basis of the types that adsorb to them into two classes—F pili made by F factors and

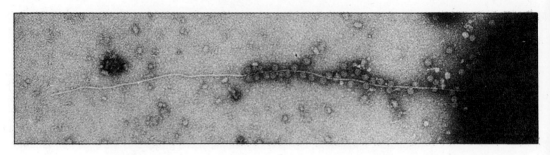

FIGURE 20-10
F pili of an E. coli *cell with the long helical phage f1 attached to its tip and the spherical phage f2 attached all along the sides. (Courtesy of Lucien G. Caro.)*

their relatives, and I pili made by Col I and its relatives. The androviruses probably inject their RNA through the pilus, but the inoviruses may act like female cells in a mating; that is, they may induce retraction of the pilus back into the male (by depolymerization of the protein, perhaps) so the whole virion is pulled inside the host cell. This is the only known case in which the capsid protein of a phage gets inside its host.

Animal cells have thin, flexible surfaces, with relatively little polysaccharide covering the membrane. They can take in many little particles by endocytosis, and animal viruses take advantage of this fact by using the old "Trojan horse" trick to gain entrance. Most animal viruses seem to adsorb to the cell surface like any food particle; they induce endocytosis and are carried into an endocytic vesicle (Fig. 20-11). However, once inside they behave like Greeks rather than food particles, and take over the cell. Invasion in this way is called *viropexis.*

The specificity of infection by bacterial and animal viruses, at least, resides primarily in the adsorption of the virion to specific cellular receptor sites. Phage like T4 form a specific association between their tail fibers and receptor sites on the bacterial surface; this positions the phage for injection of its DNA. Adsorption is generally dependent upon divalent cations, and occasional phage require other factors; for example, one strain of T4 requires tryptophan. Under the proper conditions, adsorption occurs rapidly and irreversibly. However, under other conditions of temperature and concentrations of critical ions it can be shown that there is a less specific, reversible adsorption condition that may precede irreversible adsorption or compete with it.

Animal viruses also recognize their host cells by means of specific sites. Differences between the surfaces of different cells in an animal are important factors in determining the type of tissue that will be infected, and it is not hard to understand why some viruses attack only nervous tissue while others attack the epithelium of the respiratory tract. It is probably sufficient that an animal virus find a site where it can induce viropexis; it does not have to adsorb in any specific position like a phage. The concentrations of divalent cations are again important. The adsorption sites on animal cells frequently contain a neuraminic acid, and pretreatment of sensitive cells with neuraminidase will often make them insensitive. Receptor-destroying enzymes are found in many fluids, such as egg albumin and urine.

The specific association between a virus and a sensitive cell is a kind or paratactic recognition that is not basically different from the association of antigen and antibody. Many viruses can be identified and characterized by using this property if some type of red blood cell can be found to which they are paratactic. When viruses and appropriate red blood cells are mixed, they will associate and precipitate, a process called **hemag-**

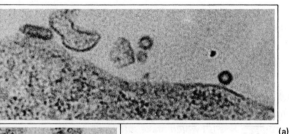

(a)

FIGURE 20-11
Rabies virus invading its host cell. The virions attach to the cell surface (a) and cause an invagination (b) which deepens (c) and finally pinches off (d), leaving the viruses in an endocytic vesicle (e) where they will be uncoated and can start their multiplication. (Courtesy of Klaus Hummeler, reprinted by permission from J. Virology, 1, 1967.)

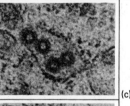

(b)

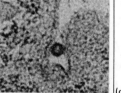

(c)

(d)

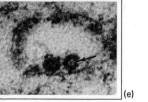

(e)

glutination. If two viruses agglutinate the same type of red blood cells, this may be taken as evidence of their relatedness. The specific protein on the virion that contains the paratypic site is called a **hemagglutinin** (HA).

exercise 20-5 Myxoviruses carry both a hemagglutinin and a neuraminidase. What would you expect to see if myxoviruses are mixed with appropriate red blood cells and the suspension is observed over a period of several hours?

exercise 20-6 How can you use the hemagglutination test to determine if an animal has antibodies to a hemagglutinating virus in its blood serum? Assume that at least some of the antibodies are paratactic to the HA epitopes of the virus.

The nucleic acid injected by a bacteriophage is naked and presumably capable of beginning its own functions immediately. However, a plant or animal virus is still largely intact when it enters the host cell, and the first events in their multiplication must *uncoat* them and release an infectious nucleic acid. Uncoating generally takes a few hours; where it has been possible to initiate an infection with a purified nucleic acid derived from the whole virus, the typical patterns of viral biosynthesis begin a few hours earlier, since the initial step of uncoating has been bypassed.

Werner K. Joklik has studied the uncoating of pox viruses (vaccinia) very thoroughly. The virus gets into its host cell by endocytosis, and its peplos is soon removed by host enzymes. This releases some kind of inducer present in the virion, which migrates into the nucleus and induces the synthesis of an uncoating protein, informed by host genes. This protein migrates back to the virion and releases the genophore from its nucleocapsid. As you would expect, actinomycin D completely inhibits uncoating. However, this complicated uncoating process is unique, and most virions are probably opened up much more simply by the action of lysosomal enzymes.

Once the nucleic acid is exposed inside the cell, it has to replicate and start to direct the synthesis of new proteins. We can identify three general patterns of replication and biosynthesis associated with the three principal types of genophore found in viruses: single- and double-stranded DNA and single-stranded RNA. (The double-stranded RNA viruses are still rather obscure.) We will now outline these patterns. Two generalizations can be made here. Every DNA virus apparently requires some special DNA polymerase for its replication, and it may also require a special

FIGURE 20-12

The pattern of multiplication in ΦX 174. An infective plus strain of DNA (1) is converted into a replicative form (RF) and attached to a site, probably on the cell membrane. This RF produces secondary RFs (2) which make mRNA (3) for the synthesis of phage proteins. The unique RF synthesizes new plus strands (4), which are wrapped up in protein to make new virions.

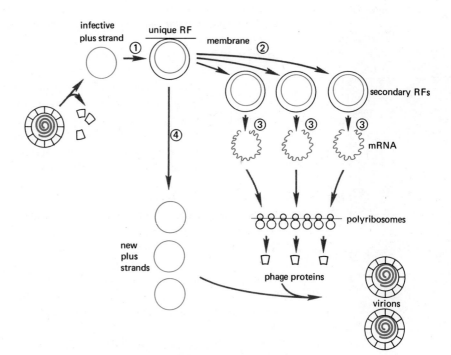

DNA-dependent RNA polymerase. An RNA virus also requires at least one new polymerase for its replication, and we now know that certain RNA viruses which cause tumors make an RNA-dependent DNA polymerase to make copies of their genomes in DNA. However, these tumor viruses are exceptional; in most RNA viruses, there is only RNA, which serves both as a genophore and as a messenger for protein synthesis.

20-4 THE SINGLE-STRANDED DNA VIRUS PATTERN

We will first consider the pattern of synthesis in viruses like ΦX 174 (Fig. 20-12), which has been studied very extensively by Robert Sinsheimer and his group. The virus has a circular, single-stranded DNA which we will call the *plus* strand; it is infectious by itself on *E. coli* spheroplasts, and during a normal infection it somehow is injected into the host cell.

The plus strand is immediately converted into a double-stranded, circular **replicative form** (RF) by synthesis of a complementary *minus* strand. This must be done by a host enzyme, since the conversion is not blocked by inhibitors of protein synthesis. This RF then attaches to a site inside the cell, presumably on the cell membrane, and begins to replicate in the usual semiconservative way to make a small population of RFs. However, the one RF containing the original plus strand always remains attached to its site, and this one is unique and important.

RF synthesis ceases after a few minutes, and at about the same time several phage proteins are evidently made, because host DNA synthesis stops and the infected cell becomes immune to superinfection by more phage. It appears that only the unique RF attached to the site can make other RFs; the latter, which must be free-floating, seem to have no function except to make mRNA. Among the proteins informed by these messengers there is one polymerase that can make single-stranded DNA from a double-stranded template.

Using the latter enzyme, the unique, attached RF begins to make a population of new plus strands of DNA, which are the genophores of new phages. As fast as these plus strands are made, they are attached to new phage capsid protein and become mature phage particles.

exercise 20-7 Pulse-labeled RNA in ΦX-infected cells will hybridize with denatured RF DNA but not with viral DNA. Which strand is the mRNA made from?

20-5 THE SINGLE-STRANDED RNA VIRUS PATTERN

Many studies on the small RNA phage, the picornaviruses, and even on the larger myxoviruses all lead to a fairly consistent pattern of replication, although, of course, some variations must be expected. A general model is shown in Fig. 20-13.

The injected plus strand of RNA must play two roles: messenger and genophore. Host cells have no polymerases that can act on single-stranded RNA, so the plus strand must first attach to host ribosomes and form a polyribosome which makes an RNA polymerase and certain other proteins, including inhibitors of cellular RNA and protein synthesis.

We will call this initial enzyme a type-I RNA polymerase; its function is to make a

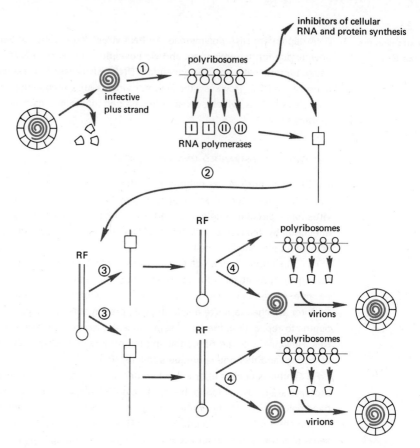

FIGURE 20-13

The pattern of multiplication in a single-stranded RNA virus. An infective plus strand of RNA (1) serves as mRNA for the synthesis of virus proteins, including RNA polymerases. Polymerase I converts the plus strand into an RF (2), from which additional plus strands are made by polymerase II (3). These new plus strands may also be made into RFs. Some of the plus strands are again used as messengers (4) while others are wrapped up as the genomes of new virions.

minus strand on the plus strand and thereby convert it into a replicative form (RF). Whether the same plus strand can serve first as a messenger and then later become part of an RF, or whether several plus strands must infect together, we don't know. In any case, there must be a second polymerase, which we will call type II; its function is to make new plus strands from the RF. The greatest controversy in this field lies right here, and at present it is unresolved. Synthesis of new plus strands from the RF is certainly asymmetrical (in that no new minus strands are made), but it is hard to tell whether it is conservative or semiconservative. The experiments designed to answer this question generally depend upon the transfer of density labels from one fraction to another in replication, à la Meselson and Stahl. Some of the results are very hard to interpret.

These experiments are complicated by the existence of a **replicative intermediate** form (Fig. 20-14); it seems clear that several polymerase II molecules may be working along an RF simultaneously, so they create a form that has "tails" sticking off. (Remember Fig. 12-17, showing the tails of rRNA coming off the nucleolar organizer.) It is hard to know, at this point, whether one plus strand remains with the minus strand of the RF and is merely displaced slightly (conservative replication) or whether each molecule displaces the one ahead of it (semiconservative replication).

Regardless of the details, many new plus strands are made; they serve as messengers for making new proteins and as genophores, which combine with these proteins to make new virions.

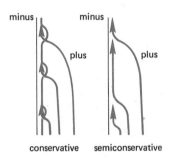

minus   minus

plus   plus

conservative   semiconservative

FIGURE 20-14
The presumed structures of replicative intermediates, on the semiconservative or conservative models.

20-6 THE DOUBLE-STRANDED DNA VIRUS PATTERN

Because the extensively studied T-even phage T2, T4, and T6 fall into this class, we probably know more about this pattern of replication than about any other. Of course, this also makes it necessary to pick and choose major topics from among an enormous amount of published information.

We have ignored the fact, until now, that the DNA of the T-even phage contains 5-hydroxymethylcytosine (5-hmC) instead of ordinary cytosine and that most of the 5-hmC residues in the DNA are bonded to glucose molecules. There are other phage that have strange bases such as 5-hydroxymethyluracil linked to mannose residues. Furthermore, all of these phage require special polymerases, both for DNA and RNA,

α or β linkage

dhmCMP   glucose

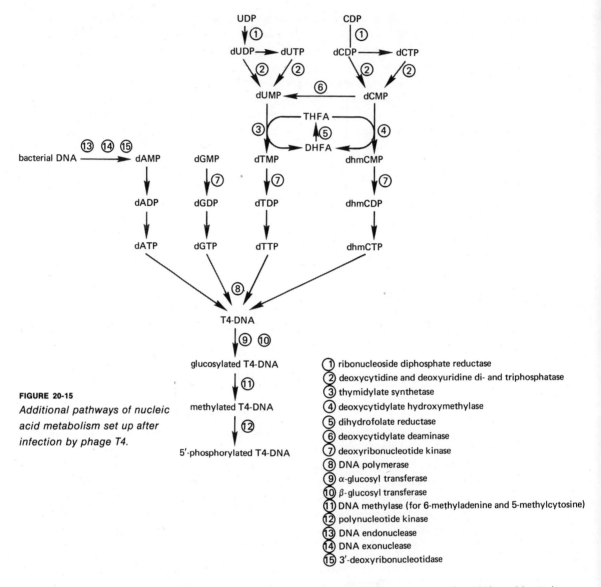

FIGURE 20-15

Additional pathways of nucleic acid metabolism set up after infection by phage T4.

① ribonucleoside diphosphate reductase
② deoxycytidine and deoxyuridine di- and triphosphatase
③ thymidylate synthetase
④ deoxycytidylate hydroxymethylase
⑤ dihydrofolate reductase
⑥ deoxycytidylate deaminase
⑦ deoxyribonucleotide kinase
⑧ DNA polymerase
⑨ α-glucosyl transferase
⑩ β-glucosyl transferase
⑪ DNA methylase (for 6-methyladenine and 5-methylcytosine)
⑫ polynucleotide kinase
⑬ DNA endonuclease
⑭ DNA exonuclease
⑮ 3′-deoxyribonucleotidase

and they must enhance certain enzyme activities not present in uninfected hosts in sufficient amounts to support their biosynthesis. Figure 20-15 shows the new and enhanced pathways that phage T4 must set up before any DNA replication can begin.

Other viruses with ordinary DNA also must make new enzymes. For example, many animal viruses invade host cells that are not making DNA and have no thymine kinases; they must therefore carry information for making their own. In every case, new DNA polymerases seem to be essential.

The synthesis of these new enzymes begins as soon as the viral DNA is injected, in the phage, or is uncoated, in other viruses. Presumably cellular RNA polymerases are used for mRNA synthesis. However, in every case the level of the new enzymes rises to a maximum and then levels off. In T4-infected *E. coli,* all of these early enzymes are turned off at about 12 minutes; in the animal viruses, their synthesis ceases after a few hours. The reasons for this pattern are unknown, but it now appears that during phage T4 multiplication the host RNA polymerase is modified in various ways and that new sigma factors are made in a regular sequence to restrict the polymerase to transcription of limited sets of phage genes at different times.

Following this period of early enzyme synthesis there is a synthesis of late proteins, mostly capsid proteins. This also seems to depend upon DNA synthesis, at least in T4; this fact suggests that among the late proteins there is some inhibitor of early enzyme synthesis. In any case, the cell rapidly accumulates a pool of capsid proteins and a pool of new DNA. Maturation of new phage virions begins soon, as new DNA molecules are pulled out of their replicating pool and combined with new proteins. This maturation process is among the most fascinating of all topics in virus research, and we will now turn to it.

20-7 NUCLEOCAPSID ASSEMBLY

The first demonstration of virus self-assembly was achieved by Heinz Fraenkel-Conrat and Robley C. Williams, who dissociated tobacco mosaic virus into its RNA and protein and then showed that they could recreate the whole infectious virus by incubating the two fractions together. They also made the interesting observation that the protein by itself will form rods that look like ordinary virus nucleocapsids, but have all different lengths. The reconstituted virus, on the other hand, consists of uniform rods. It is assumed that the RNA serves as a measuring stick; protomers associate along the RNA until it is entirely covered and then polymerization stops, so if the RNAs are uniform, the nucleocapsids will be uniform.

The principal product of dissociation induced by mild alkali (pH 10) is the A protein, which is a trimer of the elementary protomer. It can be dissociated into its subunits at pH 13 or at very high dilutions. As the pH is lowered (and in general as the ionic strength of the solution increases), polymerization is enhanced. The A protein appears to be the first intermediate stage in nucleocapsid assembly; from the trimer, larger units are built up, including one of about 32 subunits that is a disc of two layers, a form often observed in microscopic studies. Polymerization is also enhanced by raising the temperature; this suggests that the process is strongly entropy driven (as the term $T \Delta S$ increases), but this appears to be true of most self-assembly processes. Several theories about the precise bonding patterns between protomers have been evolved, but the problem is still too unsettled for us to consider these in more detail.

The assembly of several other viruses has now been studied. One interesting case is that of the RNA phage R17. This little virus has a small genome with only three cistrons; one of them (B) codes for the major capsid protein (13,800 daltons) and another (A) for a special protein (35,000–40,000 daltons). Capsids are formed by amber mutants of the A cistron, but they are noninfectious and do not adsorb to host cells. These noninfectious capsids can also be reconstituted from B protein alone; when A protein is added to the mixture, fully infectious viruses are formed. There is probably only one A protein per capsid.

John P. Bancroft has done an interesting study of cowpea chlorotic mottle virus (CCMV) and brome mosaic virus (BMV), which are the same size but infect very different hosts. The two can be dissociated and reconstituted by themselves, and hybrids can be made from the protein of one and the RNA of the other. The hybrids always have the host specificity of their RNA; thus, the attachment of the virus to its host cell membrane is not critical in these viruses, whereas intracellular events afterward are very critical. (Remember that plant viruses apparently always infect via wounds in their host cells.)

One of the most interesting problems in virus assembly is being studied with phage T4. Figure 20-16 shows the present map of T4 with genes defined primarily by amber and temperature-sensitive mutants. Most of these appear to be "late" genes whose function is to specify capsid structures or to aid in capsid assembly. After Edgar's group had assembled all of these mutants, William A. Wood tried the daring experiment of combining an extract from cells infected with a "headless" mutant with one from cells infected with a "tailless" mutant. He found a rapid assembly of mature, infectious virions, and thus opened the door to a program of *in vitro* complementation studies. Wood, Edgar, and their students have been able to draw out a biosynthetic pathway for T4 assembly (Fig. 20-17) by combining extracts from bacteria infected with different defective phage. Relatively little of this pathway has been studied in detail. Jonathan King has shown that the core-endplate unit is converted into

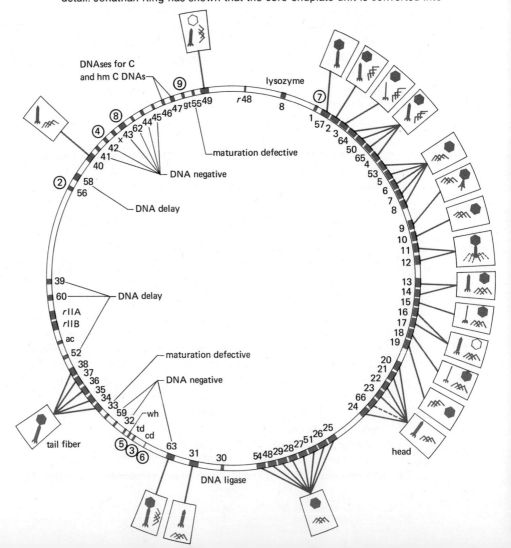

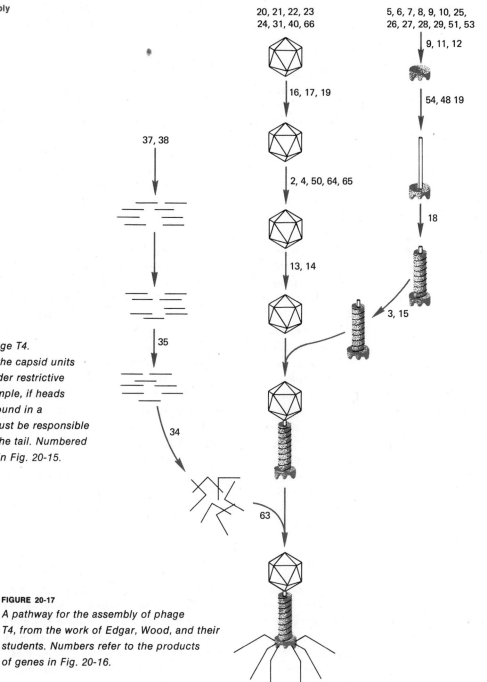

20, 21, 22, 23
24, 31, 40, 66

5, 6, 7, 8, 9, 10, 25,
26, 27, 28, 29, 51, 53

9, 11, 12

16, 17, 19

54, 48 19

37, 38

2, 4, 50, 64, 65

18

13, 14

3, 15

FIGURE 20-16
Edgar's map of phage T4.
The pictures show the capsid units
found in lysates under restrictive
conditions; for example, if heads
and tail fibers are found in a
mutant, the gene must be responsible
for making part of the tail. Numbered
enzymes are those in Fig. 20-15.

35

34

63

FIGURE 20-17
A pathway for the assembly of phage
T4, from the work of Edgar, Wood, and their
students. Numbers refer to the products
of genes in Fig. 20-16.

a mature tail by the products of genes 18, 15, and 3. Gene 18 apparently informs the structural gene for the sheath protein; genes 15 and 3 must then act to stabilize the sheath on the core.

This raises the interesting question of how assembly of such complicated structures proceeds. Edouard Kellenberger has distinguished several orders of morphopoiesis (assembly). The simplest phage, as described above, apparently exhibit only *first-order* interactions; their components contain all of the necessary information within their own structure, and under the right conditions they simply fall into place. However, phage like T4 may require *second-order* interactions and special factors; for example, a critical gene may not specify any protein that becomes a part of the finished capsid, but its product may be an enzyme that has to modify one capsid unit before it can attach properly. Other gene products may form some rather small, nonprotein unit that fits between two proteins and holds them together. The T4 system presents an opportunity to explore some of these possibilities; we can expect some very exciting results to come from it.

One characteristic of nucleocapsid assembly is its sensitivity to errors. Bancroft has shown that changing the medium in which his nucleocapsids assemble can make them take on a variety of abnormal forms, such as elongate spheres that contain too many structure units but still conform to the basic rules about the architecture of closed shells. In making models with hexad and pentad units, it is not hard to make forms that look just like the abnormal objects seen in the electron microscope. Edgar and Epstein also observed that in some of their mutants abnormal structures form intra-cellularly. In some mutants, "polyhead" structures that look like enormously long tubes of head protein can be seen; other mutants make long "polysheaths." In both cases there is presumably something missing that would normally terminate the assembly of these structures once they have reached a certain length.

Finally, abnormal, noninfectious capsids are found in many normal infections. When the viruses are centrifuged through a sucrose gradient, the infectious virions typically form a single band while one or two "top components," which move more slowly, can be separated out. These often look like empty capsids that have formed without any nucleic acid core. Most virus infections are inefficient in the sense that many capsid units are always produced that are never assembled or that assemble themselves into useless structures.

20-8 VIRUS MATURATION AND RELEASE

If a virus particle is simply a nucleocapsid, then it is "mature" once it has been assembled. The accumulation of mature virions inside a cell looks very much like crystallization; Fig. 20-18 shows one stage of the process in T4 maturation, where dense DNA cores have been pulled from the replicating pool and crystallized as immature heads. It looks as if capsid head protein forms around these cores; as Fig. 20-17 indicated, the tails are apparently attached later. After the cell is largely filled with mature particles, its lipids start to break down, a lysozyme begins to attack the murein wall structure, and soon the cell lyses.

The maturation of a naked plant or animal virus is no more complicated. Figure 20-19 shows a polyomavirus forming a similar crystalline mass in the nucleus of an animal

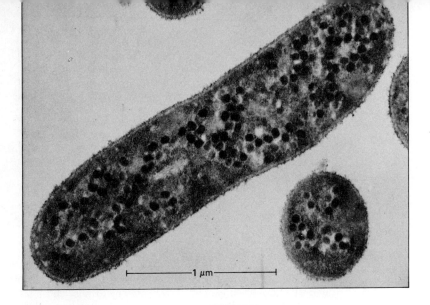

1 μm

FIGURE 20-18
E. coli *cells about 10 minutes after infection
with phage T4. The phage DNA is
crystallizing out and starting to form mature
virions. (Courtesy of Janine Sechaud.)*

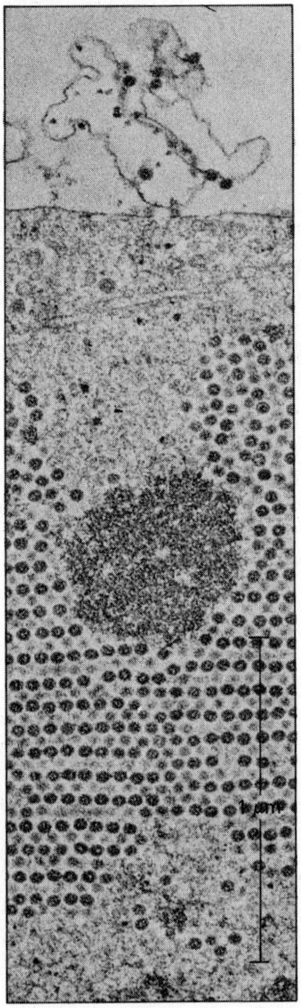

FIGURE 20-19
*Crystalline arrays of the papillomavirus SV 40
forming inside the nucleus of a tissue culture
cell. (Courtesy of Councilman Morgan, reprinted
by permission from J. Virology,* **1,** *1967.)*

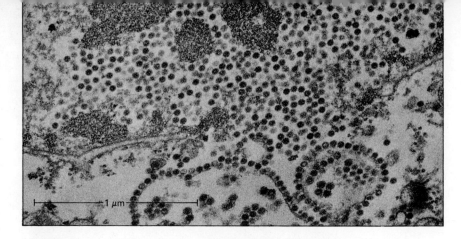

FIGURE 20-20

*A later stage in SV 40 infection. The nuclear envelope has broken and virions are pouring out into the cytosol where some are aligning on fragments of membrane. (Courtesy of Councilman Morgan, reprinted by permission from J. Virology, **1**, 1967.)*

cell. Eventually, the nuclear envelope ruptures and mature virions start to pour out (Fig. 20-20); the cell membrane also disintegrates, and infectious virus particles can be found in the surrounding fluid or the blood.

It should be pointed out that virus protein synthesis in plant and animal cells always occurs on the ribosomes of the ER, just as in uninfected cells. However, every virus has a characteristic place, either nuclear or extranuclear, where its nucleic acid replicates. The virus proteins then migrate to that point and assembly of virions begins; each virus typically assembles itself on some specific cell structure (nuclear envelope, ER, microtubules, etc.) and the complex of virions and other materials makes a characteristic **inclusion body** that can be seen with the light microscope and can often be used to identify the virus.

If the virus has a peplos, then maturation becomes a little more interesting. The nucleocapsid is assembled as usual; it must then acquire one or more layers of membrane from the host, and it always does so by pushing through one of the host

FIGURE 20-21

*Maturation of parainfluenza virus type 2 at the surface of an infected cell. The black dots on the membrane surface are ferritin-conjugated antibodies. Cross sections through helical nucleocapsids can be seen inside some of the budding viruses. (Courtesy of Councilman Morgan, reprinted by permission from J. Virology, **1**, 1967.)*

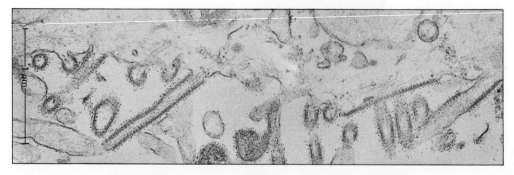

membranes in a process like endo- or exocytosis. One of the simplest cases of this is seen in the maturation of myxoviruses, as shown in Fig. 20-21. The nucleocapsids have already been formed; they are probably long, random coils with little definitive structure aside from their internal helical form, and in this condition they migrate toward the cell membrane. The membrane has already been changed significantly; remember that myxoviruses contain specific antigens on their surfaces, including a hemagglutinin and a neuraminidase. These new proteins have already been added to the cell membrane during the time of viral protein synthesis; the significance of the

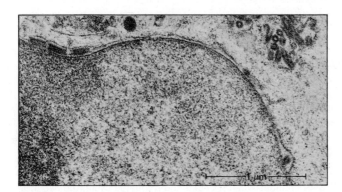

FIGURE 20-22
*Rabies virus particles lying between the layers of a nuclear envelope. (Courtesy of Klaus Hummeler, reprinted by permission from J. Virology, **1**, 1967.)*

subunit structure of membranes becomes apparent here, for it is easy to see now how the cell membrane can be altered by interpolating new virus-informed subunits within the existing structure, just as the membranes of bacteria can be changed when different cytochromes are induced. In the figure, these new antigens have been made visible by using antibodies bonded to ferritin, an iron protein that is very electron dense; these antibodies have made paratactic contacts with antigens on the cell surface, where nucleocapsids can be seen pushing out fingers of membrane to make mature virions. These fingers eventually pinch off and round up, and the virions coming off in the picture are already coated with antibody.

exercise 20-8 The fact that myxovirus antigens appear on the surface well before virions are released has been made the basis of the **hemadsorption** method for following virus synthesis. What would you see if you infected cells with myxovirus and at various times mixed a sample of cells with red blood cells that could be hemagglutinated by the virus? (Note: The red blood cells are much smaller than the host cells in these experiments.)

Other enveloped viruses acquire their peplos from internal membranes. The stomatoviruses, including rabies virus, characteristically mature in association with the nuclear envelope; virions can sometimes be seen between the two layers of the envelope (Fig. 20-22). The morphology of the virus is still not entirely clear, but it has at least one membrane and possibly two; it may therefore require two passages through the host membrane to complete its maturation. One passage may occur through the nuclear membranes; viruses are also seen passing through the membranes that surround vacuoles and accumulating there (Fig. 20-23), and they may mature at the surface like myxoviruses.

Viruses that mature by budding, rather than by causing the total breakdown of cellular membranes, may really do little damage to their host cells. The cells may continue to metabolize and proliferate normally for a long time. The inoviruses that infect male or piliated bacteria are somehow released continually from their hosts at the cell surface without any apparent damage, for the bacteria continue to grow. The virus may be assembled as it passes through the membrane, for no intracellular pool of mature virions can be detected.

C. Lysogeny and latency

20-9 LYSOGENY AND ITS REGULATION

A long and fascinating chapter could be written now about the regulation of temperate viruses, particularly because of recent research on phage λ. We will very briefly review some of this work here because of the light it sheds on the more complex regulatory systems we can expect to find in both viruses and uninfected cells.

Figure 20-24 shows the map of λ as it now stands. It is drawn as a circle to show how the injected DNA changes when it begins the lysogenic cycle; the two strands have short, complementary regions at their ends which pair with one another and are then sealed by a ligase that closes the DNA into a double-stranded, covalently bonded circle. The circle attaches at its *att* site to the corresponding bacterial site, and after the crossovers we drew in Fig. 9-6, the prophage becomes integrated.

When ordinary λ is plated on sensitive bacteria, it makes turbid plaques because many of the bacteria in the plaque region are lysogenized and do not lyse. Dale Kaiser isolated clear (*c*) mutants and showed that they fall into three cistrons (c_I, c_{II}, c_{III}) that all map close together in what is now called the **immunity region** (*imm,* around 3 o'clock on the map). Kaiser found that c_I mutants could not lysogenize at all, whereas c_{II} and c_{III} mutants could establish stable lysogeny at a low frequency. Moreover, it appeared that the products made by the latter two cistrons are required very early in lysogenization, but that the c_I product was required later. It is therefore assumed that the c_{II} and c_{III} products must act immediately after DNA injection to stop the phage from going into the lytic cycle and the c_I product is required for maintenance of stable lysogeny later. It is now clear the c_I codes for a repressor—what we called the "immunity substance" in Chapter 9. Mark Ptashne has succeeded in separating the repressor protein from other λ proteins and characterizing it.

A series of other temperate phages similar to λ can be separated into those that are **coimmune** with λ or **heteroimmune** with it. If two phage are coimmune, then the repressor made by each can repress lytic growth of the other; in other words, their repressors mutually recognize one another's "operators." If they are heteroimmune with one another, then each one has its own repressor-operator system. In principle, one might find that the A repressor represses B, while the B repressor does not repress A, but this never occurs. Kaiser and Jacob crossed λc mutants with the heteroimmune phage 434. They obtained phage with the immunity of λ when they used λc_{II} or λc_{III}

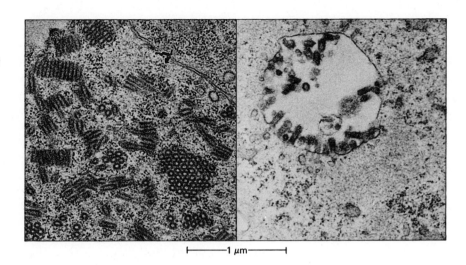

FIGURE 20-23
(a) *Rabies virus nucleocapsids filling a cell as they assemble themselves.* (b) *The viruses maturing by passing through a vacuole membrane.* (*Courtesy of Klaus Hummeler, reprinted by permission from J. Virology,* **1,** *1967.*)

—————1 µm—————

FIGURE 20-24
A map of phage λ. Arrows point in the direction of mRNA synthesis. b2 is the extent of a density mutation (deletion) that cannot lysogenize.

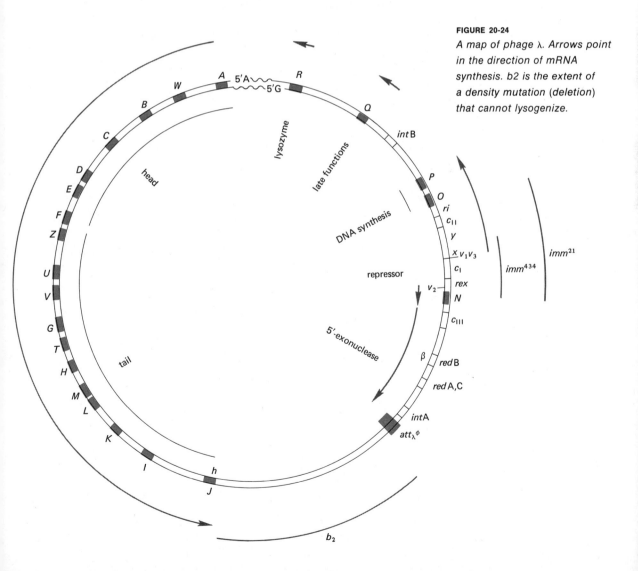

A 5′A
R 5′G
W
B
C
D
E
F
Z
U
V
G
T
H
M
L
K
I
J
h

head

tail

lysozyme

late functions

DNA synthesis

repressor

5′-exonuclease

Q
int B
P
O
ri
c_{II}
y
x $v_1 v_3$
c_I
rex
N
c_{III}
v_2
β
red B
red A,C
int A
$att_λ^{\phi}$

imm^{434}
imm^{21}

b_2

mutants, but not with λc_i mutants, indicating that the c_i region controls immunity. By a series of crosses, they obtained a hybrid phage, λhy or λimm^{434} that has all of the λ genes except for the one critical region where it carries the immunity of 434. This region is shown on the map, as is a similar immunity region for phage 21.

Jacob and Wollman described a virulent mutant of λ, λvir, that cannot lysogenize. There are actually three (apparent) mutations that can be separated into two sites, v_2 and $v_1 v_3$. Mark Ptashne and Nancy Hopkins have shown that the λ repressor binds poorly to DNA from these mutants. These are therefore presumed to be the operators on which the repressor normally acts.

Several genes (A through R, reading around the map) have been defined by amber mutants (originally called *sus* mutants—*s*uppressor *s*ensitive—in λ). By performing experiments similar to Wood and Edgar's, Jean Weigle showed that genes A through F inform the phage head units and genes Z through J inform the tail units; J itself informs the tail fibers. The interesting mutants associated with regulation of multiplication all map close to the immunity region. Gene N function is apparently required for synthesis of DNA and phage proteins. Genes O and P are required for DNA synthesis, and gene Q is required for the "late proteins"—the capsid proteins, lysozyme, etc.

The two strands of phage λ can be separated from one another, perhaps most easily by centrifugation in a CsCl gradient containing poly G, which binds to cytosine-rich regions of the DNA. Labeled messengers can be made at various times after infection and from mutants that lack various functions, and by hybridizing these messengers with appropriate DNA preparations one can determine which strands are transcribed in each region of the genome and at various times after the onset of lytic multiplication. (In stable lysogeny, only c_i appears to be transcribed.) For convenience, the DNA strands are designated L and R. The L strand is transcribed toward the *left* on the standard map, it is *light* in CsCl with poly G, and it has a 5'G nucleotide at its *left* end. The R strand is transcribed toward the *right*, it is *heavy* in the CsCl + poly G gradient, and it has a 5'A nucleotide on its *right* end (and the R gene on the right end is probably transcribed from it).

The principal genes expressed early during a lytic infection are in the region from

FIGURE 20-25

Conversion of Salmonella O-antigens by prophages. The nonlysogenic cell bears the epitope at the left. Infection by phage ϵ^{15} leads to a change from an α-linked to a β-linked galactose. If the cell is later infected by phage ϵ^{34}, a molecule of glucose is added to the epitope.

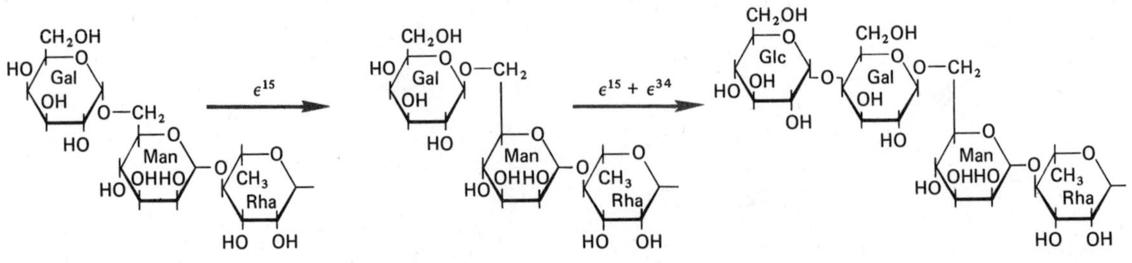

c_i to P and in the region from N toward *att.* As you can see, these regions are transcribed in opposite directions, and from the positions of the operator mutants v_2 and $v_1 v_3$ these might be considered simply operons under control of the repressor. Things are not quite this simple; the N gene appears to be a critical intermediate, for without N function there is no synthesis of anything else. A more appropriate model postulates that the v_2 operator controls N and that the N product, in turn, stimulates the transcription of the other regions.

Gene Q is somehow required for the transcription of all the late proteins; this may simply be a case of positive control by a Q activator. The importance of genes O and P may lie primarily in providing enough DNA to produce sufficient messenger for all the capsid protein needed to make a large burst. However, the details of regulation are still being determined. We present this here not to give the false impression that everything is settled—because the subject is really in a very exciting state of activity—but to illustrate the sort of complexity in regulation that one can expect in this kind of situation, where the choice between lysis and lysogeny is very critical.

20-10   POLYOMA AND LYSOGENIC CONVERSION

It is now well known that cells carrying stable episomes undergo a **conversion** to a condition in which they show many properties that are different from cells without episomes, often in ways not obviously related to the function of the episome itself. For example, Fig. 20-25 shows the changes that occur in the surface polysaccharides of *Salmonella* cells carrying certain temperate phages. There is no obvious reason why an episome should carry the information for such a conversion (immunity is an intracellular function, after all, not an inability of viruses to adsorb and inject). Another well-known example is that of *Corynebacterium diphtheriae,* the cause of diphtheria in man. The disease is produced only by strains that make a specific toxin, and these are strains that carry a temperate phage, which produces the same kind of conversion shown here for *Salmonella.*

Animal viruses such as polyoma behave very much like temperate phage. Polyoma can infect lytically and productively, yielding a normal burst of infectious virus after several hours. It can also infect nonproductively, so that instead of yielding viruses the host cells are **transformed** into a neoplastic or malignant condition. Monolayers of cells in tissue culture lose their contact inhibition and begin to proliferate, so each point of infection appears as a disorganized little pile of cells. In susceptible animals, of course, transformation produces tumors.

The interesting correlate of these facts is that the polyoma and papilloma viruses all have small, circular DNA genomes. Extending the analogy of phage lysogeny just a trifle, it is tempting to say that in transformed cells the viral genome has integrated itself into the host chromosomes just like a prophage. To probe this hypothesis, Thomas L. Benjamin measured the amount of labeled RNA in both productive and transformed cells that would hybridize with polyoma DNA. In productively infected cells, the amount of virus RNA increases substantially in parallel with multiplication, just as one would expect. In transformed cells, he detected a small amount of virus-specific RNA. Similarly, in cells transformed by the related virus SV 40 he found RNA that would hybridize with SV 40 DNA, but neither of these RNAs would hybridize with

DNA of the opposite virus. The amount of RNA is what would be expected from about 1 to 5 virus genomes per cell.

Virus-induced surface antigens can be detected in cells infected by polyoma virus, both productively and in transformation. These changes in surface structure may be important in releasing the cells from contact inhibition by their neighbors, although there are presumably intracellular correlates that permit the cells to start proliferating again.

The transformation of cells from a nonproliferating to a proliferating state is a common effect of viruses; many leukemias and other neoplasms can be attributed to specific viruses. We also know that some viruses can maintain themselves in a "latent" state for long times; for example, most people probably carry herpes simplex virus in a hidden condition, but they don't know about it until they undergo some kind of stress—including emotional stress—that allows the virus to come out of hiding and produce painful cold sores in and around the mouth. Now if viruses can exist in a state that is indistinguishable from that of a prophage in a bacterium, it becomes interesting to ask the more general question: How many proviruses is each of us carrying around in his own genome? It is well known that the tendency to develop certain types of cancer is hereditary; is this because families with these tendencies are really carrying proviruses that are inherited along with the rest of their genomes? These questions, in one form or another, are being asked seriously by people concerned with cancer. The answers are very likely to be affirmative.

Many agents have been linked to the production of tumors; these include x rays, ultraviolet light, and a long list of complex hydrocarbons. It seems quite clear that some agents found in cigarette smoke are also carcinogenic. Because of the similarity of some of these materials to those that cause induction of prophages, it is reasonable to ask whether they produce carcinomas by inducing specific proviruses in the sensitive cells. This is another promising line of inquiry.

The solutions to many problems of human biology will undoubtedly come from an application of techniques and concepts developed through research on very simple systems. The evolutionary link between bacteria and man may be rather remote, but if we are really establishing general principles of biology through research on primitive systems, we must expect these principles to apply to human biology as well.

20-11 SIGMA: A LATENT RIBOVIRUS

When most *Drosophila* are given moderate doses of CO_2, they become anesthetized for a while and then recover. Ph. L'Heritier and G. Teissier discovered a strain of *Drosophila* that is unusually sensitive to CO_2; these flies never recover from anesthesia. Sensitivity is now known to be associated with a virus called *sigma*, which is one of the stomatoviruses.

Sigma virus is transmitted from one generation of flies to the next rather like a nonchromosomal factor. A CO_2-sensitive strain may be *nonstabilized*, in which case the virus is only inherited through females, or *stabilized*, in which case a male can transmit the virus to a certain percentage of his progeny. Flies that receive sigma from their fathers are identical to those that receive the virus through an artificial injection. The males cannot transmit the virus to their progeny and the females produce only nonstabilized lines.

Injected sigma virus grows much more poorly in stabilized flies than in nonstabilized flies, as if cells carrying sigma are somewhat immune to further infection. Moreover, L'Heritier found a strain called *rho* in which injected sigma virus grows very poorly; rho flies produce only a very small amount of virus, and they rarely become sensitive to CO_2.

Robert L. Seecof has drawn attention to the analogy between sigma and temperate phage. He suggests that sigma is capable of existing in forms analogous to the states of an episome. In a stabilized line, the virus—or its genome—would exist in something approximating the nonintegrated state of an episome, within the sperm and eggs. The rho line would be analogous to the integrated state of an episome, and the genome would have assumed a provirus condition.

A possible solution to the problem of sigma virus lies in observations first reported in 1970 by Howard Temin and by David Baltimore, who found that several RNA tumor viruses carry the information for synthesis of a DNA polymerase which makes copies of their RNA genomes. This polymerase has now been found in several tumor viruses which have structures of the myxovirus-stomatovirus type. These exciting results permit us to extend the concept of a provirus to the RNA viruses, without invoking any unusual types of stable RNA. Sigma virus, tumor viruses, and latent RNA viruses could easily be carried in cell lines as DNA copies which require an inductive step to become RNA genomes again, and there are many ways in which these viruses could affect normal cell activities. This is certain to be a fruitful line of investigation in the near future.

origins of biotic systems

21

We live on a small planet about 8000 miles in diameter. In fact, we occupy very little of the planet—a surface layer at most a few miles thick, even though we sometimes venture into the depths of the sea or the fringes of space. This whole planet is a mere dust speck; it circles a medium-size sun at a distance of 93,000,000 miles along with eight other planets and many smaller bits of dust, and this whole solar system lies in an obscure position about 25,000 light years from the center of a huge galaxy containing at least 10^{11} other stars. As far as we can see in all directions there are other galaxies with similar numbers of stars. So here we are, bits of self-reproducing organic matter clinging to a speck in space. Where did we come from? What are we doing here? And where are we going?

21-1 STAR SEQUENCES AND BIOTIC LIMITATIONS

While the opening paragraphs of this book were meant primarily to set a scene, they were also intended to convey a picture of stellar origins as cosmologists currently picture them. Somehow enormous, diffuse clouds of dust and gases begin to condense under internal gravitational forces to form galaxies containing millions or billions of star clouds. These star clouds become much hotter as they condense until eventually thermonuclear processes are initiated that lead to the formation of heavier elements out of hydrogen.

Every star can be placed in a **spectral class** on the basis of its color, which is a function of its temperature; the classes are designated O, B, A, F, G, K, M, R, N, and S [standing, some say, for Oh Be A Fine Girl, Kiss Me Right Now, Sweet], and each of them has subclasses 1 through 9. Figure 21-1 shows the positions of stars on the **Russell–Hertzsprung diagram**, a graph of the spectral class of a star against its **luminosity** (its rate of energy output per unit mass). Some stars can be classified as huge red giants and others are tiny white dwarfs, but most of them fall into the **main sequence**. Stellar evolution can be understood in relation to this diagram.

In the earliest stages of its evolution, when it is still a mass of contracting gas and dust, the star begins to glow red because of energy released during contraction. As it becomes more dense it gets hotter, and finally hot enough to initiate thermonuclear reactions in its core that convert hydrogen into helium:

$$\begin{aligned}
{}_1^1\text{H} + {}_1^1\text{H} &\rightarrow {}_2^2\text{He} \rightarrow {}_1^2\text{H} + e^+ \\
{}_1^2\text{H} + {}_1^1\text{H} &\rightarrow {}_2^3\text{He} \\
{}_2^3\text{He} + {}_1^1\text{H} &\rightarrow {}_3^4\text{Li} \rightarrow {}_2^4\text{He} + e^+
\end{aligned}$$

At this point, the star moves downward and to the left on the

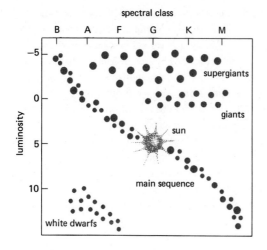

FIGURE 21-1

The Russell-Hertzsprung diagram.

Russell-Hertzsprung diagram and enters the main sequence. It has now achieved a steady-state condition in which its energy output is just balanced by its energy production in such reactions and in others, such as the carbon cycle shown in Fig. 21-2, in which hydrogen is continually transformed into heavier elements. The star remains in the main sequence until about 10–20% of its hydrogen has been exhausted; it then leaves the main sequence and expands into a red giant. It may remain in this state for many years, but it finally burns out completely and becomes a white dwarf, which consists of a strange form of matter far different from anything we know on earth, with a density of about 10^4 g/cc.

FIGURE 21-2

The carbon cycle of stellar reactions.

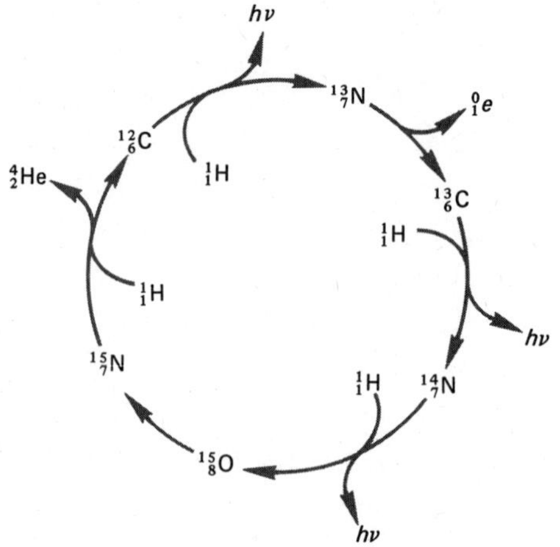

However, it is the time the star resides in the main sequence that concerns us, for during this time there is the possibility that biotic systems can evolve on some of its planets. Planets are probably formed more or less as outlined by G. P. Kuiper. During the formation of the star cloud, up to 10% of its mass may remain outside the main body as a kind of halo; the halo breaks up into protoplanets, consisting of perhaps 1% of the total mass, and these condense along with the star itself. By this time the star has gotten hot enough to glow, and the radiation it emits sweeps out across the protoplanets and carries their lighter gasses, particularly hydrogen and helium, out into space. The remaining heavier elements condense further into planets, although the planets farthest from the sun retain much of their light elements because the radiation pressure is too weak at these distances. On this model, planetary systems should not be particularly rare in the universe. (It was once thought that planets were formed only when two stars nearly collided and pulled bits of themselves off into space; this theory is no longer tenable.) Furthermore, since a primitive mass starts to burn only when it achieves a critical mass, the protoplanets never get anywhere near the temperature of the sun and we do not have to picture a long, fiery period in the life of a planet from which it must cool down so it can support organisms.

Su-Shu Huang has made some useful analyses of the conditions for the evolution of biotic systems. The requirements are simply that a star have planets, that it exist long enough to allow the evolution of biotic systems, that it have a zone of the right temperature, and that it not be a double star system whose planets would have an extremely erratic motion. Conveniently, several of these requirements seem to be met together. It has taken the earth's biota about 3×10^9 years to get where they are, and less than a billion years of this has been spent in evolution from quite simple cells to our present forms; all of the previous time was spent in evolving structures that could be called organisms. If we take 2–3×10^9 years as a minimal time, then we must exclude all stars of the O, B, A, and early F classes because they do not reside in the main sequence long enough. We are confined to stars from about F5 downward. However, we also require a star that has sufficient luminosity to create a zone of the right temperature, where a habitable planet can exist; Huang calculates that stars below K5 are not bright enough to have such zones, and so we are confined to stars between about F5 and K5.

About 1930, Otto L. Struve reported that stars rotate with equatorial velocities ranging from about 300 to 500 km/second for the O and B types to only a few km/second for the coolest ones. The remarkable thing he noted is a break at about F5; stars below this point rotate much more slowly than those above it. The obvious explanation is that those below F5 have planets that take up much of the angular momentum of the system; the sun, for example, has only 2% of the angular momentum of the entire solar system and rotates quite slowly. It is an interesting coincidence that the stars that last long enough for a significant biotic evolution are also those that have planets on which organisms could evolve.

The limitation that the star must be single eliminates about half the stars in the proper spectral range. However, about 3–5% of the stars still meet all the requirements. This means that our galaxy alone might well contain a few hundred million planetary systems on which some organisms are growing! This is a fantastic realization; it means that we are probably not alone in even our tiny corner of the universe. Of course, in making such assertions we are really running ahead of our story, because so far you have no good reason to believe that biotic systems are likely to originate on warm planets that last a few billion years. We will take up this problem next.

21-2 ORIGINS OF BIOTIC SYSTEMS

One of the longest controversies in the history of biology has centered around the question of the spontaneous origin of biotic systems. For a long time, it was simply a commonplace "fact of life" that frogs grew out of the mud and that rats and flies came from garbage; only a fool would doubt what was so obvious. The first fool to back up his doubts with a serious experiment was Francesco Redi, who showed in 1668 that fly maggots only grew out of meat accessible to adult flies, while meat covered with gauze attracted flies but remained unmaggoty. This settled a major part of the controversy until bacteria were discovered; the proof that even bacteria do not arise spontaneously in a pure culture medium required the genius of Louis Pasteur, who showed that a carefully sterilized flask of medium remained sterile if bacteria could not fall in it, even though it was open to the air (Fig. 21-3). Aside from an occasional crackpot, no one today seriously believes that new organisms arise out of totally abiotic materials. However, this has no bearing on the origin of the first organisms on this (or any other) planet billions of years ago; it is clear that at some very early time, there were no organisms on earth, so the first organisms must have come out of abiotic surroundings. Some people have suggested silly solutions to the problem, such as imagining that at some time in the remote past some space travellers from another planet stopped here and contaminated the earth with their "bacteria," which then evolved into all the other terrestrial organisms, but this just begs the real question. We really want to ask how organisms could arise on any planet out of abiotic surroundings, considering that this is apparently not possible according to present knowledge.

As far back as the nineteenth century, a few clever organic chemists were able to produce simple organic compounds out of even simpler materials, such as CO_2 and ammonia, but apparently no one considered this work in relation to the problem of the abiogenic origin of organisms on the primitive earth. The first serious discussion of the problem was apparently the work of J. B. S. Haldane, in 1929; Haldane simply sug-

FIGURE 21-3
Pasteur's "swan-neck" flask. While air can diffuse freely through the neck, bacteria are trapped so they cannot fall into the nutrient medium. When the flask is tipped so medium runs into the neck and then back into the bowl, bacteria are carried back and they grow into a dense culture.

gested that in a "hot dilute soup" of inorganic compounds, ultraviolet irradiation might produce simple organic compounds which could eventually aggregate themselves into real organisms. He recognized that this process would only be possible in the absence of free oxygen, and this is the really critical point. It was picked up and emphasized by A. I. Oparin about 1936; Oparin pointed out that the primitive earth must have had a *reducing* atmosphere rather than our present *oxidizing* atmosphere, and that it consisted primarily of CH_4, NH_3, N_2, and H_2O. (There may also have been some CO_2 — one of the major unsettled questions today is the amount of CO_2 or methane.) In fact, a variety of CH and CN radicals are responsible for some of the important lines in the spectra of stars like our sun, which is a G type. Then the energy of uv light and electrical discharges would drive the synthesis of various organic compounds. Thus Haldane, Oparin, and their contemporaries laid down the essential ideas upon which all later work is based. There are relatively few ideas that we would want to change today. As we have seen, there is no reason to assume that the primitive earth was terribly hot, so the "soup" was merely dilute (although Harold Urey has suggested that the early oceans may have contained as much as 10% organic compounds). Fred Hoyle has also pointed out that the proper atmosphere existed even in the solar cloud before any planets had condensed, so organic compounds might have formed on bits of interplanetary dust and then settled out on all of the protoplanets. The further evolution of these compounds, of course, took place only on those planets that fell into the right temperature belt. This is a very attractive notion for several reasons; a dust particle could provide a favorable surface that would catalyze many reactions, and the high concentration of organic compounds on such a particle could increase the probability that energetically unfavorable reactions would occur. Furthermore, the hypothesis is testable; the eventual discovery of primitive organic complexes frozen in the ice of some of the more distant planets would lend great support to this idea and would also give us valuable information about the course of primitive evolution.

Around 1951–1953, experiments were initiated in Melvin Calvin's laboratory and by Stanley Miller, in Urey's laboratory, to test the Haldane–Oparin hypothesis. These experiments showed clearly that a variety of gas mixtures that might have existed on the primitive earth will form organic compounds, with the energy of uv light and electrical discharges. These experiments have now been extended considerably. From compounds such as methane, ammonia, and water, simple organic molecules such as formaldehyde ($H_2C{=}O$), formate (HCOOH), $HC{\equiv}N$, and cyanoacetylene ($HC{\equiv}C{-}C{\equiv}N$) are easily produced, and these react further to make amino acids such as glycine, alanine, aspartate, and glutamate, intermediate metabolites such as acetate, lactate, succinate, and propionate, and at least the purine adenine.

There is little doubt that polypeptides and polynucleotides — the most critical macromolecules — will form from this primitive mixture. There is still plenty of room for speculation about ways in which the informational relationships between these molecules could develop; an outline of one sequence, based largely on the ideas of Crick and Orgel, is shown in Fig. 21-4. The earliest stages were probably a kind of "doodling" in which random polynucleotides and polypeptides were produced, and the only selection was for molecules that could replicate themselves well. Eventually, an informational coupling between the two types of macromolecules began to develop, and then selection could begin for polynucleotides that could direct the synthesis of

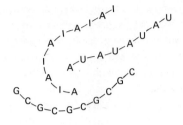
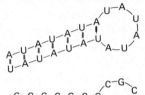
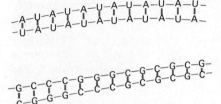

| In the primitive soup, certain polynucleotides arise with sequences that | can be extended easily by internal replication and can be replicated easily. | These polynucleotides proliferate more rapidly than others and form a pool of "protogenomes." |

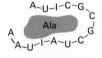

Other polynucleotides occur that do not replicate themselves so well, but have some affinities for amino acids. These form a pool of "proto-tRNA."

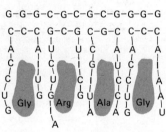

Proto-tRNAs may line up along protogenomes

Gly—Arg—Ala—Gly

and catalyze the synthesis of certain polypeptides.

FIGURE 21-4

A general sequence of events in polynucleotide evolution that could have led to our present coding mechanism.

Gradually, protogenomes are selected which inform useful polypeptides, particularly those with polymerase activity which can speed replication.

useful polypeptides—in general, those with enzymatic activity. Eventually a specialization occurred, with DNA becoming a stable genome and RNA becoming the more transient messenger. Of course, at some point all of these systems had to get wrapped up in primitive cellular complexes. Some structures with the right properties have been made in artificial reaction mixtures, and it does not seem to be a major problem, since membrane-like structures form so easily out of a variety of protein and lipid mixtures.

The evolution of metabolism is another interesting and important problem. The earliest cells were *heterotrophs* which made themselves out of the monomers available in the primitive soup, which must have been a veritable Garden of Eden in which everything was provided. However, the first crisis eventually occurred in Eden—the supply of some monomer began to run out. This created a selective stress, and the organisms that survived were those that managed to acquire a little more knowledge—that is, information. Norman Horowitz has pointed out that biosynthetic pathways would evolve more or less as shown in Fig. 21-5, by working *backward* from the end product. Suppose some amino acid becomes scarce. A cell will then be selected that happens to have an enzyme that will convert a common compound in the soup into that amino acid. (Initially, all the cells were auxotrophs, and now this one partial prototroph has a selective advantage.) But eventually the supply of this precursor runs out; now a mutant will be selected that can transform a second precursor into the first precursor. This process continues until all the biosynthetic pathways are created back to some common intermediates; it is easy to see why our present biosynthetic pathways form a series of families, for there is obviously great selection for efficient pathways that use common precursors for several end products. Notice, also, that evolution proceeds by the "duplicate, then diversify" tactic. Therefore, there will be a strong tendency for duplications to be selected just as the figure shows so all the genes for this pathway will lie next to one another, often in the same order as the biosynthetic sequence. Of course, this is what we observe in much of the bacterial genome.

FIGURE 21-5

Evolution of a hypothetical enzyme sequence for biosynthesis of histidine.

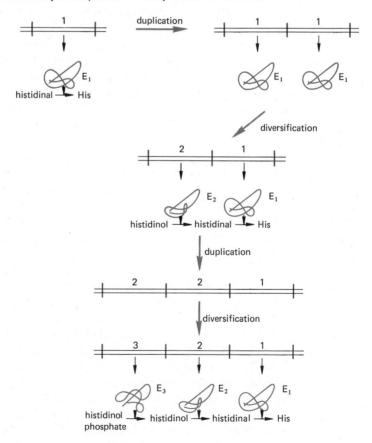

We are still a long way from understanding the probable course of early evolution; we may never understand it all just because it is so hard to obtain good evidence for processes that occurred so long ago. However, there is no great, unfathomable step from an abiotic world to a biotic one, and eventually we hope to have outlined such a gradual series of steps that it would be hard to get biologists to agree that any one of them is the point where life began. Of course, some of the steps may have been rather improbable, but even unlikely events become more likely to occur as time goes on, and there was certainly enough time.

The earth was probably formed about 4.8 billion years ago and the oldest rocks that contain identifiable remnants of primitive organisms are about 3–3.5 billion years old; there was at least a billion years for the evolution of simple cells. In contrast, it has only taken about half a billion years for the most primitive plants and animals to evolve into the variety we see today. The only reason these events seem impossibly fast is that we are viewing the whole panorama of events at once. We cannot really conceive of a million years, let alone a billion.

The first functioning cells must have been procaryotic, and they were certainly heterotrophs. Some of them may have become chemolithotrophs by discovering mechanisms for getting energy out of reduced inorganic compounds. The primitive atmosphere certainly contained very little oxygen, but the oxygen concentration increased gradually through the escape of oxygen from rocks and the photodissociation of water in the upper atmosphere (with the hydrogen created in this process escaping into space). Some of this oxygen was trapped by iron and precipitated as ferric oxide or ferric hydroxide. The primitive organisms probably also used iron to catalyze their oxidative processes, since it is an abundant and adequate electron carrier; Preston Cloud has pointed out that very extensive banded iron formations were laid down between 3.4 and 1.8 billion years ago, probably by primitive organisms at the time when they were evolving their basic metabolic mechanisms. However, any increase in oxygen concentration at this time created a problem for the primitive organisms since oxygen is a poison in biotic systems—specifically, it is a mutagen. The organisms of this period responded to the problem by evolving pathways for the synthesis of porphyrins; the iron porphyrins and enzymes which use them for prosthetic groups (catalases and peroxidases) must have evolved as mechanisms for efficiently reducing molecular oxygen to water. This not only eliminated the problem of free oxygen, but also opened up a more efficient type of electron-transport system, with oxygen as the terminal electron acceptor and hemes as the electron carriers. This system became dominant because of the abundance of the gas and the fact that it has a higher potential than other acceptors.

However, the evolution of the tetrapyrrole ring must have led to the synthesis of the magnesium tetrapyrroles, the chlorophylls, and thus to primitive photosynthesis. Eventually the blue-green algal type of photosynthesis evolved, in which oxygen is produced from water, and at this point the oxygen concentration must have increased dramatically in a relatively short time. The primitive organisms themselves were therefore responsible for the greatest part of the transition from a reducing to an oxidizing atmosphere, an event which occurred on the order of 1.2 billion years ago.

The creation of an oxygen atmosphere had several consequences. It ended the period of abiogenic synthesis, since uv light is absorbed by oxygen, thus removing the energy source, and since any new compounds synthesized abiogenically would have been immediately oxidized. However, the high concentration of oxygen in the atmosphere also demanded that all organisms either adapt to oxygen tolerance or be limited to special niches where there is no oxygen. Lynn S. Margulis has suggested that this demand for adaptation to oxygen was largely responsible for the evolution of the primitive eucaryotes. There may have been procaryotes which lived by fermenting carbohydrates and perhaps by ingesting smaller cells. Some of them may have ingested cells which they retained as intracellular symbionts, instead of digesting them for food; a number of organisms exist today which maintain such a relationship with small intracellular algae. Then internal symbionts which lived chemosynthetically evolved into mitochondria and those which were phototrophs became chloroplasts; some such origin for these structures is certainly suggested by their size, their content of a circular DNA, and their generally procaryotic internal ribosomes. The mitotic mechanism itself is much harder to account for, but it could have evolved in association with some internal symbionts which were rather like basal bodies or centrioles. In any case, the transition from procaryotes to eucaryotes probably occurred around 1.2–0.6 billion years ago, at the time when it became critical to adapt to a high concentration of oxygen in the atmosphere. The enormous radiation from the first eucaryotes into all of the plants and animals we know today has occurred in the relatively short span of the last 600 million years.

21-3 EPILOGUE

Biotic systems on this planet have come a long way from primitive procaryotes to creatures like ourselves. But we must now seriously ask ourselves where we are going, because there is a real possibility that in the next few decades we can destroy everything that has taken a billion years or more to evolve. A cynical view of the "purpose of life" could easily run like this: "You start with a primitive planet with a reducing atmosphere. Organisms evolve out of the carbon compounds in that atmosphere and they eventually change the planet into one with an oxidizing atmosphere. More complex organisms evolve and eventually you get an intelligent species which develops a technology that pollutes the place so badly that the atmosphere becomes reducing again and then the cycle repeats itself."

This cynical viewpoint may be very realistic. Surely our problems are largely the result of a technology that has been used unwisely and has gotten out of control; surely science and technology have changed our world so rapidly and dramatically that we are hardly able to understand the changes and cope with them psychologically. But we have in our hands not merely a weapon for our own destruction; we have, in fact, the potential to make for all of humanity a life that is healthier, longer, and more satisfying than at any time in the past. Our troubles do not simply come from too much knowledge, but rather from too little knowledge about some important matters and from a failure to use our knowledge wisely. Thus, we have the medical knowledge to extend life and make it reasonably comfortable and carefree; but we have failed to acquire some important information about birth control and we are in danger of proliferating ourselves out of existence. Furthermore, we are unable to offer the benefits of modern medicine to a major part of the world and even to many of our own citizens in rural areas or at the lower end of the economic scale. We have the agricultural knowledge to provide ourselves with a bountiful feast; but in our haste and greed we have destroyed wildlife and forests and drained wetlands for agriculture, and the land strikes back by flooding our cities with uncontrolled rivers filled with valuable topsoil. We can make a million things that make life more pleasant, but we have failed to cope with the byproducts of our technology — with beer cans and garbage, with an atmosphere that is becoming unbreathable, and with rivers so polluted they can catch fire.

Though the future looks gloomy, our fate is not sealed. We do have alternatives; we do have the knowledge and technology to repair the damage we have already done, and there is still time to act. The commitment to act must come from all of us — or at least from enough of us to have some political effect and to initiate positive programs. But the knowledge and technology must come from a smaller number of people who are trained in biology, chemistry, geography, economics, and some other areas. We now demand a corps of people who have the necessary training and who are committed to applying their skills to these problems. We cannot afford to remain uneducated. We cannot afford to remain ignorant of those aspects of biology, for example, that may provide solutions to our problems. This book has been an attempt to relate some of the most fundamental concepts of biology; its purpose will be accomplished if it helps to train a generation of biologists who can solve our medical and environmental problems and who can do some significant research in biology. You still have a great deal to learn about more complex biotic systems, particularly at the level of multicellular organisms and populations of organisms, but the information and orientation you have acquired here should serve as a foundation for all of your future work in biology.

APPENDIX

If you don't know the meaning of a biological term, the terse definition of a standard glossary will not help you much. You should consult the index for a full explanation in context. However, a large share of the biological vocabulary is derived from a few common roots, and if you understand these you will have a basis for learning new words when you first encounter them and for jogging your memory if you forget.

1-1 Common roots every educated person should know

a-, an-, in-, im- all indicate negation: not or without

ad-, af- = toward	*meta-* = after, beyond
ex-, ef- = away from	*pro-* = before
auto- = self, same	*post-* = after
allo- = other, different	*pre-* = before
inter- = between	*levo-* = left
intra- = within, inside	*dextro-* = right
extra- = outside	*amphi-* = both
infra- = below	*eu-* = true, real
supra- = above	*pseudo-* = false
hyper- = over, more than	*iso-* = equal
hypo- = under, less than	*aniso-* = unequal
para- = beside, near	*homo-* = like, similar
peri- = around	*hetero-* = unlike, dissimilar
epi- = above, upon	*sym-, syn-* = with, together
sub- = under, below	*anti-* = against, opposite
macro- = large	*archi-* = old
micro- = small	*neo-* = new
trans- = across	*co-* = with
per- = through	*apo-* = away from
exo-, ecto- = outer	*prot-, proto-* = first, beginning
endo-, ento- = inner	*tel-, telo-* = last, end, far
meso- = middle	
ana- = up	*andro-* = male
cata- = down	*gyn-* = female
	phyto- = plant
	zoo- = animal

653

1-2 Numbers

1 = *uni-, mono-*	*hemi-* = half
2 = *di-, bi-*	*holo-* = whole
3 = *tri-*	
4 = *tetra-, quadra-*	*haplo-* = simple
5 = *penta-, quinque-*	*diplo-* = double
6 = *hexa-*	*triplo-* = triple
7 = *hepta-*	
8 = *octo-*	*oligo-* = few
9 = *nona-*	*poly-* = many
10 = *deca-*	*multi-* = many
20 = *icosa-*	*pleio-* = more

There is now a standard set of prefixes to denote multiples and fractions of all physical units; you should know these prefixes and their standard abbreviations. In this book, distances and lengths are particularly important; for cells and subcellular objects, these are usually given in micrometers (μm, formerly called microns, μ) and nanometers (nm, formerly called millimicrons, mμ). Atomic distances are more generally given in angstrom units, abbreviated Å, which are 0.1 nm.

PREFIX	SYMBOL	VALUE
atto-	a	10^{-18}
femto-	f	10^{-15}
pico-	p	10^{-12}
nano-	n	10^{-9}
micro-	μ	10^{-6}
milli-	m	10^{-3}
centi-	c	10^{-2} .
deci-	d	10^{-1}
deka-	dk	10
hecto-	h	10^{2}
kilo-	k	10^{3}
mega-	M	10^{6}
giga-	G	10^{9}
tera-	T	10^{12}

1-3 Colors

chloro- = green (chlorophyll, chloroplast).
cyano- = blue (cyanophyte, anthocyanin, cyanosis).
flavo- = yellow (flavoprotein, acriflavin).
iodo- = violet (iodine, iodopsin).
leuco- = white (leucocyte, leucine).
nigr- = black (Negro, nigrosin, nigrescent).
rhodo-, erythro- = red (erythrocyte, Rhodophyta, erythema).

1-4 The adjective-noun transforms

Biologists commonly change adjectival descriptions into nouns; the transformations are so regular and common that it is easiest to describe them here, rather than each time a term is introduced.

1. *-morphic* → *-morph* and *-morphous* → *-morph:* Greek, *morphe* = form or shape. If the form or shape of an object is described adjectivally as, for example, allomorphic, the object may also be called an allomorph.

2. *-obic* → *-obe:* Greek, *bios* = life. An aerobic organism is frequently called an aerobe; both forms describe its way of life.

3. *-otic* → *-ote* and *-ous* → *-ote.* An organism that is heterozygous is also called a heterozygote; a syngenote is syngenotic.

4. *-somal* → *-some:* Greek, *soma* = body. An object described as allosomal might also be called an allosome.

5. *-trophic* → *-troph:* Greek, *trophe* = nourishment. An organism's way of obtaining food or energy may be described as, for example, allotrophic; the organism may then be called an allotroph.

1-5 Words relating to cells

blast = that which will generate something else (Greek, *blastos* = bud). An osteoblast is a cell that produces bone; the blastoderm is a tissue that will become a large part of a developing embryo.

cary, kary, nucl- (Greek *karyon* = Latin *nucleus* = kernel) refer to the cell nucleus.

cyt = cell (Greek, *kytos* = hollow). The ending *-cyte* indicates a particular kind of cell.

hist- = tissue (Greek, *histos*). Histology is the study of tissues.

plasm (Greek, *plasma* = form) refers to the body or form of a cell, cell part, etc., mostly in obsolete words.

chondr (Greek, *chondros*) = cartilage.

derm (Greek, *derma*) = skin (or layer of cells).

hem (Greek, *haima*) = blood (Note: *haem* in British spelling).

hepa (Greek, *hepar*) = liver.

myo (Greek, *mys*)
sarc (Greek, *sarx* = flesh) } = muscle.

neuro (Greek, *neuron*) = nerve.

oste (Greek, *osteon*) = bone.

rhiz (Greek, *rhiza*) = root.

xylo (Greek, *xylon*) = wood.

1-6 Words relating to reproduction

gamo (Greek, *gamos* = marriage) refers to reproductive cells, fertilization, and reproduction in general.

gen (Greek, *genos* = race) refers to descent, heredity, and any other aspect of the genetic process.

gon (Greek, *gone* = birth or seed) refers to birth, sexual reproduction, reproductive structures, etc.

oo- and *ovi-* (Greek *oon* = Latin *ovum* = egg) refer to eggs and to female reproductive structures in general.

sperm (Greek, *sperma* = seed) refers to male reproductive cells and related structures.

sporo- (Greek, *sporos* = seed) refers to seeds, spores, or similar reproductive units.

1-7   Miscellaneous words

acro- (Greek, *akros* = tip) refers to the end of an object.

aer- (Greek, *aer-* = air) means air.

brachi- (Greek, *brachion* = arm) refers to an arm-like appendage or an arm-like shape.

chondr- (Greek, *chondros* = grain) refers to a granule or small particle, as in chondriosome or mitochondrion.

enter- (Greek, *enteron* = gut) refers to the intestine, as in enteric, coelenterate.

fer (Latin, *ferre* = to carry) means bear, carry, or possess: Porifera, transferrin.

ferr- (Latin, *ferrum* = iron) means containing iron: transferrin, ferroheme.

fuco- (Latin, *fucus* = seaweed) means characteristic of seaweeds or originally isolated from them.

gastr- (Greek, *gaster* = stomach) refers to the stomach, intestine, or anything shaped like a stomach or pouch.

gluc-, glyc- (Greek, *glykys* = sweet) means sugar or carbohydrate.

gymno- (Greek, *gymnos* = naked) means naked, not covered.

halo- (Greek, *hals* = salt) means salt or salty.

hydro- (Greek, hydros = water) means water or aqueous.

id-, idio- (Greek, *idios* = distinct) means distinct, characteristic, private, or peculiar.

kin- (Greek, *kinein* = to move) means move, moving or dynamic.

lipo- (Greek, *lipos* = fat) means fatty or lipid.

lys- (Greek, *lysis* = loosening) means loosening, splitting, or opening.

mer- (Greek, *meros* = part) means part, piece, unit, or segment.

mito- (Greek, *mitos* = thread) means thread or thread-like.

myc- (Greek, *mykes* = fungus) means fungus, characteristic of fungi, or originally isolated from fungi.

muco- (Greek, *mucus*) = mucous, slimy.

nema- (Greek, *nema* = thread) means thread or thread-like.

ortho- (Greek, *orthos* = straight) means straight or direct.

pachy- (Greek, *pachys* = thick) means thick or heavy.

phag- (Greek, *phagein* = to eat) refers to a method of eating or obtaining nourishment.

phor (Greek, *pherein* = to bear) means bear, carry, or possess.

photo- (Greek, *phos* = light) means light.

pycno- (Greek, *pykhos* = dense) means dense, compressed, or compact.

rhin- (Greek, *rhis* = nose) refers to the nose.

schizo- (Greek, *schizein* = to cleave) means cut, cleaved, divided, fragmented, or split.
scler- (Greek, *skleros*) means hard or tough.
stom- (Greek, *stoma* = mouth) means mouth or opening.
theca (Greek, *theke* = case) means a sheath or covering.
tom (Greek, *tome* = cut) means cut, cutting, piece, or section.
trop (Greek, *tropos* = turn) means to move or turn toward.
zyg- (Greek, *zygon* = yoke) means together, brought together, paired, or joined.

1-8 Words used mostly in zoology

branch (Greek, *branchia*) refers to gills.
cephal- (Greek, *kephale*) refers to the head.
coel- (Greek, *koiloma*) means hollow.
gloss (Greek, *glossa*) means tongue.
gnath (Greek, *gnathos*) means jaws.
maxilla (Greek, *maxilla*) means jaws.
nephro- (Greek, *nephros*) refers to the kidneys.
noto- (Greek, *noton*) refers to the back.
pod, ped (Greek, *pous*) refers to the foot.

There is a set of general directions that are most useful in describing the body of an animal, but have occasional use throughout biology:

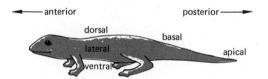

We also commonly use the terms *distal,* meaning *away from* some reference point, and *proximal,* meaning *near* some reference point.

1-9 Biological nomenclature

We will refer to organisms by their technical names in this book, and there may be a few people who need to be reminded how this system works, even though it would probably be hard to find a college student who does not know that organisms are given a binomial designation. Almost everyone knows that man is called *Homo sapiens* and has shared the great joke about "Homo Sap." The modern binomial system was developed by Carolus Linnaeus (Karl von Linne) in 1758. The first or *generic* name defines the small group of closely related species (*genus,* plural *genera*) to which the one in question belongs. The second, or *trivial,* name designates the species within the genus. For example, *Felis* is a genus of cats; *Felis catus* is the common house cat, *F. leo* is the lion, and *F. pardus* is the leopard. Notice that the technical name is always italicized and that only the generic name is capitalized.

In contemporary biology we have to perform many calculations of the number of ways in which objects can be combined and the probabilities that certain combinations will occur. Since this is not a mathematics that is uniformly taught in beginning college courses, we will discuss it here briefly.

appendix 2 probability and combinations

2-1 Elementary probability

In spite of some real biological factors, the numbers of male and female births in a human population are just about equal. Let us assume a 50–50 chance ($p = \frac{1}{2}$) that a given child will be a boy. We can then calculate the probability that a family will have a given sequence of births and the probability that it will have a given combination of boys and girls, without regard to sequence.

For a family of two children, imagine that there are two spaces to be filled. The probability that the first is filled by a boy is $\frac{1}{2}$ and the probability that the second is filled by a boy is also $\frac{1}{2}$, so the probability is $\frac{1}{2} \times \frac{1}{2} = \frac{1}{4}$ that both children will be boys. The same is true of the other combinations, so the four possible orders,

$$\text{BB, BG, GB, GG}$$

are all equally probable ($p = \frac{1}{4}$). However, suppose that the probabilities were 0.6 for a boy and 0.4 for a girl. Then the probabilities of these sequences would be

$$p(\text{BB}) = 0.6 \times 0.6 = 0.36$$
$$p(\text{BG}) = 0.6 \times 0.4 = 0.24$$
$$p(\text{GB}) = 0.4 \times 0.6 = 0.24$$
$$p(\text{GG}) = 0.4 \times 0.4 = 0.16$$

In general, for sequences of any length containing j boys and k girls,

$$p(\text{B}_j\text{G}_k) = [p(\text{B})]^j [p(\text{G})]^k$$

Since we have been assuming that $p = \frac{1}{2}$ for either sex, the probability for any particular sequence of n human births is just $(\frac{1}{2})^n$.

Now let us consider the probabilities of certain combinations of boys and girls, without regard to sequence. We begin by calculating the probability of a single sequence and then count the number of sequences that will satisfy the terms of the problem. For example, what is the probability that a family of three children will consist of two boys and a girl? The probability of any one sequence, such as BBG, is just $(\frac{1}{2})^2(\frac{1}{2}) = \frac{1}{8}$. However, the girl can be placed in any of three positions in the family, so we multiply by 3 to obtain the answer $\frac{3}{8}$. In other words, out of the

659

eight possible sequences of three children in a family, three will consist of two boys and a girl.

exercise 2-1 What is the probability that a family of three children consists of
(a) three boys; (b) two girls and a boy?

exercise 2-2 What is the probability that a family of four children consists of (a) four girls; (b) one girl and three boys?

A more interesting and complex case arises when we consider the following. What is the probability that of five children, three will be boys and two girls? The probability of any particular sequence of five children is $(\frac{1}{2})^5 = \frac{1}{32}$. Now notice that the first girl can go into any of five positions and her sister into any of the remaining four; the other three must then be occupied by boys. However, we are not asked to distinguish the girls from one another: the sequences Jane-June and June-Jane are identical, so we must divide by 2. The number of combinations of two girls and three boys is therefore $(5 \times 4)/2 = 10$, and the probability that a family of this type will occur is $\frac{10}{32} = \frac{5}{16}$.

exercise 2-3 Repeat this calculation by considering ways to arrange the three boys; obviously you must get the same answer.

exercise 2-4 What is the probability that a family of six will consist of four boys and two girls?

exercise 2-5 If you flip a coin 10 times, what is the probability of getting (a) all tails; (b) seven heads and three tails; (c) four heads and six tails?

2-2 A mathematical formulation

It is useful to generalize this approach to probabilities for other cases. As a model, we consider an enormous pot containing a large number of black and white marbles, all thoroughly mixed. Let the fraction of black marbles be b and the fraction of white marbles be $a = 1 - b$. Now we pick out N marbles and lay them in a row. The probability that any one is black is b and that any one is white is a, so the probability that n of them are black and $N - n$ are white is $b^n a^{(N-n)}$.

Now we must calculate the number of combinations of n black marbles and $(N - n)$ white ones, denoted by $C(N, n)$. We calculate the number of ways in which the n black marbles can be drawn, because the remaining marbles must then all be white. If there are N spaces to be filled, the first black marble can go into any of them; the second can go into any of the remaining $N - 1$, the third into any of $N - 2$, and finally the last one can go into any of $N - n + 1$ spaces. The total number of arrangements is therefore

$$N(N - 1)(N - 2) \cdots (N - n + 1)$$

Now we must recognize that we cannot tell the difference between the marbles, so we can rearrange all of the black marbles among themselves in all possible ways. Considering only the n spaces filled with black marbles, the first marble could go into any of these n positions, the second into any of $n - 1$, and so on, until the last marble must go into the one remaining space. The number of such internal rearrangements is therefore $n(n-1)(n-2) \cdots 1 = n!$ Combining these two series, the number of distinguishable combinations is

$$C(N, n) = \frac{N(N-1)(N-2) \cdots (N-n+1)}{n!} \tag{2-1}$$

$$= \frac{N!}{(N-n)!n!}$$

and the probability of drawing n black marbles out of N is

$$p_n = C(N, n)b^n a^{(N-n)} \tag{2-2}$$

2-3   Binomial expansion and the normal distribution

It is interesting to tabulate the values of $C(N, n)$ for all values of N and n in the following way:

```
N = 0                          1
    1                      1    1
    2                   1   2   1
    3                 1   3   3   1
    4              1   4   6   4   1
    5            1   5  10  10   5   1
    6          1  6  15  20  15   6   1
    7        1  7  21  35  35  21   7   1
```

This pyramid, which could continue indefinitely, is called Pascal's triangle. Notice that every number is the sum of the two numbers above it to the right and left, and by following this rule you can always reconstruct the triangle even if you forget the formula for $C(N, n)$. Incidentally, these numbers are also called the binomial coefficients, since they are the coefficients that appear when the expression $(a + b)^N$ is expanded.

Examine the numbers in each row of Pascal's triangle; plot some of them on graph paper around some arbitrary center point. Notice that as N increases, the curve of binomial coefficients becomes more and more like the bell-shaped curve usually called the *normal distribution function;* in fact, it can be shown that this curve is the limit of the binomial distribution as N approaches infinity. This is very significant in any quantitative science, for it means that the normal distribution of values in any measurement is simply the summation of many factors that produce small variations from a mean value. This can be seen by looking at a device that separates steel balls binomially (Fig. 2-1); as a ball falls on a point, it is assumed to have equal probabilities of going in either direction. As this process is repeated at each level with many balls, the balls accumulate in a pattern which approximates Pascal's triangle and, eventually, the normal distribution curve.

Suppose each ball represents a measurement to be made on a sensitive instrument. Each layer of points represents one of the many, tiny bits of interference that can produce errors in the measurement, such as fluctuations in electric current or

FIGURE 2-1
*Balls which have equal probabilities of going left or right at each choice point fall in a
normal distribution curve.*

vibrations in the building. When many independent measurements are made, they
center around some modal value in a normal distribution; we generally take the
modal (or mean) value for the purposes of the experiment, and we make several
measurements whenever possible to get the best mean value.

Similarly, if we sample a population of organisms or other variable objects and find
a normal distribution in some characteristic, we generally assume that this distribution
is due to the summation of many small variations, perhaps mostly unknown. For
example, the population of adult, male, American humans has a normal distribution of
height. You can imagine many factors that contribute to this distribution, including
nutrition, medical history, and hereditary factors.

2-4 The poisson distribution

We will now consider a very important variation of the binomial distribution function
for cases in which (1) N is very large compared to n and (2) b is much less than 1. In
other words, the event we are studying is rare, but there are many opportunities for it
to occur. The problem is to calculate the probability that it will occur a certain number
of times ($n = 0, 1, 2, 3$, rarely more) in a given situation. For example, if the probability that an amateur archer will hit a target is small, but there are many targets, how
many of them will he hit zero times, once, twice, and so on?

If b is the probability that the event will occur at each opportunity and N is the number of opportunities, then $m = Nb$ is the average number of times it will occur. We ask for the probability that it will occur n times, and from equations 2-1 and 2-2,

$$p_n = \frac{N(N-1)(N-2)\cdots(N-n+1)}{n!} b^n a^{(N-n)}$$

Since $N \gg n$, the numerator is approximately N^n and the equation becomes

$$p_n = \frac{N^n b^n}{n!} a^{(N-n)} = \frac{m^n}{n!} a^{(N-n)}$$

where the second simplification is allowed because $m = Nb$. Since $a = 1 - b$, $a^{(N-n)}$ can be expanded as

$$a^{(N-n)} = (1-b)^{(N-n)}$$

$$= 1 - b(N-n) + \frac{(N-n)(N-n-1)b^2}{2!} - \cdots$$

from the formula for binomial expansion. Again, since $N \gg n$ and $m = Nb$,

$$a^{(N-n)} = 1 - Nb + \frac{N^2 b^2}{2!} - \frac{N^3 b^3}{3!} + \cdots$$

$$= 1 - m + \frac{m^2}{2!} - \frac{m^3}{3!} + \cdots$$

$$= e^{-m}$$

since the infinite series in m is just a definition of e^{-m}. Combining all of these expressions,

$$p_n = \frac{m^n e^{-m}}{n!} \tag{2-3}$$

For example, suppose an event occurs on the average of once a month ($m = 1$). The probability that it will not occur in a given month is $e^{-1} = 0.37$ (see Table 2-1)

TABLE 2-1

values of e^{-x}

x	0.0	0.1	0.2	0.3	0.4	0.5	0.6	0.7	0.8	0.9
0.	1.00000	0.90484	0.81873	0.74082	0.67032	0.60653	0.54881	0.49658	0.44933	0.40657
1.	0.36788	0.33287	0.30119	0.27253	0.24660	0.22313	0.20190	0.18268	0.16530	0.14957
2.	0.13534	0.12246	0.11080	0.10026	0.09072	0.08208	0.07427	0.06721	0.06081	0.05502
3.	0.04979	0.04505	0.04076	0.03688	0.03337	0.03020	0.02732	0.02472	0.02237	0.02024
4.	0.01832	0.01657	0.01500	0.01357	0.01228	0.01111	0.01005	0.00910	0.00823	0.00745
5.	0.00674	0.00610	0.00552	0.00499	0.00452	0.00409	0.00370	0.00335	0.00303	0.00274
6.	0.00248	0.00224	0.00203	0.00184	0.00166	0.00150	0.00136	0.00123	0.00111	0.00101
7.	0.00091	0.00082	0.00075	0.00068	0.00061	0.00055	0.00050	0.00045	0.00041	0.00037
8.	0.00034	0.00030	0.00027	0.00025	0.00022	0.00020	0.00018	0.00017	0.00015	0.00014
9.	0.00012	0.00011	0.00010	0.000091	0.000083	0.000075	0.000068	0.000061	0.000056	0.000050
10.	0.000045									

the probability that it will occur twice is $e^{-1}/2! = 0.185$, and so on. There are many important applications of this formula in biology.

exercise 2-6 Suppose you have a bucket with a large number of marbles, 80% of them black and 20% white. What is the probability that you will get the following combinations?
(a) $N = 2$, both black;
(b) $N = 3$, two black and one white;
(c) $N = 5$, two black and three white;
(d) $N = 6$, three black and three white;
(e) $N = 5$, all white.

exercise 2-7 Now suppose that 10^{-3} of the marbles are black and the rest are white. What is the probability of the following?
(a) $N = 10^3$, no black marbles;
(b) $N = 10^3$, two black marbles;
(c) $N = 10^2$, no black marbles;
(d) $N = 10^2$, one black marble;
(e) $N = 10^4$, five black marbles.

2-5 Goodness of fit and chi square

Suppose you are told that a large urn contains equal numbers of black and white marbles, and you decide to test this statement by drawing 100 marbles at random. You certainly would not expect to get exactly 50 black and 50 white marbles in every trial. If you got 45 of one and 55 of the other you would probably say that the statement was true. You might begin to wonder if you got 42 of one and 58 of the other, but your intuition still says that this is not an unreasonable result. However, suppose the agreement between actual and expected values was worse than this; at what point would you be inclined to say that the deviations from the expected results are too large and the statement is probably false?

By convention, statisticians start to question a hypothesis in a situation like this one when the experimental results are so far from the expected that they would be obtained in only 5% of the trials. Such a deviation from the expected is called *significant;* if the deviation would be expected in only 1% of the trials, it is *highly significant.* In order to make judgments of this kind, they have developed simple measures of the **goodness of fit** between data and hypothesis, and here we will briefly discuss one of the most useful of these, the **chi-square** test invented by Karl Pearson.

If you were content to test the hypothesis about the numbers of marbles in the urn by drawing out small samples, you could easily calculate the probability that you would get a given distribution. For example, if you drew only two marbles out you would expect them both to be black or both to be white half of the time, if the ratio of black to white is really 1:1. However, this is not a very good test of the hypothesis simply because there is such a high probability of getting grossly unequal numbers of black and white marbles in your sample. To really test the hypothesis you must draw a large sample, but as the size of the sample increases it becomes very tedious to use the formulas we have developed to calculate the probability that a given distribution would be obtained. The chi-square test is a simple way to obtain this probability.

To apply this test, we calculate the results to be expected on the basis of the hypothesis we are testing. In this example, we would expect 50 black and 50 white marbles in a sample of 100. Suppose we then perform one experiment and get 46 black and 54 white marbles. We have two classes of data (number of black marbles, number of white marbles), and in each class we then calculate the deviation, d, of the observed results from the expected: $d = $ (expected - observed). We then define chi square as

$$\chi^2 = \sum_n \frac{d^2}{\text{expected}} \qquad (2\text{-}4)$$

where the sum is taken over all n classes. In this case, $d = 4$ and expected $= 50$ in each case, so $\chi^2 = 16/50 + 16/50 = 32/50 = 0.64$.

This number is clearly a measure of the deviations from the expected results and it will increase as the data depart from the expected results, but it cannot be interpreted by itself. To obtain the probability that this deviation would be obtained by chance we consult a graph or a table (Fig. 2-2) giving this number as a function of χ^2 and the

FIGURE 2-2
*Table of χ^2. Read along
the appropriate line for
degrees of freedom until
you come to the
calculated value for χ^2;
diagonal lines give the
probability of this value.*

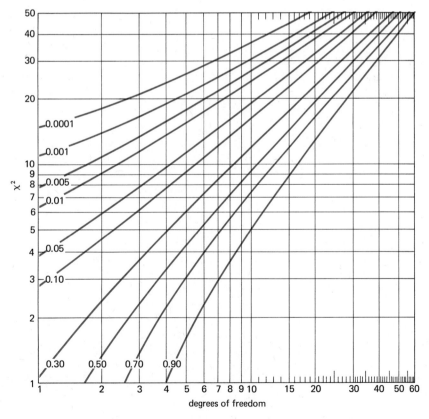

number of **degrees of freedom.** In the cases we will consider, degrees of freedom $=$ $n - 1$, since the last class is determined once the other $n - 1$ classes are determined. Consulting the chart for one degree of freedom, we find that a χ^2 of 0.64 would be obtained in 40% of the trials, and the results are certainly consistent with the hypothesis. However, if we had obtained 40 black marbles and 60 white ones, we would have $d = 10$ in each case and $\chi^2 = 100/50 + 100/50 = 4.0$. For one degree of freedom we find a probability of slightly less than 5%, and so with deviations of this size we would have to question the hypothesis.

exercise 2-8   Someone tells you that he has filled an urn with 250 red marbles, 250 white marbles, and 500 black marbles. In each of the following trials, calculate χ^2 and consider whether you would seriously question his statement.

(a) 22 red, 21 white, 57 black in 100;

(b) 20 red, 25 white, 55 black in 100;

(c) 27 red, 20 white, 53 black in 100;

(d) 31 red, 28 white, 41 black in 100;

(e) 30 red, 22 white, 48 black in 100.

Our understanding of biotic systems depends very heavily on the ability to determine the properties of large molecules such as proteins and nucleic acids. Many large, complex instruments are now available for this work; we will try to understand some of the principles these instruments employ.

3-1 Sedimentation velocity

The analytical ultracentrifuge, which was originally developed by The Svedberg and K. O. Pedersen, is certainly one of the most useful devices for analyzing the properties of macromolecules. It employs the principle that heavy particles will fall through a solvent faster than light ones, so the rate at which a substance sediments in a gravitational field is a measure of its weight (and shape).

If macromolecules of volume v and density ρ_s are suspended in a solvent of density ρ, the particles will be buoyed up by a force equal to the weight of the liquid they displace, and will have an effective mass $m' = v(\rho_s - \rho)$. If the solution is now spun in an ultracentrifuge at a radial distance x from the center of the rotor (about 7.5 cm in the common commercial machine) and with a radial velocity ω, the solute molecules will sediment with a velocity dx/dt, abbreviated $\dot{x}$. Let f be the coefficient of friction between the solvent and solute molecules. Then when the solute molecules are falling with a constant terminal velocity, the forces on them will be

$$F = m'\omega^2 x - f\dot{x} = 0$$

and we may define the sedimentation coefficient of the solute molecules, s, as their terminal velocity per unit gravitational field,

$$s = \frac{\dot{x}}{\omega^2 x} = \frac{m'}{f} = \frac{v}{f}\rho_s\left(1 - \frac{1}{\rho_s}\rho\right)$$

where $v\rho_s$ is the true mass of the particles; this is equal to their molecular weight divided by Avogadro's number, M/N. The value $1/\rho_s$ is called the partial specific volume of the solution, $\bar{v}$, and this is relatively easy to measure. Therefore,

$$s = \frac{M(1 - \bar{v}\rho)}{Nf}$$

However, since the frictional coefficient f is hard to measure, we replace it by an expression in terms of D, the diffusion constant of the molecules, which can also be determined easily; the two are related by the Einstein equation $D = kT/f$, where k is

Boltzmann's constant and T is the absolute temperature. This gives the useful expression

$$s = \frac{MD(1 - \bar{v})}{RT}$$ (3-2)

since $Nk = R$, the gas constant. By measuring the diffusion and sedimentation constants of a molecule, its molecular weight can be determined.

FIGURE 3-1

Distribution of molecules in an ultracentrifuge cell. The black spaces are merely position references; the space at the left is an air pocket left when the cell is filled. At first, the molecules are uniformly distributed throughout the cell; those that were initially near the bottom have sedimented and those near the top have pulled away from the meniscus, leaving a region of pure solvent. Gradually, more and more molecules sediment and the boundary moves farther down the cell. Below are two photographs taken at times corresponding to the drawings above (Courtesy of Pauline Yang). The optical system measures the derivative of concentration, so the horizontal line shows a peak at the position of the boundary. The peak is sharper initially because molecules at the boundary diffuse away during the course of centrifugation.

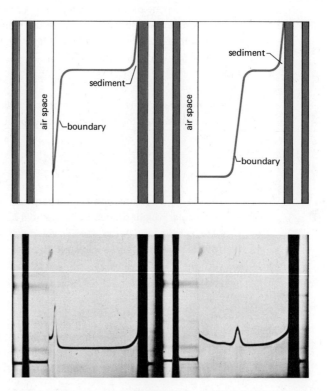

In practice, a relatively concentrated solution (about 1 mg/ml) of the substance is put into a small cell whose top and bottom are made of strong, quartz lenses. This cell is placed in a rotor in an evacuated chamber and spun at high speed; the cell is positioned so it passes through a light beam leading into an optical system with a camera. Suppose there is only one kind of solute molecule; as Fig. 3-1 shows, these molecules will have sedimented to a certain point after a given time, leaving a region of pure solvent near the top of the cell. Since the refractive index of the solution is proportional to its concentration, light passing through different regions of the cell will be refracted to different degrees. The schlieren optical system converts part of the light passing through it into a line that measures the rate of change of refractive index across the cell from top to bottom. Since this rate of change is highest right at the boundary between pure solvent and solution, the line forms a peak at the boundary. By taking photographs every few minutes, the movement of this peak down the cell can be followed and the sedimentation velocity can be calculated. As a dividend, the solute concentration can be determined, because it is proportional to the area under the peak.

A slightly different optical system built into the instrument uses uv light. The absorption of this light, particularly by nucleic acids and proteins, creates a pattern of density on uv-sensitive film so the sedimentation of small amounts of these substances can be followed. If desired, the pattern of density on the film can be converted into a simple line function by another instrument, a microdensitometer.

Values of s for macromolecules fall in a range near 10^{-13} seconds; this inconveniently small number is called a **svedberg unit, S,** and sedimentation constants are then given as multiples of S. Small proteins fall at around 2–4 S, while larger cell components may have constants of a few hundred S. You must remember that s is not strictly proportional to mass alone, so if two 10-S particles aggregate into a large unit its sedimentation constant will generally not be 20 S; S values are not additive.

The sedimentation coefficient clearly depends on the viscosity of the solution and on temperature, so it is customary to report the value $s_{20,w}$ of a substance, which is the sedimentation coefficient corrected to 20°C for a solution with the viscosity of water at that temperature. Measurements made on solutions at various concentrations are extrapolated back to a value for infinite dilution.

There are many other methods for measuring sedimentation coefficients and molecular weights with the ultracentrifuge. We have also ignored many of the complications that always creep into this kind of analysis. The important lesson to carry away is a feeling for the meaning of sedimentation coefficients and the way they can be measured.

3-2 Viscosity

A liquid will resist any shearing force which tends to move one region faster than another; the viscosity of the liquid, η, is a measure of the resistive force. Viscosity is measured in poises; a solution has a viscosity of 1 poise if a force of 1 dyne is required to move one layer of solution 1 cm/second faster than another layer 1 cm away. It is useful to know that the viscosity of pure water at 20°C is almost exactly 1 centipoise. Viscosity may be determined by measuring the rate at which a liquid flows through a narrow tube of radius r and length l; if v ml flow through in t seconds, Poiseuille's law gives

$$\eta = \frac{r^4 \pi P t}{8 v l} \tag{3-3}$$

where P is the driving force in dynes/cm^2. In the Ostwald viscometer, liquid is allowed to fall through a long tube past two marks several centimeters apart. The instrument is calibrated with liquid 1, which has a viscosity η_1, a density ρ_1, and takes t_1 seconds to fall from one mark to the other. If an unknown liquid 2, with density ρ_2, takes t_2 seconds, its viscosity is obtained from

$$\frac{\eta_1}{\eta_2} = \frac{\rho_1 t_1}{\rho_2 t_2} \tag{3-4}$$

The addition of macromolecules to a solvent with low molecular weight will cause an increase in viscosity. If the pure solvent has a viscosity η_0 and the viscosity of the solution is η, the *relative viscosity* is defined as $\eta_r = \eta/\eta_0$. Another equation derived by Einstein gives the relative viscosity as a function of ϕ, the volume fraction of solute molecules:

$$\eta_r = 1 + 2.5 + 4\phi^2 + 5.5\phi^3 + \cdots \tag{3-5}$$

This equation is valid at small flow rates for very dilute solutions of rigid, spherical macromolecules that are large compared to the solvent molecules and are not wet by them. However, even with all these restrictions the higher-order terms lose significance, and we can replace equation 3-5 by $\eta_r = 1 + 2.5\phi$.

We now define the *specific viscosity* as $\eta_s = \eta_r - 1$ and the *volume intrinsic viscosity* as

$$[\eta] = \lim_{\phi \to 0} \frac{\eta_s}{\phi} \tag{3-6}$$

(Alternatively, we can define the *weight intrinsic viscosity* as the corresponding limit for solutions where we measure concentration in grams per 100 ml of solution.) Evidently, the intrinsic viscosity for a solution of rigid spheres should be 2.5; some typical values are given in Table 3-1, which shows that this value is rarely obtained for real

TABLE 3-1

specific viscosities of representative suspensions

SUSPENSION	η_s
Gum gamboge in water	0.02–0.03
Sulfur in water	0.02–0.03
Glass spheres in water	0.018–0.027
Polystyrene in toluene	0.5–2.0
Rubber in benzene	0.8–3.5
Cotton in water	3.0–5.0

solutions. This means that real macromolecules do not fit the ideal model for which the equations were derived, either because they are not just rigid spheres or because they interact strongly with the solvent (for example, they may bind water molecules strongly). We will now try to derive some correction factors that allow us to interpret viscosity data in terms of the actual shape of the macromolecules.

3-3 Shape and hydration factors

From all of the attempts to derive equations that will permit us to interpret viscosity data in a more meaningful way, we will pick one general approach devised by R. Simha. Instead of considering the molecules as perfect spheres, we will idealize them as either prolate or oblate spheroids (Fig. 3-2). These figures are made by rotating an ellipse around one of its axes, and we let a be the semiaxis of revolution and b be the equatorial semiaxis, so the *axial ratio* is $j = b/a$.

FIGURE 3-2

Oblate and prolate spheroids of revolution.

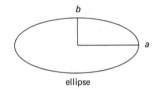

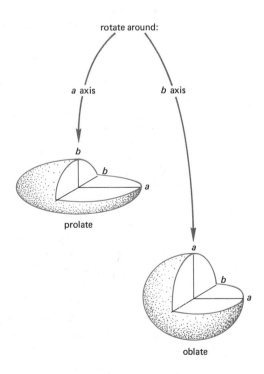

In the Einstein equation for viscosity, we replace the constant 2.5 by a variable ν, which is a function of j given by Table 3-2:

$$\eta_r = 1 + \nu\phi \qquad (3\text{-}7)$$

If one assumes that all deviations from the Einstein equation are due to a shape factor, the approximate shape of the molecules can be determined by consulting the table. However, we must also consider the factor of hydration of the macromolecules by solvent molecules which produce an additional viscous drag.

TABLE 3-2

values of ν as a function of axial ratio

AXIAL RATIO	PROLATE	OBLATE
1.0	2.50	2.50
1.5	2.63	2.62
2.0	2.91	2.85
3.0	3.68	3.43
4.0	4.66	4.06
5.0	5.81	4.71
6.0	7.10	5.36
8.0	10.10	6.70
10.0	13.63	8.04
12.0	17.76	9.39
15.0	24.8	11.42
20.0	38.6	14.8
25.0	55.2	18.2
30.0	74.5	21.6
40.0	120.8	28.3
50.0	176.5	35.0
60.0	242.0	41.7
80.0	400.0	55.1
100.0	593.0	68.6
150.0	1222.0	102.3
200.0	2051.0	136.2
300.0	4278.0	204.1

Assume that w g of water are bound per gram of solute and that the bound and free water have the same density, ρ. If the partial specific volume of the solute is $\bar{v}$, then a hydration factor $H = 1 + w/\bar{v}\rho$ will account for increases in viscosity due to this factor. In practice, one cannot separate the factors H and ν; J. L. Oncley has derived a chart (Fig. 3-3) showing the values of j and w consistent with various values of ν.

You must remember that the shapes calculated for various macromolecules are just idealizations. They may be good approximations, but no protein molecule, for example, really has the shape of a perfect spheroid of any kind. Harold Sheraga and Leo

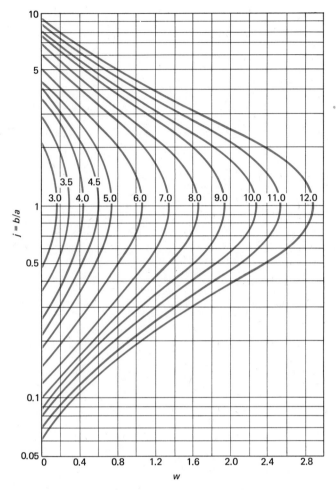

FIGURE 3-3

Values of ν (on the curves) as a function of axial ratio and w.

Mandelkern have taken the view that these ellipsoids of revolution are complete fictions and that the calculated shapes may have little relation to actual shapes. By making a fresh start, they have obtained a very useful relationship between viscosity and sedimentation coefficient:

$$M = \frac{4690(s_{20,w})^{3/2}[\eta]^{1/2}}{(1 - \bar{v}\rho)^{3/2}}$$

(3-8)

The constant 4690 is an average for a variety of molecules with different shapes, and in fact better values can be used for molecules that are known to be extremely long, such as extended nucleic acid chains.

references

References for this book have been compiled under the assumption that most readers will be interested in a fuller explanation of major concepts or in a guide to further exploration of some topic, rather than the experimental detail of original research papers. Many other textbooks cover the same ground as this book at a comparable level, with no more depth or attention to detail (and, hopefully, no better explanation of concepts); only a few of these are listed which have some special virtues. To help in understanding basic ideas, your best reference may be *Scientific American,* which consistently manages to convey the latest scientific news in well-written, superbly illustrated articles. Many of these are now available as offprints from W. H. Freeman and Co., San Francisco, which also publishes paperback collections of the most useful articles.

To get into a field a little deeper, try some of the texts listed here. An excellent way to begin an exploration in many fields is through articles in the *Annual Reviews* series (Annual Reviews, Inc., Palo Alto, California), particularly *Annual Reviews of Biochemistry, Microbiology, Genetics, Plant Physiology,* and *Physiology.* Also consult review journals, such as *Bacteriological Reviews* (which is not as narrow as its title implies), *Biological Reviews,* and *The Quarterly Review of Biology.*

SCIENTIFIC AMERICAN ARTICLES
Numbers refer to offprints of the original articles.

CHAPTER 4

5   F. H. C. Crick, The Structure of the Hereditary Material (October, 1954).
18   Rollin D. Hotchkiss and Esther Weiss, Transformed Bacteria (November, 1956)
22   David Lack, Darwin's Finches (April, 1953)
42   E. O. P. Thompson, The Insulin Molecule (May, 1955)
74   L. S. Penrose, Self-reproducing Machines (June, 1959)
1148   Margaret O. Dayhoff, Computer Analysis of Protein Evolution (July, 1969)

CHAPTER 5

120   Seymour Benzer, The Fine Structure of the Gene (January, 1962)

CHAPTER 6

121   John C. Kendrew, The Three-Dimensional Structure of a Protein Molecule (December, 1961)
1155   R. D. B. Fraser, Keratins (August, 1969)

CHAPTER 8

1030   John Cairns, The Bacterial Chromosome (January, 1966)
1124   Arthur Kornberg, The Synthesis of DNA (October, 1968)

CHAPTER 9

50   Elie L. Wollman and Francois Jacob, Sexuality in Bacteria (July, 1956)
89   Francois Jacob and Elie L. Wollman, Viruses and Genes (June, 1961)
106   Norton D. Zinder, "Transduction" in Bacteria (November, 1958)

CHAPTER 10

119   Jerard Hurwitz and J. J. Furth, Messenger RNA (February, 1962)
153   Marshall W. Nirenberg, The Genetic Code: II (March, 1963)
183   S. Spiegelman, Hybrid Nucleic Acids (May, 1964)
1033   Robert W. Holley, The Nucleotide Sequence of a Nucleic Acid (February, 1966)
1041   Luigi Gorini, Antibiotics and the Genetic Code (April, 1966)
1052   F. H. C. Crick, The Genetic Code: III (October, 1966)
1092   Brian F. C. Clark and Kjeld A. Marcker, How Proteins Start (January, 1962)

CHAPTER 11

79   Peter Satir, Cilia (February, 1961)
156   Christian de Duve, The Lysosome (May, 1963)
182   Robert D. Allen, Amoeboid Movement (February, 1962)
1026   H. E. Huxley, The Mechanism of Muscular Contraction (December, 1965)
1085   Anthony Allison, Lysosomes and Disease (November, 1967)
1134   Marian Neutra and C. P. Leblond, The Golgi Apparatus (February, 1969)

CHAPTER 14

1002   Ruth Sager, Genes Outside the Chromosomes (January, 1965)

CHAPTER 15

84   Curtis A. Williams, Jr., Immunoelectrophoresis (March, 1960)
131   Arthur K. Solomon, Pumps in the Living Cell (August, 1962)
196   M. F. Perutz, The Hemoglobin Molecule (November, 1964)
199   G. J. V. Nossal, How Cells Make Antibodies (December, 1964)
1008   Jean-Pierre Changeux, The Control of Biochemical Reactions (April, 1965)
1055   David C. Phillips, The Three-Dimensional Structure of an Enzyme Molecule
       (November, 1966)
1083   R. R. Porter, The Structure of Antibodies (October, 1967)

CHAPTER 16

20   Bernhard Katz, The Nerve Impulse (November, 1952)
122   J. A. Bassham, The Path of Carbon in Photosynthesis (June, 1962)
1007   Keith R. Porter and Clara Franzini-Armstrong, The Sarcoplasmic Reticulum
       (March, 1965)
1016   Eugene I. Rabinowitch and Govindjee, The Role of Chlorophyll in Photosynthesis
       (July, 1965)
1101   Efraim Racker, The Membrane of the Mitochondrion (February, 1968)
1163   R. P. Levine, The Mechanism of Photosynthesis (December, 1969)

CHAPTER 18

157   Edward O. Wilson, Pheromones (May, 1963)
180   Wolfgang Beerman and Ulrich Clever, Chromosome Puffs (April, 1964)
1013   Eric H. Davidson, Hormones and Genes (June, 1965)

CHAPTER 19

95   A. A. Moscona, How Cells Associate (September, 1961)

1142   Nathan Sharon, The Bacterial Cell Wall (May, 1969)

88   Jerome Gross, Collagen (May, 1961)

CHAPTER 20

9   Heinz Fraenkel-Conrat, Rebuilding a Virus (June, 1956)

128   Robert L. Sinsheimer, Single-Stranded DNA (July, 1962)

147   R. W. Horne, The Structure of Viruses (January, 1963)

185   Harry Rubin, A Defective Cancer Virus (June, 1964)

1004   R. S. Edgar and R. H. Epstein, The Genetics of a Bacterial Virus (February, 1965)

1037   Fred Rapp and Joseph L. Melnick, The Footprints of Tumor Viruses (March, 1966)

1058   Edouard Kellenberger, The Genetic Control of the Shape of a Virus (December, 1966)

1069   Renato Dulbecco, The Induction of Cancer by Viruses (April, 1967)

1079   William B. Wood and R. S. Edgar, Building a Bacterial Virus (July, 1967)

GENERAL REFERENCES FOR PART 1

Hawkins, David, *The Language of Nature,* W. H. Freeman and Co., San Francisco, 1964. Stimulating essays in the philosophy of science, including valuable discussions of information theory and the thermodynamics of biotic systems.

Howland, John L., *Introduction to Cell Physiology: Information and Control,* Macmillan Co., New York, 1968. A short, lucid textbook.

Lwoff, Andre, *Biological Order,* M. I. T. Press, Cambridge, Mass., 1962. A fine series of lectures—not to be missed.

Stahl, F. W., *The Mechanics of Inheritance,* Prentice-Hall, Inc., Englewood Cliffs, N.J., 1969, 2nd ed. An excellent supplement for much of this book; try the exercises in it.

Schrodinger, Erwin, *What is Life?* Doubleday Anchor Books, Doubleday and Co., Inc., Garden City, N.Y., 1956. A reprint of lectures given in 1943 and therefore somewhat out of date, but interesting reading nevertheless.

CHAPTER 1

"The Biosphere," *Scientific American,* September, 1970. Appeared too late to be consulted, but *you* should read it.

Bradbury, S., *The Microscope, Past and Present,* Pergamon Press, Ltd., Oxford, 1968. An interesting history.

Stehle, Georg, *The Microscope—and How to Use It,* Sterling Publishing Co., New York, 1960. A bit old, but full of fun.

Thomas, M. D., "Photosynthesis (Carbon Assimilation): Environmental and Metabolic Relationships," in *Plant Physiology,* vol. IVA (F. C. Steward, ed.), Academic Press, New York, 1965.

CHAPTER 3

Klotz, Irving M., *Some Principles of Energetics in Biochemical Reactions,* Academic Press, New York, 1957. Short, easy to read, and valuable.

Pardee, A. B., and L. L. Ingraham, "Free Energy and Entropy in Metabolism," in *Metabolic Pathways* (D. M. Greenberg, ed.), vol. 1. Academic Press, New York, 1960.

Lehninger, Albert L., *Bioenergetics: The Molecular Basis of Biological Energy Transformations,* W. A. Benjamin, Inc., New York, 1965. A good reference for many chapters in this book.

CHAPTER 4

Bennett, Thomas P., *The Mechanism of Protein Synthesis: An Instructional Model,* W. A. Freeman and Co., San Francisco, 1969. Highly recommended; learn the essentials of protein synthesis in five minutes by making your own polypeptides. The accompanying booklet is nice and has additional exercises in the spirit of those in this book.

Watson, J. D., *The Double Helix,* Atheneum Press, New York, 1968. Watson's own story of the discovery of DNA structure; recommended for an enjoyable evening of reading.

CHAPTER 6

Dickerson, Richard E., and Irving Geis, *The Structure and Action of Proteins,* Harper & Row, New York, 1969. A lucid, beautiful book resulting from the perfect collaboration between artist and chemist; not to be missed if you want further insight into protein structure.

Crick, F. H. C., and J. C. Kendrew, "x-Ray Analysis and Protein Structure," *Advances in Protein Chemistry* **12,** 133 (1957).

Dickerson, Richard E., "x-Ray Analysis and Protein Structure," in *The Proteins,* vol. 2, Academic Press, New York, 1964. These two articles present the best explanations available of x-ray crystallography and its use in protein structure analysis.

Wolstenholme, G. E. W., and M. O'Connor (eds.), *Principles of Biomolecular Organization* (Ciba Foundation Symposium), Little, Brown and Co., Boston, 1966. The articles by Caspar, Liquori, and Kendrew and Watson are particularly recommended.

CHAPTER 7

Needham, G. H., *The Practical Use of the Microscope,* Charles C. Thomas, Springfield, Ill., 1958.

Branton, Daniel, and Roderic B. Park, *Papers on Biological Membrane Structure,* Little, Brown, and Co., Boston, 1968. Contains some classics in the field, including several that are broad enough to be useful to a beginning student.

CHAPTER 8

Krebs, H. A., and H. L. Kornberg, *Energy Transformations in Living Matter,* Springer-Verlag, Berlin, 1957. A concise, valuable survey (in English), available as an inexpensive paperback pamphlet, worth more than its price.

Maaløe, Ole, and Niels O. Kjeldgaard, *Control of Macromolecular Synthesis,* W. A.

Benjamin, Inc., New York, 1966. Highly recommended after you have grasped some of the basic ideas of cellular regulation.

CHAPTER 9

Hayes, William, *The Genetics of Bacteria and Their Viruses,* John Wiley and Sons, New York, 1968, 2nd ed. Probably the best general reference.

Jacob, Francois, and Elie L. Wollman, *Sexuality and the Genetics of Bacteria,* Academic Press, New York, 1961. Two founders of the field write a clear, often exciting account which gives some important historical insights.

CHAPTER 10

Bennett's model (see references for Chapter 4) now becomes even more valuable.

Synthesis and Structure of Macromolecules. Cold Spring Harbor Symposium on Quantitative Biology, vol. 28, 1963. Contains some important papers, including a review by Jacob and Monod.

Ingram, Vernon, *The Biosynthesis of Macromolecules,* W. A. Benjamin, Inc., New York, 1965.

Hartman, Philip E., and Sigmund R. Suskind, *Gene Action,* Prentice-Hall, Inc., Englewood Cliffs, N.J., 1969, 2nd ed.

CHAPTER 11 AND 12

Bourne, Geoffrey H., *Division of Labor in Cells,* Academic Press, New York, 1962.

DuPraw, Ernest J., *Cell and Molecular Biology,* Academic Press, New York, 1968.

Fawcett, Don W., *The Cell: An Atlas of Fine Structure,* W. B. Saunders Co., Philadelphia, 1966. Recommended for its beautiful micrographs.

Allen, R. D., and N. Kamiya (eds.), *Primitive Motile Systems in Cell Biology,* Academic Press, New York, 1964. Contains several important articles on cell movement.

Wolpert, L., "Cytoplasmic Streaming and Amoeboid Movement," in *Function and Structure in Microorganisms,* 15th Symposium of the Society for General Microbiology, Cambridge University Press, 1965.

Padilla, G. M., G. L. Whitson, and I. L. Cameron (eds.), *The Cell Cycle: Gene-Enzyme Interactions,* Academic Press, New York, 1969.

CHAPTER 13

Hadzi, J., *The Evolution of the Metazoa,* Macmillan Co., New York, 1963.

Klein, R. M., and A. Cronquist, "A Consideration of the Evolutionary and Taxonomic Significance of Some . . . Characteristics in the Thallophyta," *Quart. Rev. Biol.* **42,** 105 (1967).

Scagel, R. F., et al., *An Evolutionary Survey of the Plant Kingdom,* Wadsworth Publishing Co., Belmont, Calif., 1965.

CHAPTER 14

Fincham, J. R. S., and P. R. Day, *Fungal Genetics,* F. A. Davis Co., Philadelphia, 1965, 2nd ed.

Jinks, John L., *Extrachromosomal Inheritance,* Prentice-Hall, Inc., Englewood Cliffs, N.J., 1964.

Stansfield, William D., *Theory and Problems of Genetics,* Schaum's Outline Series, McGraw-Hill Book Co., New York, 1969. A very useful guide to problem solving in genetics.

Chapters 6 and 9 of Stahl's book (see references to Part I) are strongly recommended here.

Taylor, J. H., "Patterns and Mechanisms of Genetic Recombination," in *Molecular Genetics* (J. H. Taylor, ed.), part II, Academic Press, New York, 1967.

CHAPTER 15

Bernhard, Sidney, *The Structure and Function of Enzymes,* W. A. Benjamin, Inc., New York, 1968.

Burnet, F. M. *The Integrity of the Body,* Harvard University Press, Cambridge, Mass., 1962. An interesting discussion by a grand master of the field.

Cohen, E. P., "On the Mechanism of Immunity: In Defense of Evolution," *Ann. Rev. Microbiology* **22,** 283 (1968).

Fincham, J. R. S., *Genetic Complementation,* W. A. Benjamin, Inc., New York, 1966.

Weiser, R. S., Q. N. Myrvik, and N. N. Pearsall, *Fundamentals of Immunology,* Lea and Febiger, Philadelphia, 1969. A good, concise introduction to the field.

CHAPTER 16

Calvin, M., and J. A. Bassham, *The Photosynthesis of Carbon Compounds,* W. A. Benjamin, Inc., New York, 1962.

Greville, G. D., "A Scrutiny of Mitchell's Chemiosmotic Hypothesis of Respiratory Chain and Photosynthetic Phosphorylation," *Current Topics in Bioenergetics,* vol. 3, Academic Press, New York, 1968.

Green, D. E., and H. Baum, *Energy and the Mitochondrion,* Academic Press, New York, 1970. Controversial; to be taken with some salt.

Mitchell, Peter, *Chemiosmotic Coupling in Oxidative and Photosynthetic Phosphorylation,* Glynn Research Ltd., Bodmin, England, 1966. See also *Biological Reviews,* August, 1966. Thought provoking, but have some more salt handy.

CHAPTER 17

Cohen, Georges N., *The Regulation of Cell Metabolism,* Holt, Rinehart and Winston, Inc., New York, 1968.

Wagner, Robert P., and Herschel K. Mitchell, *Genetics and Metabolism,* John Wiley & Sons, Inc., New York, 1964, 2nd ed.

CHAPTER 18

Robison, G. A., R. W. Butcher, and E. W. Sutherland, "Cyclic AMP," *Ann. Rev. Biochem.* **37** (1968).

Atkinson, Daniel, "Regulation of Enzyme Activity," *Ann. Rev. Biochem.* **35** (1966).

CHAPTER 19

Nikaido, H., "Biosynthesis of Cell Wall Lipopolysaccharides in Gram Negative Enteric Bacteria," *Adv. Enzymology* **31,** 77 (1968). Another field where genetic and metabolic studies have yielded an interesting picture.

Salton, M. R. J., *The Bacterial Cell Wall,* Elsevier Publishing Co., Amsterdam, 1964.

Steinberg, M., "The Problem of Adhesive Selectivity in Cellular Interactions," in *Cellular Membranes in Development* (M. Locke, ed.), Academic Press, New York, 1963.

CHAPTER 20

Dove, W. F., "Genetics of the Lambdoid Phages," *Ann. Rev. Genetics,* vol. 2, 1968.

Fraenkel-Conrat, H. (ed.), *Molecular Basis of Virology,* Van Nostrand Reinhold Company; New York, 1968. Contains several excellent reviews.

Luria, S. E., and J. E. Darnell, Jr., *General Virology,* John Wiley & Sons, Inc., New York, 1967, 2nd ed. The best introduction to the field.

Lwoff, A., and P. Tournier, "The Classification of Viruses," *Ann. Rev. Microbiol.* **20** (1966).

Many articles in *Advances in Virus Research,* vols. 1–15, Academic Press, New York, 1953–1969, can also be recommended as the starting points for further inquiry.

CHAPTER 21

Cameron, A. G. W. (ed.), *Interstellar Communication: The Search for Extraterrestrial Life,* W. A. Benjamin, Inc., New York, 1963. A collection of very stimulating articles.

Kenyon, Dean H., and Gary Steinman, *Biochemical Predestination,* McGraw-Hill Book Co., New York, 1969.

Oparin, A. I., *Genesis and Evolutionary Development of Life,* Academic Press, New York, 1968.

1-1 (a) Linear homopolymer.

(b) Linear heteropolymer (or copolymer).

(c) Branched polymer.

(d) Branched or network polymer.

(e) Nothing.

1-2 (a) 3 and 4.

(b) 3.

(c) 2, 3, 4, and 7.

1-3 (a) $4^2 = 16$.

(b) $5^3 = 125$.

(c) 10^{100}.

1-4 (a) 7.85 μm^2.

(b) 7.85×10^6 (since each monomer is 1 nm^2 and 10^6 $nm^2 = 1$ μm^2).

1-5 The microscope *resolves* the bits of green color.

1-6 Probably not, if you consider that each organism can use the amino acids in its food directly, without changing them, if all organisms in the food chain use the same type.

1-7 About 5%.

1-8 (a) 2.5×10^{-13} g.

(b) 2.5×10^{-13} g $\times 6 \times 10^{23}$ daltons/g $= 15 \times 10^{10}$ daltons.

(c) $7.5 \times 10^{10} \div 3 \times 10^4 = 2.5 \times 10^6$ molecules.

2-1 (a) 0.66 hour^{-1}.

(b) 0.85 hour^{-1}.

2-2 $\tau = t_D/0.693$.

2-3 (a) 2.4 hours.

(b) 1.6 hours.

(c) 0.5 hours.

(d) 0.67 hours.

2-4 2.3 hours.

2-5 $g = 0.36$.

2-6 3.68×10^8 cells/ml.

2-7 Two mutations per tube; 5×10^{-7} per cell

2-8 about 6520; $.87 \times 10^{-6}$/cell

2-9 2.2 hours.

2-10 10^{-8}/hour; 1.3×10^{-7}/hour.

2-11 24 years; 55 years; late exponential; almost stationary; and do you really need to be told the last answer?

2-12 Plate bacteria on agar containing small amounts of penicillin.

2-13 Continual chemical warfare; "lower" bacteria will always be selected that are resistant to the antibiotics and "higher" bacteria and fungi will be selected that can make more effective antibiotics.

2-14 For example, grow bacteria in a chemostat with limited carbon source and small amounts of various hydrocarbons.

2-15 Killing coyotes allows larger rodent populations and so less grass for the sheep. It is better to let the coyotes live.

2-16 The original type has diverged into a long, thin line and a short, fat line.

2-17 1800 million years.

2-18 1960 million years.

3-1 100.

3-2 $W = nRT \ln (V_2/V_1)$; 2.76×10^3 J.

3-3 1/36; 6/36 = 1/6; 1/36.

3-4 −691.09 kcal/mole. (Did you use 6 moles of each product?)

3-5 (a) 6.69 kcal/mole.

(b) 1.4×10^{-5}.

(c) 1.16 kcal/mole.

3-6 0.026; 0.060.

3-7 −5.5 kcal; −2.73 kcal; 1.38 kcal; 0 kcal; 2.7 kcal; 5.5 kcal.

3-9 25.7 bits; 2.45×10^{-15} cal/deg; 3.18×10^{-15} cal/deg.

3-10 (a) 1000 lb of deer, 100 lb of puma; 10 lb of condor.

4-1 Asp-Gly-Pro-Pro-Ser-Thr-Gly-Lys-Met-Arg-Gly-Cys-Asp-Lys-Glu-Ilu-Thr.

4-4 Plectonemic.

4-5 Half dense, half light; one-fourth dense, three-fourths light.

4-6 dRib: 134; $H_3PO_4^=$: 96; A: 135; G: 151; C: 111, T: 126; dAdo: 251; dGuo: 267; dCyd: 227; dThd: 242; dAMP: 329; dGMP: 345; dCMP: 305; dTMP: 320; A-T: 613; GC: 614; 61350 per 100 nucleotides. (Note: You must subtract additional waters at each step.)

4-7 61725; 1.22%.

4-9 AUU CCG GUA CGU UCG GAA AUC.

4-10 Ilu-Pro-Val-Arg-Ser-Glu-Ilu.

4-11 UCG → CCG, Ser → Pro; mis-sense.

4-12 Two.

5-1

lng⁺ thn⁺ tls⁺ flg⁺	*lng⁺ thn⁺ tls flg⁺*
lng thn⁺ tls⁺ flg⁺	*lng⁺ thn⁺ tls flg⁻*
lng thn tls⁺ flg⁺	*lng⁺ thn tls⁺ flg⁺*
lng⁺ thn tls flg⁻	*lng⁺ thn tls flg⁺*
lng⁺ thn⁺ tls⁺ flg⁻	*lng⁺ thn tls⁺ flg⁻*

5-2 markers are unlinked; $\chi^2 = 2.65$, for three degrees of freedom. *P* is between 40 and 50%.

5-3

```
m           tu          h          r
|           |           |          |
    5       |   6.1     |   7.1     |
|           |        10.8           |
|           |           |          |
|         10.0          |          |
```

5-4 10.5; S = 1.8.

5-5 (a) *a-c-b*.

(b) 1.3; 0.7; 1.6.

(c) 22.2.

5-6 *flg*   *tls*   *thn*

	15.3		9.3	
		21.8		

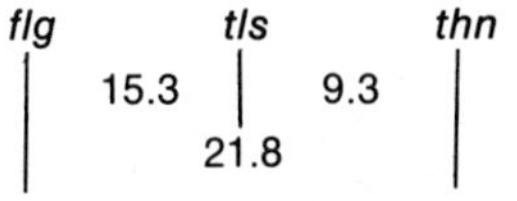

5-7 No;

	1	2	3	4	5	6
1	0	0	0	0	0	1
2	0	0	0	1	1	1
3	0	0	0	1	0	1
4	0	1	1	0	0	1
5	0	1	0	0	0	0
6	1	1	1	1	0	0

5-8
  36
  711
  54
  304
  72
  19

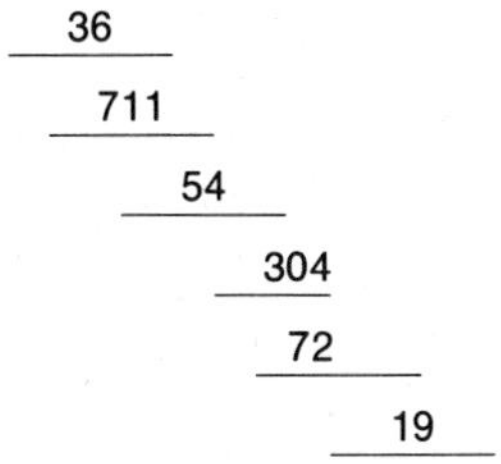

5-9 11-(3,5,8)-1-12-9-13-(2,7)-6-(4,10).

6-1 (a) 0.01.
  (b) Zero.
  (c) 0.105.
6-2 Total: 126.4; minus 18, equals 108.4
6-3 (a) 277
  (b) 461
6-4 (a) 277 nm; 2.74×10^5 daltons
  (b) 461 nm; 4.56×10^5 daltons
6-5 1.2 μm

7-1 (a) 0.0725.
  (b) 4.2 μm.
7-2 31.4×10^{-12} g.

7-3

0.9	0.6	0.4	0.6
1.33	0.67	0.33	0.67
1.125	0.375	0.125	0.375
0.576	0.524	0.476	0.524

7-4 All ratios are identical for each set of initial conditions; there is no potential.
7-5 The envelope has ruptured; $M/10$.

7-6

edge	surface	volume	surface/volume
1	6	1	6.0
2	24	8	3.0
5	150	125	1.2
10	600	1000	0.6
20	2400	8000	0.3
50	15000	125000	0.12
100	600000	1000000	0.06

8-1 (a) 9.6×10^5 daltons; 1.6×10^{-18} g.

           16.4×10^5 daltons; 2.7×10^{-18} g.

     (b) About 38.

     (c) 1.6×10^{-14} g.

8-2 (a) 1.5×10^{11} daltons.

     (b) 5×10^9 daltons; 8×10^6 pairs; 2770 μm.

     (c) 10,000 rpm!

8-3 15; 18.

8-4 47%.

8-8 C-4.

8-9 (a) Asymmetrically.

     (b) C-1 and C-6.

8-10 Citrate: C-1 is 1, C-5 is $\frac{1}{2}$, C-6 is $\frac{1}{2}$. Ketoglutarate: C-1 is 1, C-5 is $\frac{1}{2}$. Succinate: C-1 and C-4 are $\frac{1}{2}$.

8-11 1.86 kcal/mole; 0.243 kcal/mole.

9-1 $\rightarrow$ germanate $\rightarrow$ italine $\rightarrow$ francine $\rightarrow$ swedate.

      D   A   C   B

     B would feed francine to A.

9-2 A and C would feed B.

     $\rightarrow$ X $\rightarrow$ sarabasol $\rightarrow$ γ-saraboside

     B   C   A

9-3 (a) 10————(14, 16, 18)————12.

     (b) establishes the order 14-16-18.

9-4 *Pro*-4-5-7-1-6-3-2-8-*Ade*

9-5 | 1 | 3 5 | 6 2 4 | *wht.*

9-6 (a) 3.25×10^6.

     (b) about 10^6.

     (c) about 2000.

     (d) 11%.

10-1 mRNA weighs 10 times the protein.

10-2

200	200	1	1
100	100	100	100
<0.1	100	<0.1	100
<0.1	<0.1	<0.1	<0.1

10-3

Trp	Tyr	Asn		Ala	Val	Arg

$$\begin{array}{ccc} \text{UGG} & \text{UAU} & \text{AA}^{\text{U}}_{\text{C}} \\ & \downarrow & \end{array}$$

$$\begin{array}{ccc} \text{GCU} & \text{GUG} & ^{\text{A}}_{\text{C}}\text{GG} \\ & \downarrow & \end{array}$$

AUG	GUA	UAA		UGC	UGU	GGG
Met	Val	Tyr		Cys	Cys	Gly

10-4

$$\text{GG}^{\text{A}}_{\text{G}} \longrightarrow \begin{array}{c} \text{C}^{\text{A}}_{\text{G}}\text{G}^{\text{A}} \quad (\text{A23}) \\ \text{GA}^{\text{A}}_{\text{G}} \quad (\text{A46}) \end{array}$$

For example:

$$\begin{array}{c} \text{A G A} \\ \text{T C T} \quad (\text{A23}) \\ \text{G | A A} \\ \text{C | T T} \quad (\text{A46}) \end{array} \longrightarrow \begin{array}{c} \text{G G A} \\ \text{C C T} \end{array}$$

10-5

$$\text{AA}^{\text{U}}_{\text{C}} \longrightarrow \text{AG}^{\text{U}}_{\text{C}} \qquad \text{UC}^{\text{A}}_{\text{G}} \longrightarrow \text{UU}^{\text{A}}_{\text{G}}$$

Asn   Ser   Ser   Leu

$$\text{GA}^{\text{U}}_{\text{C}} \begin{array}{c} \nearrow \text{GC}^{\text{U}}_{\text{C}} \\ \searrow \text{GG}^{\text{U}}_{\text{C}} \end{array}$$

$$\text{AC}^{\text{U}}_{\text{C}} \begin{array}{c} \nearrow \text{AUC}^{\text{U}}_{\text{A}} \\ \longrightarrow \text{AUG} \\ \searrow \text{AG}^{\text{U}}_{\text{C}} \text{ or } \text{UC}^{\text{U}}_{\text{C}} \end{array}$$

You can easily identify the mutations in DNA; not all are transitions.

10-6 (a) 1700.

    (b) 4.4.

    (c) 1.4 seconds; a little less than a millisecond.

10-7 Genes for A, B, and D may be in one operon; C is not.

10-8 Constitutive mutants for β-galactosidase.

10-9 *lac⁻* and inducible strains; that is, you select *against* constitutives.

10-10 In a medium containing E but not D the operon will be repressed but there will be no D for protein synthesis. Because they could not grow under this condition, they are not very well adapted; you are not likely to find such a regulatory mechanism.

10-11 This showed that tRNA is a true adaptor. An enzyme must recognize the right tRNA and attach the appropriate amino acid; and the anticodon must recognize the proper codon. But once the amino acid has been attached, the tRNA cannot know what it is carrying and so it can be fooled.

10-12 (a) An operator.

    (b) $R^s > R^+$, $R^s > R^-$, $R^+ > R^-$.

    (c) *Ara* constitutive (as if a promotor has been formed there).

11-1 A, mitochondria; B, nucleus; C, nuclear envelope; D, lumen of ER; E, channels to the cell surface; F, mitochondria; G, Golgi membranes; H, pore in nuclear envelope.

11-2 Perhaps two mitochondria fusing, or one dividing.

11-3 Some of these may form channels from the ER lumen to the outside.

12-1   73 protomers/second.

12-2   Yes; yes, and this is not compatible with the data; after another round, only half the chromatids are labeled.

12-3 0.65×10^{-15} g; 6.33×10^5 pairs.

12-4   8000; about 80.

14-1   8.1 map units.

14-2   13.0 map units.

14-3   30.5 map units.

14-4   Very close to centromere.

14-5

14-7 $\chi^2 = 0.492$, P close to 50%.

14-11 $\chi^2 = 0.456$, P about 90–95%.

14-12   Round, yellow; $R/r, Y/y$; 9:3:3:1 distribution.

14-13   (a)   Round, yellow, colored; $R/r\ Y/y\ C/c$.

      (b)   27/64, 9/64. 1/64, 9/64, 3/64.

14-14   16; 81; 16.

14-15   (a)   3/16; 1/16.

      (b)   1/2; 0.

14-16   9/16 sepia, 3/16 brown, 1/4 red-yellow.

14-17 $c^k = 40\%$, $c^d = 20\%$; $c^r = 5\%$, $c^a = 0\%$.

14-18   0.00476.

14-19

14-20   Parent is $d\ F/D\ f$, E/e; 8.33 map units between D and F.

14-21   Fatherkin is XXX, motherkin XXY, unclekin XYY.

14-22   Male is XY, shemale X′Y, female XX or XX′ or X′X′ (X and X′ must produce the same result in the absence of a Y chromosome; otherwise there will be some difference between the three types of female, some selection for one kind or another, and the system will be unstable).

14-23

	A^1B^1	A^1B^2	A^2B^1	A^2B^2	
A^1B^1				C	
A^1B^2			C		only C types are
A^2B^1		C			compatible
A^2B^2	C				

14-24   Nondisjunction could occur at M_I only, producing XY sperm and XX eggs; it could at both M_I and M_{II}, producing XXYY sperm and XXXX eggs. You can put these combinations together yourself, along with normal gametes.

14-25 (a) 1/200; 1/200.

      (b) 1/4; 0.

      (c) 1; 0.

14-26 (a) Yellow.

      (b) Two—black and yellow.

14-27 9.5 map units; half male, half female.

14-28 9.25 map units.

14-29 Mother is X^+/X^c, daughter is X^c/O.

14-30 (a) Autosomal recessive.

      (b) Sex-linked recessive; heterozygous woman gives the mutation to half her sons and half her daughters; a man with the mutation makes all of his daughters heterozygotes.

      (c) Sex-linked dominant; a woman gives the trait to half her children; a man gives it to all his daughters.

14-31 36.7 map units.

15-1 $V = 57.2$ for both; $K_m = 1.9 \times 10^{-5}$ M for glucose and 1.32×10^{-5} M for mannose; mannose has greater affinity.

15-2 Growth temperature is limited by enzyme stability. Strains adapted to higher temperatures have selected for more heat-stable structures— perhaps with less hydrophobic bonding and more covalent bonding.

15-3 Suggests that threonine not only inhibits enzymes for threonine biosynthesis, but also that it is a precursor of isoleucine. Similarly, glutamate inhibits enzymes for glutamate biosynthesis and is a precursor of the other amino acids.

15-4 Enzyme I and HPr are common to all; but each sugar has its own specific enzyme II.

15-5 6-P-sugars are common early metabolites in sugar catabolism, as glucose-6-P in the Embden–Meyerhof pathway. It makes sense for the cell to use a single phosphorylation for catabolism and translocation simultaneously.

15-6 One step of methionine biosynthesis must be coupled to K^+ translocation; a single protein does both jobs.

16-1 About 0.2 V.

16-2 (a) 0.254 V, 1 ATP.

      (b) 0.566 V, 2 ATP.

      (c) 0.785 V, 3 ATP.

      (d) 1.098 V, 5 ATP.

      (Assuming 100% efficiency.)

16-3 51%.

16-4 50%.

16-5 PSII → b-559 → "M" → c-553 → plastocyanin → P700 → PSI

17-1 This is *bivalent* repression—both Thr and Ilu must bind to make an active repressor.

17-2 Inhibition must be less than 100%, to allow for Met biosynthesis.

17-3 37%.

17-4 The two pools don't mix; how many explanations for this can you devise?

17-5 Feedback resistant; in gene *hisG,* which informs the first enzyme in the pathway.

18-1 Acetate $\rightarrow$ acetyl-CoA, which uses up oxalacetate to make citrate, thus lowering the PEP level.

18-2 To ensure continued citrate synthesis, the oxalacetate pool must be large, and therefore the PEP pool is also large.

18-3 Fatty acids are oxidized to acetyl-CoA, which goes to malate, to oxalacetate, to PEP, and to carbohydrates.

19-1 (a) Le^a.

    (b) A and some H.

    (c) B, some Le^a, and Le^b.

19-2 (a) A, B, AB.

    (b) A, B, AB, O.

    (c) B, O.

    (d) A, B.

19-3 Where did the child get its B allele?

20-1 1, 3, 4, 7, 9, 13, 16, 19, 21, 25, 31, 36, 37, 39.

20-2 1, 3, 4, 7, 9, 12, 13, 16, 19, 21, 25.

20-3 $10T + 2$.

20-4 $T = 1$, 12 pentamers; $T = 3$, 12 pentamers $+$ 20 hexamers; $T = 4$, 12 pentamers $+$ 30 hexamers; $T = 9$, 12 pentamers $+$ 80 hexamers.

20-5 First you see agglutination, but then the neuraminidase starts to destroy receptor sites and you see disaggregation.

20-6 Blood serum containing antibodies will block the viruses and prevent hemagglutination.

20-7 The minus strand.

20-8 As epitopes appear on the cell surface, erythrocytes will adsorb directly to the cells. The number adsorbed per cell increases regularly and is a good measure of virus synthesis.

2-1 (a) $\frac{1}{8}$.

    (b) $\frac{3}{8}$.

2-2 (a) $\frac{1}{16}$.

    (b) $\frac{1}{4}$.

2-4 15/32.

2-5 (a) 1/1024.

    (b) 15/128.

    (c) 105/516.

2-6 (a) 0.64.

(b) 0.384.

(c) 0.0512.

(d) 0.088.

(e) 0.00032.

2-7 (a) 0.37.

(b) 0.185.

(c) 0.905.

(d) 0.0045.

(e) 0.0379.

2-8 (a) $\chi^2 = 2$, $P \sim 0.4$.

(b) $\chi^2 = 1.5$, $P \sim 0.5$.

(c) $\chi^2 = 1.34$, $P \sim 0.5$.

(d) $\chi^2 = 3.42$, $P \sim 0.2$.

(e) $\chi^2 = 1.44$, $P \sim 0.5$.

ATOMIC MASSES OF ELEMENTS
REFERRED TO $^{12}C = 12.0000$

NAME	SYMBOL	ATOMIC NUMBER	ATOMIC WEIGHT	NAME	SYMBOL	ATOMIC NUMBER	ATOMIC WEIGHT
Actinium	Ac	89	(227)	Gallium	Ga	31	69.72
Aluminum	Al	13	26.9815	Germanium	Ge	32	72.59
Americium	Am	95	(243)	Gold	Au	79	196.967
Antimony	Sb	51	121.75	Hafnium	Hf	72	178.49
Argon	Ar	18	39.948	Helium	He	2	4.0026
Arsenic	As	33	74.9216	Holmium	Ho	67	164.930
Astatine	At	85	(210)	Hydrogen	H	1	1.00797
Barium	Ba	56	137.34	Indium	In	49	114.82
Berkelium	Bk	97	(249)	Iodine	I	53	126.9044
Beryllium	Be	4	9.0122	Iridium	Ir	77	192.2
Bismuth	Bi	83	208.980	Iron	Fe	26	55.347
Boron	B	5	10.811	Krypton	Kr	36	83.80
Bromine	Br	35	79.909	Lanthanum	La	57	138.91
Cadmium	Cd	48	112.40	Lawrencium	Lw	103	(257)
Calcium	Ca	20	40.08	Lead	Pb	82	207.19
Californium	Cf	98	(251)	Lithium	Li	3	6.939
Carbon	C	6	12.01115	Lutetium	Lu	71	174.97
Cerium	Ce	58	140.12	Magnesium	Mg	12	24.312
Cesium	Cs	55	132.905	Manganese	Mn	25	54.9380
Chlorine	Cl	17	35.453	Mendelevium	Md	101	(256)
Chromium	Cr	24	51.996	Mercury	Hg	80	200.59
Cobalt	Co	27	58.9332	Molybdenum	Mo	42	95.94
Copper	Cu	29	63.54	Neodymium	Nd	60	144.24
Curium	Cm	96	(247)	Neon	Ne	10	20.183
Dysprosium	Dy	66	162.50	Neptunium	Np	93	(237)
Einsteinium	Es	99	(254)	Nickel	Ni	28	58.71
Erbium	Er	68	167.26	Niobium	Nb	41	92.906
Europium	Eu	63	151.96	Nitrogen	N	7	14.0067
Fermium	Fm	100	(253)	Nobelium	No	102	(253)
Fluorine	F	9	18.9984	Osmium	Os	76	190.2
Francium	Fr	87	(223)	Oxygen	O	8	15.9994
Gadolinium	Gd	64	157.25	Palladium	Pd	46	106.4